Geophysical Monograph Series

Including

IUGG Volumes
Maurice Ewing Volumes
Mineral Physics Volumes

Geophysical Monograph Series

91 **Seafloor Hydrothermal Systems: Physical, Chemical, Biological, and Geological Interactions** *Susan E. Humphris, Robert A. Zierenberg, Lauren S. Mullineaux, and Richard E. Thomson (Eds.)*

92 **Mauna Loa Revealed: Structure, Composition, History, and Hazards** *J. M. Rhodes and John P. Lockwood (Eds.)*

93 **Cross-Scale Coupling in Space Plasmas** *James L. Horowitz, Nagendra Singh, and James L. Burch (Eds.)*

94 **Double-Diffusive Convection** *Alan Brandt and H. J. S. Fernando (Eds.)*

95 **Earth Processes: Reading the Isotopic Code** *Asish Basu and Stan Hart (Eds.)*

96 **Subduction Top to Bottom** *Gray E. Bebout, David Scholl, Stephen Kirby, and John Platt (Eds.)*

97 **Radiation Belts: Models and Standards** *J. F. Lemaire, D. Heynderickx, and D. N. Baker (Eds.)*

98 **Magnetic Storms** *Bruce T. Tsurutani, Walter D. Gonzalez, Yohsuke Kamide, and John K. Arballo (Eds.)*

99 **Coronal Mass Ejections** *Nancy Crooker, Jo Ann Joselyn, and Joan Feynman (Eds.)*

100 **Large Igneous Provinces** *John J. Mahoney and Millard F. Coffin (Eds.)*

101 **Properties of Earth and Planetary Materials at High Pressure and Temperature** *Murli Manghnani and Takehiki Yagi (Eds.)*

102 **Measurement Techniques in Space Plasmas: Particles** *Robert F. Pfaff, Joseph E. Borovsky, and David T. Young (Eds.)*

103 **Measurement Techniques in Space Plasmas: Fields** *Robert F. Pfaff, Joseph E. Borovsky, and David T. Young (Eds.)*

104 **Geospace Mass and Energy Flow: Results From the International Solar-Terrestrial Physics Program** *James L. Horwitz, Dennis L. Gallagher, and William K. Peterson (Eds.)*

105 **New Perspectives on the Earth's Magnetotail** *A. Nishida, D. N. Baker, and S. W. H. Cowley (Eds.)*

106 **Faulting and Magmatism at Mid-Ocean Ridges** *W. Roger Buck, Paul T. Delaney, Jeffrey A. Karson, and Yves Lagabrielle (Eds.)*

107 **Rivers Over Rock: Fluvial Processes in Bedrock Channels** *Keith J. Tinkler and Ellen E. Wohl (Eds.)*

108 **Assessment of Non-Point Source Pollution in the Vadose Zone** *Dennis L. Corwin, Keith Loague, and Timothy R. Ellsworth (Eds.)*

109 **Sun-Earth Plasma Interactions** *J. L. Burch, R. L. Carovillano, and S. K. Antiochos (Eds.)*

110 **The Controlled Flood in Grand Canyon** *Robert H. Webb, John C. Schmidt, G. Richard Marzolf, and Richard A. Valdez (Eds.)*

111 **Magnetic Helicity in Space and Laboratory Plasmas** *Michael R. Brown, Richard C. Canfield, and Alexei A. Pevtsov (Eds.)*

112 **Mechanisms of Global Climate Change at Millennial Time Scales** *Peter U. Clark, Robert S. Webb, and Lloyd D. Keigwin (Eds.)*

113 **Faults and Subsurface Fluid Flow in the Shallow Crust** *William C. Haneberg, Peter S. Mozley, J. Casey Moore, and Laurel B. Goodwin (Eds.)*

114 **Inverse Methods in Global Biogeochemical Cycles** *Prasad Kasibhatla, Martin Heimann, Peter Rayner, Natalie Mahowald, Ronald G. Prinn, and Dana E. Hartley (Eds.)*

115 **Atlantic Rifts and Continental Margins** *Webster Mohriak and Manik Talwani (Eds.)*

116 **Remote Sensing of Active Volcanism** *Peter J. Mouginis-Mark, Joy A. Crisp, and Jonathan H. Fink (Eds.)*

117 **Earth's Deep Interior: Mineral Physics and Tomography From the Atomic to the Global Scale** *Shun-ichiro Karato, Alessandro Forte, Robert Liebermann, Guy Masters, Lars Stixrude (Eds.)*

118 **Magnetospheric Current Systems** *Shin-ichi Ohtani, Ryoichi Fujii, Michael Hesse, and Robert L. Lysak (Eds.)*

119 **Radio Astronomy at Long Wavelengths** *Robert G. Stone, Kurt W. Weiler, Melvyn L. Goldstein, and Jean-Louis Bougeret (Eds.)*

120 **GeoComplexity and the Physics of Earthquakes** *John B. Rundle, Donald L. Turcotte, and William Klein (Eds.)*

121 **The History and Dynamics of Global Plate Motions** *Mark A. Richards, Richard G. Gordon, Rob D. van der Hilst (Eds.)*

122 **Dynamics of Fluids in Fractured Rock** *Boris Faybishenko, Paul A. Witherspoon, and Sally M. Benson (Eds.)*

123 **Atmospheric Science Across the Stratopause** *David E. Siskind, Stephen D. Eckerman, and Michael E. Summers (Eds.)*

124 **Natural Gas Hydrates: Occurrence, Distribution, and Detection** *Charles K. Paull and Willam P. Dillon (Eds.)*

125 **Space Weather** *Paul Song, Howard J. Singer, George L. Siscoe (Eds.)*

126 **The Oceans and Rapid Climate Change: Past, Present, and Future** *Dan Seidov, Bernd J. Haupt, and Mark Maslin (Eds.)*

Geophysical Monograph 127

Gas Transfer at Water Surfaces

M. A. Donelan
W. M. Drennan
E. S. Saltzman
R. Wanninkhof
Editors

American Geophysical Union
Washington, DC

Published under the aegis of the AGU Books Board

John E. Costa, Chair; David Bercovici, Andrew Dessler, Jeffrey M. Forbes, W. Rockwell Geyer, Rebecca Lange, Douglas S. Luther, Darrell Strobel, and R. Eugene Turner, members.

Library of Congress Cataloging-in-Publication Data
Gas transfer at water surfaces / M.A. Donelan ... [et al.], editors.
      p. cm.-- (Geophysical monograph ; 127)
   Includes bibliographical references.
   ISBN 0-87590-986-8
   1. Diffusion. 2. Mass transfer. 3. Gas dynamics. 4. Fluid dynamics. 5. Transport theory. 6. Ocean-atmosphere interaction. I. Donelan, M. A. (Mark A.), 1942-. II. Series.

QC185.G32 2002
530.4'27--dc21   20011053770

ISBN 0-87590-986-8
ISSN 0065-8448

Copyright 2002 by the American Geophysical Union
2000 Florida Avenue, N.W.
Washington, DC 20009

Figures, tables, and short excerpts may be reprinted in scientific books and journals if the source is properly cited.

Authorization to photocopy items for internal or personal use, or the internal or personal use of specific clients, is granted by the American Geophysical Union for libraries and other users registered with the Copyright Clearance Center (CCC) Transactional Reporting Service, provided that the base fee of $1.50 per copy plus $0.35 per page is paid directly to CCC, 222 Rosewood Dr., Danvers, MA 01923. 0065-8448/02/$01.50+0.35.

This consent does not extend to other kinds of copying, such as copying for creating new collective works or for resale. The reproduction of multiple copies and the use of full articles or the use of extracts, including figures and tables, for commercial purposes requires permission from the American Geophysical Union.

Printed in the United States of America.

CONTENTS

Preface
Mark A. Donelan, William M. Drennan, Eric S. Saltzman, and Rik Wanninkhof ix

Introduction

Gas Transfer at Water Surfaces——Concepts and Issues
Mark A. Donelan and Rik Wanninkhof 1

Section 1: Effects of Surface Waves and Turbulence on Gas Transfer

Turbulence Generated by Microscale Breaking Waves and Its Influence on Air-Water Gas Transfer
Muhammad H. K. Siddiqui, Mark R. Loewen, Christine Richardson, William E. Asher, and Andrew T. Jessup 11

On the Surface Kinematics of Microscale Breaking Wind Waves
William L. Peirson and Michael L. Banner 17

Effect of Microscale Wave Breaking on Air-Water Gas Transfer
C. J. Zappa, W. E. Asher, A. T. Jessup, J. Klinke, and S. R. Long 23

Statistics of Geometric Properties of Breaking Wind Waves Observed in Laboratory
Guillemette Caulliez 31

Overview of the CoOP Experiments: Physical and Chemical Measurements Parameterizing Air-Sea Heat Exchange
Erik John Bock, James Bearer Edson, Nelson M. Frew, Tetsu Hara, Horst Haussecker, Bernd Jähne, Wade R. McGillis, Sean P. McKenna, Robert K. Nelson, Uwe Schimpf, and Mete Uz 39

Surface Wave Observations During CoOP Experiments and Their Relation to Air-Sea Gas Transfer
Tetsu Hara, B. Mete Uz, Hua Wei, James B. Edson, Nelson M. Frew, Wade R. McGillis, Sean P. McKenna, Erik J. Bock, Horst Haußecker, and Uwe Schimpf 45

On the Investigations of Statistical Properties of the Micro Turbulence at the Ocean Surface
Uwe Schimpf, Bernd Jähne, and Horst Haußecker 51

On the Skewness of the Sea Slope Probability Distribution
Bertrand Chapron, D. Vandemark, and T. Elfouhaily 59

Directional Distributions and Mean Square Slopes of Surface Waves
Paul A. Hwang and David W. Wang 65

Numerical Simulation of Hydrodynamic Processes Beneath a Wind-Driven Water Surface
Wu-Ting Tsai 71

Direct Numerical Simulation of Turbulent Free Surface Flow with Carbon-Dioxide Gas Absorption
Tomoaki Kunugi and Shin-ichi Satake 77

LIF Measurements of Oxygen Concentration Gradients Along Flat and Wavy Air-Water Interfaces
Philip T. Woodrow, Jr. and Steve R. Duke 83

CONTENTS

A Direct Visualization Method of CO_2 Gas Transfer at Water Surface Driven by Wind Waves
Kohsei Takehara and Goji T. Etoh 89

Measurements of Wind-Driven Current Near the Water Surface by the Use of the PTV Method
Hajime Kato, Hisamichi Nobuoka, and Naoki Ooshima 97

Physics from IR Image Sequences: Quantitative Analysis of Transport Models and Parameters of Air-Sea Gas Transfer
Horst W. Haussecker, Uwe Schimpf, Christoph S. Garbe, and Bernd Jähne 103

Measuring the Sea Surface Heat Flux and Probability Distribution of Surface Renewal Events
Christoph S. Garbe, Bernd Jähne, and Horst Haußecker 109

Effects of Counter-Swell on Both the Mean Current and Turbulent Structure Below the Wind-Waves
Shinjiro Mizuno 115

A Closer Look at Short Waves Generated by Wave Interactions With Adverse Currents
Steven R. Long and Jochen Klinke 121

Surface Divergence and Air-Water Gas Transfer
Sean P. McKenna and Wade R. McGillis 129

Section 2: Effects of Buoyancy and Surfactants on Gas Transfer

The Critical Importance of Buoyancy Flux for Gas Flux Across the Air-Water Interface
Sally MacIntyre, Werner Eugster, and George W. Kling 135

A Model of Air-Sea Gas Exchange Incorporating the Physics of the Turbulent Boundary Layer and the Properties of the Sea Surface
Alexander Soloviev and Peter Schluessel 141

Air-Lake Interaction and Surface Layer Processes
G. N. Panin, A. E. Nasonov, and S. G. Sarkisian 147

Spatial Variations in Surface Microlayer Surfactants and Their Role in Modulating Air-Sea Exchange
Nelson M. Frew, Robert K. Nelson, Wade R. McGillis, James B. Edson, Erik J. Bock, and Tetsu Hara 153

Thermal Profiling of the Sea Surface Skin Layer Using FTIR Measurements
Jennifer A. Hanafin and Peter J. Minnett 161

An Autonomous Profiler for Near Surface Temperature Measurements
Brian Ward and Peter J. Minnett 167

Water Column CO_2 Measurements During the Gas Ex-98 Expedition
R. A. Feely, R. Wanninkhof, D. A. Hansell, M. F. Lamb, D. Greeley, and K. Lee 173

Fine Thermohaline Structure and Gas-Exchange in the Near-Surface Layer of the Ocean During GasEx-98
Alexander Soloviev, Jim Edson, Wade McGillis, Peter Schluessel, and Rik Wanninkhof 181

CONTENTS

Section 3: Gas Transfer in Strong Turbulent Flows

A Field Study of Whitecap Coverage and Its Modulations by Energy Containing Surface Waves
V. A. Dulov, V. N. Kudryavtsev, and A. N. Bol'shakov 187

The Physical and Practical Implications of a CO_2 Gas Transfer Coefficient That Varies as the Cube of the Wind Speed
Edward C. Monahan 193

Fractional Area Whitecap Coverage and Air-Sea Gas Transfer Velocities Measured During GasEx-98
William Asher, James Edson, Wade McGillis, Rik Wanninkhof, David T. Ho, and Trina Litchendorf 199

Gas Transfer in Energetic Conditions
David Kevin Woolf 205

Estimation of Whitecap Coverage Percentage Using Shallow Grazing-Angle Video and FMICW Radar
Craig L. Stevens, M. J. Smith, and J. A. McGregor 213

ASGAMAGE, the Air-Sea Gas Exchange/MAGE Experiment
Wiebe Oost and ASAMAGE participants 219

Comparison of the Deliberate Tracer Method and Eddy Covariance Measurements to Determine the Air/Sea Transfer Velocity of CO_2
Cor Jacobs, Phil Nightingale, Rob Upstill-Goddard, Jørgen Friis Kjeld, Søren Larsen, and Wiebe Oost 225

Gas Transfer Velocities for ^{3}He in a Lake at High Wind Speeds
Philippe Jean-Baptiste, Elise Fourré, and Alain Poisson 233

Measurement Uncertainty in Gas Exchange Coefficients
J. S. Gulliver, B. Erickson, A. J. Zaske, and K. S. Shimon 239

Measurements of Free Surface Turbulence
Joseph J. Orlins and John S. Gulliver 247

Gas Transfer Across a Zero-Shear Surface: A Local Approach
Mohamed A. Atmane and Jacques George 255

Atmosphere-Ocean Gas Exchange Due to Bubbles Generated by Wind Wave Breaking
Roman S. Bortkovskii 261

The Effect of Bubbles on Air-Water Oxygen Transfer in the Breaker Zone
Shohachi Kakuno, Douglas B. Moog, Tetsuya Tatekawa, Kenji Takemura, and Tatsuya Yamagishi 265

Bubble Size Distributions on the North Atlantic and North Sea
Gerrit de Leeuw and Leo H. Cohen 271

Measurements of Large Bubbles in Open-Ocean Whitecaps
Dale Stokes, Grant Deane, Svein Vagle, and David Farmer 279

The Effects of Bubbles on Mass Transfer Across the Breaking Air-Water Interface
Satoru Komori and Ryuta Misumi 285

CONTENTS

LUMINY – An Overview
G. de Leeuw, G. J. Kunz, G. Caulliez, D. K. Woolf, P. Bowyer, I. Leifer, P. Nightingale,
M. Liddicoat, T. S. Rhee, M. O. Andreae, S. E. Larsen, F. A. Hansen, and S. Lund 291

Bubbles Outside the Plume During the LUMINY Wind-Wave Experiment
Gerrit de Leeuw and Ira Leifer 295

Bubble Measurements in Breaking-Wave Generated Bubble Plumes During the LUMINY Wind-Wave Experiment
Ira Leifer and Gerrit de Leeuw 303

An Experimental Study of Bubble Mediated Gas Exchange for a Single Bubble
Nobuhito Mori, Masahiro Imamura, and Ryosuke Yamamoto 311

Better Bubble Process Modeling: Improved Bubble Hydrodynamics Parameterization
Ranjan Patro, Ira Leifer, and Peter Bowyer 315

Development and Testing of an Eddy Accumulator
W. John Cooper, Mehran Alaee, and Mark Donelan 321

Section 4: Remote Sensing for Large-Scale Gas Transfer

A Multi-Year Time Series of Global Gas Transfer Velocity from the TOPEX Dual Frequency, Normalized Radar Backscatter Algorithm
David M. Glover, Nelson M. Frew, Scott J. McCue, and Erik J. Bock 325

A Global, High Resolution, Satellite-Based Model of Air-Sea Isoprene Flux
David J. Erickson III and Jose L. Hernandez 333

Daily Surface Wind Fields Produced by Merged Satellite Data
Abderrahim Bentamy, Kristina B. Katsaros, William M. Drennan, and Evan B. Forde 343

The Effect of Using Time-Averaged Winds on Regional Air-Sea CO_2 Fluxes
Rik Wanninkhof, Scott C. Doney, Taro Takahashi, and Wade R. McGillis 351

Section 5: Gas Transfer by Aeration

A Dynamic Method to Estimate the Reoxygenation Rate at Hydraulic Structures
Badre E. Boumansour, Olivier Dufayt, Jean L. Vasel, and Jean M. Hiver 357

Mass Transfer in Bubbly Flows: Influence of Physico-Chemical and Hydrodynamic Conditions
Cornelia Lang 363

Air-Water Gas Transfer in Uniform Flows With Large Gravel-Bed Roughness
Douglas B. Moog and Gerhard H. Jirka 371

Large Scale Laboratory Experiments for Water Oxygenation Under Breaking Waves
V. K. Tsoukala, E. I. Daniil, and C. I. Moutzouris 377

PREFACE

The transfer of gases across the air-water interface has received much attention over the past two decades, particularly in light of increased societal interest in the exchange of greenhouse gases and pollutants between natural water bodies and the atmosphere. Gas transfer at the interface between liquids and gases holds great fascination for a wide range of researchers, from fluid dynamicists to biogeochemists. However, the phenomena of gas transfer, and the problems we face in understanding them, involve daunting issues, including multi-phase flows over a wide range of spatial and temporal scales. Such complexity is increased by the presence of surface films of both natural and anthropogenic origin, which can modify the physical and chemical nature of the interface. As a result, the challenge of working on gas transfer has stimulated the development of multidisciplinary, collaborative efforts and the development of a variety of innovative experimental and observational techniques.

Certainly, one is challenged by problems that involve multi-phase flow, which start with interactions at microscopic scales yet ultimately depend on large-scale flows. Equally challenging are the effects of surface chemistry, which involves a diverse array of natural and man-made compounds whose chemical and physical properties are complex and poorly characterized. The air-water interface is also a region of intense gradients that impose interesting constraints on the lives and productivity of microscopic organisms. Research in this area requires working at scales far smaller than those normally associated with the bulk processes on either side of the interface, and requires new experimental and theoretical approaches. The multi-disciplinary interest in the topic of gas transfer at water surfaces has led to important new experimental techniques such as laser-induced fluorescence and controlled flux application. In the geophysical context, the sun's diurnal and seasonal radiation pattern leads to important effects on the stability of the fluids and to biochemical changes brought about by organisms. Riverine flows experience turbulence at the air-water interface propagated from interaction of water flow with the bottom, while wind-generated waves introduce bubbles to the liquid phase and droplets to the gas phase. All of these modify the flow in the boundary layers and, generally, enhance the exchange of heat and mass across the interface.

This monograph represents the current state-of-the-art in research on the unifying topic: gas transfer at water surfaces. The papers are arranged in five sections representing the fundamental processes underlying the flux of mass across the air-water interface or the methodology involved in estimation of that flux:

Section 1	Effects of surface waves and turbulence on gas transfer
Section 2	Effects of buoyancy and surfactants on gas transfer
Section 3	Gas transfer in strong turbulent flows
Section 4	Remote sensing for large-scale gas transfer
Section 5	Gas transfer by aeration

Section 1 is devoted to the effects of surface waves and turbulence on mass transfer. It includes detailed laboratory experiments, field observations, and numerical simulation. During the first symposium held on the subject, in 1983 at Cornell University, it was postulated that wave breaking and its attendant turbulence production and bubble generation were fundamental to understanding gas transfer at water surfaces. Since then many laboratory experiments and field studies have left little doubt that there is a causal connection between wave breaking and enhancement of gas transfer, although the responsible mechanisms have not yet been fully quantified. Although complete understanding of the effects of surface waves and turbulence on wave breaking cannot be claimed, the papers in this section greatly aid in clarifying the issues.

Section 2 deals with the effects of buoyancy and surfactants on gas transfer. Buoyancy differences arise from heating, evaporation, or mixing of different water bodies and can suppress or enhance turbulence near the surface. Surfactants affect gas transfer primarily by inhibition of near-surface turbulence due to the presence of natural or man-made slightly soluble and insoluble organic material that tends to concentrate at the interface between liquid and gas phases, both at the surface of the liquid and on the "skins" of bubbles.

There is continued interest in the processes of air-sea interaction in strong winds, in part fueled by a hypothesized increase in the intensity of some meteorological phenomena due to climate change. In the third section, we focus on strongly forced wind-generated seas or their laboratory simulants as well as interfaces attacked from below by intense turbulent eddies. In many of these papers, bubbles are central to flux enhancement, and attempts to measure and model their behavior are prominent.

Ultimately, we will need to apply our knowledge and understanding of the microprocesses at the air-water interface to larger-scale general surface and boundary layer flow properties. The interpretation of satellite retrievals in terms of surface properties and fluxes is clearly of great importance for assessing the impact of air-sea exchange on biogeochemical budgets, and is undoubtedly an area of future growth for the field. Section 4 presents a quartet of papers that lead in that direction.

Section 5, which concludes the monograph, acknowledges the special problems of intense mixing at flow control structures as well as mixing in flows with significant entrained air or bottom-induced turbulence in riverine flows. Here the production of turbulence and the attendant interfacial mixing is continuous, as contrasted with the characteristic intermittency of wave breaking.

Altogether, these five sections of 57 papers represent the current thinking of the gas transfer research community and provide a resilient springboard for further advances. An introductory paper clarifies the expansive context of the work as a whole and the wider implications of work on gas transfer at water surfaces.

We are pleased to offer this volume and trust that you will find it informative and that it will join its predecessors in your library of valuable references on the important topic of gas transfer at water surfaces.

A series of symposia over the years has led to the present work. As mentioned briefly, the first symposium devoted entirely to the subject was held in 1983 at Cornell University, Ithaca, New York. It attracted a wide range of researchers devoted to the study of the natural environment as well as engineers concerned mainly with the industrial problem of mass transfer between liquids and gases. Seven years later the second symposium, held in Minneapolis, Minnesota, in September 1990, focused on "improvements in the ability to describe the physical and chemical processes associated with air-water mass transfer and applications of that knowledge to solve engineering problems." The third symposium, held in Heidelberg, Germany, in 1995, reflected the increasing awareness of the role of breaking waves and entrained bubbles in the transfer of gases at the air-water interface. By then it had become apparent that the increasing societal interest in the consequences of mass transfer at air-water interfaces, the size and productivity of the research community, and the wide geographic distribution of active researchers argued for a regular series of international symposia with a periodicity of about five years. Thus the Fourth International Symposium on Gas Transfer at Water Surfaces was held under the auspices of the University of Miami, Miami, Florida, in June 2000, with 110 researchers presenting a total of 72 papers and 38 posters. Of these, 57 were selected, through a rigorous review process, for inclusion in this volume.

We wish to acknowledge the considerable efforts of the Organizing and Program Committee members in coordinating the meeting and making this Fourth Symposium interesting, informative, and memorable.

Organizing Committee: M. Donelan, University of Miami, USA (Chair); W. Drennan, University of Miami, USA; E. Monahan, University of Connecticut, USA; R. Wanninkhof, AOML/NOAA, USA.

Program Committee: W. Asher, University of Washington, USA; E. Bock, University of Heidelberg, Germany; G. Caulliez, IRPHE, France; D. Erickson, Oak Ridge National Laboratory, USA; J. George, Institut de Mecanique des Fluides, France; J. Gulliver, University of Minnesota, USA; T. Hanratty, University of Illinois, USA; B. Jähne, University of Heidelberg, Germany; A. Jessup, University of Washington, USA; G. Jirka, Universität Karlsruhe, Germany; K. Katsaros, AOML/NOAA, USA; B. Kerman, Canada Centre for Inland Waters; L. Merlivat, Universite Pierre et Marie Curie, France; I. Nezu, Kyoto University, Japan; P. Nightingale, Plymouth Marine Laboratory, United Kingdom; G. Panin, Russian Academy of Sciences; E. Plate, Universität Karlsruhe, Germany; D. Woolf, University of Southampton, United Kingdom.

The success of such symposia depends to a great extent on the generosity of the funding agencies. We gratefully acknowledge financial support from the following sponsors of the symposium: National Aeronautics and Space Administration, National Oceanic and Atmospheric Administration, National Science Foundation, Office of Naval Research, Scientific Committee on Oceanic Research, and the Rosenstiel School of Marine and Atmospheric Science, University of Miami.

Careful preparations and smooth running of the meetings are essential to the success of such symposia. Ms. Frances Sampedro was responsible for all aspects of meeting coordination, including assisting in the production of this volume. Her skill, efficiency, and dedication facilitated the jobs of the organizing committee and the editors of this volume. During the symposium, Ms. Patricia Archuleta assisted with registration and the day-to-day organization.

Now that the value of these symposia has been established and their regularity deemed desirable, plans are already in place for the Fifth Symposium, to be held in 2005 in Europe. We look forward to continuing our discussions on gas transfer at water surfaces there.

Mark A. Donelan
William M. Drennan
Eric S. Saltzman
Rik Wanninkhof

DEDICATION

This volume is dedicated to the memory of Dr. Erik J. Bock. Erik was torn out of the middle of his life at the age of 39 by a tragic bicycle accident near his home in Obrigheim on June 25, 2001. Our community has suffered a great loss.

Erik participated in the 4th International Symposium Gas Transfer at Water Surfaces, where he presented two papers and is an author of four that appear in these pages. All of us who had the pleasure to work with Erik remember his dedication to science, his broad interdisciplinary knowledge and his subtle humor, which he kept up even in hard times. Erik supervised a number of students at the Woods Hole Oceanographic Institution and Heidelberg University. They remember him as a dedicated and competent mentor always having an ear and hand for them. His untimely death reminds us all of the fleetingness of life and we will all miss Erik greatly.

Erik was born on November 2, 1961 in Buffalo, New York. He obtained his B.S. (Magna cum laude in 1984), M.S. (1986); and Ph.D. (1987) degrees in Chemistry from the Rensselaer Polytechnic Institute in Troy, New York. In 1990, he joined the Woods Hole Oceanographic Institution as an Assistant Scientist in the Department of Applied Ocean Physics and Engineering and became an Associate Scientist in 1994. In 1998, Erik went on to work at the University of Heidelberg. At the time of his death, he was about to accept a professorship at the Baltic Sea Research Institute of the University of Rostock, Germany.

Erik's research centered on the interfacial properties of the ocean surface. He made significant contributions in this interdisciplinary field by quantifying how the chemistry of the air-sea interface affects the dynamics of short waves, near surface flows, and fluxes of heat, mass, and momentum across the interface. He was a leading expert in the measurement of gravity-capillary wind waves and ocean surface films and slicks and the design of optical and electronic instruments for laboratory and field measurements. Erik pioneered the field measurement of short wind waves with a scanning laser slope gauge. He was a major player in many of the recent large field experiments within the ONR Marine Boundary Layer Initiative and the NSF Coastal Oceanography Programme (CoOP). Shortly before his death, he had returned from the very successful Equatorial Pacific Air-Sea CO_2 Exchange Experiment.

Bernd Jähne
Horst Haussecker
Uwe Schimpf

Gas Transfer at Water Surfaces—Concepts and Issues

Mark A. Donelan and Rik Wanninkhof

This introductory paper puts the technical articles to follow in the context of the need to understand gas transfer at water surfaces and to apply improved methods to the estimation of the exchange of gases between air and water. We summarize the physical and chemical background to processes of interfacial gas transfer, discuss field and laboratory approaches to measuring the gas exchange rate and to elucidating its causes. Finally, we illustrate the application of acquired understanding in gas transfer to the global flux of carbon dioxide. This issue is of societal relevance in predicting and possibly reducing anthropogenic causes of climate change.

SOCIETAL IMPORTANCE

The importance of air-sea gas transfer processes has recently been highlighted by the ocean's role in taking up a large fraction (30-40 %) of fossil fuel-produced carbon dioxide. CO_2, of course, is a greenhouse gas that is increasing by about 0.5 % yr^{-1} due to burning of fossil fuels. In the early nineteenth century, atmospheric concentration was approximately 280 ppm; today it stands at 370 ppm -- and is projected to reach 560 ppm by the middle of this century. This projected increase is expected to significantly affect the earth's radiation balance and contribute to climate change [*Houghton et al.*, 1995]. Nonetheless, current constraints on the spatial and temporal variability of the oceanic uptake are poor, in large part because of uncertainty in regional air-sea CO_2 fluxes. Certainly, an improved understanding of processes controlling the gas transfer, which should lead to better quantification of fluxes, is critical to address the "CO_2 question." Current estimates of global CO_2 uptake using simple parameterizations of gas exchange with wind speed yield estimates ranging from 1 to 3 Pg C yr^{-1} (Table 1).

PHYSICAL BACKGROUND

The exchange of gases across air-water interfaces occurs by way of molecular and turbulent diffusion. Turbulent diffusivity is determined by the scale and rapidity of turbulent motions, while molecular diffusivity depends on molecular characteristics and operates at very small scales. In the body of air or water the dominant agency for diffusion is turbulent mixing. Close to the interface the turbulent motion is suppressed and the exchange of mass depends on molecular diffusivity there. The enormously greater mixing efficiency of turbulent flow leads to weak gradients away from the interface and much stronger gradients very close to the surface where the turbulence is suppressed. This view has led to conceptual models of the process of gas transfer in which the boundary layers on either side of the interface are further divided into a turbulent outer layer and an inner "diffusive sub-layer" (see Figure 1). Of course, there is a continuous transition from one layer to another, but the concept of diffusive sub-layers is useful because it is these thin layers that provide the largest resistance to the transfer of gases across the interface. Indeed, the diffusivities of gases in air are generally much larger than in water, so that, unless they are very soluble in, or reactive with water, their flow across the interface is limited by the resistance in the diffusive aqueous boundary layer. In this class of "water phase limited" gases fall all the major constituents of dry air including CO_2 and O_2; the preeminent biological gases. As a "greenhouse gas" CO_2 has received a lot of attention in the climate dynamics and prediction communities and in public forums.

The 800-fold difference in density of water and air has led to the consideration of the air-water boundary in analogy to flow of fluids over solid porous walls. This allows one to draw on a wealth of well established work on heat and mass transfer at such interfaces. In fact, when the wind

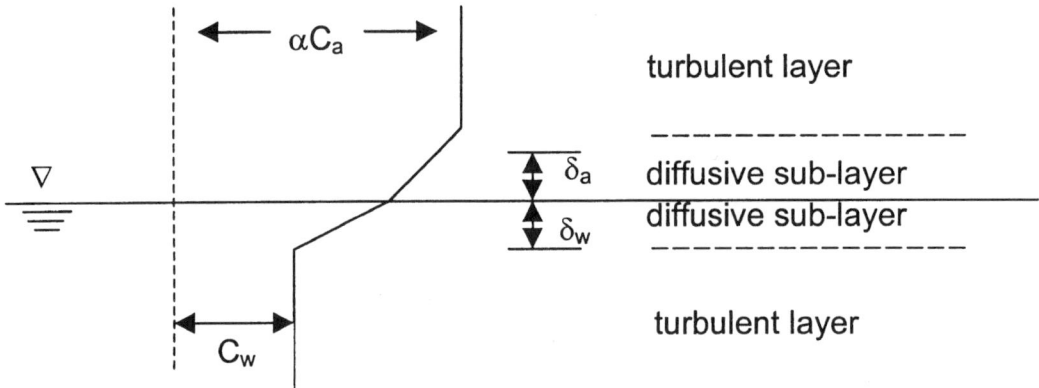

Figure 1. The four layer concept of resistance to gas transfer at water surfaces.

is light and the surface unbroken, this would seem to be a reasonable analogy. However, the laboratory experiments of Ocampo-Torres et al. [1994] show that the transfer of H_2O (unequivocally air-phase limited) is slower by a factor of 2 in light winds compared to the well established solid porous wall evaporation rates. An explanation for this still seems to be lacking. On the other hand, one would expect that the direct mixing of the diffusive aqueous sub-layer, brought about by wave breaking, would greatly enhance the transfer of water phase limited gases like CO_2. As long ago as 1987, Jähne et al. demonstrated this enhancement, and the field measurements of the mass transfer of slightly soluble gases show a strong wind speed dependence (as strong as cubic) that is consistent with enhancement related to the onset of breaking (e.g. Wanninkhof and McGillis [1999]).

CONCEPTUAL MODELS OF BOUNDARY LAYER MIXING

The concept of a resistive diffusive sub-layer provides the underpinning for the models of gas transfer at water surfaces. If this layer is stretched or eroded in some way the result will be an increase of the gas flux, at least locally. For simplicity we restrict the following discussion to the aqueous sub-layer, although many of the ideas apply equally well to both sides of the interface. The models differ in the way in which turbulence produced outside the layer acts to erode it or to mix it altogether. The turbulence may be produced by interaction with the bottom, as in river flows; by the breakdown of the wind-driven shear layer; by the rupture of the surface caused by breaking waves or the return of bubbles to the surface; or by the interaction of injected momentum from wave breaking into the relatively vorticity free environment beneath the waves. The choice of appropriate model depends on which of these agents for thinning or rupturing the diffusive sub-layer is considered to be dominant. The possibility that many act together with varying degrees of efficiency suggests a composite model. However, the attempts so far concentrate on one process at a time.

Mixing Length Models

If the source of turbulence is the breakdown of the shear flow imposed by a tangential stress at the surface, the simplest model is the well-established mixing length model first introduced by Prandtl [1925], which leads to a turbulent diffusivity, D_t:

$$D_t = \ell^2 \frac{\partial U}{\partial z} \qquad (1)$$

where ℓ is Prandtl's mixing length and $\frac{\partial U}{\partial z}$ the local velocity gradient (shear). Prandtl hypothesized that this length, characteristic of the turbulent scales, should be related to the distance from the boundary: $\ell = \kappa z$. Experiments yield: $\kappa = 0.4$, von Karman's constant. Van Driest [1956] generalized the mixing length concept to allow a smooth transition from fully turbulent flow away from the wall to fully laminar flow at the wall:

$$\ell = \kappa z[1 - \exp(-zu_*/26\nu)] \qquad (2)$$

where u_* is the friction velocity and ν the kinematic viscosity. This important advance deals only with flow over smooth surfaces, i.e. where the momentum transfer to the surface is entirely by molecular processes and there is no "form drag" or momentum transfer to the roughness elements on the surface through windward/leeward differences in pressure caused by flow around the roughness elements. In air-sea interaction the air flow is considered to be entirely rough, in which case virtually all momentum transfer occurs through pressure forces, for wind speeds in excess of 7.5 m/s. This is roughly the global average for marine winds, so a suitable model has to be able to deal with transitional and fully rough conditions as well as smooth. Riley et al. [1982] have proposed a variant of (2) to deal with the full range of roughness:

$$\ell = \ell_s + \kappa z[1 - \exp(-zu_*/13\nu)]^2 \quad (3)$$

where ℓ_s is the surface value of the mixing length, which is proportional to the heights of the roughness elements for rough flow and vanishes for smooth flow. The roughness length for gas transfer is then [Donelan, 1990]:

$$\ell_g = 1.24 Re_*^{-0.5} Sc^{-0.39} \ell_s + Sc_t^{-0.5} \kappa z[1 - \exp(-zu_*/13\nu)]^2 \quad (4)$$

where ℓ_g is the roughness length for gas transfer, Re_* is the roughness Reynolds number, Sc is the molecular Schmidt number and Sc_t the turbulent Schmidt number, which is 0.85 according to Kader and Yaglom [1972]. The exponents have been determined empirically.

Surface Renewal Models

Surface renewal models, first introduced by Dankwerts [1951] have gained wide acceptance in the gas transfer community. They are based on the concept that turbulent eddies bring fluid from the bulk to the surface periodically and it is this "renewal rate", r, that determines the rate of gas exchange rather than the thickness of the diffusive layer. The renewal events are assumed to occur randomly with a frequency and intensity that are related to the turbulence properties in the bulk or to the degree of wave breaking. The corresponding gas transfer velocity, k is given by [Gulliver, 1991]:

$$k = \sqrt{Dr} \quad (5)$$

where D is the molecular diffusivity of the gas.

This leads to an exponential profile of concentration in the diffusive sub-layer that has to be matched to the well-established logarithmic profile in the fully turbulent layer away from the wall [Kraus and Businger, 1994].

CHEMICAL BACKGROUND

Solubility

Air-sea gas fluxes are governed by physical forcing at the interface as described above and the chemical potential gradient of the gas across the interface. This gradient, which for slightly soluble gases is in the water boundary layer, is controlled by several factors. These include the composition of the water, the reactivity of the exchanging gas of interest, and the physical characteristics of the gas, such as its diffusion coefficient and solubility [Liss and Merlivat, 1986].

The bulk flux estimate is commonly expressed as $F = k \Delta\mu$ where $\Delta\mu$ is the chemical potential gradient of the gas between water and air. The chemical potential gradient can be expressed in different ways involving different expressions of solubility. In terms of concentrations it can be described as $(C_w - \alpha C_a)$ where α is the dimensionless Ostwald solubility coefficient defined as the volume of gas (at T and P) dissolved per unit volume of solvent, and subscripts "w" and "a" stand for water and air. The flux can be expressed as $(C_w - K_s p_a)$ where p_a is the partial pressure/fugacity of the gas in air and K_s is the solubility coefficient [mol l^{-1} atm^{-1}]. The gradient is also often measured in terms of partial pressure (or fugacity) of the gas in air and aqueous phases: $K_s (p_w - p_a)$. The solubility of a gas is gas specific and inversely related to temperature. It is of the order of -2 % °C^{-1} for slightly soluble gases [Wilhelm et al., 1977]. The solubility of Helium is a notable exception in that it hardly shows any dependence on temperature. The solubility of gas in seawater is about 20% less then in fresh water. The salinity dependence can be expressed as the Setschenow salting out coefficient: $b = \exp(-A(t) S)$ where $A(t)$ is gas and temperature dependent.

Diffusion Coefficient

The aqueous diffusion coefficient of the gas affects its rate of transfer. It is gas specific and a function temperature. For most gases the diffusion coefficients are within a factor of two (1-2 10^{-5} cm^2 s^{-1} @20 °C) except for He and H which have coefficients about six times higher than other gases. Diffusion coefficients increase by about a factor of three over ambient temperature ranges with the diffusion in salt water being about 5 % lower then in fresh water. For gas transfer studies the effect of diffusion is often expressed in terms of the Schmidt number, defined as the kinematic viscosity of the water divided by the diffusion coefficient. Gas transfer rates of different gases and under different conditions can be related through the Schmidt number (Sc): $k_a = k_b (Sc_a/Sc_b)^x$ where x is the

Schmidt number dependence that is -2/3 for smooth surfaces and -1/2 for rough surfaces.

Air/water Phase Limited

The solubility and aqueous reactivity of the gas determines if the gas transfer will be limited in the aqueous or air boundary layer. For gases with low solubility ($\alpha < 5$), which include all the constituents of dry air down to N_2O at the 0.5 ppm level, the resistance to gas transfer is predominantly in the liquid boundary layer. The exchange of gases with high solubility ($\alpha > 15$) and reactivity are limited by transfer through the air boundary. These gases include SO_2, O_3, and H_2O. Gases that fall between this range, such as DMS, will experience part liquid and part airside resistance, and the predominance will depend, in part, on the turbulence conditions and temperature. At higher winds the air side resistance will increase in importance [*McGillis et al.*, 2000].

Surfactants

Surfactants can affect air-water gas fluxes in several ways. At high concentrations the surfactants can act as the third medium, in which case an added boundary layer has to be taken into account since gas will have to transfer through the water boundary layer, surfactant, and air boundary layer. The solubility of the gas in the, often heterogeneous, surfactant material will affect the rate of transfer. This condition seldom occurs in open ocean waters. More often, the surfactant layer is not continuous or little more than a mono-molecular layer such that most of its effect is related to decreasing the gas transfer velocity rather than influencing the chemical potential gradient. Surfactants damp waves and decrease near-surface turbulence [*Frew*, 1997]. The effect is pronounced even with slightly soluble surfactants and in cases where surfactant concentrations are less than required to create a mono-molecular surface. For inland and coastal waters, organic pollutants can be a significant source of surfactants. In the open ocean surfactants are believed to originate primarily as marine exudates and are poorly characterized other than being a mixture of (slightly soluble) proteins, carbohydrates and lipids. The review article by Hunter [1997] provides a good overview about the origin and composition of surfactants in natural waters. The effect of marine surfactants on gas exchange was elegantly shown by Frew [1997] using surface seawater samples, ranging from coastal water to oligotrophic ocean water. In a small wind-wave tank the waters were subjected to a constant airflow over the surface. A tenfold difference in air-sea gas transfer velocities was observed with the coastal waters, containing the highest levels of surfactants, showing the lowest gas transfer.

LABORATORY MEASUREMENTS OF GAS TRANSFER

Wind-wave tanks offer unique environments to execute controlled studies of air-water gas transfer. Much of the improvements in understanding of gas transfer originate from investigation in tanks. However, tanks also have several limitations, particularly when extrapolating tank study results to the natural environment. In particular, the limited fetch and wall effects are recognized as being different from most natural conditions. Therefore tank studies are useful to elucidate processes but the information may not be quantitatively transferred to the natural environment.

A series of studies in a variety of tanks have been performed over the past 30 years to study the effect of wind on gas exchange focusing mostly on transfer of slightly soluble gases. The study of Broecker et al. [1978] showed three regimes for air-water gas transfer, the smooth regime at low wind speeds, the wavy regime at intermediate winds, and the breaking wave regime at high winds. In each regime the gas transfer was linearly related to wind forcing. Subsequent studies confirmed this behavior but the transitions occurred at different wind speeds, in part due to different tank geometries but in large part because of varying cleanliness of the water. The tanks with higher surfactant concentrations showed the smooth surface regime extending to higher wind speeds. Limited studies with soluble gases such as H_2O show two to three orders of magnitude higher exchange and no abrupt increase in the transfer rate with the onset of wave breaking [*Liss*, 1983; *Ocampo-Torres et al.*, 1994].

By studying the exchange of several gases (and heat) simultaneously, the Schmidt number dependency has been shown to vary from -2/3 for smooth surfaces to -1/2 for wavy surfaces [*Jähne et al.*, 1987]. Exploratory studies have shown that gas exchange can be related to scatterometer return [*Wanninkhof and Bliven*, 1991], opening avenues to employ remote sensing to determine regional fluxes. Comparison of gas exchange measurements from different size tanks and tanks with different configurations, such as linear and circular "infinite fetch" tanks, show different relationships with wind speed but the results collapse to a narrow envelope when related to total mean square wave slope. Careful investigations in which the different wave number ranges of the surface wave spectrum are related to air-sea gas transfer suggest that it is the very short waves that influence the rate of gas transfer most [*Bock et al.*, 1999].

FIELD ESTIMATES OF GAS TRANSFER

Natural and Deliberate Tracers

Several different methods have been used to quantify gas transfer. In natural systems they usually involve taking advantage of a natural perturbation of gases from equilibrium by studying a gas that undergoes radio-active decay or a gas that is consumed /released by biological degradation /production. Equilibrium concentrations can also be perturbed by deliberately adding or removing gas. In all these cases the gas transfer is determined by performing a mass balance of the gas in the water column. ^{222}Radon has been used extensively as a natural tracer [Peng et al., 1979]. In a closed system it is in equilibrium with its parent ^{226}Ra. However, in surface water some of the ^{222}Rn escapes to the air phase prior to its decay ($t_{1/2}$ = 4.5 days). From the disequilibrium between the parent and daughter, along with the half-life of ^{222}Rn, the gas transfer velocity can be determined. Since the response time is directly related to the decay constant, the method lends itself best to determining average gas transfer rates over a period of 1 to 2 weeks. Entrainment of water from below and horizontal variability can bias the interpretation of the results but the method, if executed properly, can yield useful regional constraints [Roether and Kromer, 1984].

Natural variability in oxygen levels has been used as well to estimate gas transfer velocities [Redfield, 1948]. One method used in biologically productive areas in the open ocean is to monitor oxygen levels inside and outside *in situ* incubation chambers. While concentrations inside the chamber increase because of net biological productivity, the levels outside the chamber are modulated by loss through gas exchange. Assuming that there are no significant bottle artifacts, the difference in rate of change of O_2 inside and outside the chamber is a direct indication of gas transfer. Again, there are uncertainties in the results because of natural spatial and temporal variability in the water column that is difficult to quantify.

Deliberate injections of trace gases are a powerful means to determine gas transfer if the mass decrease with time can be properly accounted for. The principle is fairly simple in that the equilibrium concentration is perturbed by adding gas and following the mass decrease (dM/dt) over time. The gas transfer velocity is estimated according to:

$$k = dM/dt \ A^{-1} \ (C_w - \alpha C_a)^{-1} \quad (6)$$

For a homogeneous system the flux, dM/dt A^{-1} can be replaced by dC/dt h, where h is the average depth of the water exchanging with the atmosphere. After integration this yields:

$$k = h \ \Delta t^{-1} \ \ln((C_i - \alpha C_a)/(C_f - \alpha C_a)) \quad (7)$$

Thus the gas transfer is determined from monitoring the decrease in air-water concentration difference (($C_i - \alpha C_a$)/($C_f - \alpha C_a$)) over time Δt. This approach has been successfully used to study gas transfer rates in laboratory and lake experiments [Wanninkhof et al., 1985]. In particular, the tracer sulfur hexafluoride has been extensively used because of its low background in water (< 2 10^{-15} M), its inertness, and ease of analysis down to the sub-part per trillion range. However, for larger systems the total gas decrease in the water column is difficult to assess due to rapid changes in concentrations due to dispersion. In these cases a combination of non-volatile and volatile tracers can be used to assess the gas transfer rates by determining the rate of change in the ratio of the two tracers over time.

The earliest dual-tracer studies were performed in stream reaeration studies using the radioactive tracers ^{85}Kr and tritiated water [Tsivoglou, 1967]. Environmental concerns of adding radioactive materials to the environment have limited their use. As an alternative to the non-volatile/volatile tracer method, two volatile tracers with differing molecular diffusivity can be used as long as their relative rates of exchange are well known. The isotope ^{3}He in combination with SF_6 has been deployed in several open ocean studies [Nightingale et al., 2000; Wanninkhof et al., 1993; Watson et al., 1991]. The interpretation of the results, particularly as applied to CO_2 and other slightly soluble gases, at high winds is somewhat ambiguous because of the uncertainties of the role of bubbles in exchange processes under those conditions. In bubble mediated gas transfer solubility is a factor and simple conversions based on Schmidt number dependencies do not apply [Asher and Wanninkhof, 1998]. Fluorescent dyes and spores have been used as non-volatile tracers in a few open ocean experiments and give similar results to those based on $SF_6/^3$He suggesting that such tracers might be an attractive alternative to the dual gaseous tracer approach [Nightingale et al., 2000]. However, the long-term stability of the spores and dyes has not been fully investigated.

The deliberate tracer studies provide a means to determine gas transfer rates in the natural environment on time scales ranging from 1/2 to 4 days, which is the time required to realize a sufficiently large gas decrease in the water column to accurately determine the gas transfer. The results provide a first-order understanding of the effect of environmental forcing on gas transfer but the integration time necessary to obtain robust gas exchange measurement is longer than the time scale of variability of wind and other forcing thus preventing accurate parameterization or elucidation of processes controlling gas transfer. However, by its integrating nature it can provide robust constraints for other measurement on shorter time scales such as the direct flux measurements.

Eddy Correlation Method

The most direct estimates of the gas transfer rate are made under the assumption of stationarity and homogeneity allowing a local measurement to be representative of the flux. Near to the surface it is generally valid to assume that there is no convergence/divergence of the flux, so that the rate of passage of a gas through a chosen measurement level is representative of the air-water interfacial transfer. In these circumstances the vertical flux of a "contaminant" of concentration, C is brought about by the correlation of vertical velocities, w', with variations in the mass concentration. Thus a measurement of the fluctuations of vertical velocity and mass concentration at the same point yields the flux of the contaminant [*Donelan*, 1990]. In practice this is difficult for two reasons: 1) the vertical velocity is measured relative to a moving platform (ship or buoy) at sea and careful measurements of and corrections for these motions have to be made [*Katsaros et al.*, 1993]; 2) instruments capable of enough resolution and response speed (> 2 Hz) for the measurement of concentration are uncommon. In the air, infra-red absorption devices are often used, but the gradients of slightly soluble (water phase limited) gases are quite small in the air, as are the corresponding fluctuations. In principle these measurements could be made in either (air or water) turbulent boundary layer, but in practice measurement in the water boundary layer is extraordinarily difficult because the wave orbital velocities are orders of magnitude bigger than the turbulent velocities – the former being effectively "noise" in this context.

Inertial Dissipation Method

The difficulties associated with the direct (eddy correlation) method for the estimation of fluxes at sea has led to the use of the "inertial dissipation" method [*Fairall and Larsen*, 1986]. The method hinges on an approximate balance of the generation and dissipation of turbulent fluctuations of the concentration of the gas whose interfacial flux is to be measured. Consider the budget of concentration variance in the atmospheric boundary layer:

$$-\overline{C'w'}\left(\frac{\partial C}{\partial z}\right) - \frac{1}{2}\frac{\partial \overline{C'^2 w}}{\partial z} = \varepsilon_c \quad (8)$$

where the first term is the rate of production of concentration fluctuations arising from the action of the turbulent velocities on the gradient of concentration, the last term is the rate of destruction of these fluctuations due to molecular diffusivity and is believed to approximately balance the production term [*Edson and Fairall*, 1998]. The middle term is the divergence of fluctuations and is generally accounted for by corrections based on similarity relations between fluxes and gradients. The rate of dissipation is related to the mean square gradients of the concentration:

$$\varepsilon_c = 3D_c \overline{\left(\frac{\partial C'}{\partial x}\right)^2} = 3D_c U^{-2} \overline{\left(\frac{\partial C'}{\partial t}\right)^2} B_c \quad (9)$$

The second equation invokes Taylor's "frozen turbulence" hypothesis to relate the spatial gradients to time derivatives of the point measurements, where B_c is a correction [*Wyngaard and Clifford*, 1977] to this hypothesis.

Finally, the idea of separation of scales of production (large) from scales of dissipation (very small) due to Kolmogoroff leads to a region in the frequency spectrum of concentration fluctuations that follows a "–5/3 law", whose level reflects the cascade towards molecular dissipation at higher frequencies:

$$\varepsilon_c = \frac{1}{\alpha_c}\left(\frac{2\pi}{U}\right)^{\frac{2}{3}} \varepsilon_e^{\frac{1}{3}} f^{\frac{5}{3}} S_{cc}(f) \quad (10)$$

where ε_e is the turbulent kinetic energy dissipation rate, α_c is the Kolmogoroff constant for scalars and $S_{cc}(f)$ is the frequency spectrum of the concentration fluctuations. Then, by equating dissipation to generation and knowing the dependence of the overall gradient of concentration (see below), the gas transfer rate, $\overline{C'w'}$ may be determined.

Profile Method

This method depends on the relationship between gas flux and the gradient of concentration of the gas -- usually on the air side.

$$\frac{\partial C}{\partial z} = -\frac{\overline{C'w'}}{u_* \kappa z}\Phi_c\left(\frac{z}{L}\right) \quad (11)$$

where Φ_c is the non-dimensional gradient of concentration and is a function of the Monin-Obukhov stability index, (z/L), [*Monin and Obukhov*, 1954]. The non-dimensional gradients have been established for terrestrial boundary layers [*Businger et al.*, 1971; *Donelan*, 1990], but have yet to verified in the marine boundary layer. This method requires a measurement of the small concentration

Mean Annual Air-Sea Flux for 1995 (NCEP 41-Yr Wind, 940K, W-92)

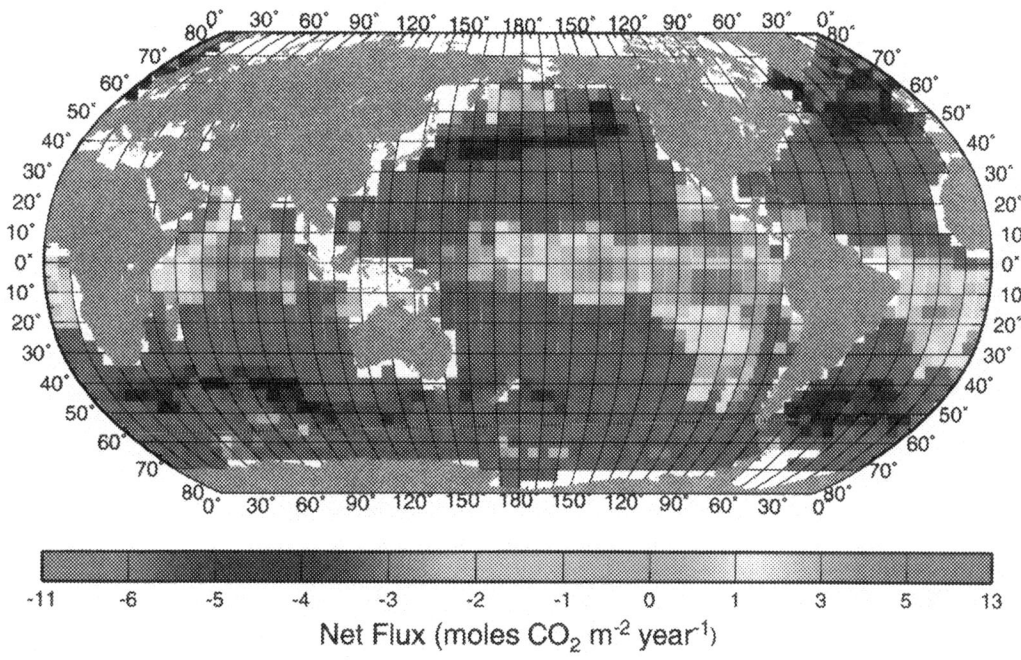

Figure 2. Annual average CO_2 fluxes based on the updated climatology of Takahashi et al. [1997, 1999]. The low latitude regions are areas where CO_2 is out gassing from the ocean and high latitude areas are where it enters the ocean.

difference between two heights and, in addition, the friction velocity, u_*, must be measured and the non-dimensional gradient, Φ_c, known. The distortion to the flow around a ship adds to the difficulties inherent in applying the profile method at sea.

PARAMETERIZATION OF GAS FLUXES

Global CO_2

The kinetics of gas transfer, often expressed as the gas transfer velocity, k, are controlled by different factors and on different time scales than the thermodynamic controls of the air water concentration difference. Therefore, the gas fluxes are often expressed as the product of the two and each factor is parameterized and/or measured separately. Because of its environmental importance as a greenhouse gas, and because the ocean is the largest natural CO_2 sink, much work has been devoted to estimating the global CO_2 flux. The first global monthly climatology of CO_2 was recently produced by applying 30 year's worth of observations of ΔpCO_2 in a surface global advection model run on a 4° x 5° grid [*Takahashi et al.*, 1997; *Takahashi et al.*, 1999] (Figure 2). The response time of CO_2 in the ocean relative to gas exchange is rather slow, because carbon dioxide in the ocean is chemically buffered, and a monthly climatology adequately captures the seasonal variability. Despite inclusion of 950,000 data points, the climatology is still data limited with roughly 1/3 of the time/space grid lacking observations and relying on extrapolations. This monthly ΔpCO_2 climatology is multiplied by a gas transfer velocity, derived from wind speed, at each grid point to estimate monthly CO_2 fluxes.

Many parameterizations of gas exchange with wind have been reported primarily based on a combination of field and laboratory results (Figure 3). There are significant differences in the relationships due to uncertainty in the field estimates and lack of field measurements at high wind speed, but primarily because other factors beside wind affect gas transfer. The different parameterizations yield global CO_2 uptakes that differ by a factor of 3 (Table 1). However, there are several independent global constraints based on the distribution of radiocarbon (^{14}C) [*Broecker et al.*, 1985], carbon isotopic ratios ($^{13}C/^{12}C$), and oxygen/nitrogen ratios (O_2/N_2) between the ocean, terrestrial biosphere and atmosphere [*Battle et al.*, 2000] that suggest

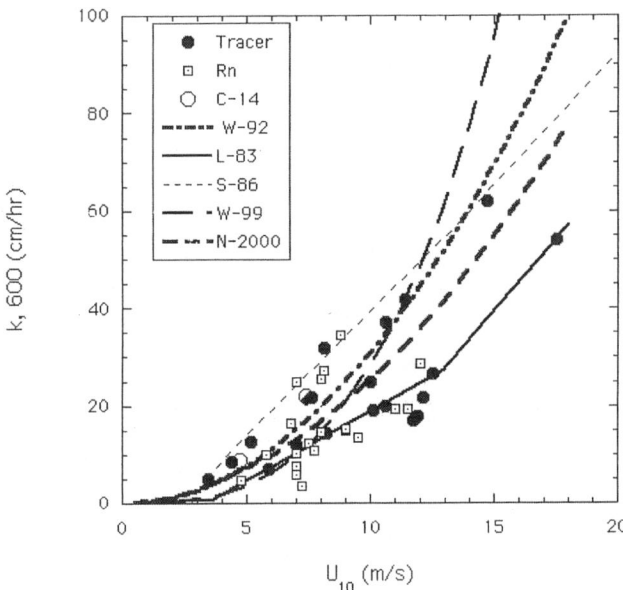

Figure 3. Summary of gas exchange results in the ocean and empirical relationships derived, in part, from this data. All data have been normalized to $Sc=600$. Dual-tracer results, Tracer; ^{222}Rn results, Rn; Global estimate based on bomb-radiocarbon, C-14. The empirical relationships are those of Liss and Merlivat [1983], L-83; Wanninkhof [1992], W-92; Smethie et al. [1986], S-86; Wanninkhof and McGillis [1999], W-99; and Nightingale et al. [2000], N-2000.

an average uptake of 2.2 ± 0.5 Pg C yr^{-1}. Year-to-year variability in oceanic CO_2 uptake remains poorly constrained but recent estimates suggest a small interannual variability (< 0.5 Pg C) that appears to be correlated with the El Niño cycle [*Le Quere et al.*, 2000; *Lee et al.*, 1998].

When averaging winds an artifact arises from the non-linear nature of the gas exchange relationship, in which the longer averaging intervals give a low bias in the gas transfer. Several relationships partially account for this by incorporating global wind speed distributions into the formulation [*Wanninkhof*, 1992]. For strongly non-linear dependency, such as the recently proposed cubic dependency of gas exchange [*Wanninkhof and McGillis*, 1999], the difference between determining the gas flux using a monthly mean wind or summing the hourly fluxes over a month can be a factor of 2.

The non-unique relationship between gas transfer and wind and the effect of variable forcing argues for an approach of determining regional fluxes over short time scales to determine the global uptake of CO_2. Satellite sensors such as scatterometers with near global coverage on daily time scales offer new opportunities to better constrain the global fluxes.

Pollutant Transfer

Aside from quantifying the flux of CO_2 and other climate relevant gases over the ocean, there is substantial

Table 1. Global oceanic CO_2 uptake estimates using different gas exchange-wind speed relationships and different wind speed products[a].

Relationship	Equation	Flux (PgC yr^{-1})
[*Liss and Merlivat*, 1986]	$k= 0.17\ U_{10}$ ($U_{10} < 3.6$ m s^{-1}) $k= 2.85\ U_{10} - 9.65$ (3.6 m s^{-1} < U_{10} < 13 m s^{-1}) $k= 5.9\ U_{10} - 49.3$ (U_{10} > 13 m s^{-1})	-1.1
[*Wanninkhof*, 1992]	$k= 0.39\ U_{10}^2$ (long term averaged winds)	-2.2
[*Wanninkhof and McGillis*, 1999]	$k= 1.09\ U_{10} - 0.333\ U_{10}^2 + 0.078\ U_{10}^3$ (long term averaged winds)	-3.3
Nightingale et al. [2000] NCEP 6-hour winds	$k= 0.333\ U_{10} + 0.222\ U_{10}^2$	-1.7
NCEP-40 year average winds[b]	$k= 0.39\ U_{10}^2$ (long term averaged winds)	-2.4
NCEP 6-hour winds[c]	$k= 0.31\ U_{10}^2$ (instantaneous winds)	-2.1
NCEP 6-hours winds[c]	$k = 0.0283\ U_{10}^3$	-2.6

a) for these calculations the monthly ΔpCO_2 climatology of Takahashi et al. (1999) was used according to: $F = k\ (Sc/660)^{-1/2}\ K_s\ \Delta pCO_2$, Sc is the Schmidt number, which is determined for each pixel from climatological SST. The solubility, K_s was determined from standard relationships with SST and salinity. Salinity is assumed to be 3.5% throughout. Unless noted the monthly mean NCEP wind speeds for 1995 were used.

b) Using the NCEP 40-year average monthly wind speed product rather than that of 1995

c) Using the NCEP 6-hour wind product. In these cases the instantaneous wind speed formulations of Wanninkhof [1992] and Wanninkhof and McGillis [1999] were used.

interest in quantifying fluxes for water quality issues. Some of the earliest work has been on quantifying oxygen uptake in rivers with high biological oxygen demand, such as rivers receiving paper mill effluents [O'Connor and Dobbins, 1958]. Gas exchange in rivers is often defined as reaeration (= k/h). Several predictive equations have been developed for oxygen exchange in rivers that include parameters such as depth, hydraulic gradient, stream flow, and bed roughness. Aside from O_2 and SF_6, propane and chloro-fluoro carbons are sometimes used to determine reaeration rates.

(Inland) water bodies can be a source of gaseous pollutants that result from industrial dumping of solid or liquid contaminants into lakes and rivers. Polychlorinated biphenyls (PCBs) are examples of industrial effluents that are in inland waters such as the Great Lakes and Hudson River [Bopp, 1983]. Large amounts of PCBs in the sediments of these water bodies slowly dissolve into the water column. The polymers that have significant vapor pressure will subsequently escape by air-water transfer. The resulting atmospheric levels, particularly near hydraulic structures that generate high aeration rates, can be significant and cause a long term health risk.

SUMMARY

We have outlined the physical and chemical background of the processes of gas transfer between air and water and discussed the methodology of "classical" field and laboratory measurements. The remaining 57 papers in this monograph deal with recent progress in the field. They are reports of independent investigations but, taken altogether, they give a balanced view of the state-of-the-art in research into gas transfer at water surfaces.

REFERENCES

Asher, W.E., and R. Wanninkhof, The effect of bubble-mediated gas transfer on purposeful dual gaseous-tracer experiments, *J. Geophys. Res.*, *103*, 10555-10560, 1998.

Battle, M., M. Bender, P. Tans, J.W.C. White, J.T. Ellis, T. Conway, and R.J. Francey, Global carbon sinks and their variability inferred from atmospheric O_2 and $\delta^{13}C$, *Science*, *287*, 2467-2470, 2000.

Bock, E.J., T. Hara, N. M. Frew, and W.R. McGillis, Relationship between air-sea gas transfer and short wind waves, *104*, 25,821-25,831, 1999.

Bopp, R.F., Revised parameters for modeling the transport of PCB components across an air water interface, *J. Geophys. Res.*, *88*, 2521-2529, 1983.

Broecker, H.C., J. Peterman, and W. Siems, The influence of wind on CO_2 exchange in a wind-wave tunnel, including the effects of mono layers, *J. Mar. Res.*, *36*, 595-610, 1978.

Broecker, W.S., T.-H. Peng, G. Östlund, and M. Stuiver, The distribution of bomb radiocarbon in the ocean, *J. Geophys. Res.*, *99*, 6953-6970, 1985.

Businger, J.A., J.C.Wyngaard, Y.K.Izumi and E.F.Bradley, Flux-profile relationships in the atmosphere surface layer, *J. Atmos. Sci.* 28, 181-189, 1971

Dankwerts, P.V., Significance of liquid-film coefficients in gas absorption, *Industrial and Engineering Chemistry*, 43(6), 1460-1467, 1951.

Donelan, M.A., Air-Sea Interaction, in *The Sea: Ocean Engineering Science 9*, edited by B. LeMehaute and D. Hanes, pp. 239-292, John Wiley and Sons, Inc., New York, 1990.

Edson, J.B., and C.W. Fairall, Similarity relationships in the marine atmospheric surface layer for terms in the TKE and scalar variance budgets. *J. Atmos. Sci.* 55, 2311-2328, 1998.

Fairall, C.W. and Larsen, S.E., Inertial-dissipation methods and turbulent fluxes at the air-ocean interface. *Bound. Layer Meteorol.* 34, 287-301, 1986.

Frew, N.M., The role of organic films in air-sea gas exchange, in *The sea surface and global change*, edited by P.S. Liss, and R.A. Duce, pp. 121-163, Cambridge University Press, Cambridge, 1997.

Gulliver, J.S., Introduction to air-water mass transfer, in *Air-Water Mass Transfer*, edited by S.C. Wihelms and J.S. Gulliver, pp. 1-7, ASCE, New York, 1991.

Houghton, J.T., L.G. Meira Filho, J. Bruce, H. Lee, B.A. Callander, E. Haites, E. Harris, and K. Maskell, *Climate Change 1994: Radiative forcing of climate change and an evaluation of the IPCC IS92 Emission Scenarios*, Cambridge University Press, Cambridge, 1995.

Hunter, K.A., The role of organic films in air-sea gas exchange, in *The sea surface and global change*, edited by P.S. Liss, and R.A. Duce, pp. 287-319, Cambridge University Press, Cambridge, 1997.

Jähne, B., K.O. Münnich, R. Bösinger, A. Dutzi, W. Huber, and P. Libner, On parameters influencing air-water gas exchange, *J. Geophys. Res.*, *92*, 1937-1949, 1987.

Katsaros, K.B., M.A. Donelan and W.M. Drennan, Flux measurements from a Swath ship in SWADE. *J. Mar. Sys.*, 4, 117-132, 1993.

Kitaigorodskii, S.A. and M.A. Donelan, Wind-wave effects on gas transfer, *in Gas transfer at the water surfaces*, edited by W. Brutsaert and G.H. Jirka, pp. 147-170, Reidel, 1984.

Kraus, E.B. and J.A. Businger, *Atmosphere-Ocean Interaction (2nd edition)*, Oxford University Press, New York, 362 pages, 1994.

Le Quere, C., J.C. Orr, P. Monfray, P. Aumont, and G. Madec, Interannual variability of the oceanic sink of CO_2 from 1979 to 1997, *Global Biogeochem. Cycles*, *14*, 1247-1265, 2000.

Lee, K., R. Wanninkhof, T. Takahashi, S. Doney, and R.A. Feely, Low interannual variability in recent oceanic uptake of atmospheric carbon dioxide, *Nature*, *396*, 155-159, 1998.

Liss, P.S., Gas transfer: experiments and geochemical implications, in *Air-sea exchange of gases and particles*, edited by P.S. Liss, and W.G. Slinn, pp. 241-299, Reidel, Dordrecht, The Netherlands, 1983.

Liss, P.S., and L. Merlivat, Air-sea gas exchange rates: Introduction and synthesis, in *The Role of Air-Sea Exchange in Geochemical Cycling*, edited by P. Buat-Menard, pp. 113-129, Reidel, Boston, 1986.

Monin, A.S., and A.M. Obukhov, Basic laws of turbulent mixing in the ground layer of the atmosphere, *Akad. Nauk. SSSR Geofiz.Inst. Tr.* 151, 163-187, 1954.

McGillis, W.R., J. W. H. Dacey, N. M. Frew, E. J. Bock, and B. K. Nelson, Water-air flux of dimethylsulfide., *J. Geophys. Res.*, 105, 1187-1193, 2000.

Nightingale, P.D., G. Malin, C.S. Law, A.J. Watson, P.S. Liss, M.I. Liddicoat, J. Boutin, and R.C. Upstill-Goddard, In situ evaluation of air-sea gas exchange parameterizations using novel conservative and volatile tracers, *Global Biogeochem. Cycles*, 14, 373-387, 2000.

Ocampo-Torres, F.J., M.A. Donelan, N. Merzi and F. Jia, Laboratory measurements of mass transfer of carbon dioxide and water vapour for smooth and rough flow conditions, *Tellus, 46B,* 16-32, 1994.

O'Connor, D.J., and W.E. Dobbins, Mechanism of reaeration in natural streams, *Trans. Am. Soc. Civ. Eng.*, 123, 641-684, 1958.

Peng, T.-H., W.S. Broecker, G.G. Mathieu, Y.H. Li, and A.E. Bainbridge, Radon evasion rates in the Atlantic and Pacific Oceans as determined during the GEOSECS program, *J. Geophys. Res.*, 84, 2471-2486, 1979.

Prandtl, L., Bericht über Untersuchungen zur ausgebildeten Turbulenz, *Z. angew. Math. U. Mech.,* **5,** 136-137, 1925.

Redfield, A.C., The exchange of oxygen across the sea surface, *J. Mar. Res.*, 7, 347-361, 1948.

Riley, D.S., M.A. Donelan and W.H. Hui, An extended Miles' theory for wave generation by wind. *Boundary-Layer Meteorol.,* 22, 209-225, 1982.

Roether, W., and B. Kromer, Optimum application of the radon deficit method to obtain air-sea gas exchange rates, in *Gas Transfer at Water Surfaces*, edited by W. Brutsaert, and G.H. Jirka, pp. 447-457, Reidel, Hingham Mass., 1984.

Smethie, W.M., T.T. Takahashi, D.W. Chipman, and J.R. Ledwell, Gas exchange and CO_2 flux in the tropical Atlantic Ocean determined from ^{222}Rn and pCO_2 measurements, *J. Geophys. Res.*, 90, 7005-7022, 1985.

Takahashi, T., R.A. Feely, R. Weiss, R. Wanninkhof, D.W. Chipman, S.C. Sutherland, and T.T. Takahashi, Global air-sea flux of CO_2: An estimate based on measurements of sea-air pCO_2 difference, *Proc. Natl. Acad. Sci. USA*, 94, 8292-8299, 1997.

Takahashi, T., R.H. Wanninkhof, R.A. Feely, R.F. Weiss, D.W. Chipman, N. Bates, J. Olafsson, C. Sabine, and S.C. Sutherland, Net sea-air CO_2 flux over the global oceans: An improved estimate based on the sea-air pCO_2 difference, in *Proceedings of the 2nd International Symposium on CO_2 in the Oceans*, edited by Y. Nojiri, pp. 9-15, Center for Global Environmental Research, NIEST, Tsukuba, JAPAN, 1999.

Tsivoglou, E.C., *Tracer measurements of stream reaeration*, Fed. Water Pollution Control Administration, U.S. Dept. of Interior, Washington D.C., 1967.

Van Driest, E.R., On turbulent flow near a wall, *J. Aero. Sci.,* 23, 1007-1001.

Wanninkhof, R., Relationship between gas exchange and wind speed over the ocean., *J. Geophys. Res.*, 97, 7373-7381, 1992.

Wanninkhof, R., W. Asher, R. Weppernig, H. Chen, P. Schlosser, C. Langdon, and R. Sambrotto, Gas transfer experiment on Georges Bank using two volatile deliberate tracers, *J. Geophys. Res.*, 98 (C11), 20237-20248, 1993.

Wanninkhof, R., and L. Bliven, Relationship between gas exchange, wind speed and radar backscatter in a large wind-wave tank, *J. Geophys. Res.*, 96, 2785-2796, 1991.

Wanninkhof, R., J.R. Ledwell, and W.S. Broecker, Gas exchange - wind speed relationship measured with sulfur hexafluoride on a lake, *Science*, 227, 1224-1226, 1985.

Wanninkhof, R., and W.M. McGillis, A cubic relationship between gas transfer and wind speed, *Geophys. Res. Let.*, 26, 1889-1893, 1999.

Watson, A.J., R.C. Upstill-Goddard, and P.S. Liss, Air-sea exchange in rough and stormy seas, measured by a dual tracer technique, *Nature*, 349, 145-147, 1991.

Wilhelm, E., R. Battino, and R.J. Wilcock, Low pressure solubility of gases in liquid water, *Chem. Rev.*, 77, 219-262, 1977.

Yaglom, A.M. and B.A. Kader, Heat and mass transfer between a rough wall and turbulent flow at high Reynolds and Peclet numbers, *J. Fluid Mech.*, 62, 601-623, 1974.

Turbulence Generated by Microscale Breaking Waves and its Influence on Air-Water Gas Transfer

Muhammad H. K. Siddiqui[1,2], Mark R. Loewen[2], Christine Richardson[3], William E. Asher[3] and Andrew T. Jessup[3]

The results from a series of wind-wave flume experiments using simultaneous DPIV (digital particle image velocimetry) measurements and IR (infrared) imagery to investigate microscale breaking waves are presented. We show that the IR signatures of microscale breaking waves are produced by a series of strong vortices that form behind the leading edge of the breakers. These strong vortices disrupt the cool skin layer and generated a thin layer of enhanced turbulence immediately below the air-water interface. In addition we used CFT (controlled flux technique) to make measurements of the local heat transfer velocity and found that the transfer velocity was correlated with the near-surface vertical turbulent velocity. We conclude that near-surface turbulence generated by microscale wave breaking determines the transfer rate at low to moderate wind speeds.

1. INTRODUCTION

Breaking of small-scale waves without air entrainment is referred to as microscale breaking [*Banner and Phillips*, 1974]. Microscale breaking waves occur at low to moderate wind speed and are O (0.1-1) m in length, a few centimeters in amplitude and have a bore-like crest directly preceded by parasitic capillary waves riding along the forward face. In the field microscale breakers are far more wide spread than whitecaps and therefore, it had been speculated that microscale breaking could be important in controlling the flux of heat and gas across the air-water interface [*Melville*, 1996]. *Zappa et al.*, [2000] recently conducted a series of wind-wave flume experiments and showed that microscale breaking is the physical process that determines the gas transfer rate at low to moderate wind speeds.

Microscale breaking waves are difficult to detect using conventional techniques and their small scale makes quantitative measurements difficult. In addition, microscale breaking waves occur randomly in time and space and the orientation and scale of each breaking event may be different. This makes it impossible to ensemble average a series of microscale breaking events making their study even more challenging.

In order to investigate the flow field beneath microscale breaking waves the first step is to properly identify which waves are microscale breaking events. *Jessup et al.*, [1997b] have shown that infrared imagery is an effective method for detecting and quantifying microscale breaking waves. Their technique is based on the fact that under most circumstances the surface or skin temperature of the ocean is a few tenths of a degree Celsius less then the bulk water temperature immediately below [*Robinson et al.*, 1984]. Breaking waves momentarily disrupt this cool skin layer and the skin temperature of the resulting turbulent wake become approximately equal to the bulk water temperature. As the wake dissipates, the skin temperature returns to its original cooler value and the skin layer is reestablished. The infrared imager measures temperature changes only in the skin layer, because the optical depth of the

[1]Department of Mechanical and Industrial Engineering, University of Toronto, Toronto, Ontario, Canada
[2]Department of Civil and Environmental Engineering, University of Alberta, Edmonton, Alberta,, Canada
[3]Applied Physics Laboratory, University of Washington, Seattle, Washington, USA.

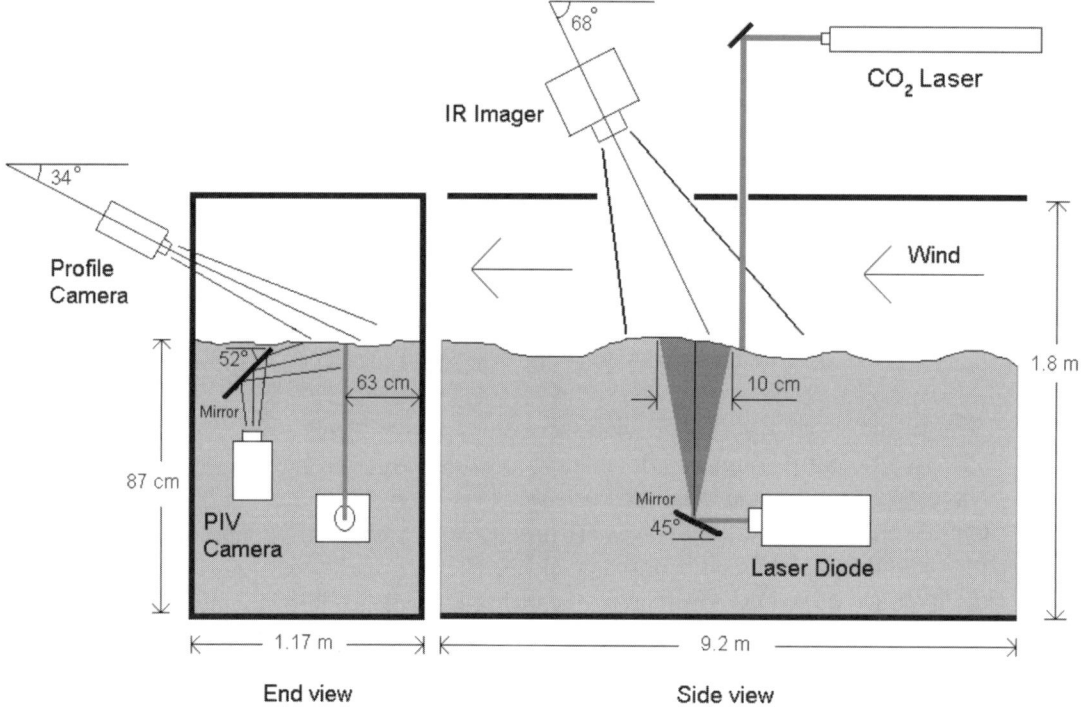

Figure 1. A schematic of the experimental setup and instrumentation in the wind-wave flume (not to scale).

infrared radiation detected is about 10 micron and this is much less then the thickness of the skin layer.

Recently, particle image velocimetry (PIV) has been used to measure the instantaneous two-dimensional flow fields beneath wind waves [*Peirson*, 1997; *Banner and Peirson*, 1998; *Hering et al.*, 1998]. However, these studies did not focus explicitly on microscale breaking waves. As a result, there is very little known about the characteristics of the turbulent flow fields produced by microscale breaking waves. We conducted a series of wind-wave flume experiments using simultaneous DPIV measurements and IR imagery to investigate the role of near-surface turbulence generated by microscale breaking waves in enhancing air-water gas transfer [*Siddiqui et al.*, 2000].

2. EXPERIMENTAL SETUP AND PROCEDURES

The experiments were conducted in a wind-wave flume at Harris Hydraulics Laboratory, University of Washington, Seattle. The flume is 9.2 m in length; 1.17 m in width and the water depth was 87 cm. A horsehair beach was placed at the downstream end of the tank to absorb wave energy. A centrifugal fan was fitted on the upstream end of the tank and it could produce wind speeds up to approximately 12 m s^{-1}. A schematic of the experimental setup and instrumentation is shown in Figure 1. Measurements were made at a fetch of 5.5 m at wind speeds ranging from 4.5 m s^{-1} to 11 m s^{-1}. The water surface was cleaned at the beginning of each day by vacuuming the surface. The water was heated prior to each experimental run and the average air-water temperature difference was maintained at approximately 10 °C. A listing of the experimental conditions is provided in Table 1.

The controlled flux technique (CFT) described by *Haußecker et al.* [1995] uses heat as a proxy tracer for gases. We implemented CFT using a CO$_2$ laser (Synrad H48-2-28S 25 Watt, 10 µm wavelength) pulsed for 40 ms to produce a 2-3 cm diameter heated circular patch on the water surface once every second. The temperature of each heated patch decays with time and the decay rate measured using IR imagery provides an estimate of the local heat transfer velocity, defined as k$_{600-H}$ [*Zappa*, 1999].

DPIV was used to measure the two-dimensional velocity in a plane parallel to the wind and bisecting the water surface. A 500mW, 678 nm wavelength diode laser was used as the light source (Magnum 500 SP, Lasiris). The diode laser was equipped with a 20° fan angle light head that created a uniform light sheet of approximately 200 µm thickness. A Pulnix TM-9701 progressive scanning full frame shutter camera with a resolution of 768 × 484 pixels^2 (cell size of 11.6µm × 13.6 µm) was used to image the

Table 1. Summary of environmental parameters for different experimental runs.

Experiment	Wind speed (m s^{-1})	Bulk air-water temperature difference (°C)	RMS wave height (mm)	Dominant wave frequency (Hz)
01	4.5	10.97	2.75	5.14
02	7.4	10.31	7.65	3.74
03	11.0	8.05	11.84	2.97

flow field. The field of view of the camera was 8.4 cm wide and 6.2 cm high. Both the laser and the DPIV camera were placed in underwater housings, see Figure 1.

Skin-layer temperature measurements were made using an Amber Radiance HS infrared imager with a resolution of 256 × 256 pixels^2. This imager is sensitive to radiation in the 3-5 μm wavelength band and its optical depth in water is approximately 10 μm. The IR imager was mounted on top of the tank looking down at an incidence angle of 22° with a field of view of approximately 34.3 × 34.3 cm^2, see Figure 1. For each experiment, 900, 12-bit digital IR images and 900, 8-bit DPIV images were acquired at the rate of 30 Hz.

In many DPIV images seed particles appear above the air-water interface due to reflections, making it impossible to locate the true position of the interface. It is very important to accurately locate the interface in the DPIV images in order to obtain reliable estimates of the near surface velocities. The interface in the DPIV images was located by first measuring the surface wave profile using a technique similar to that reported by *Banner and Peirson* [1998] and *Law et al.* [1999]. This technique is based on the fact that the laser light sheet is visible only in the water because of the high reflectivity of the seed particles. A second CCD camera (Cohu 2100 series) operating in the interlaced mode was used to image the surface wave profile, where the laser light sheet intersects the water surface. The water surface profile measurements were validated by comparing the wave properties computed from profile data with the wave properties computed from the wave gauge data [*Siddiqui et al.,* 2000].

Raw DPIV images were preprocessed and then the raw DPIV velocity fields were computed using a standard cross-correlation technique. The final velocity field was obtained by first removing spurious vectors and then interpolating the velocity data onto a regular grid. Details of the DPIV technique are presented in *Siddiqui et al.,* [2000]. The uncertainties in the velocity and vorticity measurements were estimated to be ± 0.4 cm s^{-1} and ± 2.9 s^{-1}, respectively.

The procedure used to compute the velocity field is illustrated in Figure 2. A typical profile image with the measured surface wave profile overlain on it is displayed in Figure 2a. This surface wave profile data is imported into the corresponding DPIV image, which is shown plotted in Figure 2b. Bright regions and reflected particles are clearly visible above the interface in this image. In this DPIV image the interface is visible on the forward face of the wave but in the rest of the image, it would be impossible to accurately locate the interface without the surface profile data. The corrected and interpolated velocity field is shown plotted in Figure 2c. For our experimental conditions the nearest velocity vector was located 2 mm or less from the interface and thus, on average the near-surface velocity was measured within 1-mm of the interface.

3. RESULTS AND DISCUSSION

Microscale breaking waves were identified in the IR images using a thresholding technique reported by *Zappa* [1999]. In this technique a temperature threshold was set for each IR image based on the mean and standard deviation of the skin temperature within that image. Wakes due to microscale breaking waves are identified as regions of the image that are warmer than the specified threshold. In Figure 3 a typical IR image and the corresponding instantaneous vorticity field obtained from the DPIV velocity field sampled at a wind speed of 11.0 m s^{-1} are shown plotted. The wakes associated with microscale breaking waves and the location of the DPIV field of view are delineated in the IR image with black lines. In the IR image (Figure 3a), the leading edge of a microscale breaking wave is close to the downwind end of the DPIV field of view and the wake is seen to extend upwind to the edge of the image. In the corresponding vorticity plot (Figure 3b), a series of vortices are observed upwind of the leading edge and the region of high vorticity is seen to extend to a depth of approximately 1-cm. The leading edge of the wake observed in the IR image (Figure 3a) corresponds to

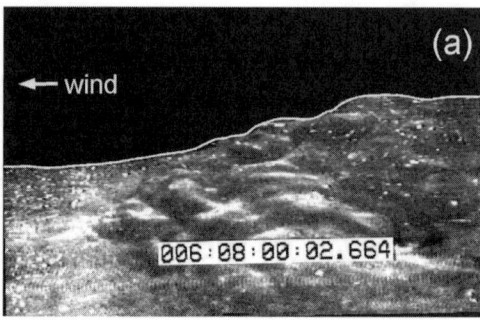

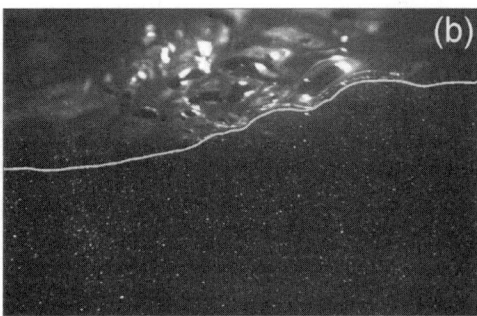

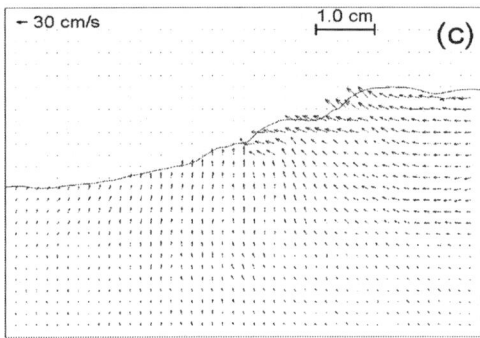

Figure 2. (a) A digital video image sampled with the profile camera at a wind speed of 11.0 m s^{-1} with the computed surface wave profile plotted as a white line. (b) The corresponding raw DPIV image with the water surface profile taken from (a) shown plotted as a white line. (c) The corresponding two-dimensional velocity field computed from the DPIV image pair.

the steep forward face of the microscale breaking wave in Figure 3b.

Analysis of the simultaneous sampled DPIV and IR data showed that a typical microscale breaking wave generates a region of high vorticity behind its leading edge that disrupts the skin layer and produces a wake that is detected simultaneously in the IR images. The vortices are strong and coherent immediately upwind of the breaking crest but as time passes, and they move from the crest region to the back face of the wave, they become weaker and less coherent.

The data obtained from the DPIV measurements is the instantaneous velocity field. The instantaneous streamwise velocity field can be decomposed into three components [*Benilov et al.*, 1974], expressed mathematically as,

$$u(t) = \bar{u} + \tilde{u}(t) + u'(t) \tag{1}$$

where $\bar{u}$, $\tilde{u}$ and u' are the mean, wave-induced and turbulent components of the instantaneous streamwise velocity field, respectively. The mean velocity component was obtained first by time averaging in a wave following coor-

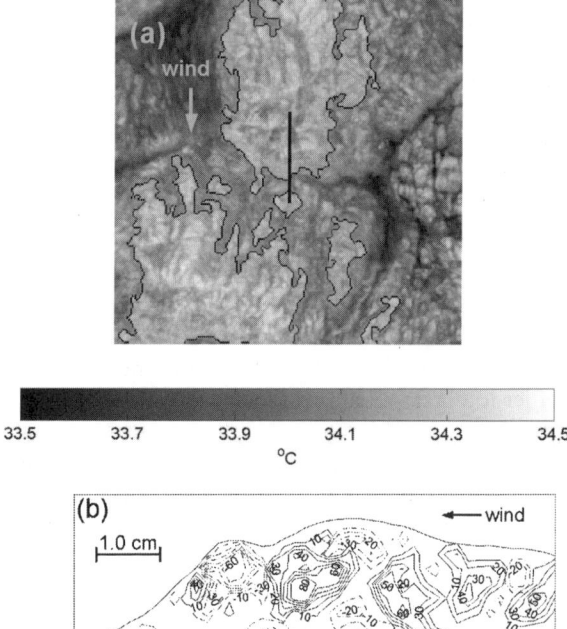

Figure 3. (a) An IR image of a typical microscale breaking wave at a wind speed of 11 m s^{-1}. The wakes generated by microscale wave breaking and the location of DPIV field of view are delineated in the IR image with black lines. (b) The corresponding instantaneous vorticity field computed from the DPIV data. Solid line contours represent counterclockwise vorticity and dashed contours clockwise vorticity.

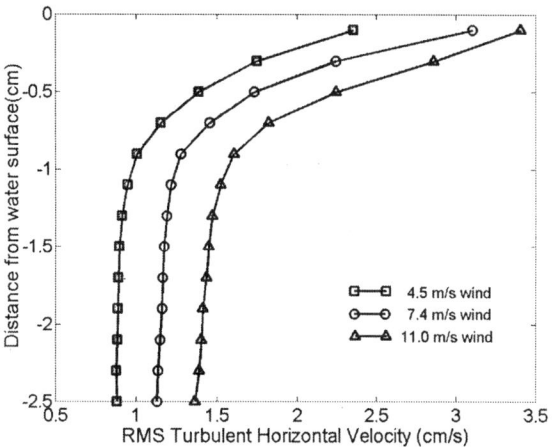

Figure 4. Variation of RMS turbulent streamwise velocity with depth below the free surface at three wind speeds.

dinate system (i.e. referenced to the fluctuating water surface). The mean component was then subtracted from the instantaneous velocity field leaving wave-induced and turbulent velocity components. The turbulent component was then extracted by spatially high-pass filtering with a 9×9 boxcar filter. Thus, the filtered turbulent velocity field is comprised of eddies from 7 to 35 mm in size.

Figures 4 and 5 show the variation of RMS (root-mean-square) turbulent streamwise and vertical velocities with depth, respectively. In both plots the maximum RMS turbulent velocities occur at the water surface and the velocities decrease with depth. As the water surface is approached there is a rapid increase in the turbulent velocities and a thin layer of enhanced turbulence approximately 1 cm thick is observed. As the wind speed increases the magnitudes of the turbulent velocities increase but interestingly the depth of this layer does not increase.

Figure 6 is a plot showing the relationship between the RMS turbulent vertical velocity and the local heat transfer velocity computed from the CFT measurements. It is evident that there is a good correlation between the transfer velocity and the vertical turbulent velocity (correlation coefficient, $r = 0.97$).

4. CONCLUSIONS

Simultaneously sampled IR images, DPIV data and CFT measurements were used to investigate the relationship between the near-surface turbulence generated by microscale breaking and air-water gas transfer. Our measurements have shown that the IR signatures of microscale breakers are typically produced by a series of strong vortices that form behind the leading edge of the breaker.

Vertical profiles of the turbulent velocity show that a thin layer of enhanced turbulence approximately 1-cm deep is formed below the air-water interface. A strong correlation was observed between the local transfer velocity and both the vertical turbulent velocity and the areal coverage of microscale breaking waves. Previous experiments by *Zappa et al.* [2000] have shown that the transfer velocity correlates with the fractional area covered by microscale breaking waves. Therefore, our results show that it is the near-surface turbulence generated by microscale wave breaking that controls the transfer rate at low to moderate wind speeds.

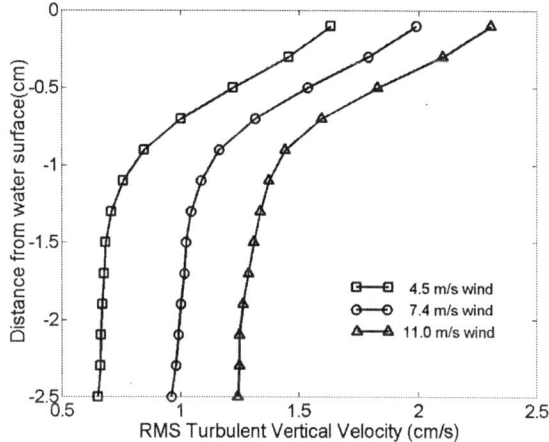

Figure 5. Variation of RMS turbulent vertical velocity with depth below the free surface at three wind speeds.

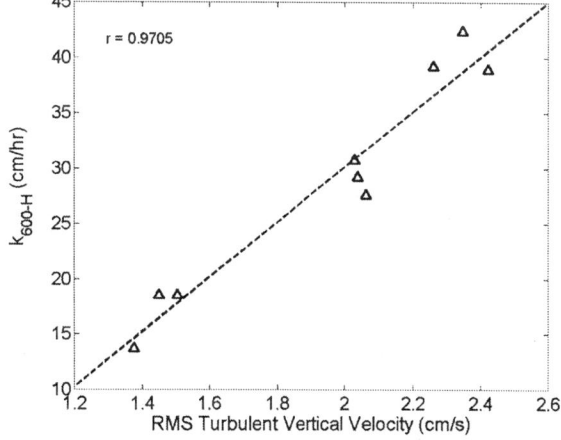

Figure 6. The local heat transfer velocity k_{600-H} referenced to at Schmidt number of 600 plotted versus the RMS turbulent vertical velocity at 1 mm below the water surface. Correlation coefficient, $r = 0.97$.

Acknowledgments. We thank Chris Zappa for his assistance with the IR measurements and Michael Marxen for developing the DPIV algorithm. This research was funded by a Natural Sciences and Engineering Research Council grant (RGPIN 138058-97) to M.R.L., and a National Science Foundation grant (OCE-9633423) to W.E.A. and A.T.J. and the office of Naval Research.

REFERENCES

Banner, M. L., W. L. Peirson, Tangential stresses beneath wind-driven air-water interface, *J. Fluid Mech.,* 364, 115-145, 1998.

Banner, M. L., and O. M. Phillips, On the incipient breaking of small scale waves, *J. Fluid Mech.,* 65, 647-656, 1974.

Benilov, A. Y., B. N. Kouznetsov, and G. N. Panin, On the analysis of wind-induced disturbances in the atmospheric turbulent surface layer, *Boundary Layer Meteorol.,* 6, 269-285, 1974.

Hau ecker, H., S. Reinelt, and B. Jähne, Heat as a Proxy for Gas Exchange Measurements in the Field: Principles and Technical Realization, in *Air-Water Gas Transfer,* edited by B Jähne, and E. C. Monahan, pp. 405-413, AEON Verlag & Studio, Hanau, Germany, 1995.

Hering, F., C, Leue, D. Wierzimok, B. Jähne, Particle tracking velocimetry beneath water waves part II: Water waves, *Exp. Fluids,* 24, 10-16, 1998.

Jessup A. T., C. J. Zappa, M. R. Loewen, V. Hesany, Infrared remote sensing of breaking waves, *Nature,* 385, 52-55, 1997a.

Jessup A. T., C. J. Zappa, H. Yeh, Defining and quantifying microscale breaking waves with infrared imagery, *J. Geophys. Res.* 102, 23145-23153, 1997b.

Law, C. N. S., B. C. Khoo, and T. C. Chew, Turbulence structure in the immediate vicinity of the shear-free interface induced by a deeply submerged jet, *Exp. Fluids,* 27, 321-331, 1999.

Melville, W. K., The role of surface-wave breaking in air-sea interactions, *Ann. Rev. Fluid Mech.,* 28, 279-321, 1996.

Peirson, W. L., Measurements of surface velocities and shear at a wavy air-water interface using particle image velocimetry, *Exp. Fluids,* 23, 427-437, 1997.

Robinson, I. S., N. C. Wells, and H. Charnock, The sea surface thermal boundary layer and its relevance to the measurement of sea surface temperature by airborne and spaceborne radiometers, *Int. J. Remote Sens.,* 5, 19-45, 1984.

Siddiqui, M. H. K., M. R. Loewen, C. Richardson, W. E. Asher, and A. T. Jessup, Simultaneous particle image velocimetry and infrared imagery of microscale breaking waves, *Phys. Fluids,* (submitted), 2000.

Zappa, C. J., Microscale wave breaking and its effect on air-water gas transfer using infrared imagery, Ph. D. dissertation, Applied Physics Laboratory, University of Washington, Seattle, 1999.

Zappa, C. J., W. E. Asher, and A. T. Jessup, Microscale wave breaking and air-water gas transfer, *J. Geophys. Res.,* (submitted), 2000.

M. H. K. Siddiqui, T. Blench Hydraulics Laboratory, University of Alberta, Edmonton, AB T6G 2N4, Canada siddiqui@mie.utoronto.ca

M. R. Loewen, Room 220 Civil/Electrical Building, University of Alberta, Edmonton, AB T6G 2G7, Canada. MRLoewen@civil.ualberta.ca

C. Richardson, W. E. Asher, and A. T. Jessup, Applied Physics Laboratory, University of Washington, 1013 NE 40th Street, Seattle, WA 98105, U.S.A.
crich@apl.washington.edu; asher@apl.washington.edu
jessup@apl.washington.edu)

On the Surface Kinematics of Microscale Breaking Wind Waves

William L. Peirson

*Water Research Laboratory, School of Civil and Environmental Engineering,
The University of New South Wales, Sydney, Australia*

Michael L. Banner

School of Mathematics, The University of New South Wales, Sydney, Australia

We have measured velocities within a few hundred micrometres of the surface of microscale breaking wind waves, using flow visualisation and particle image velocimetry (PIV) techniques. Our results elucidate the nature of the aqueous surface skin flow. We observed that in a frame of reference travelling with the waves, the transport in the aqueous surface layer is rearward along the entire surface of the wave, except in the immediate vicinity of the spilling region. Moreover, we found that transport of surface fluid forward over the crest and into the spilling region rarely occurs. This is in strong contrast with the previously envisaged structure of the wind drift layer. Our measurements demonstrate the important role of microscale wind-wave breaking in the direct transport of fluid from the surface to the highly turbulent domain below. At the toe of each microscale breaker spilling region, there is an intense and highly localised convergence of surface fluid, while on the upwind face of the wave, flow divergence processes are more diffuse. These measurements provide key insights for the development of improved models of air-water exchange of low solubility constituents at moderate wind speeds.

1. INTRODUCTION

Fundamental to a detailed understanding of the processes associated with the exchange of low-solubility constituents at air-water interfaces is an accurate assessment of the behaviour of the surface skin and turbulent mixing in immediate vicinity of the surface.

For smooth water surfaces (that is, at low wind speeds or very short fetch), the surface skin moves downwind at a speed of $0.55u_*$ (Phillips and Banner, 1974, Banner and Peirson, 1998). In this regime, Deacon, 1977 showed that exchange can be accurately parameterised by assuming smooth solid wall turbulent behaviour.

At longer fetches (>2m) and moderate wind speeds (between 4 and 12ms^{-1}), the surfaces becomes covered by microscale breaking wind waves. Even in large wind seas, the surface remains covered by microscale breaking wind waves (Banner, Jones and Trinder, 1989). Their small size means that they respond very rapidly to fluctuations in wind speed and direction and are easily forced into a breaking state.

In this paper, we examine the nature of surface skin flow and near-surface mixing beneath wind-forced microscale-breaking waves as revealed by a suite of laboratory experiments. Following Deacon (1978) who showed that

Table 1. Summary of the experimental conditions and results.

	Error	This study	Okuda (1982)
Fetch (m)	<1%	4.35	4.35
Centreline wind speed (ms^{-1})	2%	6.3	8.1
Friction velocity (ms^{-1})	10%	0.37	0.46
Air roughness length (mm)	50%	0.177	0.151
Wave energy [mm^2]	4%	14.90	30.82
Mean phase speed (ms^{-1})	2%	0.468	0.545
Characteristic mean frequency [Hz]	2%	3.74	3.17
Characteristic mean height [mm]	2%	10.77	15.69
RMS ak	1%	0.276	0.293
Std. dev. Of ak_c	4%	0.052	0.071
No. of surface velocity observations		135	135

exchange is greatly enhanced under such conditions, we present the likely implications of our observations for air-water interfacial exchange.

This paper complements the recent paper by Banner and Peirson (1998) describing direct measurements of the surface tangential stresses associated with microscale breaking wind waves for a range of wind speed and fetch conditions.

2. PREVIOUS INVESTIGATIONS

Direct field observations of microscale breaking waves are very difficult and investigations have generally been restricted to theoretical studies or suitably designed laboratory experiments. Nonetheless, observations of velocity immediately adjacent to a rapidly-moving interface with waves traversing the surface are exceedingly difficult. Only one previous investigation [Okuda, 1982] of this nature has been reported in the literature.

Okuda investigated the velocities along the surface of microscale-breaking wind waves. On the basis of earlier hydrogen bubble measurements, Okuda et al. (1977) concluded that very intense tangential stresses were induced by the wind at the crest. Okuda (1982) measured the surface velocity using ciné camera images of the motion of hydrogen bubbles and small floating beads.

For three individual waves, Okuda (1982) used hydrogen bubble measurements to derive surface velocity distributions along the surfaces of these waves (left panel of Figure 1) and concluded that the crest surface velocity was in excess of the wave celerity for a substantial proportion of waves. However, as shown in these figures, measurements for the critical cases were not possible between -45° and +20° phase and the results of Okuda et al. (1977) were used to infer that the velocity at the crest is greater than the wave speed.

The percentage of waves exceeding the height of his annotated cases II, III and IV were approximately 1, 6 and 60% of all waves, respectively. On the basis of wave statistics alone, the probability of observing transport at the crest in excess of the phase speed (his cases II and III) would be very low. Based on these experiment results, Csanady (1990) proposed a theoretical model of the enhancement of air-sea gas transfer. The roller at the crest was represented by a surface vortex. Significantly, this model yielded the correct Schmidt number dependence for interfacial exchange in the presence of wind waves.

Recently, Banner and Peirson (1998) questioned the magnitude of the very strong tangential stresses reported by Okuda et al. (1977) at the crests of wind-forced microscale breaking wind waves. Also, Okuda (1982) did not attempt to investigate the spilling region associated with these small waves and no detailed investigations have previously been attempted. The present study extends the previous investigations by addressing the influence of wind forcing on these fundamental microscale breaking wave processes. Specifically, we examined the interaction of the thin wind drift layer with the underlying microscale breaking waves.

3. EXPERIMENTAL CONDITIONS

The measurements described here were undertaken to determine the surface velocity structure of breaking microscale wind waves for comparison with Okuda (1982). This entailed probing aqueous viscous sublayers substantially less than a millimetre in thickness below a water surface that is covered in freely-propagating wind-generated waves with heights of a centimetre or more.

The measurement system is documented in detail in Peirson (1997). Here we report those measurements most directly comparable with those of Okuda, 1982. In brief, our study was conducted in a wind wave tank 220 mm wide and 8m long, with a wind tunnel section transitioning smoothly over the water surface. The water depth was 200 mm with a 420 mm deep air channel above. The maximum fetch attainable was 4.35 m although shorter fetches were achieved by shielding the water surface with a thin (<200 μm thickness) membrane. The properties of the wind wavelets generated at these very short fetches vary strongly with fetch and wind speed and were generally in a state of microscale breaking. Relevant details of the wind and wave properties are summarised in Table 1.

The local wind forcing of the surface due to a centreline wind speed was quantified in terms of the wind friction velocity ($u*$) and the aerodynamic roughness length. These

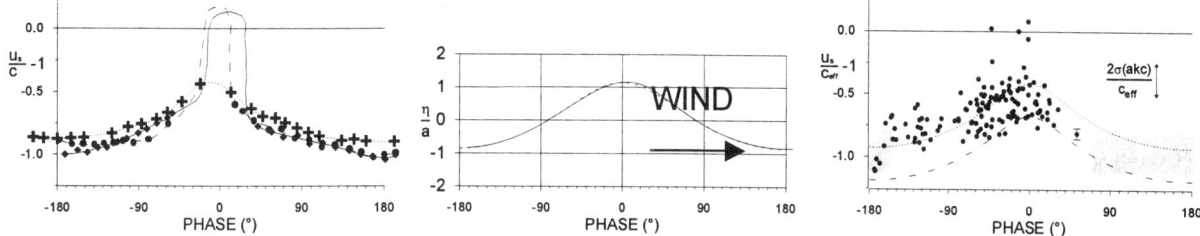

Figure 1. *Left panel:* Surface velocity distributions along the wave inferred from hydrogen bubble measurements by Okuda (1982) at a fetch of 6m and a wind speed of 5.0ms^{-1}. The measured data points are indicated by the symbols, with lines indicating the interpolations. Measurement cases and approximate relative wave heights: II ($H_{1\%}$) ♦, solid line ; III ($H_{6\%}$) ●, dashed line; IV ($H_{60\%}$) +, dotted line. ($H_{x\%}$ is the height of a wave in a population in which x% are larger.) Note the absence of measured data between –20 and +10 degrees of phase. *Centre panel*: Measured average wave form (solid line) with comparable 5th order Stokes wave (dashed line) obtained at a fetch of 4.35m and a wind speed of 6.3ms^{-1}. *Right panel*: normalised surface velocity measurements (●) as a function of wave phase with errors indicated on rightmost point. The stippled region indicates the range of velocities observed in the vicinity of the trough. Surface velocity distribution for a fifth order Stokes wave (dashed line) and with a wind drift of 0.3u∗ imposed. The vertical bar indicates the variability in crest velocity expected by applying linear theory to measured wave forms. The normalising velocity c_{eff} is the measured mean speed of the waves.

were determined for each experiment using the standard method of characterising the logarithmic region of the mean velocity profile measured in the airflow.

Measuring the properties of these short fetch waves is greatly complicated by their changing waveform and their advection by the wind drift current. By differencing the timings of zero-crossing events on the upwind faces of waves propagating past two accurately-spaced capacitance wire probes (spacing 18.00 ± 0.07 mm), the speed of individual waves relative to a fixed frame of reference could also be measured. From these records, we determined a mean speed c_{eff}. The root mean square (RMS) wave steepness (ak) is shown in Table 1.

4. SKIN VELOCITY MEASUREMENTS

The measurement system described by Peirson (1997) was developed to obtain water velocities within a few hundred micrometres of the local interface. A Kodak *Megaplus* CCD camera with an effective resolution of 1280 (horizontal) by 1024 (vertical) pixels was fitted with a low distortion lens. This system was used to capture the motion of fluorescent particles (diameters 20-60 μm, mean relative density 1.2) illuminated by pulsing a laser sheet at approximately 500 Hz within an image region of approximately 16.2 by 13.0 mm. Software was developed to enable direct displacement measurements to be made and to locate the surface beneath the relatively smooth rear faces and crests of the waves. The local surface velocity was obtained by extrapolating to the water surface using an assumed linear velocity distribution through pairs of adjacent data points. Reliable measurements were not possible within the highly turbulent spilling region nor upwind of the trough where water particle velocities were very small.

Larger-scale coincident images of the water surface profile were captured by a CCD camera viewing the water surface from above and used to determine the phase location for the high-resolution sub-surface images.

Larger-scale flow measurements were used to probe the velocity field in the vicinity of the spilling region and the field of view captured by the sub-surface camera was increased. Its lens was replaced by a Micro Nikkor lens of 60 mm focal length yielding a sub-surface field of view of approximately 45 by 37 mm. This optical arrangement yielded less than the negligible degree of distortion of the original system.

An ensemble of near-surface (<350μm depth) velocity measurements under actively breaking waves was taken with measurements registered at the crest and downwind between crest and the spilling region.

5. RESULTS

Approximately 140 surface velocity and tangential stress measurements were obtained for each experiment using the high-resolution imagery. These are presented in Figure 1 where the velocities have been normalised by c_{eff} and are shown relative to a frame of reference moving with the wave itself. Theoretical estimates obtained for a comparable 5th order Stokes wave are also shown.

Measurement of the surface velocity relied on measuring the relative motions of two particles in close proximity to the surface. This was relatively straightforward on the crests and upwind faces of the waves because of the relatively high velocities. Downwind of the spilling region and in the trough the wind drift was approximately equal to the wave orbital component yielding near stationary fluid as indicated in the right panel of Figure 1.

It was found that the measured surface velocities using this system were comparable with those obtained by Okuda (1982) (Compare the left and right panels of Figure 1). However, our technique yielded reliable measurements over the crests of the waves – a region Okuda had not been able to probe. Occasional measurements were obtained in which the surface velocity was greater than the mean wave speed (Figure 1, right panel) but this was not representative behaviour. In general, our measured surface velocity values at the crest were substantially lower than c_{eff}.

Three observations can be made:
1. The average surface velocities over the rear faces of these small waves are substantially greater that those derived from by Stokes theory by approximately (0.3 ± 0.1)$u*$ as indicated by the dashed line in Figure 1.
2. Whereas Okuda (1982) argued that the surface drift was strongly modulated at the crests of breaking waves, our measurements do not support that view. The surface drift assumes an approximately constant value over the rear faces of the waves.
3. The Stokes estimates appear to provide a natural lower bound to the velocity data but this precludes variations in the surface velocity field due to modulation of the waveforms. The wave records were processed to yield estimates of the linear crest orbital velocity akc_{eff}, the rms variability in this parameter is shown in Table 1 and its magnitude is shown in the right panel of Figure 1. The relative magnitude of 2rms(akc) to the observed variability in the surface velocity measurements and suggest that there are contributions from both the modulation of the wave field associated with wave groups and either fluctuations in the skin friction or turbulence in the wake of the spilling region.

To assess flow behaviour in the vicinity of the spilling region, the lower-resolution system was used. Analysis of these images confirmed that horizontal water velocities at the crest greater than the mean wave speed was not representative behaviour for microscale-breaking wind waves. However, there was a rapid increase in surface velocity from the crest to the spilling region.

The key kinematical features are more easily observed at higher wind speeds and therefore an image captured at 8.1ms^{-1} has been selected for presentation here. Figure 2 shows a representative flow visualisation beneath the crest and forward face of a microscale breaking wind wave with its rich flow structure. Overlaid on the images are PIV measurements of velocity and contours of vorticity within the domain. Four key kinematic aspects of the flow can be observed:
1. An increase in velocity adjacent to the surface from the crest to the spilling region. Note that horizontal velocity in the spilling region is just equal to or greater than the mean wave speed.
2. Rupture of the surface at the toe of the spilling region. This is accompanied by intense fluid shear which is highlighted by the high local vorticity measurement.
3. Turbulent flow in the wake of the toe of the spilling region. A ridge of high vorticity can be observed at a depth of 5-6mm beneath the crest.
4. Irrotational flow at depth. Below the toe wake, the vorticity magnitude is very low in comparison the values observed near the surface.

The strong surface convergence observed at the toe of the spilling region can be contrasted with observed surface convergence and divergence on the rest of the wave surface. At the toe of the spilling region, the convergence is very large (often greater than 175s^{-1}). Many (>1000) visualisations of flow at the surface of microscale breaking wind waves were examined. The only other sites of significant surface convergence or divergence were on the upwind face - presumably due to local variations in the wind-generated tangential stress. However, these were relatively small compared to values at the toe of the spilling region with no values greater than 10s^{-1} observed.

6. CONCLUSIONS

We have used PIV techniques to measure the mean wind drift velocity component on the crests, rear faces and troughs of microscale breaking wind waves as (0.3 ± 0.1)$u*$. This value is also substantially less than measured values of the total drift (0.55 ± 0.05) $u*$ determined in the presence of wind waves using floating drifters (for example, Keulegan, 1958).

Viscous sublayer flow on the windward face of a microscale breaking wave which flows over the crest and feeds into the spilling region. is not the characteristic behaviour. The spilling region downwind of the crest remains locally compact, with a strong convergence of surface fluid at the toe with relatively weak divergences near the crest and on the upwind face. Surface fluid feeds into the spilling region at a location immediately upwind of the spilling region and downwind of the crest.

Constituent exchange across the interface is enhanced in the presence of wind waves. This study has shown that the physical behaviour and surface processes of microscale breaking wind waves are profoundly different from those of smooth interfaces and clearly identifies those processes that result in failure of smooth solid wall parameterisations.

Transfer rates of low solubility gases are controlled by the behaviour of the surface viscous sublayer and such approaches do not directly address this key physical aspect.

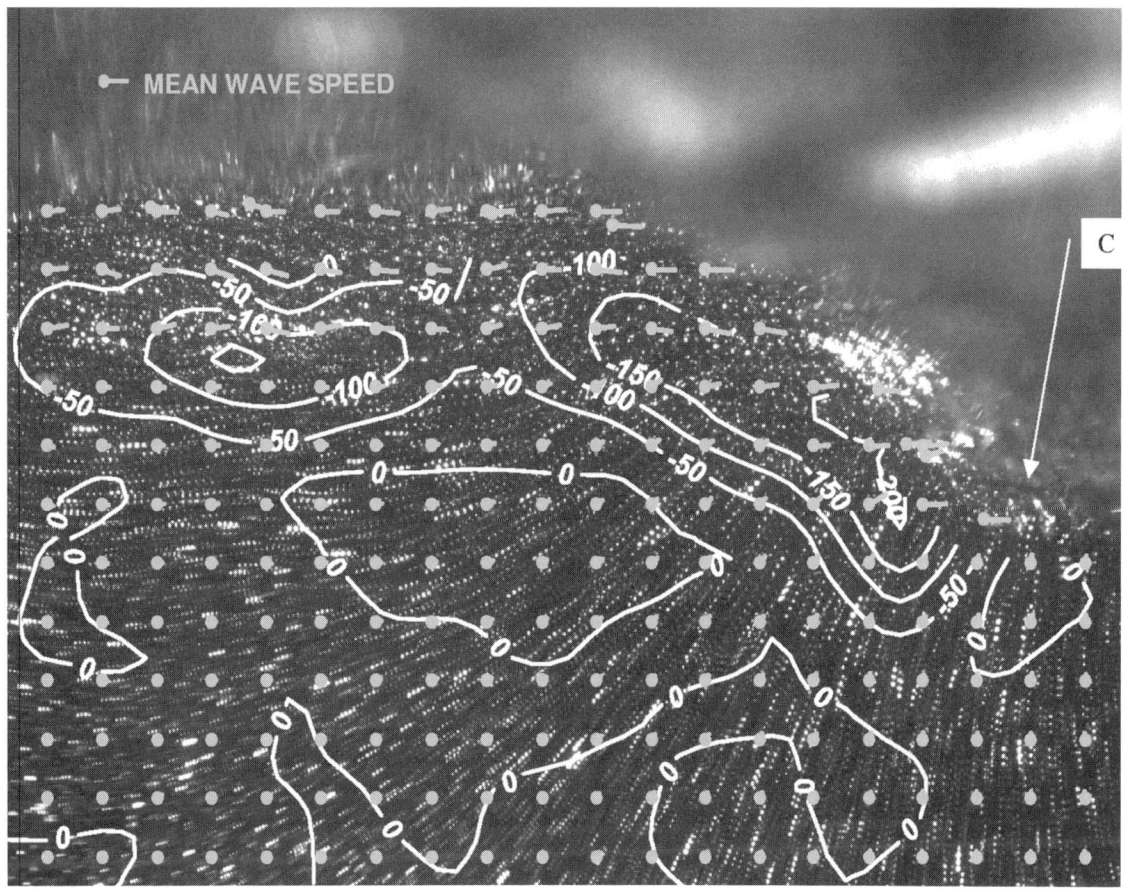

Figure 2. Flow visualisation inside a microscale-breaking wind wave crest obtained at a fetch of 4.35m and a wind speed of 8.1ms^{-1}. Approximate image size is 45.5mm by 36.4mm. Wind and wave direction is towards the right. PIV velocity vectors are imposed in light grey in laboratory co-ordinates and the mean wave speed is shown above. Vorticity contours calculated from the subsurface velocities are shown in white and their magnitudes indicated in s^{-1}. Note: the significant increase in near-surface velocity along the crest towards the spilling region; the surface rupture and intense shear at point C.

This study has examined microscale breaking wind wave processes and concludes that these processes provide plausible mechanisms for the enhancement of interfacial exchange.

In a frame of reference moving with the microscale breaking wind waves, transport is in an upwind direction in the wind-drift layer at the surface of unbroken waves. As the small gravity waves steepen, parasitic capillary wave trains form on the downwind faces that are capable of producing localised vorticity and mixing within the surface viscous sublayers. As the gravity wavelets steepen further, they break and the surface ruptures, with surface fluid subducted into an interior region of intense turbulent mixing beneath an intensively-mixed spilling region.

Acknowledgments. The authors gratefully acknowledge the support provided for this project by the Australian Research Council and the use of the facilities at the University of New South Wales Water Research Laboratory.

REFERENCES

Banner, M.L. Jones, I.S.F. and Trinder, J.C. 1989 Wavenumber spectra of short gravity waves. *J. Fluid Mech.* 189, 321-344.

Banner, M.L. and Peirson, W.L. 1998 Tangential stress beneath wind-driven air-water interfaces. *J. Fluid. Mech.* 364, 115-145

Csanady, G.T. 1990 The role of breaking wavelets in air-sea gas transfer. *J. Geophys. Res.* 95, 749-759

Deacon, E.L. 1977 Gas transfer to and across air-water interfaces. *Tellus,* 29, 363-374

Keulegan, G.H. 1951 Wind tides in small closed channels. *Journal of Research of the National Bureau of Standards*, 46, 5, Research Paper 2207

Okuda, K. 1982 Internal flow structure of short wind waves, Part I. The vorticity structure. *J. Oceanogr. Soc. Jap.* 38, 28-42.

Okuda, K, Kawai, S. and Toba, Y. 1977 Measurement of skin friction distribution along the surface of wind waves. *J. Oceanogr. Soc. Jap.* 33, 190-198.

Peirson, W.L. 1997 Measurement of surface velocities and shears at a wavy air-water interface using particle image velocimetry. *Expt. in Fluids*, 23, 427-437

Peirson, W.L. 2000 On the kinematics and dynamics of microscale-breaking wind waves. Ph.D. Thesis. University of New South Wales

Phillips, O.M. and Banner, M.L. 1974 Wave breaking in the presence of wind drift and swell. *J. Fluid Mech.* 66, 625-640.

Zappa, C.J., Jessup, A.T and Asher. W.E. 1999 Correlating microscale wave breaking with gas transfer for cleaned and surfactant-influenced water surfaces. In *The Wind-Driven Air-Sea Interface*, Ed. M.L. Banner, University of New South Wales, 357-358 (in press).

Effect of Microscale Wave Breaking on Air-Water Gas Transfer

C. J. Zappa[1,2], W. E. Asher[1], A. T. Jessup[1], J. Klinke[3], and S. R. Long[4]

Measurements of simultaneous and co-located infrared and wave slope imagery of laboratory wind waves show that the wave-related areas of thermal boundary layer disruption and renewal are the turbulent wakes of microscale breaking waves, or microbreakers. These signatures of disruption are associated with waves that have a steep forward face and a dimpled crest, and can be quantified by infrared imaging techniques. The fractional area coverage, A_B, of the surface affected by these renewal features is significant (0.25 – 0.40) and found to be linearly correlated with the transfer velocity, k. Furthermore, this correlation is insensitive to the presence of surfactants and independent of fetch. Using the controlled flux technique (CFT) to measure k locally, the renewal within the wakes of microscale breaking waves was found to enhance the transfer by a factor of 3.5 on average compared to that outside the wakes. Moreover, up to 75% of the transfer across the air-water interface under moderate wind speeds is the direct result of microbreaking. The roughness features associated with microscale breaking waves are shown to contribute significantly to the mean square slope, $<S^2>$, and may explain the observed correlation between k and $<S^2>$. The correlation between k and A_B regardless of surfactant concentration, combined with the enhanced local k and wave slope results, provides quantitative laboratory evidence that microbreaking is an important physical process contributing to gas transfer at low to moderate wind speeds.

1. INTRODUCTION

Air–sea gas exchange plays a crucial role in biogeochemical cycling, and a better understanding of ocean mixing and air-sea exchange mechanisms is needed to improve model predictions of the spatial variability of global air-sea fluxes. The gas flux is determined by the product of the gas transfer velocity, k, which characterizes the resistance to gas exchange across the aqueous boundary layer at air-sea interface for sparingly soluble gases, and the air-sea concentration difference, which is the driving potential. The transfer velocity k incorporates the dependencies of gas transfer on diffusivity and turbulent transport, which is presumed to be the driving mechanism that decreases the resistance to the transfer of gas across that air-sea interface. Since the wind stress at the ocean surface plays a central role in the generation of turbulence through the transfer of momentum to the waves and currents, considerable effort has gone into determining empirical relations between k and wind speed (see *Wanninkhof and McGillis* [1999]).

The dependence of the transfer velocity on wind speed or wind stress has been shown to be a function of the concen-

---

[1]Applied Physics Laboratory, University of Washington, Seattle, Washington

[2]Current Affiliation: Department of Applied Ocean Physics and Engineering, Woods Hole Oceanographic Institution, Woods Hole, Massachusetts

[3]Physical Oceanography Research Division, Scripps Institution of Oceanography, La Jolla, California

[4]NASA GSFC / Wallops Flight Facility, Wallops Island, Virginia

tration of surfactants [*Frew*, 1997], which are present in nature to varying degree. Laboratory measurements indicate that a wave-related mechanism may regulate gas transfer because the correlation of the transfer velocity with wave slope [*Jähne et al.*, 1987] seems to be unaffected by the presence of surfactants [*Frew*, 1997; *Bock et al.*, 1999] under pure wind-driven systems. Microscale wave breaking, or microbreaking [*Banner and Phillips*, 1974], is the breaking of very short wind waves without air entrainment, is widespread over the oceans, and has been proposed as the underlying mechanism that governs the gas transfer velocity, k, at low to moderate wind speeds [*Jähne et al.*, 1987; *Csanady*, 1990]. The specific manner by which microscale wave breaking controls k has been theorized to be the thinning of the aqueous concentration boundary layer (CBL) by the intense surface divergence generated during the breaking process (for drawing see *Zappa* [1999]).

Disruptions of the CBL associated with microscale wave breaking can be viewed as regions where the replacement of fluid within the CBL is enhanced. It is this area of enhanced turbulence generated in the wake of a microbreaker that regulates the transfer of gas across the air-water interface. If the transfer velocity within the wakes of microbreaking waves is significantly greater than that outside, k should be dependent on the fraction of the water surface covered by the wakes, which we define as A_B. A simple model for the partitioning of the contribution to k from the areas inside and outside of the wakes is

$$k_M = A_B k_B + (1 - A_B) k_{NB} \quad (1)$$

where k_B is the transfer velocity within A_B and k_{NB} is the transfer velocity outside of A_B. If the model is valid and $k_B \gg k_{NB}$, then k will be correlated with A_B and we can conclude that microbreaking contributes significantly to the gas transfer velocity.

2. EXPERIMENTS AT THE WALLOPS FLIGHT FACILITY

2.1. Facility

Experiments were performed in the wind-wave flume at the Air-Sea Interaction Research Facility at NASA Goddard Space Flight Center/Wallops Flight Facility in October–December 1998. The flume is 18.29 m long, 1.22 m high, and 0.91 m wide, the water depth is 0.76 m, and the air headspace is 0.46 m. The tank is equipped with a water heating/circulation system and a 10-cm thick water-wave-absorbing "beach" made of plastic honeycomb. The facility is instrumented to measure wind speed, friction velocity and surface displacement. In addition, for this study, measurements were made of bulk gas transfer, bulk water temperature, air temperature, and relative humidity.

2.2. Fetch and Surfactants

To investigate the dependence of gas transfer and microscale wave breaking on fetch, measurements were made at two fetches of 5.6 m and 11.1 m. The water used in the tank was filtered tap water, and the surface was skimmed before each experiment to remove accumulated surface contaminants. Biological activity was minimized by continuous bromination to 5 ppm levels using dissolving pellets. As a further precaution, the tank was drained, cleaned, and refilled between the groups of experiments performed at the two fetches. The influence of surfactants also was investigated by purposely adding 1 g m^{-3} of the soluble surfactant Triton X-100 to the water for a portion of the experiments at both fetches. During four experiments at the 11.1 m fetch, an oily putty used to seal some electronics in the tank slowly bled surfactant into the water. After observing lower transfer velocities and a Schmidt number exponent of $\frac{2}{3}$, characteristic of surfactants, the tank was drained, the oily sealant was removed, and the tank was cleaned and refilled. These surface conditions, therefore, have been denoted as adventitious surfactant cases.

2.3. IR and Wave Slope Imagery and the CFT

Infrared imagery was used to detect and quantify the fractional coverage of microscale wave breaking, A_B, and to implement the CFT. The infrared measurements were made using an Amber model Radiance HS infrared imager at a range of 1.4 m and an incidence angle of 30°. Wave slope imagery was used to complement the infrared detection of microbreaking and observe the surface-roughness characteristics of the breaking process as they relate to the mean square wave slope. The image slope system is the combination of a Sony XC-75 CCD camera at a range of 3 m to the water surface and a subsurface light box covered by an intensity absorption screen with a linear gradient in the along-tank direction. The result is a spatial map of a direct measure of the surface slope in the upwind/downwind direction that is proportional to the image intensity.

The controlled flux technique (CFT) [*Jähne and Haußecker*, 1998] uses heat as a proxy tracer for gas to obtain the remote measurement of the local water-side transfer velocity with high spatial resolution and short response times. With the CFT, the water surface is heated with a CO_2 laser to produce a spot with a measurable temperature difference that can be tracked within a sequence of infrared images. The transfer velocity, k, is determined from the surface renewal rate, which is estimated from the thermal decay of the heated spot as predicted from a surface renewal model. Previous CFT measurements by Jähne and co-workers have given reasonable estimates of k in the laboratory under conditions of minimal heat flux across the air-water interface. However, in an unanticipated result, our laboratory data for conditions of high heat flux show that, when referenced to a common Schmidt number, the CFT estimates of k are roughly 2.5 times greater than k determined by mass balance methods using He and SF_6, requiring an additional scale factor. Zappa [1999] proposed that this scale factor is dependent on heat flux. He suggested that the 1-D unsteady diffusion model used to calculate the decay of the CFT patch must incorporate the appropriate non-homogeneous heat flux boundary condition.

3. RESULTS

3.1. Infrared and Wave Slope Imagery

Simultaneous and co-located measurements of infrared and wave slope imagery were made to provide evidence that the propagating disruptions observed in the infrared imagery are produced by steep wind waves with bore-like crests. Figure 1 shows a comparison of co-located and simultaneous slope and infrared imagery, and illustrates that capillary-gravity wave packets evolve into microscale breaking waves that disrupt the diffusive boundary layers. The characteristic signature of a microscale breaking wave in the infrared imagery consists of an abrupt front of increased temperature that propagates through the image in the direction of the wind at roughly the phase speed of the dominant wave. This distinct propagating front disrupts the cool thermal boundary layer, leaving behind a decaying turbulent wake of warmer water mixed up from below (for schematic drawing see Zappa [1999]). Smaller-scale structures observed in the imagery are most likely associated with background eddies generated by shear, buoyancy (due to the high heat flux), and/or turbulence input to the near surface by microscale breaking waves.

The image slope shows the very front of the curved crest to have the steepest slope, and this steep slope corresponds directly to the leading edge of the propagating thermal front of the microscale breaking signature in the infrared. The mixing that produces the infrared signature is the direct consequence of the near-surface turbulence generated by the crest of the microscale breaking wave. The slope image also shows only extremely short capillary waves riding on the forward face at the initial moment of disruption that are barely visible and nearly disappear. However, the crest and the region behind it appear distinctly "dimpled," suggesting that an energetic process occurs right at the crest. This highly three-dimensional dimpled characteristic of the crest and of the wake diminishes in subsequent slope images, but in the infrared imagery the microscale breaker is shown to continue to disrupt the surface as it dissipates energy to smaller scales. As the dimpled features diminish, short, steeper capillary waves begin to emerge on

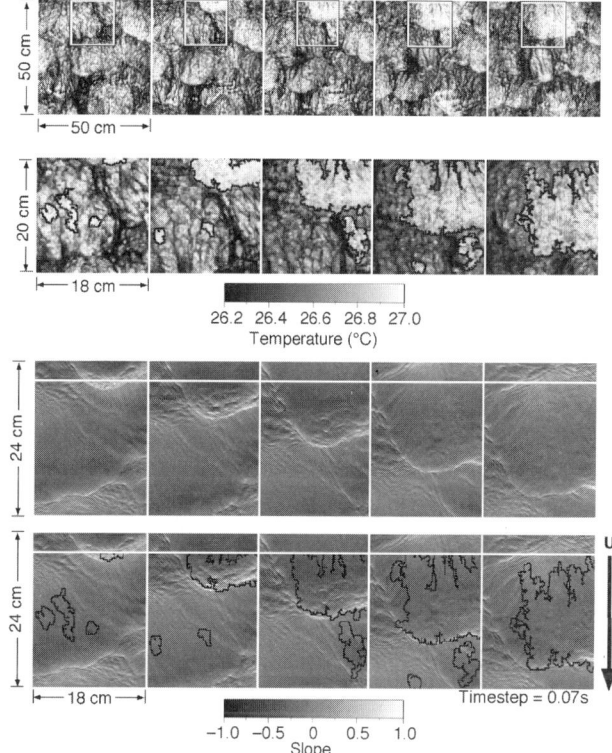

Figure 1. An example of both infrared (Top) and wave slope imagery (Bottom) at a wind speed of 6.9 m s^{-1} with a cleaned water surface depicting a microscale breaking wave; T_w = 27.2°C, ΔT_{aw} = –5.6°C, RH = 41.9%, f_p = 3.20 Hz, and Ω = 5.6 m. The disruption of the skin layer imaged in the infrared is coincident with the appearance of a "dimpled" feature on the bore-like crest in the slope imagery. Time increases from left to right in 0.07 s increments and the wind direction in the images is from top to bottom. The first sequence of infrared images shows a box drawn to highlight the overlapping slope image area. The second infrared sequence is an expanded view of this highlighted region with an outline of the measured A_B. Likewise, only the portion below the line drawn in the slope images corresponds directly to the box (highlighted region) in the infrared images. The second wave slope sequence shows an overlay of the outline of A_B.

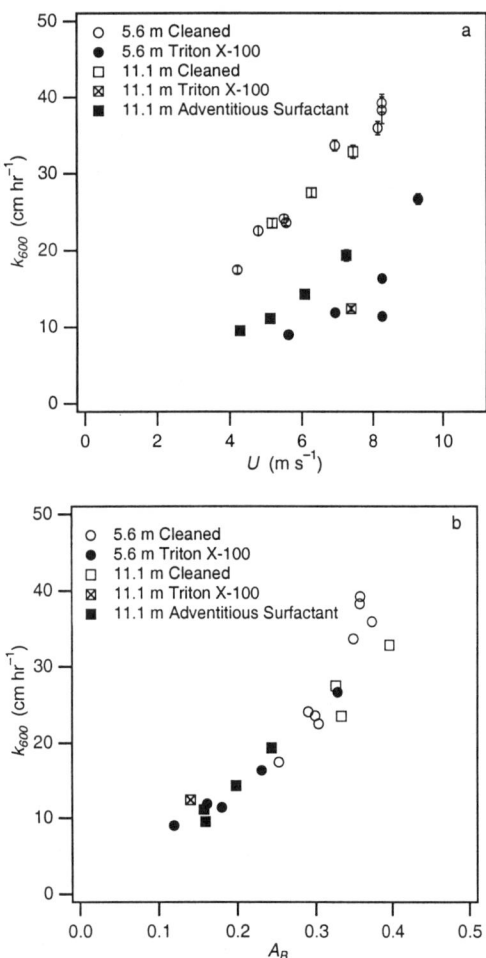

Figure 2. (a) Local transfer velocity, k_{600}, for CFT referenced to a $Sc = 600$ as a function of wind speed, U. (b) Local transfer velocity, k_{600}, for $Sc = 600$ versus A_B.

the forward face of the front of the crest and continue to grow in wavelength as the microscale breaking event evolves toward completion. These simultaneous, co-located infrared and slope observations confirm that the distinct scalloped features of skin layer disruption are produced by the propagating crests of microscale breaking waves.

3.2. Microbreaking and Transfer Velocity

The outlined features overlayed in the IR and slope imagery shown in Figure 1 correspond to A_B. A_B was computed by applying a thresholding technique to the infrared images coupled with standard morphological processing operations (for details see *Zappa* [1999]). A_B is the ratio of the area of the outlined regions to the total area of the infrared image. A_B was found to increase with wind speed for both fetches. For the cleaned surface cases, A_B increased from 0.25 to 0.40 as the wind speed increased from 4.2 to 8.3 m s^{-1}. With the addition of the surfactant Triton X-100 and for the adventitious surfactant cases observed at the 11.1 m fetch, A_B was significantly lower (nearly half) at a given wind speed, consistent with the known ability of surfactants to damp capillary-gravity waves.

Figure 2a shows the local transfer velocity, k_{600}, determined by CFT and referenced to a $Sc = 600$ as a function of wind speed, U, for both fetches. Consistent with previous results in wind-wave flumes [*Frew*, 1997], k_{600}, was lower at a given U for the Triton X-100 surfactant-influenced experiments. Experiments performed with the adventitious surfactant also showed a decrease in k_{600}. The variation in k_{600} at a given wind speed is the consequence of a variation in the degree of surface contamination. Surface contamination is known to modify the free-surface boundary condition to behave as a rigid boundary by introducing a tangential stress that works to suppress horizontal motion, and therefore near-surface turbulence. Figure 2a demonstrates the difficulty of using wind speed to parameterize gas transfer since k can be a multi-valued function of U, depending on the surface cleanliness.

Figure 2b shows that k_{600} is linearly correlated with A_B, and this correlation is independent of surfactant. The correlation coefficient is 0.94 and the linear fit passes through the origin. In contrast to the multi-valued behavior of k_{600} with respect to U, k_{600} is a single-valued function of A_B. These results corroborate the result of *Zappa et al.* [2000] and indicate that enhanced near-surface turbulence due to microbreaking may be the underlying mechanism responsible for a significant fraction of the increase in k with wind speed.

3.3. Enhanced Transfer Due to Microbreaking

The correlation between k and A_B observed in section 3.2 suggests that microscale wave breaking is the underlying mechanism controlling the gas transfer velocity. If microbreaking is the source of near-surface turbulence responsible for the correlation, the transfer velocity inside the wake of microscale breaking waves should be enhanced. Thus, based upon the model in (1), k has been partitioned into two separate transfer velocities, one within the skin-layer disruptions in the wakes of microbreakers and one outside the wakes in the background. While implementing the CFT, the individual decay rates of the actively heated patches produced by the CO_2 laser were observed to vary substantially, and these decay rates were dependent on the influence of microscale wave breaking.

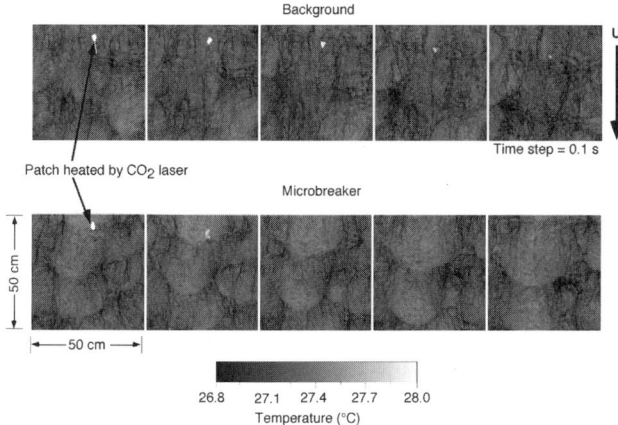

Figure 3. Comparison of CFT patch decay that was not affected at all by microbreaking (Background) and one affected directly by a microscale breaking wave (Microbreaker). The wind speed was 6.9 m s^{-1}; T_w = 28.2°C, ΔT_{aw} = –5.2°C, RH = 42.4%, and f_p = 3.17 Hz.

The objective of implementing the CFT is to quantify k_B and k_{NB} to determine the effect that microscale breaking has on the overall k. However, the patch decay will depend on the moment that breaking first affects the patch, if at all. Figure 3 shows infrared sequences illustrating the two types of decay of the actively heated CFT patches that will be used to quantify k_B and k_{NB}, and have been designated Background and Microbreaker. The top sequence shows the Background decay in which the patch is unaffected by microscale breaking. The bottom sequence shows Microbreaking decay, when a patch is laid directly in the actively breaking wave crest of a microscale breaker. As shown in these sequences, the patches directly affected by microscale wave breaking decay faster than those patches in the background. The comparison of the Microbreaker and Background examples will be used to determine the direct effect of microscale wave breaking on the gas transfer velocity.

Distinguishing microbreaker patches from background patches has been achieved by using A_B in the vicinity of the patch. Those patches that were coincident with an active microscale breaking wave and completely within A_B were classified as Microbreakers and used to determine the transfer velocity due to microscale wave breaking, k_B. Those patches that were unaffected by microscale wave breaking and never within A_B were classified as Background and used to compute k_{NB}, the transfer velocity that is unaffected by microscale breaking waves.

Figure 4 shows k_B and k_{NB} as a function of A_B. Both k_B and k_{NB} increase with A_B, but k_B is significantly greater than k_{NB}. The ratio of k_B to k_{NB} serves as an estimate of the enhancement of the transfer velocity due to microscale wave breaking. The turbulence generated by the actively breaking microscale wave crest consistently enhances the local transfer velocity by a factor of 3, with a mean enhancement factor of nearly 3.5 for all experiments. Therefore, microscale wave breaking has a significant effect on the increased gas transfer across the air-water interface observed in the presence of waves.

The ratio $R_{eff} = A_B k_B / k_M$, where k_M is found in (1), serves as an estimate of the effect of microscale wave breaking on the overall transfer velocity. At low A_B, the contribution of microscale wave breaking to gas transfer is 25%. As A_B increases to 0.4, the effect of microscale wave breaking on the total transfer velocity approaches nearly 75% and is significantly greater than the contribution of the background.

3.4. Microbreaking and Wave Slope

Laboratory measurements have indicated that statistical increases in the local wave slope are associated with microscale breaking waves [*Banner*, 1990]. The simultaneous infrared and wave slope imagery presented in section 3.1 indicates significant increases in local wave slope associated with the actively breaking waves. Therefore, the data in Figures 2b and 4 may explain why wave slope has been observed to correlate with k for a range of surfactant-influenced surface conditions [*Frew*, 1997; *Bock et al.*, 1999].

The microscale breaking process is the culminating event in the development of capillary-gravity wave packets, and

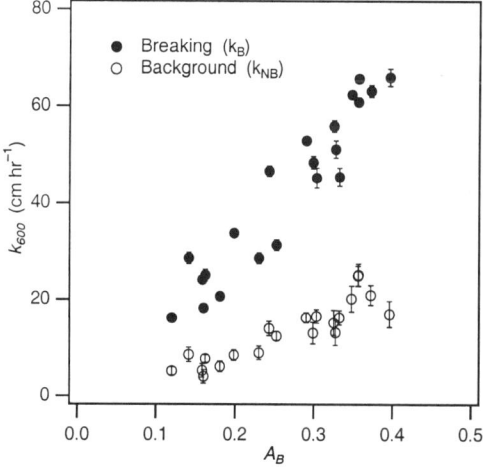

Figure 4. Local transfer velocities as determined by CFT referenced to Sc = 600 for decay patches affected directly by microscale breaking, k_B, and those in the background, k_{NB} as a function of A_B.

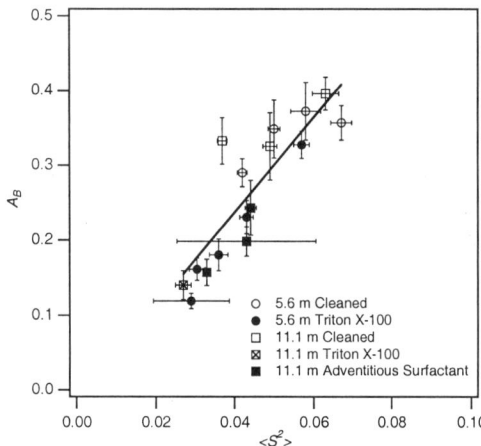

Figure 5. The fractional area coverage of microscale wave breaking, A_B, versus the total mean square slope, $<S^2>$.

it dissipates energy into near-surface turbulence that enhances k. *Jähne et al.* [1987] suggest that $<S^2>$ serves as a parameter to correlate with k as a measure of the stability of the waves and therefore near-surface turbulence, and *Frew* [1997] suggests that $<S^2>$ incorporates the effect of surfactants on the wave field and on near-surface turbulence. Microscale breaking waves are observed to develop from a gravity wave crest accompanied by parasitic capillary waves on its forward face. Since k correlates with A_B and microbreaking has been shown to directly enhance k, it is possible that microbreaking may serve as a link between the observed correlation of k with $<S^2>$, provided the surface roughness features associated with microbreaking prove to contribute significantly to $<S^2>$. Figure 5 shows A_B plotted versus $<S^2>$. The linear relationship suggests that microbreaking may be the process that links the observed correlation of gas transfer to wave slope. Specifically, the steep slope at the front of the microbreaker and the dimpled roughness features of the crest and its wake may provide a significant contribution to $<S^2>$.

To determine if these roughness features associated with microbreaking provide the link between k and $<S^2>$, the magnitude of the contribution by microbreaking to $<S^2>$ must be determined. Similar to discriminating between CFT patches affected and unaffected by microbreakers, the combination of infrared and wave slope imagery has been utilized to investigate the direct effect of microbreaking on $<S^2>$. Using the thresholded outline of A_B as a mask, the slope is calculated for the region within A_B as Microbreaking and for the region $(1 - A_B)$ as the Background. Time series show strong enhancement of slope due to microscale wave breaking signified by large spikes in mean square slope within A_B, $<S^2>_B$. Typically, as the microbreaking events appear within the infrared and slope imagery, $<S^2>_B$ spikes to roughly twice $<S^2>$. The concentrated, steep slopes during the initiation of breaking contribute significantly to $<S^2>_B$. Subsequently, the extent of the wakes left behind by the propagating microbreakers increases, and $<S^2>_B$ decreases toward $<S^2>$ as the near-surface turbulence decays. The effective contribution to $<S^2>$ due to breaking, defined as $R_{SB} = A_B <S^2>_B / <S^2>$, reached 40% to 60% for individual breaking events. The average R_{SB} for the three experiments analyzed was 30%, suggesting that a substantial contribution to $<S^2>$ was due to actively breaking waves.

4. DISCUSSION

Microscale wave breaking, and the near-surface turbulence it generates, is hypothesized to enhance gas transfer. In the infrared imagery, the microbreaking process is observed to disrupt the aqueous thermal boundary layer, producing fine-scale surface thermal structures at the crest and within its wake. In the slope imagery, microbreaking also generates the three-dimensional dimpled features. The relevance to gas exchange for the contribution of microbreaking to $<S^2>$ is the relationship between the roughness features of and the turbulence generated by the microscale breaking waves. Microscale breaking waves appear to directly contribute to $<S^2>$. The roughness features of the crest and in the wake of microscale breakers may also be indicative of the scales of turbulence produced by microbreakers as shown in the comparison of infrared and slope imagery. The detected events of microbreaking represent turbulent disruptions of the diffusive boundary layer that directly enhance gas transfer and produce surface roughness elements that contribute directly to $<S^2>$ and that may explain the correlation between k and $<S^2>$. Therefore, if near-surface turbulence is the source of these observations, the spatial scales of the eddies beneath the breaking crest should be comparable to the spatial scales of both the dimpled features and the thermal structures.

The spatial scales of the fine-scale thermal structures and the three-dimensional roughness features were observed to be of $O(10^{-2}$ m), identical to the length scale for the eddies measured by *Komori et al.* [1993]. Furthermore, the dimpled roughness features of the breaking crest correspond directly to the warmest fine-scale features of the skin-layer disruption by microbreaking. The implication is that the fine-scale thermal structures and the dimpled roughness features are directly related to the turbulence generated at the breaking crest. This suggests that microbreaking serves as the mechanism that explains the correlation between k and $<S^2>$. Since near-surface turbulence increases k, microscale breaking waves should prove to be the mechanism that enhances heat and mass transfer.

The transfer velocity was shown to be linearly correlated with A_B and this correlation was independent of surface cleanliness and fetch. The surface renewal within the wakes of microbreakers was found to be greater than that in the background. It is this area of enhanced surface renewal generated in the wakes of microbreakers that regulates the transfer of gas across the air–water interface. Therefore, the model in (1) suggests that if k_B is much greater than k_{NB}, k will correlate with A_B, and surface renewal due to microbreaking will contribute significantly to gas transfer. The results in Figure 4 indeed show that $k_B \gg k_{NB}$, and surface renewal in the wakes of microbreakers was shown to enhance gas transfer by a factor of 3.5 over that in the background. Furthermore, microbreaking was shown to directly contribute up to 75% of the transfer across the air-water interface under moderate wind speed conditions. These results show conclusively that microbreaking is an underlying mechanism that explains the observation of enhanced gas transfer in the presence of waves and may govern air-water gas transfer at low to moderate wind speeds.

Acknowledgments. This work was supported by the National Science Foundation, the Office of Naval Research, the National Aeronautics and Space Administration, and the Applied Physics Laboratory at the University of Washington.

REFERENCES

Banner, M. L., The influence of wave breaking on the surface pressure distribution in wind-wave interactions, *J. Fluid Mech., 211*, 463-495, 1990.

Banner, M. L., and O. M. Phillips, On the incipient breaking of small scale waves, *J. Fluid Mech., 65*, 647-656, 1974.

Bock, E. J., T. Hara, N. M. Frew, and W. R. McGillis, Relationship between air-sea gas transfer and short wind waves, *J. Geophys. Res., 104*, 25821–25831, 1999.

Csanady, G. T., The role of breaking wavelets in air–sea gas transfer, *J. Geophys. Res., 95*, 749-759, 1990.

Frew, N. M., The role of organic films in air-sea exchange, *The Sea Surface and Global Change*, Eds. P. S. Liss and R. A. Duce, Cambridge, Cambridge University Press, 121-172, 1997.

Jähne, B., and H. Haußecker, Air-water gas exchange, *Ann. Rev. Fluid Mech., 14*, 321-350, 1998.

Jähne, B., K. O. Munnich, R. Bosinger, A. Dutzi, W. Huber, and P. Libner, On the parameters influencing air-water gas exchange, *J. Geophys. Res., 92*, 1937-1949, 1987.

Komori, S., R. Nagaosa, and Y. Murakami, Turbulence structure and mass transfer across a sheared air-water interface in wind-driven turbulence, *J. Fluid Mech., 249*, 161-183, 1993.

Wanninkhof, R., and W. R. McGillis, A cubic relationship between air-sea CO_2 exchange and wind speed, *Geophys. Res. Lett., 26*, 1889-1892, 1999.

Zappa, C. J., Microscale wave breaking and its effect on air-water gas transfer using infrared imagery, Ph.D. Thesis, University of Washington, 1999.

Zappa, C. J., W. E. Asher, and A. T. Jessup, Microscale wave breaking and air–water gas transfer, *J. Geophys. Res., submitted*, 2000.

W. E. Asher, Applied Physics Laboratory, 1013 NE 40th St., Seattle, WA, 98105. (email: asher@apl.washington.edu)

A.T. Jessup, Applied Physics Laboratory, 1013 NE 40th St., Seattle, WA, 98105. (email: jessup@apl.washington.edu)

J. Klinke, Scripps Institution of Oceanography, La Jolla, CA 92093-0230. (email: jklinke@ucsd.edu)

S. R. Long, NASA GSFC / Wallops Flight Facility, Code 972, Wallops Island, VA 23337. (email: steve@airsea.wff.nasa.gov)

C. J. Zappa, Woods Hole Oceanographic Institution, MS #9, Woods Hole, MA, 02543. (email: czappa@whoi.edu)

Statistics of Geometric Properties of Breaking Wind Waves Observed in Laboratory

Guillemette Caulliez

Institut de Recherche sur les Phénomènes Hors Equilibre, Marseille, France

We present the results of a statistical analysis of the geometric properties of the breaking wind waves observed in the large Marseille-Luminy wind-wave facility. These investigations were performed within the framework of the LUMINY program aiming at better understanding the effects of wave breaking on gas transfer at the air-water interface [*de Leeuw et al.*, 2001]. They were concerned with both wind-generated waves of wavelength increasing with fetch and paddle-generated waves amplified by wind. Breaking was detected in time series of the wave height derivative signals using a local geometrical criterion. We distinguished two different stages of development of the breaking process: the incipient breaking and fully-developed breaking. We first show that the asymmetric profile of the wind waves at the inception of breaking is invariant with scale and, therefore, is self-similar. The statistics of the geometric properties of the breaking wind waves whatever the development of the breaking process, as the average slope distribution along the crest, the mean slope and the relative jump height of the "breaking" region, do not vary with fetch and do not depend on the way the waves were initially generated, corroborating that the shape of breaking wind waves is self-similar. The results then suggest that wave breaking at the sea surface and its induced effects both on momentum and gas transfer may be characterized by only one wave parameter as the rate of breaking.

1. INTRODUCTION

Wave breaking as source of near-surface turbulence and air entrainment by bubbles in water, is one of the most important phenomena which control gas transfer across the air-sea interface [*Merlivat and Memery*, 1983; *Melville*, 1996, *Zappa et al*, 1999]. However, our present knowledge of the basic properties of breaking wind waves remains limited due to intrinsic difficulties of experimental investigation both in laboratory or at sea. Thus, the field observations performed up to now have been mainly confined to finding the probability of occurrence of breaking as function of wind and wave parameters [*Holthuijsen and Herbers*, 1986; *Gemmrich and Farmer*, 1999]. In laboratory, the efforts were focused on mechanically-generated wave breaking [*Bonmarin*, 1989; *Rapp and Melville*, 1990; *Duncan*, 1999]. An early study paying attention to geometric properties of breaking wind waves was made by *Xu et al* [1986] and *Hwang et al.* [1989]. This work provides a statistics of breaking wave characteristics averaged over the whole lifetime of the breaking events without consideration of the various phases of the breaking process. *Koga* [1984] distinguished four different types of breaking based on the wind wave shape and the bubble occurrence at the crest. However, his

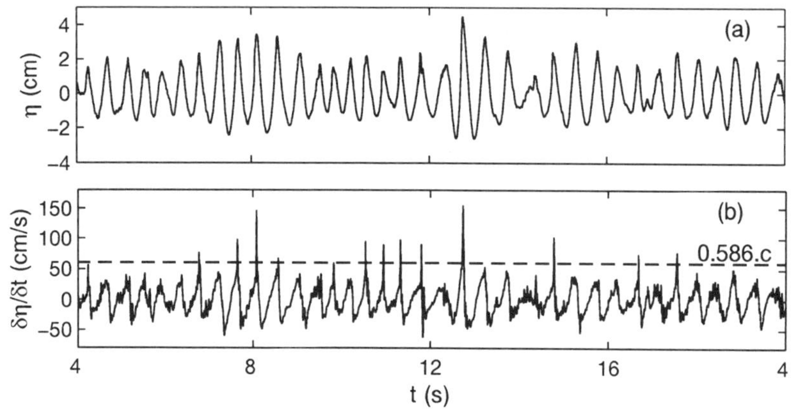

Figure 1. Time sequence of the water surface height signal (**a**) measured by a capacitance wave gauge located at 22 m fetch for a wind speed of 10 m/s, and (**b**), the corresponding time derivative signal. A wave breaking is detected everytime the time derivative signal exceeds the value 0.586c marked by the horizontal dashed line.

investigation was restricted to the analysis of a few basic properties of the breaking waves as their typical wave height or wave steepness. In this context, to built a model for predicting the gas transfer rate between the atmosphere and the ocean in which the effect of wave breaking would be included, has appeared as an uneasily tractable task. To improve this situation, during LUMINY gas transfer experiments [*de Leeuw et al.*, 2001], our focus was aimed at better quantifying wave breaking at the water surface and searching for the adequate relationships between wave breaking and global wave field and wind parameters. To reach this objective, a detailed investigation of the geometric properties of the individual breaking wind waves was performed. This enables us to find basic universal features of wind wave profiles characteristic of two distinct stages of the breaking process, and then to propose a new parameterization of wave breaking at the water surface.

2. EXPERIMENTAL PROCEDURE

The experiments were carried out in the large IRPHE-Luminy wind wave facility which consists of a 40 m long, 2.6 m wide and 0.9 m deep water tank and an air-recirculating tunnel of 1.5 m height at the test section. The facility is equipped with a submerged wavemaker made of an oscillating plate and controlled by an electro-hydraulic motor driven by a sine function. At the end of the tank, an absorbing beach damps the wave reflection. To estimate the turbulent momentum flux across the air boundary layer, measurements of the instantaneous air flow velocities at several levels above the water surface were made by means of a hot X-wire probe set up on a vertical displacement and a carriage moving along the tank. Parallel measurements of the water surface heights were performed by means of two capacitance wave gauges separated by a 5 cm along-wind distance. These probes were composed of thin sensitive wires, 0.3 mm in diameter. The time derivative signals of the water surface heights $\eta(t)$ were then obtained by use of analog derivators. Thus, the instantaneous slope of the longitudinal dominant wave profiles s(t) could be derived from these time series, using the relation:

$$s(t) = -1/c \, d\eta/dt, \qquad (1)$$

in which the phase speed c corresponding to the dominant frequency was computed from both wave signals by means of a cross-correlation method with less than 10% uncertainty. This slope-measuring method was tested by comparing this signal with direct measurements performed by a laser slope gauge [*Jähne and Riemer*, 1990]. A good agreement was found for gravity and capillary-gravity wind waves of moderate steepness, thus confirming the ability of our technique to measure water surface slopes due to high frequency disturbances. Moreover, compared to optical methods, such an indirect method has two important advantages: first, it has higher slope cut-off when the data sampling rate is large enough (200 Hz in these experiments), second, it enables one to distinguish dominant wave breaking from small-scale microbreaking, since these waves propagate at quite different wave speeds. Hereafter, to make easier the description of the wave properties, the water surface slope will be defined as the quantity -s(t).

The observations were made at 10 m/s and 13 m/s wind speed, with and without mechanically-generated waves, for fetches varying between 15 and 30 m. The dominant wave frequency decreased with fetch respectively from 2.4 to 1.8 Hz and 2.0 to 1.6 Hz for pure wind waves, and was fixed at 1.5 Hz and 1.2 Hz for wind-amplified paddle waves. The

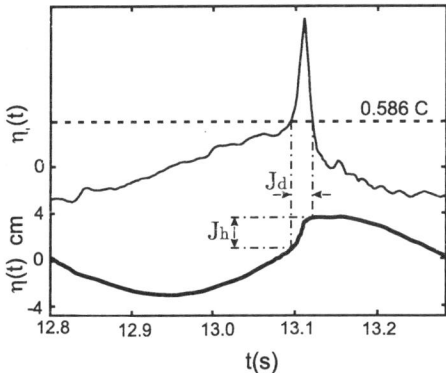
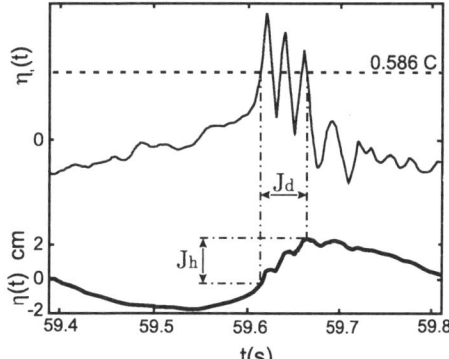

Figure 2. Typical wave amplitude signal (bold) and the corresponding time derivative signal (thin) associated respectively with a breaking wave of the first type (left) and the second type (right).

average wave steepness, of order of 0.2 to 0.3 depending on wind and paddle conditions, was found roughly constant with fetch, indicating that the respective wave fields have reached a kind of "energy saturation" by 15 m fetch.

Wave breaking was detected by a geometrical criterion as proposed by *Longuet-Higgins and Fox* [1977] based on the computations of the highest slope which a Stokes wave may reach at the crest before breaking, and then tested experimentally [*Xu et al.*, 1986; *Longuet-Higgins and Smith*, 1983]. Strictly speaking, this criterion is only justified for free surface waves, but the fact it concerns a local wave feature led these authors to extend it to wind-generated waves. Therefore, in a similar way, an individual wave has been considered as breaking when the water surface slope at the crest exceeds 0.586 in magnitude, i.e. when the time derivative wave signal exhibits a value higher than 0.586c at anytime between two successive zero down-crossings of the wave signal (Figure 1). However, a more detailed analysis of the time derivative wave signals enabled us to distinguish unambiguously two types of breaking waves associated with two distinct stages of the breaking process, as illustrated in Figure 2. The breaking waves of the first type have been identified by their steep monotonous wavefront in which η_t displays only one peak exceeding the threshold value 0.586c (Figure 2a). These waves characterized by a well-defined single-value profile can be regarded as near-breaking waves, i.e. waves at the inception of breaking. The breaking waves of the second type have been detected by their steep but corrugated wavefront in which η_t presents several peaks crossing the threshold value and separated from each other by very short time intervals (Figure 2b). These waves can be regarded as fully-developed breaking waves, i.e. waves for which a breaking phenomenon develops at the crest. In these experiments, this active phase of the breaking process was generally of enough intensity to be visualised by whitecaps.

3. RESULTS

3.1. Statistics of the Breaking Wave Crest Characteristics

To describe the shape of the breaking wind waves, we first investigated the local features of the "breaking region" of the wavefront where the instantaneous slope exceeds the chosen threshold. This study was made separately for both types of breaking waves. Following *Longuet-Higgins and Smith* [1983] approach, this region was characterised by the jump height J_h and the jump duration J_d defined respectively as the difference in the water surface elevation and the time duration between the moments when η_t rises above and falls down 0.586c (Figure 2). Focusing on the waves at the inception of breaking, the first quantity to be analysed is the mean slope of the "near-breaking region" defined as the ratio J_h/cJ_d. Its average value $<J_h/cJ_d>$ over about 40 to 130 near-breaking waves, depending on the wind and paddle conditions, was estimated for each time series recorded at the various fetches between 15 and 30 m. An illustration of this evolution with fetch for 10 m/s wind speed is given in Figure 3. Thus, for pure wind waves, quite surprisingly, $<J_h/cJ_d>$ exhibits a constant value of 1, while the wave field evolves with fetch and the dominant wavelength, ranging in the gravity domain, increases from about 40 cm to 60 cm. Besides, this quantity proves to be independent on the initial way of wave generation, remaining the same for pure wind waves and for mechanical waves amplified by wind till to the point of breaking. In the average, the "near-breaking region" of the wavefronts is thus inclined by 45° to the horizontal plane, that is an angle much larger than 30°, the value predicted for the highest Stokes wave. Similar observations were made at 13 m/s wind speed for pure wind waves of wavelength ranging between 60 and 85 cm and 130 cm for wind-amplified mechanical waves. In fact, the wavefront slope varies between 30° and the highest slope

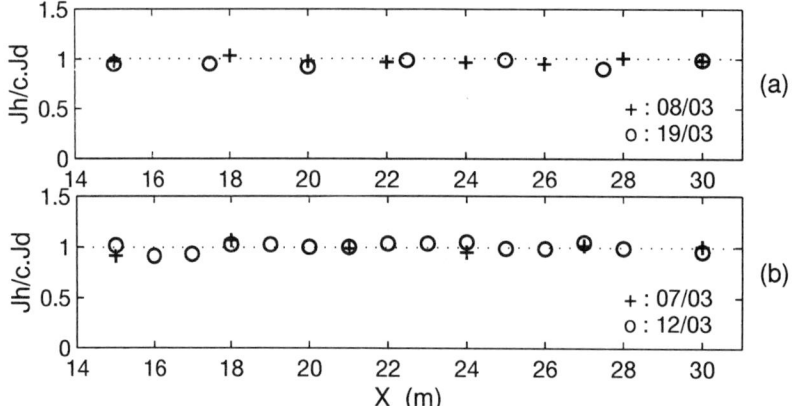

Figure 3. Evolution with fetch of the average slope of the wavefront "near-breaking" region observed at 10 m/s wind speed, respectively for: **(a)** pure wind waves and **(b)** wind- amplified mechanical waves.

measurable in these experiments with a reasonable accuracy (of order of 2), depending on the stage of the wave deformation before the crest collapse typical of this phase of breaking initiation. However, the probability for observing a given slope exceeding 30° is the same, whatever the time series and the scale of the dominant waves, as shown in Figure 4. Furthermore, from a slightly different point of view, we can mention that the average nondimensional quantities characteristic of the "near-breaking" wavefront shape such as the ratios between the jump height J_h and the related near-breaking wave height H_b, or the time duration J_d and the corresponding wave period T_b, are also independent on wave scale whatever the conditions, taking the values 0.18 and 0.03 respectively within ± 15%.

The geometric features of the wavefront breaking region of the fully-developed breaking wind waves (type II) present the same kind of invariance with respect to scale as those observed for the waves at the point of breaking, even if in this case, the fluctuations around the average values are much more pronounced. Thus, the mean angle θ_f of the breaking wavefronts to the horizontal plane is independent on fetch for such waves, but it reaches about 32° only. The latter is consistent with the fact that the crests are effectively "broken". The breaking region is much wider both in height and in the along-wind direction, the nondimensional quantities $<J_h/H_b>$ and $<J_d/T_b>$ reaching approximately 0.35 and 0.08. However, note that no significant difference in the typical values of the wave breaking height H_b is observed between both types of waves.

These results clearly show that the near-breaking wind waves have a quite specific wavefront geometry which characterises the wave crest deformation before breaking, and thus, suggest the existence of self-similar limit profiles for such waves. The histograms of the water surface slope values observed at the crest of the near-breaking wind waves were plotted in Figure 5 for each time series available at 10 m/s. In turn, these graphs exhibit the same shape, whatever the wave scale, and collapse remarkably well onto a single curve. The latter is characterised by two well-pronounced asymmetric peaks at two definite slope values (positive and negative) and a tail at high slopes due to the steep "near-breaking region" of the wavefronts. The narrow peak centred at the negative slope value of - 0.25 indicates that the rear face of the near-breaking wave crests presents a wide region of almost constant slope inclined approximately by 15° to the horizontal plane. Although at higher wind speeds, the contribution of "noise" induced by small-scale wind waves increases significantly and thus, makes the histograms less representative of the dominant

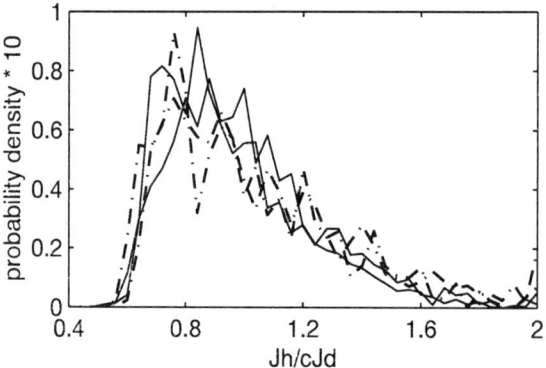

Figure 4. Probability density function of the values of the mean slope of the near-breaking region of the waves at the inception of breaking observed whatever the fetch at 10 m/s (solid) and 13 m/s (dashed) wind speed with (bold) and without (thin) mechanically-generated waves

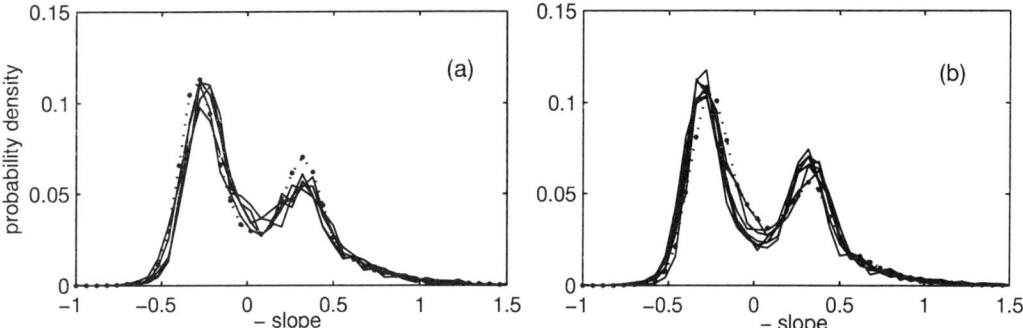

Figure 5. Histograms of the slope values of the near-breaking wind wave crests observed at 10 m/s wind speed for various fetches between 15 m and 30 m, respectively for: (a) pure wind waves and (b) wind-amplified mechanical waves. The average histogram for the wave conditions (b) is plotted in Fig. 5(a) and vice versa (dotted lines).

wave shape, the measurements performed at 13 m/s wind speed also support in the whole these observations. Then, the profile invariance with scale observed for both pure wind waves and wind-amplified mechanical waves shows undoubtedly that the shape of the incipient breaking wind waves is self-similar.

3.2. Intrinsic Relationship Between the Breaking Wave Field Characteristics

The found invariance with scale of the individual breaking wave characteristics whatever the stage of development of the breaking process enables us to infer the existence of intrinsic relationship between the different global features of breaking waves which may describe statistically the wave breaking conditions at a water surface, and consequently, such a description may be obtained from a limited number of appropriate physical quantities. To check these ideas, the first quantity to be investigated is the temporal breaking rate R_d. This quantity was introduced by *Longuet-Higgins and Smith* [1983] and defined as the ratio between the sum of the breaking jump duration ΣJ_d and the total duration D of the time sequence. An illustration of its evolution with fetch for pure wind waves is given in Figure 6. The R_d slow decrease with fetch, clearly noticeable in this figure, appears typical of the fetch-limited nature of the wind wave fields in wave tanks, as the energy balance between wind input to waves and wave dissipation by breaking is not completely achieved even at large fetches. It is however worth noting that despite these variations, the relative contributions to R_d of incipient breaking waves and fully-developed breaking waves are remarkably constant with fetch. This confirms once again the scale independence of the breaking process unfolding in time. The evolution of the wave breaking rate as function of wave steepness is also displayed in Figure 7, both quantities being estimated for each time series recorded at the various measuring fetches. Thus, it is found that for such nonhomogeneous wave fields,

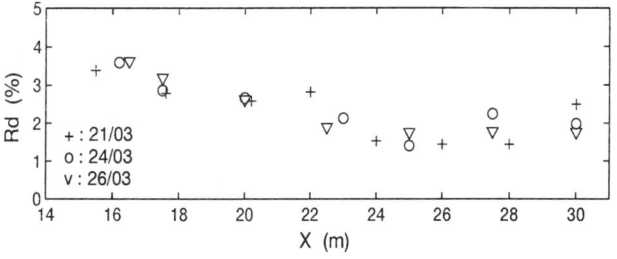

Figure 6. Evolution with fetch of the breaking rate R_d observed in the water tank when pure wind waves develop at the surface for a wind speed of 13 m/s.

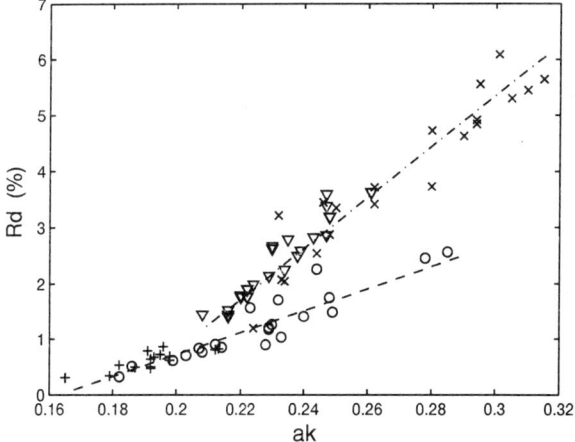

Figure 7. Evolution with wave steepness of the breaking rate R_d estimated for each time series recorded at the different fetches at 10 m/s (+; o; --) and 13 m/s (∇; x, -.-), respectively for pure wind waves (+;∇) and wind-amplified paddle waves (o, x). The wave steepness ak at a given fetch is defined as the root mean square of the associated wave slope signal.

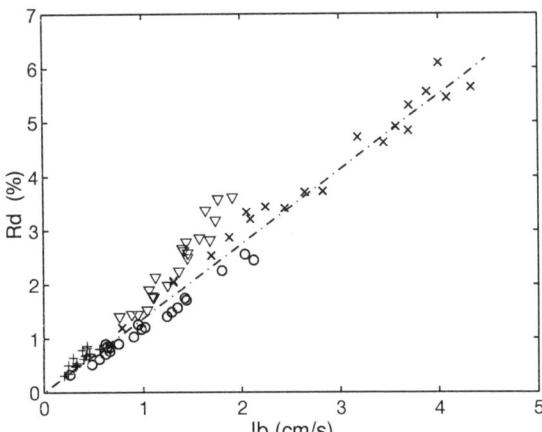

Figure 8. Coupled evolution of the wave breaking intensity I_b and the breaking rate R_d observed for the various measuring fetches at 10 m/s (+; o) and 13 m/s (∇; x), respectively for pure wind waves (+;∇) and wind-amplified paddle waves (o, x).

R_d varies linearly with the local wave steepness, but this variation is strongly wind dependent.

As it follows quite straightforwardly from the above findings, the second quantity to be introduced to better characterize wave breaking should be built upon the jump height of the breaking waves. A new quantity I_b was defined as the average jump height $<J_h>$ over a time sequence weighted by the average breaking frequency n_b (i.e. the average number of wave breaking per second) and then estimated for each time series. This quantity which is representative of the average height of the breaking regions might be a good indicator of the average wave potential energy dissipated by breaking. If confirmed, this hypothesis might be used for quantifying readily the intensity of breaking of wind wave fields.

The coupled evolution of the breaking intensity I_b and the breaking rate R_d observed for the measured wave fields is then examined in Figure 8. It is quite remarkable to observe that for so different wind and wave conditions, the respective evolutions of R_d and I_b are directly linked and can be described by a simple relation of proportionality independent on wind speed. Furthermore, we mention that a similar close relationship between the respective values of R_d and I_b is also established when fully-developed breaking waves are analyzed separately. These results of practical importance suggest that the wave breaking observed at the sea surface might be described by a single parameter, such as the temporal rate of breaking, or in space, the fractional area of breaking, whatever the wind forcing, or at least in a moderately high wind speed domain.

CONCLUDING REMARKS

The investigations of the statistical geometric properties of the breaking wind waves made during the LUMINY gas transfer experiments enable us to determine basic universal features of the breaking wind waves as a function of the stage of development of the breaking process. At the inception of breaking, the profiles of wind waves do not depend on wave scale or the way of initial wave generation. The statistics of the geometric characteristics of the fully-developed breaking waves also remains scale invariant. This robust self-similarity of the breaking wind waves in the gravity range suggests that the physical mechanisms leading to breaking, and, to a certain extent, following the breaking, are universal and weakly dependent on wind forcing at the wave crest.

This advance in describing of the breaking wind waves has direct implications for understanding and parameterization of the effect of wave breaking on momentum and gas exchanges at the sea surface. At first, for wind wave fields, wave breaking can be characterized by a single parameter, e.g., by the rate of breaking. Second, such a parameter may account for the intensity of the specific phenomena induced by wave breaking in the near-surface boundary layers, i.e. the increase of turbulent mixing in air and water and the generation of bubbles plumes in water. Consequently, as shown by *de Leeuw et al.* [2001], at high wind speeds, the evolution with wind of the gas transfer rate at the air-water interface may be modelled by various simple laws dependent on the gas solubility but function only of two dynamical parameters, namely the air friction velocity and the breaking rate.

Acknowledgments. I am particularly grateful to Prof. V.I. Shrira for fruitful discussions of this work and comments on the manuscript, Dr. S. Badulin for his contribution to the data analysis, the fellow participants of the LUMINY project "Breaking Waves and Air-Sea Gas Transfer" for their participation in the experiments and the constructive discussions which follow, and the technical staff of the laboratory for its assistance for the preparation and running of the experiments. This work was supported by the European Commission EC DG XII (contract ENV4-CT995-0080) and CNRS.

REFERENCES

Bonmarin, P., Geometric properties of deep-water breaking waves, *J. Fluid Mech.*, 209, 405-433, 1989.

De Leeuw, G., G.J. Kunz, G. Caulliez, D.K. Woolf, P. Bowyer, I.S. Leifer, P. Nightingale, M. Liddicoat, T.S. Rhee, M.O. Andreae, S.E. Larsen, F.Aa Hansen and S. Lund, LUMINY – An overview, in *Gas Transfer and water Surfaces*, edited by M.A. Donelan, W.M. Drennan, E.S. Salzman, and R. Wanninkhof, pp. xxx-xxx, AGU, this volume, 2001.

Duncan, J.H., H. Qiao, V. Philomin, A Wenz, Gentle spilling breakers: crest profile evolution, *J Fluid Mech*, 379, 191-222, 1999.

Gemmrich, J.R., D.M. Farmer, Observations of the scale and occurrence of breaking surface waves, *J. Phys. Oceanogr.*, 29, 2595-2606, 1999.

Holthuijsen, L.H., T.H. Herbers, Statistics of breaking waves observed as whitecaps in the open sea, *J. Phys. Oceanogr.*, 16, 290-297, 1986.

Hwang, P.A., Xu D., Wu J., Breaking of wind-generated waves: measurements and characteristics, *J. Fluid Mech.*, 202, 177-200, 1989.

Jähne, B., Riemer K.S., Two-dimensional wave number spectra of small-scale water surface waves, *J. Geophys. Res.*, 95, 11531-546, 1990.

Koga, M., Characteristics of a breaking wind-wave field in the light of the individual wind-wave concept, *J. Oceanogr. Soc. Japan*, 40, 105-114, 1984.

Longuet-Higgins, M.S., Fox M.J.H., 1977, Theory of the almost-highest wave: the inner solution. *J. Fluid Mech.*, 80, 721-741.

Longuet-Higgins, M.S., N. D. Smith, Measurements of breaking by a surface jump meter. *J. Geophys. Res.*, 88, 9823-9831, 1983.

Merlivat, L., Memery, L., Gas exchange across an air-water interface: experimental results and modelling of bubble contribution to transfer, . *J. Geophys. Res.*, 88, 707-724, 1983.

Melville, W.K., The role of surface-wave breaking in air-sea interaction, *Ann. Rev. Fluid Mech.*, 28, 279-321, 1996.

Rapp, R.J., W.K. Melville, Laboratory measurements of deep-water breaking waves, *Phil. Trans. R. Soc. Lond. A*, 331, 735-800, 1990.

Xu, D., Hwang P.A., Wu J., Breaking of wind-generated waves. *J. Phys. Oceanogr.*, 16, 2172-2178, 1986.

Zappa C.J., Asher W.E., Jessup A.T., Correlating microscale wave breaking with gas transfer for cleaned and surfactant–influenced water surfaces. In *The Wind-Driven Air-Sea Interface*, edited by M.L. Banner, 357-58, Univ. New South Wales, Australia, 1999.

G. Caulliez, IRPHE - IOA, 163, avenue de Luminy – case 903, 13009 Marseille, France (e-mail: guil@pollux.irphe.univ-mrs.fr)

Overview of the CoOP Experiments: Physical and Chemical Measurements Parameterizing Air-Sea Heat Exchange

Erik John Bock,[1] James Bearer Edson,[2] Nelson M. Frew,[3] Tetsu Hara,[4] Horst Haussecker,[5] Bernd Jähne,[1,6] Wade R. McGillis,[2] Sean P. McKenna,[2] Robert K. Nelson,[3] Uwe Schimpf,[1] and Mete Uz[4]

Experiments performed in the Pacific and Atlantic Oceans in 1995 and 1997 attempted to measure the short time-scale and small spatial scale variability in the air-sea gas transfer rate. Along with these measurements, physical and chemical parameters known from previous laboratory studies to influence transfer rates were also characterized. These parameters include the atmospheric forcing, the capillary and capillary-gravity wave state, the surface chemical enrichment, and the level of near-surface turbulence. In this contribution we describe the methodologies employed for the measurement campaigns and summarize some general observations resulting from them. Other contributions from the coauthors describe in more detail the specific conclusions derived from the Coastal Ocean Processes (CoOP) field program.

1. INTRODUCTION

1.1. Motivation

A review of measurements of the gas transfer rate across the air-sea interface using transient tracers [*Asher and Wanninkhof*, 1998] includes field measurements used to estimate this rate in situ. This review sites the principal disadvantage of chemical and biological tracer studies being the long time scale required for accurate assessment of the transfer rate, which requires integration times of the order of one to several days. This constraint prohibits the determination of gas transfer rates on time and space scales commensurate with fast atmospheric forcing and modulation of gas transfer rate by patchiness on the ocean surface associated with surface-active agents (surfactants). A review of the effects of surfactant films on air-sea gas exchange [*Frew*, 1997] cites examples of the extent and magnitude of these films in the oceanic environment. The conclusions indicate that coastal zones are the most influenced by the presence of both man-made (anthropogenic) and natural (phyto- and zoo- plankton exudative) films. If these films influence transfer in situ to the same extent that laboratory studies suggest [*Bock et al.*, 1999; *Frew et al.*, 1995], their impact may significantly reduce the gas transfer coefficient and this effect will be aliased by the techniques requiring integration times in excess of several hundred seconds. The need for making short time-

---

[1]Interdisciplinary Center for Scientific Computing, University of Heidelberg, 69120 Heidelberg, Germany.

[2]Department of Applied Ocean Physics and Engineering, Woods Hole Oceanographic Institution, Woods Hole, Massachusetts 02543.

[3]Department of Marine Chemistry and Geochemistry, Woods Hole Oceanographic Institution, Woods Hole, Massachusetts 02543.

[4]Graduate School of Oceanography, University of Rhode Island, Narragansett, Rhode Island 02882.

[5]Xerox Palo Alto Research Center, Palo Alto, California 94304.

[6]Institute for Environmental Physics, University of Heidelberg, 69120 Heidelberg, Germany.

Gas Transfer at Water Surfaces
Geophysical Monograph 127
Copyright 2002 by the American Geophysical Union

scale and small spatial scale estimates of gas transfer velocity, along with the physical and chemical parameters that affect it, provided a framework for the two field experiments of the Coastal Ocean Processes Program.

1.2. CoOP 95

The first field experiment of the CoOP program was conducted in April and May of 1995. A detailed description and first results of this experiment [Bock et al., 1995] demonstrated the feasibility of combining a diverse set of measurement techniques toward understanding the variability within coastal environments of atmospheric forcing, surface wave field, surface chemical enrichment and near surface turbulence. The attempt to obtain gas transfer estimates using the controlled flux technique, however, was unsuccessful owing to the inability to image a heated patch on the ocean surface for an extended period of time (see section 4, below.) As an alternative, data collected from the infrared imaging of the ocean surface was reinterpreted by incorporating a model of surface renewal based on aperiodic renewal statistics and making use of the assumption of spatial and temporal homogeneity within the space and time period of sampling. Measurements were planned to coincide with the Marine Boundary Layer Program sponsored by the Office of Naval Research, to take place in Monterrey Bay near the FLIP platform. Weather constraints limited the useful data to that obtained in the vicinity of the Channel Islands off San Diego.

1.3. CoOP 97

The second experiment of CoOP was performed in July 1997 and took place in the Northwest Atlantic. The scientific goals were essentially the same as in the first experiment, but instrumentation and techniques were refined to obtain heat transfer estimates in addition to the other physical and chemical measurements already possible in CoOP 95. As was the case in the first cruise, the measurements were obtained from an instrument suite aboard the LADAS catamaran, and instrumentation on both a bow-mounted mast and a bow-mounted extension. The mast served as a platform for atmospheric measurements of wind stress, heat flux and humidity flux. The extension served to support the infrared imaging package and to ensure that the images collected were not influenced by reflections of the ship structure in the wave signatures. The measurements were chosen to be obtained at a number of sites to include a wide variety of surface conditions, including as variables the surface roughness, surface chemical composition and interfacial stability. Cruise tracks were also selected to co-locate the instrumentation packages at places where the TOPEX/POSEIDON overflights occurred during the cruise. In doing this it was believed that ground truthing of satellite information regarding surface roughness would be obtained.

2. METHODOLOGY

2.1. LADAS Catamaran

For the 95 and 97 CoOP experiments two versions of the LADAS research catamaran were deployed. The advantages of such a platform are two-fold. Firstly, it is separated from the main research vessel so that measurements can be made in a region unaffected by flow distortions and wave fields created by the research vessel itself; and, secondly it is a small platform capable of following the longer waves and can track the air-sea interface with greater precision. In the 1995 field experiment, a towed version of the catamaran was used. This platform derived its power from the main research vessel through an umbilical cable that additionally transmitted acquired data and acted as a strength member to support the towing effort. The platform was approximately 4.9m long, weighed 1T in total, and was steered approximately 50m to the side of the R/V New Horizon using the lift from rudders to counteract the drag on the hull of the catamaran. After being lost at sea following an experiment late in 1995, LADAS was replaced with another platform. This version is also a twin pontoon platform, but has several distinct advantages over previous versions. It is slightly larger (approximately 5.4m long and weighs 2T), but it is remotely controlled, allowing operation anywhere within line of site of the research vessel. This has the advantage of being completely free of interference due to the research vessel and affords a much lower minimum speed of driving, since there is no need for generating lift to propel the platform away from the research vessel. Two maneuverable electric thrusters power the catamaran; in addition it has steerable rudders. Data is collected by onboard computer systems and an onboard wireless local area network (LAN) provides data communication between the platform and the dry lab of the research vessel. The throughput of the LAN (10Mbps) is sufficient to enable a real time assessment of data integrity and allows control of the acquisition software without boarding the platform. Steering and system power up/down is accomplished using a digital proportional radio control unit operated from the bridge of the research vessel. The onboard supply of lead-acid batteries enables the catamaran to be deployed for approximately 6 hours and afterwards it requires about the same amount of time to recharge. This version of LADAS was deployed during CoOP 97.

2.2. Scanning Laser Slope Gauge

The Scanning Laser Slope Gauge (SLSG) used in both the CoOP experiments [*Bock and Hara*, 1995] was the same used in previous field studies [*Bock et al.*, 1995; *Hara et al.*, 1998; *Hara et al.*, 1994] and laboratory studies [*Bock et al.*, 1999; *Plant et al.*, 1999; *Hara et al.*, 1995]. The operational description, data analysis and error analysis has been described in detail previously and need not be repeated here. In summary, the instrument provides estimates (in two wavenumbers and frequency) of capillary-gravity wave spectra for wavenumbers between 50 and 800 radian/m and for frequencies less than the Nyquist frequency. In CoOP 95 the Nyquist frequency was 52.1 Hz, and in CoOP 97 it was slightly higher (69.5 Hz) owing to faster acquisition hardware. In the traditional sense, Nyquist limiting comes from the fact that any sinusoidal component of a complex signal has to be sampled at least twice within it's period for a non-aliased representation of the discretely sampled signal. While it is possible to interpret the data products of the SLSG beyond this frequency, it is strictly invalid. For this reason, we limit the interpretation of the spectral results of the analysis to one-half of the rate at which we perform the repetitive scan. The slope gauge was mounted at the extreme bow of the catamaran so that it was minimally influenced by disturbances created by the pontoons. Further examples of the slope data obtained during the CoOP experiments is given elsewhere [*Hara et al.*, 2001].

2.3. Surface Microlayer Sampler

Surface chemical distributions were measured using colored dissolved organic matter (CDOM) as an indicator of surface-active organic matter adsorbed at the air-sea interface. A surface microlayer sampler (SMS) uses a rotating glass drum to continuously skim the upper 40-60 microns of the sea surface. A fluorometer measures the emission of the CDOM in the sampler flow stream at 450 nm (excitation at 350 nm). Fluorescence of a second flow stream obtained from subsurface water at a depth of 10 cm was also measured to provide bulk seawater CDOM concentration and thus, by comparison, an estimate of the degree of surface microlayer enrichment. The raw fluorescence data were calibrated with Quinine Sulfate standards and intensities were normalized to that of the water Raman band to provide normalized fluorescence. More detailed results from the use of this technique during CoOP are reported elsewhere [*Frew et al.*, 2001].

2.4. Capacitive Wave Staffs/Motion Package

A six-wire wave staff array was used in CoOP 95, and a 12-wire array was used in CoOP 97. The purpose of the wire array is to provide wave height measurements, based on the moving coordinate normalization algorithms developed in previous experiments [*Hanson et al.*, 1997]. These algorithms make use of a six-degree of freedom motion package to take into account the three linear accelerations and the three axial rotations associated with a free body. Using these motions to correct for the motion of the catamaran, it is possible to estimate wave height at a series of locations in earth coordinates. An extended version of the Data-Adaptive Spectral Estimator (DASE) algorithm [*Davies and Regier*, 1977] is used to obtain estimates of the directional spectra of waves. Using the combination of wire wave staffs and the motion package it is possible to estimate these spectra across the range of wavelengths of order 1-10m.

2.5. Acoustic Anemometer/Flux Package

A meteorological flux system mounted on the ship's bow at a height of 10m above the sea surface provided meteorological information during the 1995 cruise. The meteorological, GPS, and ADCP data were merged and used to compute the fluxes of momentum, sensible heat, and moisture (latent heat) using the inertial-dissipation [*Edson et al.*, 1991] and bulk aerodynamic methods [*Fairall et al.*, 1996]. These estimates have been used by Hara et al. [1998] to investigate the response of short wind waves to wind to wind forcing in coastal waters. During the 1997 experiment, additional instrumentation was added to the meteorological systems on both the ship and LADAS to provide direct estimates of atmospheric forcing via the direct covariance technique [*Edson et al.*, 1998].

2.6. Controlled Flux Technique

To obtain fast time- and small space- scale estimates of the gas transfer coefficient the controlled flux technique (CFT) was developed [*Haussecker*, 1996]. This technique uses heat as a proxy tracer for gas transfer. Although it is an indirect method that still requires the Schmidt number similarity to estimate gas transfer from heat transfer, it is capable of measuring heat transfer rates with a temporal resolution of less than about 1 minute. Hence the gas transfer velocity values derived from CFT-measurements can resolve even highly

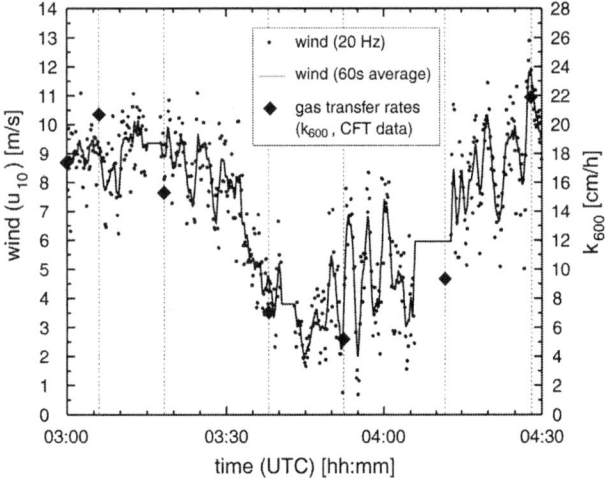

Figure 1. Measured gas transfer velocity from the infrared technique, as a function of time for a period with variable wind. Individual small points are 20Hz wind estimates along with averaged winds. Large symbols are infrared estimates of gas transfer (normalized to Schmidt number 600).

intermittent meteorological, chemical and physical influences. The CFT technique uses an infrared camera to image the spatial distribution of the sea skin surface temperature. From these images the temperature gradient ΔT, across the interface can be directly estimated, as well as the time constant t_* of the transfer process. In addition to these parameters, the infrared image sequences directly reveal the structure of microturbulence directly at the water surface. Processes contributing to gas transfer by enhancing surface turbulence, such as microscale wave breaking and instantaneous surface renewal events become directly visible. If the surface renewal events responsible for heat and mass transfer obey similarity, characterization of heat transfer allows for estimation of mass flux. The validation of this assumption is currently the topic of a series of experiments at the Heidelberg Aelotron.

3. RESULTS

The combination of data products of the CFT technique and the SLSG are presented in two time series of calculated CO_2 transfer velocities (obtained by Schmidt number scaling of surface renewal estimates) measured during the 1997 CoOP cruise in the Western North Atlantic. These are shown in Figure 1 together with 15min averages of the measured wind speed. Figure 1 shows that the gas transfer rate follows the wind speed along the sudden increase of both in the course of Year Day (YD) at about YD 190.21. Although the transfer rates match field measurements of other techniques (such as the differential tracer and the Radon deficiency method with measurement times in the order of 24 hrs) in the statistical average, some features of these series are not completely explained using wind-stress alone. Intermittent fluctuations on short time scales seem to be related to other modulating mechanisms like the occurrence of rain and the presence of sea surface films.

An example of the strong influence of rain on modulating gas exchange is demonstrated in Figure 2. During the times reported for YD192 the weather changed from moderate rain to heavy rain with a period of no rain in between. Unfortunately no rain gauge was deployed during either cruise and the intensity of the rain events is only qualitative. While the wind speed remained low - between 3 and 5 m/s - the gas transfer rates show an increase of almost a factor of 4 at the onset of heavy rain. A combined plot of wavenumber-binned mean square slope versus gas transfer rate is presented in Figure 3. The data represent laboratory studies in two separate annular wave tanks and field data obtained from the CoOP cruises. While the scatter is in some cases large, the overall trend that holds for cases including surfactant films, clean surfaces, wave scales spanning several orders of magnitude, and different environmental factors including rain is supported. While some of the more extreme outliers include the field results of CoOP, reasons that these may be the result of measurement error will be addressed below. The

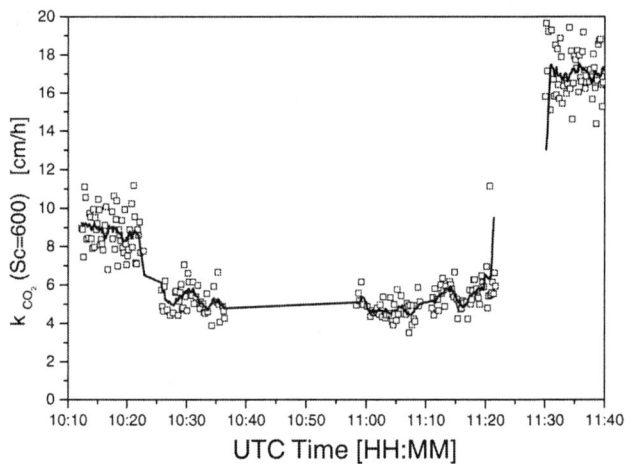

Figure 2. Measured gas transfer velocity from the infrared technique, as a function of time for a period with variable rain. Individual points are infrared estimates of gas transfer (normalized to Schmidt number 600). In the period of time after UTC 11:20 (Year Day 192), the intensity of rain increased, along with a corresponding increase of gas transfer rate. Rain rate are qualitative estimates, no rain gauge was available.

most significant result learned from this figure is that there exists a need for more field measurements so as to better constrain the appropriate parameters for gas exchange.

4. DISCUSSION

Because of the need to obtain both the temporal and spatial statistics of the ocean surface temperature structure when determining the heat transfer rate (and hence the gas transfer rate by means of Schmidt/Prandtl number scaling) it is not possible to obtain an independent estimate of the gas transfer rate from non-consecutive time series of images. Unfortunately, this was the only data product available from the infrared images recorded during the cruises. This constrained the analysis of the imagery to use the available heat estimates from the anemometer/flux package during the experiments. While in principle this is adequate, the statistics of heat flux measurements using both the direct covariance and the bulk aerodynamic methods are subject to variability by their very nature. This is the single largest source of uncertainty in the heat flux estimates obtained from the infrared, and cannot be corrected without the availability of time series of images. For this reason, modifications have been performed to the experimental apparatus and time series of images have been collected and analyzed using the 10m Heidelberg facility [*Garbe et al.*, 2001]

5. CONCLUSIONS

The CFT technique is capable of measuring gas transfer rates even under heavy rain conditions and gas transfer rates are subject to strong variations with intermittent meteorological conditions like rain. To date, all field data of gas transfer rates show that wind speed alone is not sufficient to parameterize air-sea gas transfer. Residual intermittence in the high-resolution CFT data is likely to be caused by surfactants and other surface-related properties that exhibit regional and temporal variations. Recent measurements (*Bock et al.*, 1999) show that mean square wave slope is more adequate to parameterize the combined impact of wind and surface properties which together modify the underlying mechanisms that accelerate gas transfer across the interface. This parameterization should result in a characterization that does not rely on an intermediate quantity. Currently, models for global estimates of air-sea fluxes rely on wind speed estimates obtained from space-borne radar or from model wind speeds initialized at relatively few locations. These models use the

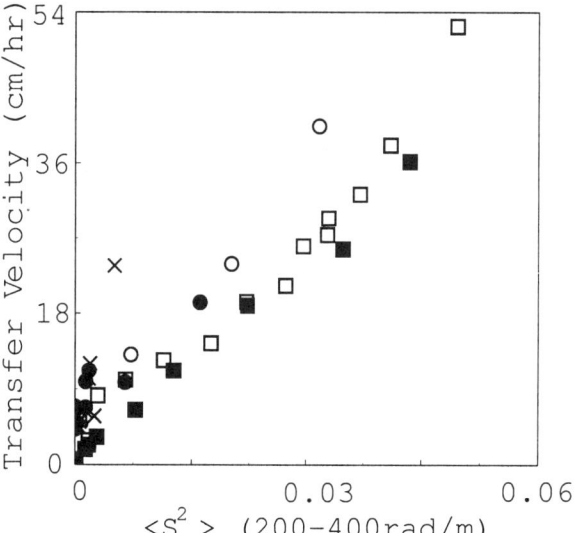

Figure 3. Composite plot of gas transfer velocity inferred from laboratory and field measurements versus wavenumber-binned mean square slope between 200 and 400 radian/m. Laboratory gas transfer rates derived from mass balance methods, field results from the infrared technique. Circles - Heidelberg 4m diameter canal data. Squares - Woods Hole 0.5m diameter canal data. Crosses - North Atlantic field data. Filled symbols represent surfaces with films, hollow symbols represent clean surfaces. The Heidelberg tank used had a channel width of 0.3m, the Woods Hole tank has a channel width of 0.1m.

accepted Liss-Merlivat relation to infer gas exchange rates from wind speed estimates. A more direct parameterization based on remote sensing of wavenumber specific sea surface roughness would be possible through the use of a relation derived from in situ measurements of short wave spectra and CFT inferred gas exchange measurements.

Finally, in both field measurement campaigns the influence of bubble mediated gas transfer has not been addressed. It is certainly clear that bubbles are a mechanism by which mass and heat transfer can be effected across the air-sea interface, and it is also clear that breaking waves make the concept of surface slope invalid. Additionally, it is not obvious what is the effect of breaking waves on the presence of the surface microlayer, since at some wind speed it is evident that the surfactant films are mixed down (presumably through solublization or micellization). The present study is limited in scope to describing the possible influences of mechanisms present at low to moderate wind speeds and serves to point out that even in these regimes, modulation of heat transfer can occur over short time and small space scales.

Acknowledgments. The authors would like to thank the National Science Foundation for the support to perform this research (OCE-9410534 and OCE-9711285). We also thank the Captains and Crews of the R/V New Horizon (Scripps Institution of Oceanography) and the R/V Oceanus (Woods Hole Oceanographic Institution) for helping carry out the at-sea deployment of the instruments. One of the authors (EJB) would like to extend thanks to Warren Witzell, Jr., and David Schroeder of the AOPE Department of WHOI for facilitating the design and reconstruction of the LADAS Air-Sea Platform.

REFERENCES

Asher, W. and R. Wanninkhof, Transient tracers and air-sea gas transfer, J. Geophys. Res., *103*, 15939-15958, 1998.

Bock, E. J., J. B. Edson, N. M. Frew, A. Karachintsev, W. R. McGillis, R. K. Nelson, K. Hanson, T. Hara, M. Uz, B. Jähne, J. Dieter, J. Klinke, and H. Haussecker, Description of the science plan for the April 1995 CoOP experiment, Gas Transfer in Coastal Waters, performed from the Research Vessel New Horizon, in Air-Water Gas Transfer, edited by B. Jähne and E.C. Monahan, pp. 801-808, AEON Verlag & Studio, Hanau, Germany 1995.

Bock, E. J. and T. Hara, Optical measurements of capillary-gravity wave spectra using a scanning laser slope gauge, J. Atmos. Oceanic Technol., *12*, 395-403, 1995.

Bock, E. J., T. Hara, N. M. Frew, and W. R. McGillis, Relationship between air-sea gas transfer and short wind waves, J. Geophys. Res., *104*, 25821-25831, 1999.

Davies, R. E. and L. A. Regier, Methods for estimating directional wave spectra from multi-element arrays, J. Mar. Res., *35*, 453-477, 1977.

Edson, J. B., C. W. Fairall, S. E. Larson, and P. G. Mestayer, A study of the inertial-dissipation technique for computing air-sea fluxes, J. Geophys. Res., *96*, 10689-10711, 1991.

Edson, J. B., A. A. Hinton, K. E. Prada, J. E. Hare, and C. W. Fairall, Direct covariance flux estimates from mobile platforms at sea, J. Atmos. Oceanic Tech., *15*, 547-562, 1998.

Fairall, C. W., E. F. Bradley, D. P. Rogers, J. B. Edson, and G. S. Young, Bulk parameterization of air-sea fluxes for TOGA COARE, J. Geophys. Res., *101*, 3747-3764, 1996.

Frew, N. M., The role of organic films in air-sea gas exchange, in The Sea Surface and Global Change, pp.121-172, edited by P. Liss and R. Duce, Cambridge University Press, Cambridge, England, 1997.

Frew, N. M., E. J. Bock, W. R. McGillis, A. V. Karachintsev, T. Hara, T. Münsterer, and B. Jähne, Variation of air-water gas transfer with wind stress and surface viscoelasticity, in Air-Water Gas Transfer, edited by B. Jähne and E. C. Monahan, pp. 529-541, AEON Verlag & Studio, Hanau, Germany, 1995.

Frew, N. M., R. K. Nelson, W. R. McGillis, J. B. Edson, E. J. Bock, and T. Hara, Spatial variations in surface microlayer surfactants and their role in modulating air-sea exchange, In Gas Transfer at Water Surfaces, edited by M.A. Donelan, W.M. Drennan, E.S. Saltzman and R. Wanninkhof, pp. xxx-xxx, AGU, this volume, 2001.

Garbe, C. S., H. Haussecker, and B. Jähne, Measuring the sea surface heat flux and probability distribution of surface renewal events, In Gas Transfer at Water Surfaces, edited by M.A. Donelan, W.M. Drennan, E.S. Saltzman and R. Wanninkhof, pp. xxx-xxx, AGU, this volume, 2001.

Hanson, K. A., T. Hara, E .J. Bock, and A. Karachintsev, Estimates of directional surface wave spectra from a towed research catamaran, J. Atmos. Ocean. Tech., *14*, 1467-1482, 1997.

Hara, T., E. J. Bock, J. B. Edson and W. R. McGillis, Observations of short wind waves in coastal waters, J. Phys. Oceanogr., *28*, 1425-1438, 1998.

Hara, T., E. J. Bock, N. M. Frew, and W. R. McGillis, Relationship between air-sea gas transfer velocity and surface roughness, in Air-Water Gas Transfer, edited by B. Jähne and E. C. Monahan, pp. 611-616, AEON Verlag & Studio, Hanau, Germany, 1995.

Hara, T., E. J. Bock, and D. Lyzenga, In situ measurements of capillary-gravity wave spectra using a scanning laser slope gauge and microwave radars, J. Geophys. Res., *99*, 12593-12602, 1994.

Hara T., B. Uz, H. Wei, J. Edson, N. Frew, W. McGillis, S. McKenna, E. Bock, H. Haussecker, and U. Schimpf, Surface wave observations and their relation to air-sea gas transfer during CoOP experiments, In Gas Transfer at Water Surfaces, edited by M.A. Donelan, W.M. Drennan, E.S. Saltzman and R. Wanninkhof, pp. xxx-xxx, AGU, this volume, 2001.

Haussecker, H., Messung und Simulation von kleinskaligen Austauschvorgngen an der Ozeanoberflche, PhD thesis, 203 pp., Interdisciplinary Center for Scientific Computing, University of Heidelberg, May 1996.

Plant, W. J., W. C. Keller, V. Hesany, T. Hara, E. Bock, and M. A. Donelan, Bound waves and Bragg scattering in a wind-wave tank. J. Geophys. Res., *104*, 3243-3263, 1999.

Surface Wave Observations During CoOP Experiments and Their Relation to air-sea gas Transfer

Tetsu Hara, B. Mete Uz, and Hua Wei

Graduate School of Oceanography, University of Rhode Island, Narragansett

James B. Edson, Nelson M. Frew, Wade R. McGillis, and Sean P. McKenna

Woods Hole Oceanographic Institution, Woods Hole, Massachusetts

Erik J. Bock, Horst Haußecker, Uwe Schimpf

Interdisciplinary Center for Scientific Computing, IWR, University of Heidelberg, Heidelberg, Germany

Gas exchange between the ocean and the atmosphere is strongly influenced by physical processes in the near-surface waters. Surface waves are particularly important for gas fluxes because they enable faster transfer of gases across the diffusive sublayer by causing more frequent renewal of the skin layer. During the CoOP air-sea gas exchange experiments (1995; 1997), we obtained one of the most comprehensive data sets of physical processes at the air-sea interface in both near-shore and off-shore waters. During these experiments simultaneous measurements of short wind waves, surface films, wind stress, and transfer velocity were made from a towed or self-propelled catamaran with a wide range of wind stress and with varying surface film conditions. The results show that the wave spectra at higher wavenumbers are significantly reduced by surfactant at wind friction velocities below 0.2 m s^{-1}. The surfactant effect may be quantified using the surface enrichment (difference between the CDOM fluorescence in microlayers and that in bulk water) with reasonable accuracy. During rain events the wave spectra are raised at higher wavenumbers (above 200 rad m^{-1}) but are not affected at 100 rad m^{-1}. The surfactant effect is also reduced during rain. The air-sea gas transfer velocity is roughly proportional to the wave spectra at higher wavenumbers but appears to be less sensitive to spectra of longer waves.

1. INTRODUCTION

Exchange of gases between the ocean and the atmosphere is strongly influenced by physical processes in the near-surface waters. Surface waves are particularly important for gas fluxes. Waves enable faster transfer of gases across the diffusive sublayer at the water surface by causing more frequent renewal of the skin layer. Even though the relevance of waves is qualitatively well established and some quantitative studies have been made in laboratory setting [e.g., *Jähne et al.*, 1987; *Bock et al.* 1999], the paucity of field data sets with concurrent wave and flux measurements has hampered parametric studies of wave effect on air-sea gas fluxes. During the CoOP air-sea gas exchange experiments off California (CoOP I 1995) and off New England (CoOP II 1997), we obtained one of the most comprehensive data sets of physical processes at the air-sea interface in both near-shore and off-shore waters. During these experiments spectral measurements of gravity-capillary waves were made from a towed or self-propelled catamaran using a scanning laser slope gauge. At the same time, wind measurements were made with sonic anemometers and surface enrichment of organic compounds was measured with an on-board surface microlayer skimmer and fluorometer. Air-sea gas transfer velocity was estimated using the controlled flux technique (CFT).

The objective of this paper is to present a comprehensive summary of wave observations under a wide range of wind and surface film conditions during the two experiments, and examine their relationship with the observed air-sea gas flux. The summary of the CoOP I experiment was previously reported by *Bock et al.*, [1995], while the overview and measurement techniques of the CoOP II experiment are given in a companion paper [*Bock et al.* 2001]. Therefore, they are not repeated in this paper.

2. SURFACE WAVES AND WIND STRESS

We first examine the relationship between surface waves and wind stress. The results from CoOP I experiment have been reported in detail by *Hara et al.* [1998]. Major findings reported in the paper are summarized below.

- Wave spectra from CoOP I are consistent with previous field observations.

- Wave spectra over clean water from field observations are much higher than laboratory results, particularly at lower wind stress and at higher wavenumbers (> 200 rad m^{-1}).

- Wave spectra at higher wavenumbers and lower wind stress show large variability in the presence of surface films.

In Figure 1 we show the results of the degree of saturation $B(k)$ from CoOP II together with the results from CoOP I at four different wavenumbers. Here, k is a wavenumber, $B(k)$ is defined as $B(k) = k^4 S(k)$, and $S(k)$ is the wavenumber spectrum integrated in all wave propagation directions. All results in this and next sections have been averaged over 5-10 min. Overall the results from both experiments are consistent. One of the striking features of the CoOP II results is that some of the low wind results show extremely low spectral values compared with CoOP I data in the presence of surface films. In fact, some results from CoOP II at wavenumber 800 rad m^{-1} are below the noise level of the instrument and are not shown in the figure. These low values are correlated with increased surface enrichment as presented in the next section. Another interesting finding from the CoOP II results is the effect of rain on surface waves. During YD192 all measurements were performed under steady moderate rain. The results from YD192 are indicated by filled squares in contrast to all the other CoOP II data shown by crosses. At wavenumber 100 rad m^{-1} the data during rain are indistinguishable from the other data. As the wavenumber increases to 400 rad m^{-1}, the data taken under rain are higher than the rest of the CoOP II data by a factor of 2-4. At 800 rad m^{-1} only rain influenced data are seen at lower wind stress because all the other data at comparable wind stress are below the noise level. These observations clearly suggest that rain preferentially enhances the spectra of very short wind waves. Unfortunately, the rain rate was not measured during the experiment, and therefore no quantitative discussion of the rain effect is possible.

3. SURFACE WAVES AND SURFACE FILMS

We next examine whether it is possible to quantify the effect of surface films on short wind waves. It has been known for some time that gravity capillary waves are suppressed in the presence of surface films. However, quantifying this effect is difficult because natural surface films consist of many chemical constituents and also because the film concentration is significantly affected by surface waves, wave breaking, and near surface turbulence. During both CoOP experiments colored dissolved organic matter (CDOM) fluorescence was measured as a proxy for surfactants in sea water (see *Frew et al.* [2001] for details). Measurements were performed both in surface microlayers and in bulk water (about 0.1 m below the surface).

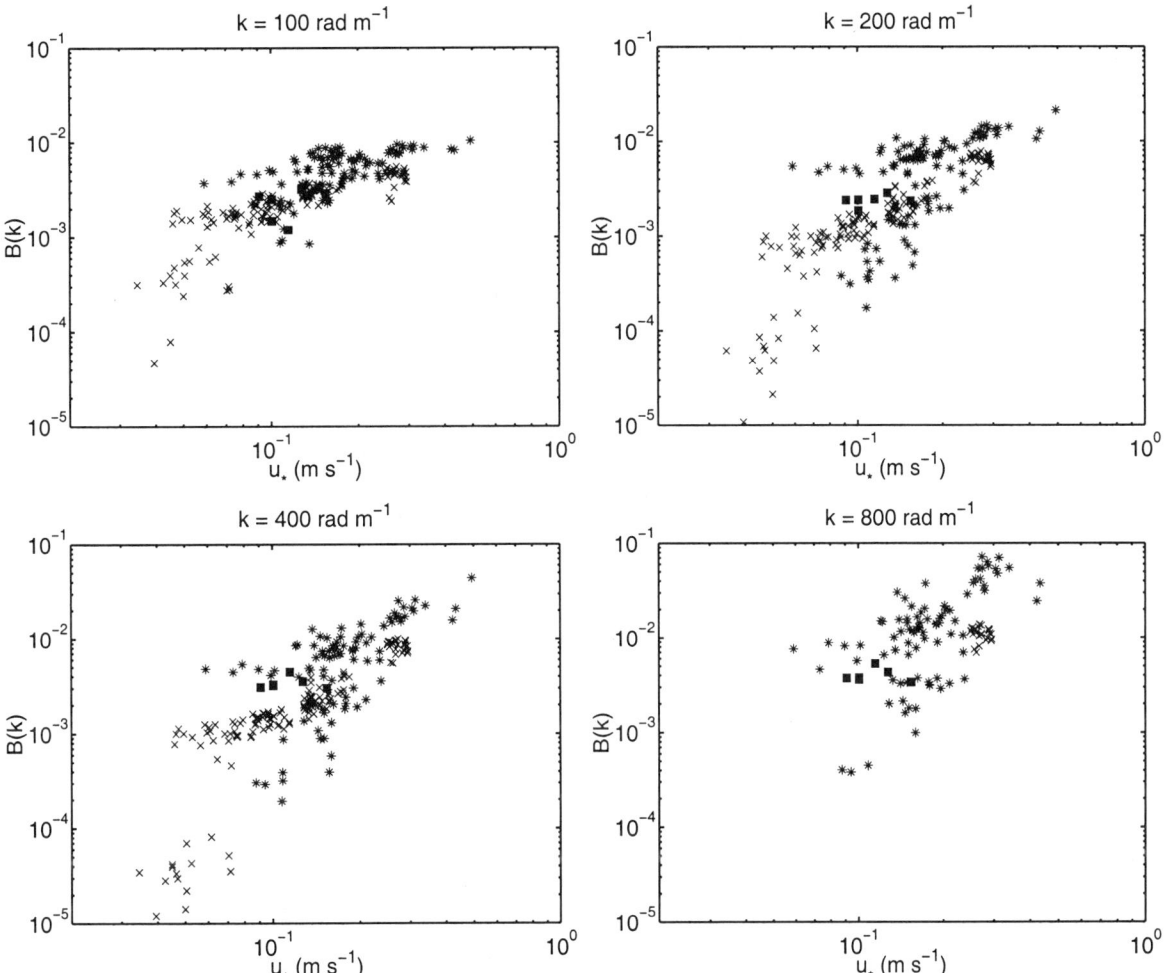

Figure 1. Degree of saturation $B(k)$ versus friction velocity u_*. Asterisks: CoOP I, off California (*Hara et al.*, 1998); crosses: CoOP II, off New England (this study); filled squares: CoOP II data during rain.

Here we present how the short wind wave spectra correlate with either the CDOM fluorescence in microlayers or the surface enrichment (difference between the CDOM fluorescence in microlayers and that in bulk water). We only show results at wavenumber 400 rad m^{-1}. The results at 200 and 800 rad m^{-1} are similar to those at 400 rad m^{-1}, while the results at 100 rad m^{-1} are difficult to interpret since the surfactant effect is much weaker. When the spectra are plotted against the microlayer CDOM fluorescence (Figure 2), there appears to be no correlation between the microlayer CDOM concentration and short wind waves. The lowest spectral values observed during CoOP II correspond to rather low values of CDOM fluorescence. On the other hand, the wave spectra and the surface enrichment show reasonable correlation in Figure 3. Overall, the spectral level reduces as the surface enrichment increases. The film effect is stronger for lower wind stress ($0.05 < u_* < 0.1$ m s^{-1}) than for slightly higher wind stress ($0.1 < u_* < 0.2$ m s^{-1}). When the friction velocity is above 0.2 m s^{-1} the surface enrichment almost disappears and the wave spectra are not influenced by surface films even if the CDOM fluorescence in microlayers is still high. This observation may be interpreted that surface films are mixed down and do not affect surface waves at wind friction velocities above 0.2 m s^{-1} under typical oceanic conditions. In contrast, we have observed surface film effects at much higher wind friction velocities in laboratory settings [e.g., *Bock et al.*, 1999]. It is clear that the surface film effect on short

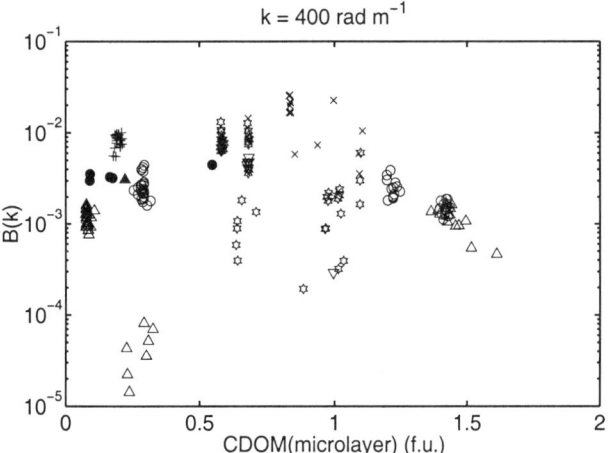

Figure 2. Degree of saturation $B(k)$ at $k = 400$ rad m^{-1} versus microlayer CDOM fluorescence. Crosses: CoOP I, $0.2 < u_* < 0.4$ m s^{-1}; pluses: CoOP II, $0.2 < u_* < 0.4$ m s^{-1}; stars: CoOP I, $0.1 < u_* < 0.2$ m s^{-1}; circles: CoOP II, $0.1 < u_* < 0.2$ m s^{-1}; downward pointing triangle: CoOP I, $0.05 < u_* < 0.1$ m s^{-1}; upward pointing triangle: CoOP II, $0.05 < u_* < 0.1$ m s^{-1}. Filled symbols indicate data during rain.

wind waves is significantly reduced when it is raining. The spectra during rain appear to be almost independent of the surface enrichment.

4. SURFACE WAVES AND GAS TRANSFER VELOCITY

From laboratory experiments *Bock et al.* [1999] find that the gas transfer velocity correlates better with the short wind waves (both the total mean square slope and the mean square slope of shorter gravity-capillary waves only ($k > 200$ rad m^{-1})) than with the wind stress. It is of interest to examine whether this is true under field conditions.

Here, the relationship between the air-sea gas transfer velocity k_{600} and surface waves is examined using the data from CoOP II. The each estimate is averaged over 2-3 hours. The gas transfer velocity is found to be roughly proportional to both the wind friction velocity and the mean square slope of gravity capillary waves as reported in the companion paper [*Bock et al.*, 2001]. In Figure 4 we present how the transfer velocity correlates with the wave spectra at different wavenumbers. Except for the lowest point of k_{600}, the wave spectra appear to increase linearly with the wave spectra at $k = 200$ and 400 rad m^{-1}. The transfer velocity increases slightly faster against the wave spectra at $k = 100$ rad m^{-1}, implying that the transfer velocity is less sensitive to longer waves than to shorter waves. This observation is qualitatively consistent with the laboratory results. However, we may not make any conclusive remarks with a very limited number of data. Finally, rain events raise the wave spectral level at higher wavenumbers but do not appear to affect the gas transfer velocity.

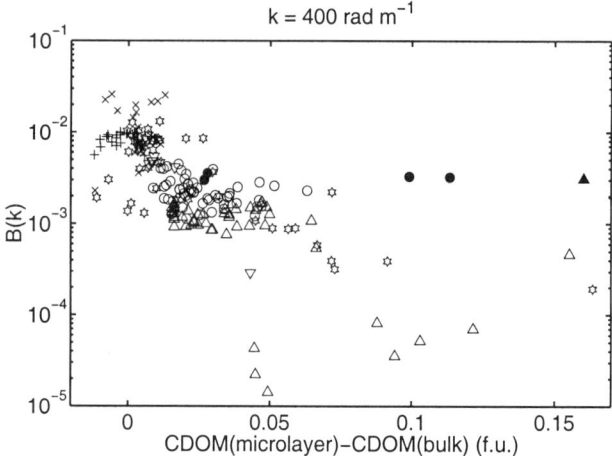

Figure 3. Degree of saturation $B(k)$ at $k = 400$ rad m^{-1} versus difference between microlayer CDOM fluorescence and bulk CDOM fluorescence. Crosses: CoOP I, $0.2 < u_* < 0.4$ m s^{-1}; pluses: CoOP II, $0.2 < u_* < 0.4$ m s^{-1}; stars: CoOP I, $0.1 < u_* < 0.2$ m s^{-1}; circles: CoOP II, $0.1 < u_* < 0.2$ m s^{-1}; downward pointing triangle: CoOP I, $0.05 < u_* < 0.1$ m s^{-1}; upward pointing triangle: CoOP II, $0.05 < u_* < 0.1$ m s^{-1}. Filled symbols indicate data during rain.

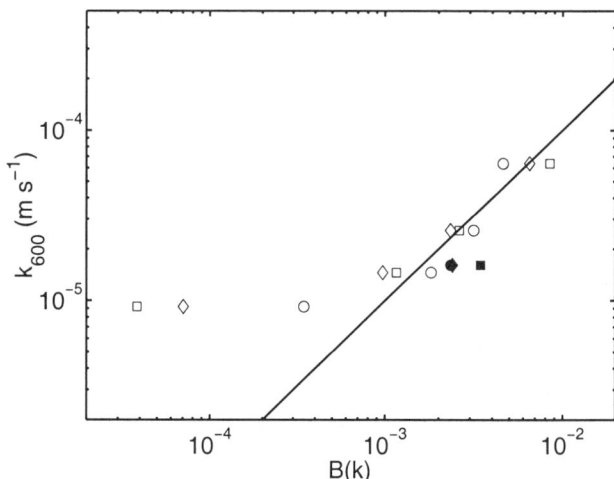

Figure 4. Degree of saturation $B(k)$ versus gas transfer velocity. Circles: $k = 100$ rad m^{-1}; diamonds: $k = 200$ rad m^{-1}; squares: $k = 400$ rad m^{-1}. Filled symbols indicate data during rain. Solid line indicates linear relationship.

5. CONCLUSION

Short wind wave spectra have been measured during CoOP I and CoOP II experiments with a wide range of wind stress and with varying surface film conditions. The spectra at higher wavenumbers are significantly reduced by surfactant at wind friction velocities below 0.2 m s^{-1}. The surfactant effect may be quantified using the surface enrichment (difference between the CDOM fluorescence in microlayers and that in bulk water) with reasonable accuracy. During rain the wave spectra are raised at higher wavenumbers (above 200 rad m^{-1}) but are not affected at 100 rad m^{-1}. The surfactant effect on waves is also reduced during rain. The air-sea gas transfer velocity is roughly proportional to the wave spectra at higher wavenumbers but appears to be less sensitive to the spectra of longer waves.

Although the CoOP II was the first experiment to measure both the air-sea gas transfer velocity and related physical and chemical variables, the results from our correlation studies are far from conclusive because of the limited number of available data. Further accumulation of similar data is definitely needed to further our understanding of air-sea gas exchange processes and their relationship with physical and chemical processes at the air-sea interface.

Acknowledgments. This work was supported by NSF grants OCE-9409222 and OCE-9711391.

REFERENCES

Bock, E. J., et al., Description of the science plan for the April 1995 CoOP experiment, 'Gas Transfer in Coastal Waters', performed from the research vessel New Horizon, in *Air-Water Gas Transfer,* edited by B. Jähne and E. C. Monahan, pp. 801-810, AEON Verlag & Studio, Hanau, Germany, 1995.

Bock, E. J., T. Hara, N. M. Frew, and W. R. McGillis, Relationship between air-sea gas transfer and short wind waves, *J. Geophys. Res., 104,* 25821-25831, 1999.

Bock, E. J., et al., Overview of the CoOP experiments: physical and chemical measurements parameterizing air-sea gas exchange, in *Gas Transfer at Water Surfaces,* edited by M. A. Donelan, W. M. Drennan, E. S. Saltzman and R. Wanninkhof, pp. xxx-xxx, AGU, this volume, 2001.

Frew, N. M., E. J. Bock, R. K. Nelson, W. R. McGillis, J. B. Edson, and T. Hara, Spatial variations in surface microlayer surfactants and their role in modulating air-sea exchange, in *Gas Transfer at Water Surfaces,* edited by M. A. Donelan, W. M. Drennan, E. S. Saltzman and R. Wanninkhof, pp. xxx-xxx, AGU, this volume, 2001.

Hara, T, E. J. Bock, J. B. Edson, and W. R. McGillis, Observation of short wind waves in coastal waters, *J. Phys. Oceanogr., 28,* 1425-1438, 1998.

Jähne, B. K., O. Münnich, R. Bösinger, A. Dutzi, W. Huber, and P. Libner, On the parameters influencing air-water gas exchange, *J. Geophys. Res., 92*, 1937-1949, 1987.

T. Hara, B. M. Uz, and H. Wei, Graduate School of Oceanography, University of Rhode Island, South Ferry Road, Narragansett, RI 02882. (e-mail: thara@uri.edu)

J. B. Edson, N. M. Frew, W. R. McGillis, and S. P. McKenna, Woods Hole Oceanographic Institution, Woods Hole, MA 02543.

E. J. Bock, H. Haußecker, and U. Schimpf, Interdisciplinary Center for Scientific Computing, IWR, University of Heidelberg, Im Neuenheimer Feld 368, D-69120 Heidelberg, Germany.

On the Investigations of Statistical Properties of the Micro Turbulence at the Ocean Surface

Uwe Schimpf[1] and Bernd Jähne[1,2]

Interdisciplinary Center for Scientific Computing, University of Heidelberg, Germany

Horst Haußecker

Xerox Palo Alto Research Center, Palo Alto, California

Using heat as a proxy tracer for gases the exchange process at the air/water interface and the micro turbulence at the water surface is investigated. The analysis of the infrared image sequences allow the determination of the heat transfer velocity, provided that the net heat flux density at the water surface is known. Furthermore, infrared image sequences of the water surface allow a detailed study of near surface turbulence with respect to scale and orientation. Laboratory studies in the circular Heidelberg wind wave flume and a field study during the 1997 *CoOP* (Coastal Ocean Processes) experiment in the Atlantic Ocean have been conducted to measure the gas transfer velocity and to investigate the relationship between the exchange process and the micro-scale temperature fluctuations at the water surface. The measured gas transfer rates in the field, as well as in the laboratory, agree with the wind speed dependence of the *Wanninkhof* relationship. The micro-scale temperature fluctuations showed a similar behavior in the field and laboratory. At low wind speeds larger scales are dominant, whereas at high wind speeds the small scales dominate the transport. At moderate speed all scales contribute equally to the temperature fluctuations.

1. INTRODUCTION

Knowledge about the parameter which influence the mass transfer across the aqueous boundary layer increased considerably in recent years. The parameterization has become better but still is not much known about the key processes which control the exchange across the air/sea interface. Gas exchange models (e.g. surface renewal model, small eddy model) differ in the prediction of the concentration profiles, but not in dependence of crucial parameters (e.g. Schmidt number Sc) of the transfer velocity. *Deacon* [1977] showed, that the transfer rate k is proportional to the friction velocity u_*:

$$k = \beta^{-1} u_* Sc^{-n}, \qquad (1)$$

with the dimensionless transfer resistance β and the Schmidt number exponent n. Further progress in this

---

[1] Also at Institute for Environmental Physics, University of Heidelberg, Germany.
[2] Also at Scripps Institution of Oceanography, UCSD, California.

research area therefore requires techniques which are able to measure concentration profiles or directly give an insight into the key processes of the gas exchange. The controlled flux technique [*Jähne et all.*, 1989] is able to measure the local temperature difference across the thermal boundary layer with high temporal resolution. Given the net heat flux density at the surface the transfer rate for heat (k_h), respective the gas transfer velocity (k_g) is determined [*Haußecker et all.*, 1998].

Near surface turbulence is one of the specifying key processes of gas exchange [*Jähne*, 1987]. The analysis of the surface temperature fluctuations, which are associated with the interplay of diffusive and turbulent transport, give direct insight into the mechanisms of gas transfer. Using digital image processing techniques the spatial and temporal structure of the micro turbulence at the ocean surface is investigated. The relation between the statistical properties of the temperature fluctuations and the mean temperature difference across the aqueous thermal boundary layer was modeled and the results strongly sustain a surface renewal model [*Schimpf et all.*, 1999a]. Even under low wind speed conditions surface renewal events were directly observable in the IR image sequences supporting the hypothesis that surface renewal seems to be the dominant turbulent transport mechanism.

2. EXPERIMENTAL METHODS

Detailed laboratory studies in the circular Heidelberg wind wave flume of the temporal and spatial statistics of the temperature fluctuations at the water surface and the relation to the exchange process have been carried out. In this facility the temperature difference across the interface was controlled by periodically switching the heat flux density at the water surface on and off. This could be achieved by controlling the ventilation system of the gastight facility. If the ventilation system is in closed mode, the air temperature instantaneously adjusts to the water temperature and the humidity in the airspace of the flume rises to 100%, so that no heat flux density at the water surface is present [*Schimpf et all.*, 1999b]. Flushing the air space of the flume with dry air establishes an evaporative heat flux. From the temperature drop of the water body the net heat flux j_h was calculated. The infrared imaging system measured temperature difference ΔT across the interface and the transfer velocity for heat k_h was determined by

$$k_h = \frac{j_h}{\rho c_p \Delta T}. \qquad (2)$$

Using Schmidt number scaling

$$\frac{k_g}{k_h} = \left(\frac{Sc_h}{Sc_g}\right)^n, \qquad (3)$$

where Sc_h denotes the Schmidt number for heat in water, and Sc_g the Schmidt number for an arbitrary gas, the gas transfer rate k_g was calculated.

During the 1997 *Coastal Ocean Processes* experiment in the Atlantic Ocean the CFT-instrument was mounted on a boom at the bow of the research vessel *Oceanus*. The instrument consists basically of an infrared camera, a temperature calibration device and a remote controlled PC for image acquisition and temperature calibration. Micrometeorological measurements have been carried out by *Edson* [1999] and provided e.g. temperature, humidity, heat flux, friction velocity, wind speed/direction. *Haußecker* [1996] developed a method to estimate the temperature difference ΔT across the thermal boundary layer from the temperature distribution at the ocean surface. Assuming a surface renewal model [*Danckwerts*, 1970] the theoretically calculated distribution of the sea surface temperature proved to fit the measured distribution and predicted the measured temperature difference across the interface very well. Additionally, the statistical method was applied to the data sets of the laboratory experiments. It could be shown that the temperature difference, measured by periodical switching the heat flux on/off, agrees within the error with the theoretical prediction [*Schimpf*, 2000].

3. RESULTS AND DISSCUSSION

3.1. Gas Transfer Velocities in the Field and Laboratory

The results of the laboratory measurements showed that the temperature gradient and thus the transfer rate is reliably determined by fitting the theoretical temperature distribution to the measured histograms. In Figure 1 the measured transfer velocities during the *CoOP* cruise are plotted versus wind speed and compared with the empirical relationships from *Liss* [1986] and *Wanninkhof* [1992]. A further comparison with field data of other authors shows that the CFT data fit the general wind speed dependence [*Haußecker*, 1995]. The large scatter in the data mirrors the temporal and spatial fluctuations of the meteorological parameters influencing the transfer rate.

In Figure 2 the transfer rates in the circular wind wave flume and the during the 1997 *CoOP* cruise are plotted

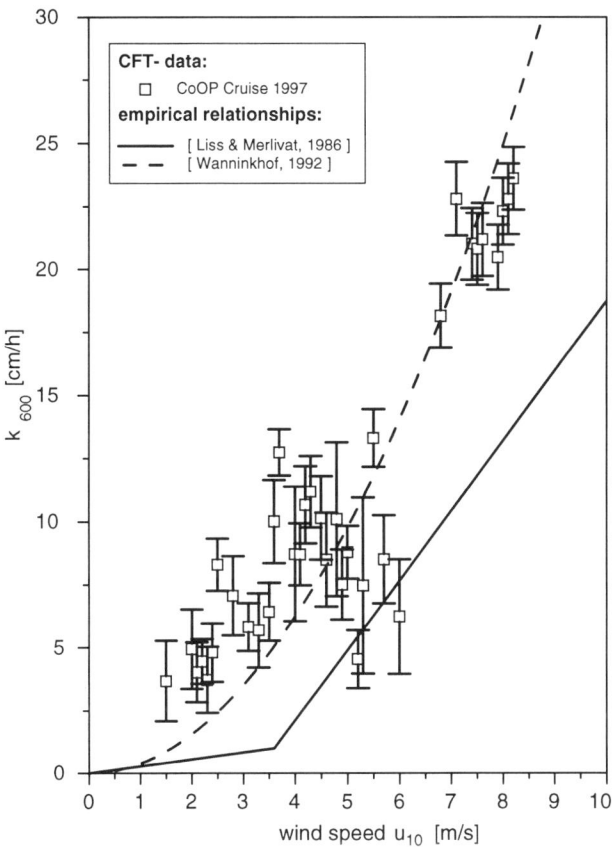

Figure 1. Transfer velocities versus wind speed during the *CoOP* cruise. The wind speed dependence of the transfer velocity follows the *Wanninkhof* relationship, whereas the difference decreases at higher wind speeds. At very low wind speeds the data shows an offset in comparison with the empirical relationships.

as a function of friction velocity. Also included are the theoretical curves for a smooth surface based on a diffusion model [*Deacon*, 1977] and for a rough surface based on a surface renewal model [*Danckwerts*, 1970]. The curve for a smooth surface represents the lower limit for the exchange rate at the ocean. Up to a friction velocity of $u_* \approx 0.5\text{-}0.7$ cm/s the transfer rates are close to the theoretical curve for a rough surface. At higher friction velocities a constant offset of the gas exchange rate is observed, i.e. the transfer resistance across the interface was reduced. This transition agrees with the results of the scale analysis, where at a wind speed of $u_{10} \approx 4.5\text{-}5.5$ m/s a change in the for the transport process dominant scales occur.

The controlled flux technique is capable to measure the local transfer velocity with high temporal resolution in the field. In Figure 3 the transfer rates are plotted versus time for different runs. The time series (a) and (b), as well as (c) and (d) are taken under nearly the same wind speed conditions. While in (a) the transfer rate is nearly constant during one hour, there is a large fluctuations of the rate in (b). The same behavior was observed in (c) and (d) at much higher wind speeds. The high fluctuations in the data can only partly be attributed to uncertainties and systematic errors in the measurements, most of it is certainly due to other factors that influence air-sea gas transfer. Consequently, the gas transfer velocity cannot simply be parameterized only by wind speed and more complex models for air-sea gas exchange are required.

3.2. Scale Analysis of Micro-Scale Temperature Fluctuations

The concept of scale analysis is based on the pyramid approach by means of digital image processing [*Jähne*, 1997]. From the infrared image, the *Gaussian* and *Laplacian* pyramids are calculated (see Figure 4). These multigrid data structures constitute a powerful tool for investigations of micro turbulence at the ocean surface. Only a few levels of the pyramid are necessary to span a wide range of structure sizes and allow a detailed study of the scales to be made. From each image sequence the standard deviation of the temperature on every level of the Laplacian pyramid was calculated. The temperature variance of a level is a measure for the dominance of a certain scale. The spatial resolution of the different levels are shown in Table 1. The micro-scale temperature fluctuations showed a similar behavior in the field and laboratory. With increasing wind speed the occurrence of small scales increases (Figure 5a), medium scales stay about at the same level (Figure 5b), and the large scales decline (Figure 5c). To investigate which scales dominate at different wind speed conditions the transport process, the ratio of the standard deviation of the temperature fluctuations for the different scales was calculated. At low wind speeds, large scales dominate the temperature distribution by a factor of about 8 (Figure 5d). At moderate wind speeds all scales contribute about equally to the temperature distribution. Finally, at higher wind speeds the smallest scales dominate the temperature distribution by a factor of about 6 (Figure 5f).

4. CONCLUSIONS

The measured transfer velocities in the laboratory and in the field are consistent with measurements of other authors [*Jähne and Haußecker*, 1998]. The CFT-

54 STATISTICAL PROPERTIES OF MICRO TURBULENCE

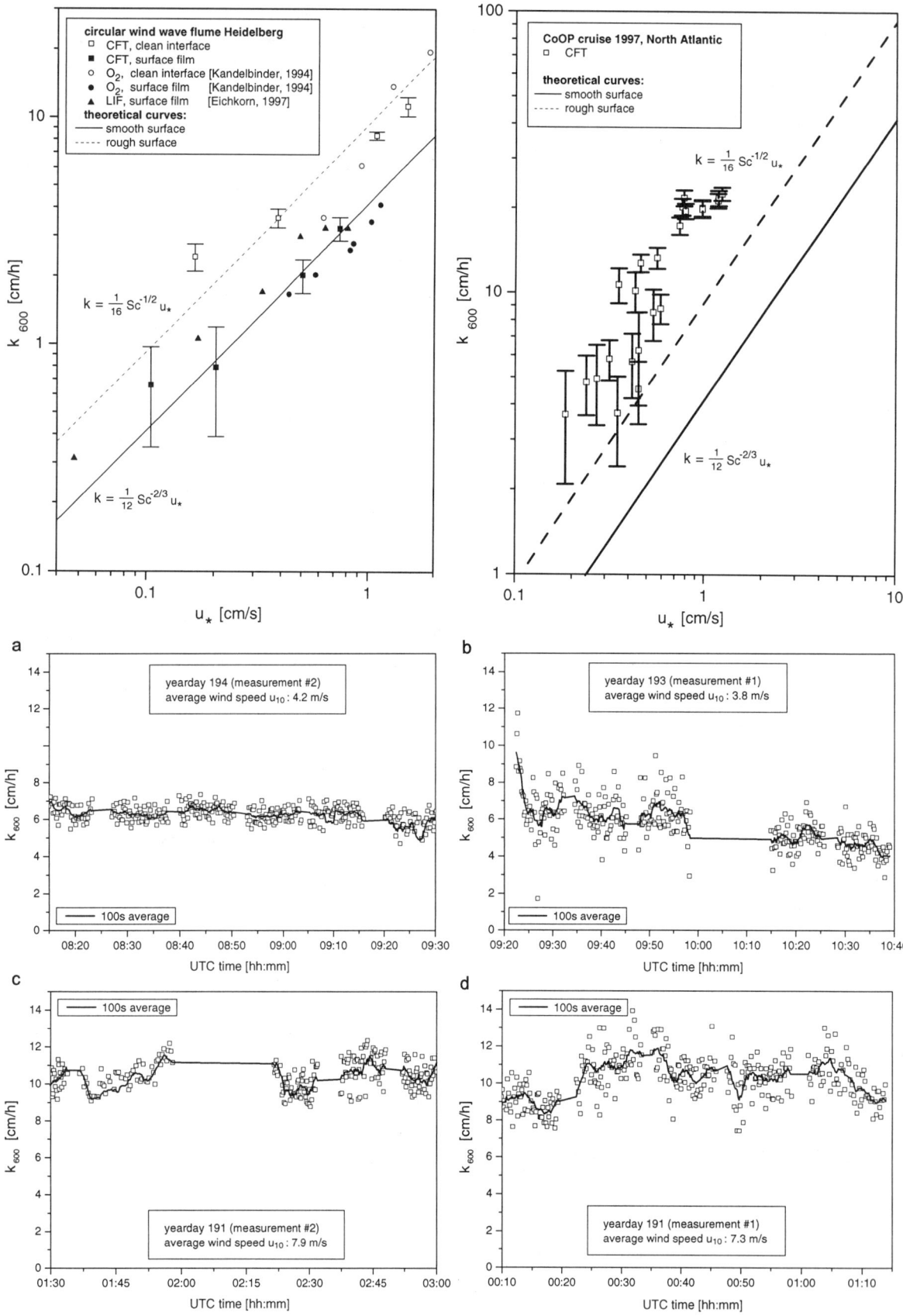

Figure 2. Transfer velocities in the circular wind wave flume and during the *CoOP* cruise as a function of friction velocity. The transfer velocities are normalized to a Schmidt number of $Sc=600$ (CO_2 at 20°). Also included are the theoretical curves for a smooth surface based on a diffusion model [*Deacon*, 1977] and for a rough surface based on a surface renewal model [*Danckwerts*, 1970].

Figure 3. Transfer velocities versus time for different runs. The time series (a) and (b), as well as (c) and (d) are taken under nearly the same wind speed conditions. While in (a) the transfer rate is nearly constant during one hour, there is a large fluctuations of the rate in (b). The same behavior was observed in (c) and (d) at much higher wind speeds.

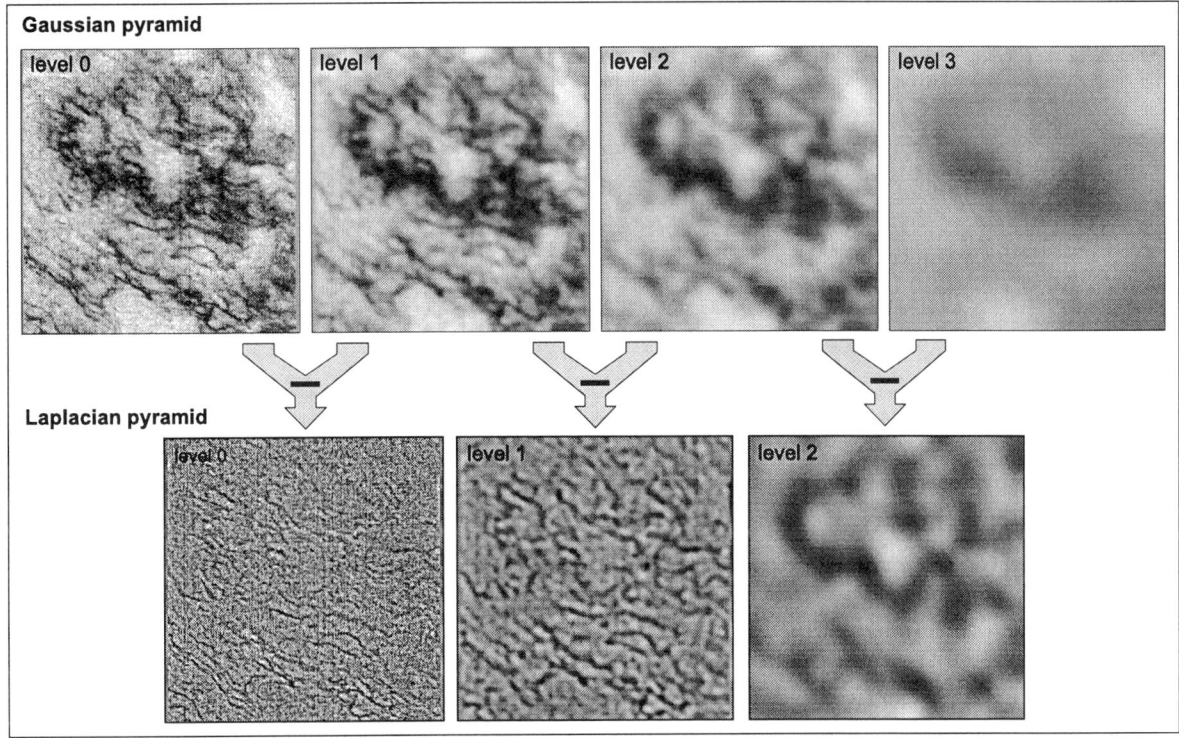

Figure 4. The *Gaussian* pyramid constitutes a series of low pass filtered images in such a way that more and more coarse details remain in the image. By subtracting the images between two consecutive levels the so called *Laplacian* pyramid is obtained. Only fine scales, removed by the smoothing operation to compute the next level of the Gaussian pyramid, remain in the finer level. The Laplacian pyramid represents an effective scheme for a band pass decomposition of an image.

technique is a method to measure locally the transfer velocity with high temporal resolution. The CFT technique is insensitive to bubble mediated gas transfer, i.e. it measures the transfer rate of a gas with very high solubility. In conjunction with conventional mass balance methods and eddy correlation techniques the different mechanisms which contribute to the exchange process could be separated. The visually observed surface renewal events and the statistical properties of the microscale temperature fluctuations support the hypothesis that surface renewal seems to be the dominant turbulent transport mechanism. It could be demonstrated

Table 1. Laplacian pyramid scale sizes and resolution.

level	center wave number	structure size	remark
0	574.5 rad/m	0.85 - 1.45 cm	small scales
1	287.2 rad/m	1.68 - 3.13 cm	...
2	143.6 rad/m	3.37 - 6.25 cm	med. scales
3	71.8 rad/m	6.75 - 12.5 cm	...
4	35.9 rad/m	13.5 - 25.0 cm	large scales

Center wave numbers and structure sizes on level 0 to 4 of the Laplacian pyramid based on a footprint of the image at the water surface of 70×70 cm.

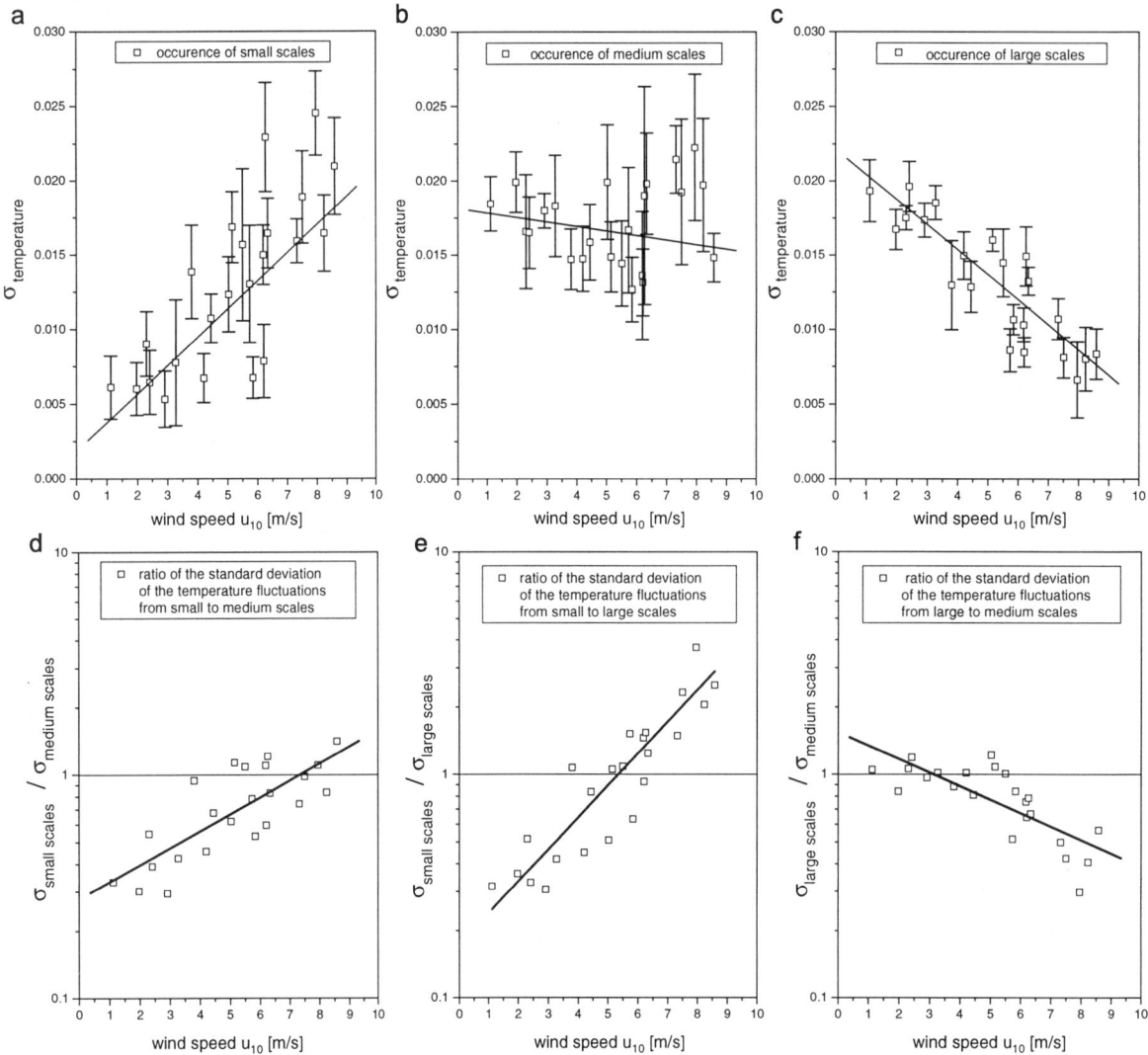

Figure 5. Occurrence of small scales (a), medium scales (b), and large scales (c) versus wind speed. Ratio of the standard deviation of the temperature fluctuations from small to medium scales (d), small to large scales (e), and large to medium scales (f).

that the digital image processing techniques are suitable to investigate the scale dependence of the transport process. Furthermore, the micro-scale temperature fluctuations reveal details of near surface turbulence which could be useful for a better understanding of the transport processes at the air-water interface.

Acknowledgments. We gratefully acknowledge financial support by the Office of Naval Research, ONR, and the National Science Foundation, NSF, within the frame of the Marine Boundary Layer (MBL), and Coastal Ocean Processes (CoOP) research initiatives, respectively.
Jim Edson from WHOI for the supplied micrometeorological data.

REFERENCES

Danckwerts, P.V., Gas-liquid reactions, McGraw-Hill, New York, 1970.

Deacon, E.L., Gas Transfer to and across an Air-Water Interface, Tellus *29*, 363-374, 1977.

Edson, J. B., personal communication and data exchange, Woods Hole Oceanographic Institution (WHOI), MA, June 1999.

Eichkorn, S., Visualisierung und Quantifizierung des CO_2 Gasaustausches mittels laserinduzierter Fluoreszenz (LIF), M.A. thesis, 121 pp., Institution of Environmental Physics, Univ. of Heidelberg, 1997.

Haußecker H., and B. Jähne, In Situ Measurements of the Air-Sea Gas Transfer Rate during the MBL/CoOP West Coast Experiment, in *Air-water gas transfer, selected pa-*

pers from the third international symposium on air-water gas transfer, edited by B. Jähne and E.C. Monahan, pp. 775-784, Aeon Verlag & Studio, Hanau, Germany, 1995.

Haußecker, H., Messung und Simulation von kleinskaligen Austauschvorgängen an der Ozeanoberfläche, PhD thesis, 203 pp., Interdisciplinary Center for Scientific Computing, Univ. of Heidelberg, May 1996.

Haußecker, H., U. Schimpf, and B. Jähne, Measurements of the Air-Sea Gas Transfer and its Mechanisms by Active and Passive Thermography, *Proc. IGARSS '98, sensing and managing the environment, IEEE, Seattle*, 1998.

Jähne, B., K. Münnich, R. Bösinger, A. Duzti, W. Huber, and P. Libner, On the Parameters Influencing Air-Water Gas Exchange, Journal of Geophysical Research *92*, 1937-1949, 1987.

Jähne, B., P. Libner, R. Fischer, T. Billen, and E. Plate, Investigating the Transfer Process Across the Free Aqueous Viscous Boundary Layer by the Controlled Flux Method, Tellus *41B*, 177-195, 1989.

Jähne, B., *Practical Handbook on Image Processing for Scientific Applications*, 358 pp., CRC Press, Boca Raton, Fl, 1997.

Jähne, B., and H. Haußecker, Air-Water Gas Exchange, Annual Review Fluid Mechanics *30*, 443-468, 1998.

Kandelbinder, T., Gasaustauschmessungen mit Sauerstoff, M.A. thesis, 92 pp., Institution of Environmental Physics, Univ. of eidelberg, 1994.

Liss, P.S., and L. Merlivat, Air-Sea Gas Exchange Rates: Introduction and Synthesis, in *he Role of Air Sea Exchange in Geochemical Cycling*, edited by Buat-Menard, pp. 113-127, Reidel, Dordrecht, 1986.

Schimpf, U., H. Haußecker, and B. Jähne, Studies of Air-Sea Gas Transfer and Micro Turbulence at the Ocean Surface using Passive Thermography, in *The Wind-Driven Air-Sea Interface: Electromagnetic and Acoustic Sensing, Wave Dynamics and Turbulent Fluxes*, edited by M. L. Banner, pp. 345-352, School of Mathematics, University of New South Wales, Sydney, Australia, 1999a.

Schimpf, U., H. Haußecker, and B. Jähne, Air-Sea Gas Transfer and Micro Turbulence at the Ocean Surface using Infrared Image Processing, *Remote Sensing of the System Earth - A challenge for the 21st Century, IEEE, Hamburg*, 1999b.

Schimpf, U., Untersuchung des Gasaustausches und der Mikroturblulenz an der Meeresoberfläche mittels Thermographie, PhD thesis, 142 pp., Interdisciplinary Center for Scientific Computing, Univ. of Heidelberg, February 2000.

Wanninkhof, R.W., Relationship between Gas Exchange and Wind Speed over the Ocean, Journal of Geophysical Research *97*, 7373-7381, 1992.

U. Schimpf, B. Jähne, Interdisciplinary Center for Scientific Computing, Im Neuenheimer Feld 368, 69120 Heidelberg, Germany. (e-mail: uwe.schimpf@iwr.uni-heidelberg.de; bernd.jaehne@iwr.uni-heidelberg.de)

H. Haußecker, Xerox Palo Alto Research Center, Palo Alto, CA 94304. (e-mail: hhausses@parc.xerox.com)

On the Skewness of the Sea Slope Probability Distribution

Bertrand Chapron

IFREMER, Département Océanographie Spatiale, Plouzané, France

D. Vandemark

NASA/Goddard Space Flight Center, Wallops Island, Virginia USA

T. Elfouhaily

Johns Hopkins University, Applied Physics Laboratory, Laurel, Mariland USA

We revisit skewness observations for the sea slope probability density function as deduced using optical glitter measurements collected in the 1950s. This slope skewness was addressed by Longuet-Higgins who concluded that localization of short steep wavelets along the phase of underlying longer waves was the most likely physical cause for the phenomena. The present work suggests that the actual wave form under near-breaking conditions, along with the varying population and length scales for these breaking events, should also contribute to the skewness. This latter component is a larger factor for moderate to high winds while the former should dominate at light winds. It is suggested that Cox and Munk's skewness estimate can be related to the modeling and remote sensing of wave breaking probability and perhaps to quantifying the critical role that wave dissipation plays within air-sea gas transfer.

1. INTRODUCTION

An area of unfilled promise in ocean remote sensing is the development of a consistent inversion of sea surface short wave characteristics via the ever-increasing complement of microwave and optical techniques. Such an interest stems largely from the potential of these measurements to provide information on geometrical and kinematic properties of the sea surface that can be related to physical ocean surface processes, such as the rate of gas exchange across the sea surface.

For instance, there is increasing experimental evidence that breaking waves, which impact exchanges of momentum, heat and gas, do contribute strongly to certain remote sensing measurements. From a statistical viewpoint, wave breaking events occur infrequently. These events are termed sporadic or intermittent, and are difficult to quantify. Geometrically, the wave profile under breaking conditions is typically characterized with asymmetry and steepness approaching a large critical value. The tools necessary for precise field determinations remain in development, nonetheless individual breaking events at a multitude of length scales are likely contributors to distinct signatures in optical and microwave remote sensors.

This note is primarily focused on issues pertaining to the inference of the ocean surface slope statistics using techniques that rely upon the quasi-specular reflection mechanism. The initial assumption often invoked to interpret experimental data in such studies is Gaussianity in the surface slope probability distribution function (PDF). However, it is quite clear from the seminal Cox and Munk sun glitter measurements and their reported parameters [Cox and Munk, 1956] that the slope PDF exhibits substantial third and fourth order corrections with respect to a Normal distribution [Cox and Munk, 1956; Longuet-Higgins, 1982]. One immediate impact of non-Gaussian wave slope statistics is its place in modifying the predicted occurrence of exceedingly steep and/or asymmetric slope values. A better identification of these statistical moment signatures within remote sensing measurements [e.g. Chapron et al., 2000; Liu et al., 1997] should help to clarify the relation between these parameters and air-sea gas exchange, as well as the development of possible strategies to parameterize these exchanges in terms of remote sea surface geometrical inferences.

Recently, even-order moments of the slope PDF were addressed [Chapron et al., 2000] to show that observed distribution kurtosis is wholly consistent with a two scale model for the ocean surface. This study indicated that the commonly invoked 'long-wave short-wave interaction' concept readily leads to a non-Gaussian, peaked distribution. The present study concentrates on the skewness of the sea slope distribution and revisits the work of Longuet-Higgins (1982) which in turn starts from Cox and Munk's (1956) observations. The primary objective is to take another look at the physical explanation for the observed skewness.

2. OPTICALLY-DEDUCED SEA SURFACE SLOPE DISTRIBUTIONS

The intensity of sun glitter versus the viewing angle was used by Cox and Munk for their slope PDF estimations. The technical limitations of that study have been well documented. As recently re-evaluated [Chapron et al., 2000], the authors could not estimate the true moments of the distributions due to lack of normalization. They were left to approximate the distribution statistics using observations with an angular limit that left uncertainty in those estimates [Wentz, 1976]. That analysis utilized a polynomial fit to capture the non-Gaussianity and directionality they observed in the data. The regression was performed on the natural logarithm of their intensity observations versus view angle, written as

$$\begin{aligned}\log(p) = &\; a_o - a'_o s^2 + a''_o s^4 \\ &+ s(a_1 - a'_1 s^2)\cos\alpha' \\ &+ s^2(a_2 + a'_2 s^2)\cos 2\alpha' \\ &+ a_3 s^3 \cos 3\alpha' + a_4 s^4 \cos 4\alpha' \\ &\text{and } a'_o s^2 \leq 4 \end{aligned} \quad (1)$$

with s the slope value, α' the azimuthal angle according to principal axis (the wind direction), and a_n^m the fit coefficients. These coefficients were then related to the expansion coefficients pertaining to the cumulants of the slope distribution under a Gram-Charlier approximation. Note that a_o in Eq. 1 is an arbitrary offset value due to the inability to fully resolve the steepest angles.

As developed, two key parameters emerge with respect to non-Gaussianity. These are the ratio a''_o/a'^2_o and the odd-order coefficient a_1. A non-zero value for the former indicates that the slope PDF exhibits a substantial departure from a parabolic fit approximation corresponding to a pure Gaussian distribution. Cox and Munk did find this term to be non-negligible and indicative of a distribution peakedness. We refer the reader to Chapron et al. [2000] for a recent re-interpretation of these results and their physical relevance. For the present paper we wish to recall that the peakedness can be directly attributed to the modulation of short-scale wavelets by a random underlying longer-scale processes such as the long wave field, its varying steepness, wind gustiness etc... Moreover, that paper notes the simple, yet perhaps subtle, issues involved when converting from Cox and Munk's observed polynomial coefficients to distribution moments for the case where exact values for variances and higher-order moments are of interest.

A non-zero value for the odd-order a_1 coefficient indicates an angular shift of the most-likely specular point. The study of Longuet-Higgins (1982) provides a robust examination of this optically-derived evidence for skewness of the surface slopes based upon the case study data of Cox and Munk (1956). As observed, the mode of the slope distribution did not occur at zero slope for the case of wind-driven slope data collected looking along the wind direction. As reported, the most-likely slope value is of the order $3-4°$ towards the downwind. This apparent 'skewness', or shift angle (Δ) was not observed looking in the crosswind direction or for the case where oil was used to damp out cm-scale waves.

Longuet-Higgins computed a first-order development using the Gram-Charlier approximation that permits a link between observed Δ values and the coefficient of

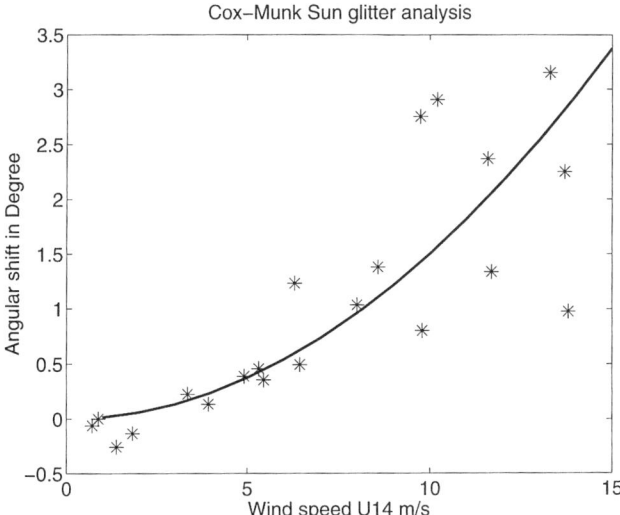

Figure 1. Most-likely specular points as deduced from Cox and Munk fit parameter.

skewness (λ_3). We would propose a slightly modified, but nearly equivalent, approach where according to Eq. 1, the maximum intensity would approximately occur at an angle given by

$$\Delta \simeq \frac{a_1}{2(a_o' - a_2)} \quad (2)$$

We suggest that since the total variance is not directly estimated, an interpretation of this shift angle in terms of the skewness statistical parameter is not straightforward. A more precise rendering normalizes a_1 by the along-wind incomplete variance, $1/(a_o' - a_2)$. This will result in somewhat smaller values compared to the initial study conducted by Longuet-Higgins (1982). Regardless, the non-zero shift angle and its dependence on the mean-square-slope (i.e. wind speed for the Cox and Munk case) are still clearly in evidence as shown in Figure 1.

3. MODELING NON-GAUSSIAN SEA SURFACE SLOPES

Several questions are posed within Longuet-Higgins (1982). The majority of this effort deals with modeling related to this question: What is the physical explanation for the observed skewness? The study examines several possibilities to explain the observed angular shift. The first approach is to consider the skewness of the individual waves where one assumes the existence of a shifted, bound second harmonic that distorts the linear form. Such a geometry should only occur in a transient state, just before breaking. The author concludes that this effect may be relevant to longer gravity waves but its direct application to the range of high-frequency waves that dominate the slope PDF is somewhat unclear. A second possibility of skewness in waves when considering viscous damping is dismissed due to the very weak computed impact. The study's search for physical explanation ends with the assumption of a deterministic two scale model in which capillary-gravity waves, contributing the most to the slope variance, are modulated in phase and amplitude by a fast-propagating but slowly varying longer wave. The author concludes that this last phenomenon leads to plausible agreement with the data in terms of magnitude and sign for Δ. Explicit in this choice is the realization that slope skewness is directly associated with long wave steepness and propagation direction and thus there is no necessary fundamental relation between the wind stress and distribution skewness. Moreover, the sign can change if the angle between wind waves and swell were to approach π.

We postulate that this deterministic and hydrodynamic model for the skew effect should be rejoined with that study's first possibility - that the wave's themselves can be skewed under near-breaking conditions. Moreover, the source of short-wave modulation can be of both hydrodynamical and aerodynamical origins. Theoretically and experimentally, it is known that modulational and direct wind input effects produce shortening and steepening of the shorter waves. The two main physical explanations mentioned above can thus both contribute to the resulting non-Gaussian slope statistics.

Undoubtedly, near-breaking wavelets can locally dominate the slope variance (e.g. sea spikes). Consequently, the non-Gaussian statistics can be hypothesized to result from the modulated occurrence of steep asymmetric short waves along longer wave profiles. This modulated occurrence contributes to an apparent modulation of the slope variance (between non-breaking and breaking short wave packets). If such intermittencies and groupiness are sufficiently sampled, a compound model may be considered to describe the surface slope statistics [Chapron et al., 2000]. Following such a model, all sources of non-stationarities and modulations, including wave-wave interactions, wind input modulation, local breaking, leads one to consider the slope variance as a random variable. Mathematically, such a compound model leads at the lowest even orders to an expression congruent with the empirical shape analysis proposed by Cox and Munk, i.e. Eq. 1. Under such a development, the departure from Gaussianity is explicitly associated with the variance of the mean square slope fluctuations, and will thus control the value of the ratio $a_o''/a_o'^2$.

To augment the hydrodynamical approach of Longuet-Higgins, one can then follow the phenomenological approach introduced by Phillips [Phillips, 1985] to consider the skewness contribution associated with near-breaking waves. The evaluation can be made by considering the average total length of breaking fronts at a given scale, $\Lambda(\mathbf{k})$. The breaking fronts are conceptually associated with isolated irregular line segments with discontinuous asymmetric slope changes. To simplify the description, the sharp crested near-breaking waves are further assumed to have self-similar asymmetric shapes (bound phase-shifted higher harmonics). Larger breaking waves are then simply magnified copies of smaller breakers.

The total average surface covered by sharp crested waves at a given scale per unit surface will be propotional to

$$dA(k) \propto \frac{\Lambda(k)}{k} dk \qquad (3)$$

leading to an expected total slope asymmetry proportional to

$$\Delta \propto \Delta_i \int \frac{\Lambda(k)}{k} dk \qquad (4)$$

where Δ_i is the self-similar asymmetry of an individual self-similar transient. As suggested by Phillips, the integration covers the gravity range. This range is limited at high wavenumber to scale whose phase speed is not small compared with the friction velocity $u_\star$, i.e.

$$k_{max} \leq \min\{(g/\gamma)^{1/2}, g/u_\star^2\} \qquad (5)$$

with γ the surface tension. Indeed, in the range of small gravity waves ($\leq$15-20 cm and $\geq$3-5 cm), the dissipation will be dominated by the generation of parasitic capillaries [Kudryavtsev et al., 1999].

According to Phillips, $\Lambda(k)$ should vary as $u_\star^3 k^{1/2}$, leading to $\Delta \propto u_\star^2$ which is consistent with the Cox and Munk's very short fetch glitter observations (see Figure 1). Under this phenomenological approach, it can also be postulated that over the total average surface covered by active wave-breaking events, the surface slope variance along the wind direction is necessarily limited to a critical constant maximum value. The along-wind slope variance should thus also vary as $u_\star^2$.

However, under open sea conditions, $\Lambda(k)$ and the total fraction of the surface covered by sporadic breaking events will also depend upon be non-wind environmental conditions (e.g. swell amplitude and directionality, current, atmospheric stability, slicks). These effects will likely limit a direct identification between the shift angle and the wind stress.

The combination of the two-scale hydrodynamic model with the multi-scale breaking phenomena provides two skewness-generating processes that are in play over most of the ocean most of the time. Their relative contribution or precedence will vary and they are also, in some sense, inseparable. However, one anticipates that the direct long wave modulation effect will be most evident at lighter winds where field observations of the modulation transfer function show a trend running from a maximum at lightest winds, decreasing with $u_\star$. Wave breaking and thus actual wave geometry skew will generally increase in proportion to the population of breaking events, and corresponding increase of the range of length scales involved in breaking. We reiterate that some combination of hydrodynamic and aerodynamic modulation along the longer waves is likely fundamental to the breaking environment. Nonetheless, as described above one can expect that, to first-order, the average fraction of the surface covered by active wave breaking events will scale with $u_\star^n$ in a manner that may likely follow field-observed trends for wave breaking at micro and macro-scales.

4. CONCLUSION

With an overall objective to refine links between ocean surface processes and remote sensing measurements, it is necessary to have a precise sea surface slope statistical description. In many studies, Gaussian statistics are a leading assumption. This first-order approximation is numerically attractive and requires only the variance for a complete model. However, optical scattering measurements are shown to consistently indicate the need to report at least two parameters in an even-order statistical analysis, in addition to a systematic angular displacement of the most intense glitter reflection. A physical interpretation for observing infrequent steep asymmetric slopes can be given via considering the modulated occurrence of near breaking waves.

Numerous studies (e.g. Longuet-Higgins, 1982; Walsh et al. 1998; Shaw and Churnside, 1997; Liu et al., 1997) suggest that high frequency radar and/or optical techniques applied at near-nadir incidence angles should lead to further clarifications of the physics and correlatives involved in determining the slope distribution. But from the presnt ananlysis, it is stressed that the data-fitting procedures must include up to fourth order corrections. This is necessary to properly determine, as a minimum, the slope variance.

We anticipate that one end goal for such data should be a better determination of wave breaking statistics, and improved quantification of the role of sharply crested waves in gas transfer across the sea surface.

REFERENCES

Chapron B., V. Kerbaol, D. Vandemark and T. Elfouhaily, Importance of peakedness in sea surface slope measurements and applications, *J. Geophys. Res*, *105*, 17195-17202, 2000.

Cox, C. and W. Munk, Slopes of the sea surface deduced from photographs of sun glitter, *Bull. Scripps Inst. Ocean.*, *6*, *9*, 401-488, 1956.

Kudryavtsev, V.N., V.K. Makin and B. Chapron, Coupled sea surface-atmosphere model 2. Spectrum of short wind waves, *J. Geophys. Res.*, *104*, 7625-7639, 1999.

Liu Y., X.-H. Yan, W.T. Liu, and P.A. Hwang, The probability density function of ocean surface slopes an its effect on radar backscatter, *J. Phys. Oceanogr.*, *25*, 782-797, 1997.

Longuet-Higgins, M.S., On the skewness of sea surface slopes, *J. Phys. Ocean.*, *12*, 1283-1291, 1982.

Phillips, O.M., Spectral and statistical properties of the equilibrium range in wind-generated gravity waves, *J. Fluid Mech.*, *156*, 505-531, 1985.

Shaw, J. and J. Churnside, Scanning laser glint measurements of sea-surface slope statistics, *Appl. Optics*, *36*, 4202-4213, 1997.

Walsh, E. D., D. Vandemark, C. Friehe, S. Burns, and D. Khelif, Measuring sea surface mean square slope with a 36-GHz scanning radar altimeter, *J. Geophys. Res.*, *103*, 12613-12628, 1998.

Wentz, F., Cox and Munk's sea surface slope variance, *J. Geophys. Res.*, *81*, 1607-1608, 1976.

B.Chapron, IFREMER, Département Océanographie Spatiale, 29280 Plouzané, France (e-mail: bchapron@ifremer.fr)

D. Vandemark, NASA Goddard Space Flight Center, Laboratory for Hydrospheric Processes, Wallops Island, 23337, VA, USA (e-mail: vandemark@gsfc.nasa.gov)

T. Elfouhaily, The Johns Hopkins University, Applied Physics Laboratory, 11100 Johns Hpkins road, Laurel, MD 20723-6099, USA (e-mail: Elfouhaily@jhuapl.edu)

Directional Distributions and Mean Square Slopes of Surface Waves

Paul A. Hwang and David W. Wang

Oceanography Division, Naval Research Laboratory, Stennis Space Center, Mississippi

Field observations show that the crosswind component constitutes a significant portion of the ocean surface mean square slope. The average ratio between the crosswind and upwind mean square slope components is 0.88 in slick-covered ocean surfaces. This large crosswind slope component cannot be explained satisfactorily based on our present models of ocean wave directional distributions. Two-dimensional spectral analysis of 3D ocean surface topography reveals that bimodal directional distribution is a common feature for wave components shorter than the peak wavelength. The calculated result of the upwind and crosswind mean square slope components using a bimodal directional distribution function is in very good agreement with field measurements.

1. INTRODUCTION

Ocean surface waves are the roughness element of the air-water interface. The directional distribution of the ocean surface roughness is an important parameter in air-sea interaction studies and ocean remote sensing applications. Examples include magnitude and direction of energy and momentum fluxes between air and water, scatterometer wind velocity measurement, synthetic aperture radar (SAR) imaging of the ocean wave field, and optical detection of surface and subsurface features. Close correlation between the gas transfer velocity and surface wave parameters, especially the mean square slope, has been reported extensively in the literature (e.g., papers presented in the Air-Water Gas Transfer Symposia, *Brutsaert and Jirka*, 1984; *Wilhelms and Gulliver*, 1991; *Jahne and Monahan*, 1995; *Donelan et al.*, 2001).

The preferred representation of the ocean surface roughness properties is the directional wavenumber spectrum of the ocean waves. Such data become available only recently. The results on the roughness wavenumber spectrum come in small pieces because of difficulties in data acquisition and analysis. Several field measurements of wavenumber spectra of short waves (from several millimeters to several decimeters) obtained by scanning laser slope sensing have been reported [e.g., *Bock and Hara*, 1995; *Hara et al.*, 1994, 1998; *Hwang et al.*, 1996]. In principle, scanning slope data include two orthogonal components of ocean surface slopes measured in the space domain and detailed information on the directional distributions of individual spectral components of ocean waves are available, but only limited results on the directional properties of short waves over a small range of wind speeds have been reported [*Hwang*, 1995; *Klinke and Jahne*, 1995; *Hara et al.*, 1998].

Most ocean wave directional spectra are obtained from temporal measurements acquired by directional buoys [e.g., *Longuet-Higgins et al.*, 1963; *Mitsuyasu et al.*, 1975; *Hasselmann et al.*, 1973, 1980] or wave gauge arrays [e.g., *Donelan et al.*, 1985] and frequency spectra are derived. The transformation from frequency to wavenumber domain is carried out using the dispersion relationship. This approach is rather successful for long gravity waves. The application to shorter waves is not very satisfactory because of the large Doppler frequency shift caused by convection of short waves by surface currents. The large frequency shift introduces large uncertainties in the interpretation of the length scales of the measured (apparent) wave frequency [e.g., *Phillips*, 1985; *Hwang et al.*, 1996; *Donelan et al.*, 1999].

For surface roughness investigations, the surface slope data of Cox and Munk (1954) remain the most comprehensive in terms of the range of wind speeds encountered and the extent of their statistical analysis. Based on their measurements, the ocean surface roughness has a significant crosswind component. In most cases, the ratio of the

Gas Transfer at Water Surfaces
Geophysical Monograph 127
This paper not subject to U.S. copyright
Published in 2002 by the American Geophysical Union

crosswind and upwind mean square slope components is greater than 0.7. The large crosswind to upwind ratio is also found in later measurements of high frequency ocean wave spectra using laser slope gauges [e.g., *Hughes et al.*, 1977; *Tang and Shemdin*, 1983; *Hwang and Shemdin*, 1988]. This ratio is much larger than what's expected from applying unimodal directional distribution functions currently established in the ocean wave spectral models.

Banner and Young [1994] emphasize that bimodality is a robust feature of the directional distribution of wind-generated waves. The dynamic process that produces a bimodal directional distribution is the nonlinear wave-wave interaction mechanism. Although bimodal directional distributions have been obtained from directional buoy data using analysis techniques such as maximum entropy method (MEM) or maximum likelihood method (MLM) [*Young*, 1994; *Young et al.*, 1995; *Ewans*, 1998; *Wang and Hwang*, 2000], there are continuous disagreements on the existence of directional bimodality, mainly because the quantitative results of bimodal directional distribution differ significantly depending on the processing method employed. *Hwang et al.* [2000a,b] report 2D wavenumber spectral analysis of 3D surface topography measured by an airborne topographic mapper (ATM, an airborne scanning lidar system). The directional resolution derived from 3D topographic data is excellent. The analysis presented in *Hwang et al.* [2000b] illustrates that the directional resolution for a wave field generated by a steady 10 m/s wind field is better than 10° for wavenumber components higher than the peak wavenumber. The directional spectra of the ATM data show clear directional bimodality [*Hwang et al.*, 2000b]. They also notice that cases of bimodal directional distributions are commonly found in the directional wavenumber spectra derived from similar 3D spatial measurements, such as those acquired by aerial stereo photography [*Phillips*, 1958; *Cote et al.*, 1960; *Holthuijsen*, 1983], airborne imaging radar (*Jackson et al.* 1985] and land-based imaging radar [*Wyatt*, 1995].

In this paper, we present computations of the upwind and crosswind mean square slope components using four different directional distribution models. These calculations are compared with the measurement of mean square slopes by *Cox and Munk* [1954]. Due to the fact that there remain major uncertainties on the spectral properties in the short wave regime, we limit our investigation to gravity wave components in the equilibrium and saturation ranges of the surface wave spectrum. Quantitative results of the ratio of the crosswind and upwind slope components calculated using four directional distribution functions are presented in Section 2. Further discussions on the spectral function and directional distributions are presented in Section 3. A summary of the study is given in Section 4.

2. DIRECTIONAL DISTRIBUTIONS AND MEAN SQUARE SLOPE COMPONENTS

2.1. Field Data

The mean square slope results reported in *Cox and Munk* [1954] are derived from analyzing the sun glitter patterns of the ocean surface obtained from an aircraft. The area of coverage for each image of glitter patterns is typically on the order of one-half square kilometer. The results, therefore, yield a high degree of statistical confidence. Based on these data, the total mean square slopes of the ocean surface increase linearly with wind speed, and the following two formulas are given:

$$s_{clean}^2 = 5.12 \times 10^{-3} U + (3 \pm 4) \times 10^{-3}, \quad (1)$$

and

$$s_{slick}^2 = 1.56 \times 10^{-3} U + (8 \pm 4) \times 10^{-3}. \quad (2)$$

In (1) and (2), the wind velocity, U, is measured at 12.5-m elevation. Combining their results with other later datasets and making corrections for cases that are affected by background swell, a slightly different formula in terms of U_{10}, the neutral wind speed at 10-m elevation, is given in *Hwang* [1997]

$$s_{clean}^2 = 5.12 \times 10^{-3} U_{10} + 1.25 \times 10^{-3}. \quad (3)$$

The data of special interest to this paper are those collected from slick covered surfaces. Altogether, *Cox and Munk* [1954] report 9 slick cases (one natural slick and 8 man-made slicks) with wind speeds ranging from 1.6 to 10.6 m/s. The ratio between crosswind and upwind components is quite large. For the 9 slick cases, the mean value with one standard derivation is 0.88±0.097, and none of the cases has a ratio less than 0.75. This large fraction of the crosswind mean square slope component has never been explained satisfactorily. As will be shown below, calculations using established unimodal directional distribution functions under-predict the crosswind component considerably.

2.2. Directional Distribution

For a given directional spectrum, $\chi(k,\theta)=\chi(k)D(k,\theta)$, the upwind and crosswind slope spectra can be expressed as

$$\chi_{1u}(k,\theta) = k_u^2 \chi(k) D(k,\theta), \quad (4)$$

and

$$\chi_{1c}(k,\theta) = k_c^2 \chi(k) D(k,\theta), \quad (5)$$

where k is wavenumber, subscripts u and c denote the upwind and crosswind components, $\chi_1(k)$ is the slope spectrum, relating to the displacement spectrum $\chi(k)$ by $\chi_1(k)=k^2\chi(k)$, and $D(k,\theta)$ is the directional distribution function (assuming wind direction is at $\theta = 0$, thus $k_u=k\cos\theta$, and $k_c=k\sin\theta$). In the following, the displacement spectral function is assumed to be

$$\chi(k) = \begin{cases} bu_* g^{-0.5} k^{-2.5}, & k \leq k_i \\ Bk^{-3}, & k > k_i \end{cases}. \quad (6)$$

where $B=4.6\times10^{-3}$, $b=5.2\times10^{-2}$, and k_i is the matching wavenumber separating the equilibrium and the saturation spectra; the magnitude of k_i is in the neighborhood of $6.5k_p$ [e.g., *Phillips*, 1977, 1985; *Hwang et al.*, 2000a; *Hwang and Wang* 2000]. The ratio of crosswind and upwind mean square slope components integrated to an upper limit wavenumber, k_I, can be written as

$$r(k_I) = \frac{\int_{k_p}^{k_I} k^2 \chi(k) \left[\int_{-\pi}^{\pi} \sin^2\theta D(k,\theta)d\theta \right] dk}{\int_{k_p}^{k_I} k^2 \chi(k) \left[\int_{-\pi}^{\pi} \cos^2\theta D(k,\theta)d\theta \right] dk}. \quad (7)$$

For individual wavenumber components, the ratio is

$$d(k) = \frac{\int_{-\pi}^{\pi} \sin^2\theta D(k,\theta)d\theta}{\int_{-\pi}^{\pi} \cos^2\theta D(k,\theta)d\theta}. \quad (8)$$

Clearly, the specific functional form of the directional distribution plays an important role on the magnitudes of $r(k_I)$ and $d(k)$. Four different directional distribution models reported in the literature are investigated here. The first three models [*Mitsuyasu et al.*, 1975; *Hasselmann et al.*, 1980; *Donelan et al.*, 1985, with modification by *Banner*, 1990] are unimodal. The key features of these models include: (i) The dominant propagation directions of all wave components are in the wind direction. (ii) The directional beamwidth is narrowest near the spectral peak and becomes broader as the wavelength of an individual spectral component increases or decreases from the peak wavelength. Of these three unimodal directional distribution models, the *Donelan et al.* distribution is narrower, and the *Hasselmann et al.* distribution is broader, than the *Mitsuyasu et al.* distribution. The directional distributions of shorter wave components ($k \geq 1.3k_p$) of the fourth model [*Hwang et al.*, 2000b, Appendix A] are bimodal. As stated earlier, bimodality is a robust feature of nonlinear wave-wave interaction [*Banner and Young*, 1994]. The beamwidth (an integrated property of the directional distribution) predicted by all four models are comparable [*Hwang et al.*, 2000b].

Figures 1a and b show the computed $d(k)$ and $r(k_I)$ for a wave field generated by 10 m/s wind. The ratio between crosswind and upwind slope components, $d(k)$ calculated using the bimodal function is usually larger than those calculated using unimodal directional distributions, especially in the lower range of k/k_p. The integrated ratio, $r(k_I)$, based on the bimodal function is much closer to the observed value. Following the calculation that waves shorter then 0.3 m are suppressed by slicks [*Cox and Munk*, 1954; *Phillips*, 1977], the corresponding wavenumber is $k_s=2\pi/0.3$ rad/m. For this wind speed, $k_s=214k_p$, and $r(k_s)\approx0.8$. In contrast, $r(k_s)$ is approximately 0.7 based on the models of *Mitsuyasu et al.* [1975] and *Hasselmann et al.* [1980], and is less than 0.6 based on the model of *Donelan et al.* [1985]. *Hwang and Wang* [2000] examine the effect of cutoff wavenumber on the integrated mean square slope, and show that the effect is minor. Sensitivity tests on other spectral parameters are also described in *Hwang and Wang* [2000].

The ratio $r(k_s)$ as a function of wind speed is plotted in Fig. 1c. For comparison, the field data of *Cox and Munk* [1954] are also shown. Of the four directional models, the results of bimodal directional distribution are in best agreement with the field data, but still underestimate the crosswind to upwind ratio by approximately 14%, the other three (unimodal) models underestimate the ratio by 26 to 39%.

3. DISCUSSIONS

As noted in Section 2, the crosswind to upwind ratio of mean square slopes calculated using a bimodal directional distribution is still 14% lower than the field data. Several factors may contribute to the observed larger crosswind surface slope component in the ocean. Firstly, the computation is based on the assumption of steady wind forcing. Fluctuations in the wind field, especially the wind direction, may contribute significantly to a higher magnitude of $r(k_s)$. Secondly, the derivation of the bimodal directional function is based on data from a mature wave field where the dominant wave component aligns with the wind vector. In a young sea, the dominant wave may propagate in oblique angles with respect to the wind vector in order for the wave field to maintain in resonant condition with the forcing wind field. According to *Phillips* [1957] resonance mechanism of wind-wave generation, two wave systems propagating at oblique angles symmetric to the wind are generated. Spatial measurements of 3D ocean surface topography using airborne scanning radar [*Walsh et al.*, 1985, 1989] and

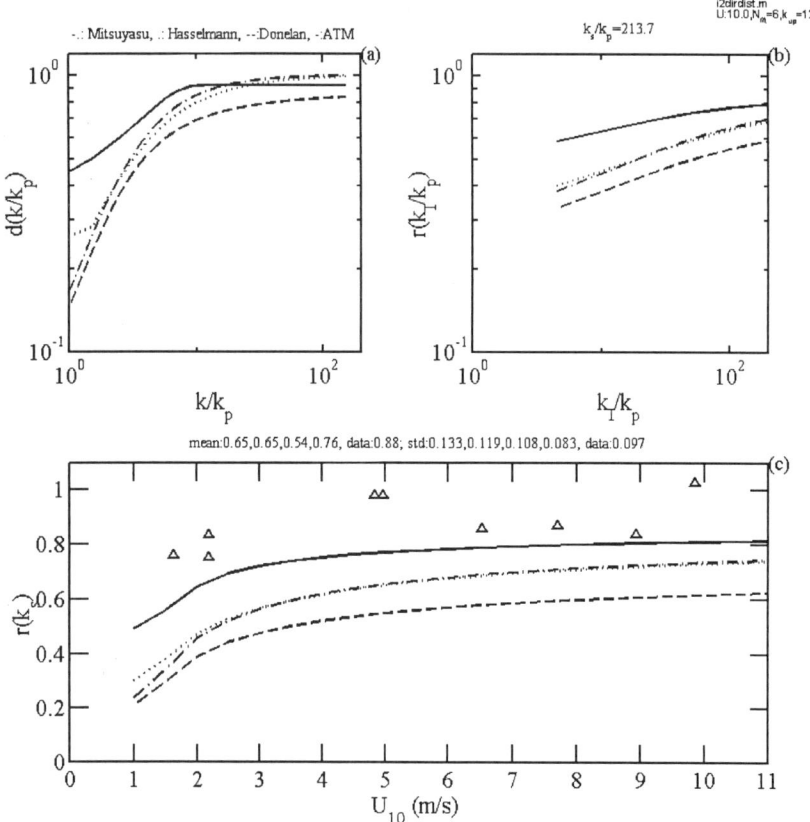

Figure 1. The ratios (a) $d(k)$ (Eq. 8), and (b) $r(k_l)$ (Eq. 7), calculated for a wave field generated by 10 m/s wind. (c) The ratio $r(k_s)$ as a function of wind speed, where the upper limit of integration wavenumber, $k_s=2\pi/0.3$ rad/m, representing the wavenumber above which the waves are damped by the slicks. The results of slick cases of *Cox and Munk* [1954] are also shown as triangles in the plot. Four different directional distribution models are used in the calculation: dashed-and-dotted curves: *Mitsuyasu et al.* [1975], dotted curves: *Hasselmann et al.* [1980], dashed curves: *Donelan et al.* [1985], and solid curves: *Hwang et al.* [2000b].

airborne scanning lidar [*Hwang and Wang*, 2000] have obtained examples of waves propagating in oblique angles with respect to the wind vector under steady wind forcing. In both cases presented by *Walsh et al.* [1985, 1989], most of the reported directional spectra have one wind-generated wave system. At shorter fetches (less than approximately 100 km for 10 m/s wind forcing), the direction of the wind-wave system deviates significantly from the wind vector. In the case reported by *Hwang and Wang* [2000], two wave systems straddling the wind vector exist. The swell condition in the dataset of *Hwang and Wang* [2000] is much milder than those encountered in *Walsh et al.* [1985, 1989], and may contribute to the observed differences. In any case, one expects a significant increase in the ratio between crosswind and upwind mean square slope components when waves are propagating at an oblique angle with respect to wind.

4. SUMMARY

The mean square slope dataset of *Cox and Munk* [1954] is regarded a masterpiece in the study of ocean surface roughness. It has been used as a major calibration reference for many areas of research, ranging from air-sea interaction and wave dynamics, to acoustic and electromagnetic remote sensing applications. One aspect of the dataset remains quite puzzling over the last half century or so is the large value of the crosswind slope component. For slick covered cases, the ratio between crosswind and upwind mean square slope components ranges from 0.75 to 1.03. The mean value with one standard deviation is 0.88±0.097. The large crosswind slope component cannot be explained by the unimodal directional functions established in modern wave spectral models (Fig. 1c). The mean ratio from three unimodal directional distribution functions [*Donelan et al.*, 1985, with modification by *Banner*, 1990; *Hasselmann et al.*, 1980; *Mitsuyasu et al.*, 1975] are 0.54±0.11, 0.65±0.12 and 0.65±0.13, respectively.

Directional spectral analysis of 3D ocean surface topography collected by an airborne scanning lidar system shows that in a mature sea, unimodal directional distribution exists only in a narrow wavenumber range near the spectral peak. For wave components shorter than the dominant wavelength, bimodality is a robust feature of the directional distribution. Nonlinear wave-wave interaction is the mechanism that generates the bimodal feature [*Banner and Young*, 1994; *Hwang and Wang*, 2000]. Using the bimodal directional distribution function, the calculated average ratio between crosswind and upwind slope components is 0.76±0.083, which is in much better agreement with field measurements. This mean value is still about 14% lower than the field data. Possible factors contributing to the observed larger crosswind slope components in the field data include fluctuation in the wind field and less-than-mature stage of the wave field. In both situations, wave components traveling at oblique angles from the wind vector contribute to the observed larger magnitude of the average crosswind slope component.

APPENDIX. BIMODAL DIRECTIONAL DISTRIBUTION FUNCTION

Hwang et al. (2000b) acquire 3D ocean surface topography using an airborne scanning lidar system. Based on data obtained under a quasi-steady wind field the directional distribution function at each wavenumber of the measured 2D spectrum is expressed in Fourier series,

$$D(k,\theta) = \frac{1}{\pi}\left[1 + \sum_{n=1}^{N} A_n(k)\cos 2n\theta\right], \quad -\pi/2 \leq \theta \leq \pi/2 \quad (A1)$$

Coefficients for the third order polynomial fitting to each of the first 9 Fourier components, A_n, $n=1, 2, ..., 9$, are tabulated for reconstructing the bimodal function. The database that establishes the polynomial coefficients is limited to $k \leq 10\, k_p$ so extrapolation too far beyond $10\, k_p$ may produce large excursions in the directional function. The computation presented in this paper is based on extrapolation of the directional coefficients to $12k_p$, and the coefficients at $12k_p$ are used for the remaining higher wavenumber components in the computation range. The coefficients of the third order polynomial fitting of the first 9 Fourier components are tabulated below.

Acknowledgements. This work is sponsored by the Office of Naval Research (NRL "Phase-resolved nonlinear shoaling waves," PE62435). [NRL contribution PP/7332—00-0013.]

REFERENCES

Banner, M. L., Equilibrium spectra of wind waves, *J. Phys. Oceanogr.*, 20, 966-984, 1990.

Banner, M. L., and I. R. Young, Modeling spectral dissipation in the evolution of wind waves. Part I: Assessment of existing model performance, *J. Phys. Oceanogr.*, 24, 1550-1571, 1994.

Bock, E. J., and T. Hara, Optical measurements of capillary-gravity wave spectra using a scanning laser slope gauge, *J. Atm. Oceanic Tech.*, 12, 395-403, 1995.

Brutsaert, W., and G. H. Jirka (eds.), *Gas Transfer at Air-Water Surfaces*, D. Reidel Publ. Co., Dordrecht, Holland, 1984.

Cote, L. J., et al., The directional spectrum of a wind generated sea as determined from data obtained by the Stereo Wave Observation Project, *Meteor. Papers, New York Univ.*, 2, W. J. Pierson (ed.), 88 pp, 1960.

Cox, C. S., and W. Munk, Statistics of the sea surface derived from sun glitter, *J. Mar. Res.*, 13, 198-227, 1954.

Donelan, M. A., J. Hamilton, and W. H. Hui, Directional spectra of wind-generated waves, *Phil. Trans. Roy. Soc. Lond.*, A315, 509-562, 1985.

Donelan, M. A., W. M. Drennan, and E. A. Terray, Wavenumber spectra of wind waves in the range of 1-50 m, in *The wind-driven air-sea interface*, ed. M. L. Banner, ADFA Document Producton Center, Camberra. Australis, 35-42, 1999.

Table A1. Third-order polynomial fitting ($y = c_1x^3 + c_2x^2 + c_3x + c_4$, where y is $A_1, A_2, ... A_9$, and x is k/k_p) of the Fourier components, $A_1, A_2, ... A_9$, of the bimodal directional distribution (A1).

	c_1	c_2	c_3	c_4
A_1	-6.83×10^{-4}	2.20×10^{-2}	-2.42×10^{-1}	9.87×10^{-1}
A_2	-2.66×10^{-3}	5.32×10^{-2}	-3.82×10^{-1}	7.83×10^{-1}
A_3	-1.44×10^{-3}	3.29×10^{-2}	-2.08×10^{-1}	3.26×10^{-1}
A_4	-1.13×10^{-3}	2.15×10^{-2}	-1.01×10^{-1}	1.17×10^{-1}
A_5	-7.22×10^{-4}	1.09×10^{-2}	-4.70×10^{-2}	5.96×10^{-2}
A_6	-9.04×10^{-4}	1.21×10^{-2}	-4.92×10^{-2}	7.40×10^{-2}
A_7	5.92×10^{-4}	-8.34×10^{-3}	2.75×10^{-2}	-9.78×10^{-3}
A_8	-1.10×10^{-3}	1.57×10^{-2}	-7.13×10^{-2}	9.80×10^{-2}
A_9	4.33×10^{-4}	-5.93×10^{-3}	2.06×10^{-2}	-1.52×10^{-2}

Donelan, M. A., W.M. Drennan, E.S. Saltzman and R. Wanninkhof (eds.), *Gas Transfer at Water Surfaces,* AGU, this volume.

Ewans, K. C., Observations of the directional spectrum of fetch-limited waves, *J. Phys. Oceanogr.,* 28, 495-512, 1998.

Hara, T., E. J. Bock, and D. Lyzenga, In situ measurements of capillary-gravity wave spectra using a scanning laser slope gauge and microwave radars, *J. Geophys. Res.,* 99, 12593-12602, 1994.

Hara, T., E. J. Bock, J. Edson, and W. McGillis, Observation of short wind waves in coastal waters, *J. Phys. Oceanogr.,* 28, 1425-1438, 1998.

Hasselmann, D. E., M. Dunckel, and J. A. Ewing, Directional wave spectra observed during JONSWAP 1973, *J. Phys. Oceanogr.,* 10, 1264-1280, 1980.

Hasselmann, K., et al., Measurements of wind wave growth and swell decay during the Joint North Sea Wave Project (JONSWAP), *Herausgegeben vom Deutsch. Hydrograph. Institut.,* Reihe A, no. 12, 95 pp, 1973.

Holthuijsen, L. H., Observations of the directional distribution of ocean-wave energy in fetch-limited conditions, *J. Phys. Oceanogr.,* 13, 191-207, 1983.

Hughes, B. A., H. L. Grant, and R. W. Chappell, A fast response surface-wave slope meter and measured wind-wave moment, *Deep-Sea Res.,* 24, 1211-1223, 1977

Hwang, P. A., Spatial measurements of small-scale ocean waves, in *Air-Water Gas Transfer,* B. Jahne and E. C. Monahan (eds.), AEON Verlag & Studio, Hanau, Germany, 153-164, 1995.

Hwang, P. A., A study of the wavenumber spectra of short water waves in the ocean. Part II: Spectral model and mean square slope, *J. Atm. and Oceanic Tech.,* 14, 1174-1186, 1997.

Hwang, P. A., and O. H. Shemdin, The dependence of sea surface slope on atmospheric stability and swell condition, *J. Geophys. Res.,* 93, 13903-13912, 1988.

Hwang, P. A. and D. W. Wang, Directional distributions and mean square slopes in the equilibrium and saturation ranges of the wave spectrum, *J. Phys. Oceanogr.,* 2000 (in press).

Hwang, P. A., S. Atakturk, M. A. Sletten, and, D. B. Trizna, A study of the wavenumber spectra of short water waves in the ocean, *J. Phys. Oceanogr.,* 26, 1266-1285, 1996.

Hwang, P. A., D. W. Wang, E. J. Walsh, W. B. Krabill, and R. N. Swift, Airborne measurements of the wavenumber spectra of ocean surface waves. Part 1. Spectral slope and dimensionless spectral coefficient, *J. Phys. Oceanogr.,* 30, 2753-2767, 2000a.

Hwang, P. A., D. W. Wang, E. J. Walsh, W. B. Krabill, and R. N. Swift, Airborne measurements of the wavenumber spectra of ocean surface waves. Part 2. Directional distribution, *J. Phys. Oceanogr.,* 30, 2768-2787, 2000b.

Jackson, F. C., W. T. Walton, and C. Y. Peng, A comparison of in situ and airborne radar observations of ocean wave directionality, *J. Geophys. Res.,* 90, 1005-1018, 1985.

Jahne, B., and E. C. Monahan (eds.), *Air-Water Gas Transfer,* Aeon Verlag & Studio, Hanau, Germany, 1995.

Klinke, J., and B. Jahne, Measurements of short ocean waves during the MBL ARI West Coast Experiment, in *Air-Water Gas Transfer,* B. Jahne and E. C. Monahan (eds.), AEON Verlag & Studio, Hanau, Germany, 165-173, 1995.

Longuet-Higgins, M. S., D. E. Cartwright, and N.D. Smith, Observations of the directional spectrum of sea waves using the motions of a floating buoy, in *Ocean Wave Spectra,* Prentice Hall, Englewood Cliffs, N. J., 111-136, 1963.

Mitsuyasu, H. *et al.*, Observation of the directional wave spectra of ocean waves using a cloverleaf buoy, *J. Phys. Oceanogr.,* 5, 750-760, 1975.

Phillips, O. M., On the generation of waves by turbulent wind, *J. Fluid Mech.,* 2, 417-445, 1957.

Phillips, O. M., On some properties of the spectrum of wind-generated ocean waves, *J. Mar. Res.,* 16, 231-240, 1958.

Phillips, O. M., *The dynamics of the upper ocean* (2nd ed.), Cambridge U. Press, Cambridge, England, 1977.

Phillips, O. M., Spectral and statistical properties of the equilibrium range in wind-generated gravity waves, *J. Fluid Mech.,* 156, 505-531, 1985.

Tang, S., and O. H. Shemdin, Measurement of high frequency waves using a wave follower, *J. Geophys. Res.,* 88, 9832-9840, 1983.

Wang, D. W., and P. A. Hwang, Transient evolution of the directional distribution of ocean waves, *J. Phys. Oceanogr.,* 2000 (in press).

Wilhelms, S. C., and J. S. Gulliver (eds.), *Air-Water Mass Transfer,* ASCE, New York, 1991.

Wyatt, L. R., The effect of fetch on the directional spectrum of Celtic Sea storm waves, *J. Phys. Oceanogr.,* 25, 1550-1559, 1995.

Young, I. R., On the measurement of directional wave spectra, *Appl. Ocean Res.,* 16, 283-294, 1994.

Young, I. R., L. A. Verhagen, and M. L. Banner, A note on the bimodal directional spreading of fetch-limited wind waves, *J. Geophys. Res.,* 100, 773-778, 1995.

Paul A. Hwang and David W. Wang, Oceanography Division, Naval Research Laboratory, Bldg. 1009, Stennis Space Center, MS 39529-5004.

Numerical Simulation of Hydrodynamic Processes Beneath a Wind-Driven Water Surface

Wu-ting Tsai

Department of Civil Engineering, Chiao Tung University, Hsinchu, Taiwan

Turbulent flow driven by a constant wind stress acting at the water surface was simulated numerically to gain a better understanding of the hydrodynamic processes governing the transfer of slightly soluble gases across the atmosphere-water interfaces. Simulation results show that two distinct flow features, attributed to subsurface surface renewal eddies, appear at the water surface. The first characteristic feature is surface streaming, which consists of high-speed streaks aligned with the wind stress. Floating Lagrangian particles, which are distributed uniformly at the water surface, merge to the predominantly high-speed streaks and form elongated streets immediately after they are released. The second characteristic surface signatures are localized low-speed spots which emerge randomly at the water surface. A high-speed streak bifurcates and forms a dividing flow when it encounters a low-speed surface spot. These coherent surface flow structures are qualitatively identical to those observed in the experiment of *Melville et al.* [1998]. The persistence of these surface features also suggests that there must exist organized subsurface vortical structures that undergo autonomous generation cycles maintained by self-sustaining mechanisms. These coherent vortical flows serve as the renewal eddies that pump the submerged fluids toward the water surface and bring down the upper fluids, and therefore enhance the scalar exchange between the atmosphere and the water body.

1. INTRODUCTION

The most direct and dominant exchange of gases between the atmosphere and the ocean is through the air-sea interface. For slightly soluble gases, such as CO_2, the resistance of the transfer process occurs within a layer of water adjoining and including the interface. The transfer process, which is liquid-phase controlled, is determined by the near-surface turbulence in the ocean, of which the dominant generation mechanism is the wind stress on the interface. Field experiments of near-surface turbulence in the ocean, however, are extremely difficult and have rarely been performed to shed light on the hydrodynamic process of gas exchange.

To our knowledge, *Gemmrich and Hasse* [1992] performed the only such investigation, in which small-scale motions on the ocean surface that might enhance air-sea gas exchange by surface renewal were sought. The study of Gemmrich and Hasse was motivated by the pioneering experiment of *Woodcock* [1941]. Woodcock visualized the motion of the water surface using lycopodium spores. He observed that when winds blow, high-speed surface water movements occur that subsequently form surface streaming in lines and streaks roughly parallel to the wind direction. The water in the streams moves with a cross spacing of the order of centimeters were intuitively interpreted either as a result of wave breaking or as an indication of the existence of counter-rotating helical vortex rolls similar to those in Langmuir circulation, which have spacing of the order of meters to kilometers [*Langmuir*, 1938; or see the review by *Leibovich*, 1983].

In a related study but with a contradictory conclusion, *Kenney* [1993] similarly observed small-scale regularly spaced streaks aligned with local wind on the surface of lakes. According to his results, a shear layer 1 to 2 cm thick forms immediately beneath the water surface under glassy calm condition. Within this surface shear layer, bands of algae were generated. The distance between bands typically ranges from 5 to 10 cm. Furthermore, the algal filaments are neutrally buoyant and move horizontally in the shear layer without significant vertical motions. In contrast, the motions of filaments below the surface layer are strongly three-dimensional, and individual algal filaments fall and rise as they cross beneath several bands. This occurrence indicates that the flow within the surface layer seems to be uncoupled from the underlying turbulent flow. No evidence was found of an array of counter-rotating vortex rolls with axes aligned with the algal bands.

In a recent laboratory experiment by *Melville et al.* [1998], predominant streaks aligned in the wind direction were also observed at a wind-driven water surface. The streaky surface undergoes several stages in its evolution beginning with a quiescent water surface at the initiation of the wind. The appearance of the surface streaks was shown to closely follow the initial growth of the surface waves and the inception of the Langmuir circulations. In addition to the surface streaks, local upwellings and divergence of the velocity field at the surface were also observed, which cause bifurcation or dislocation of the streaks. The generation and evolution of these surface features and the possible Langmuir circulations were attributed to an instability in the wind-driven surface shear layer.

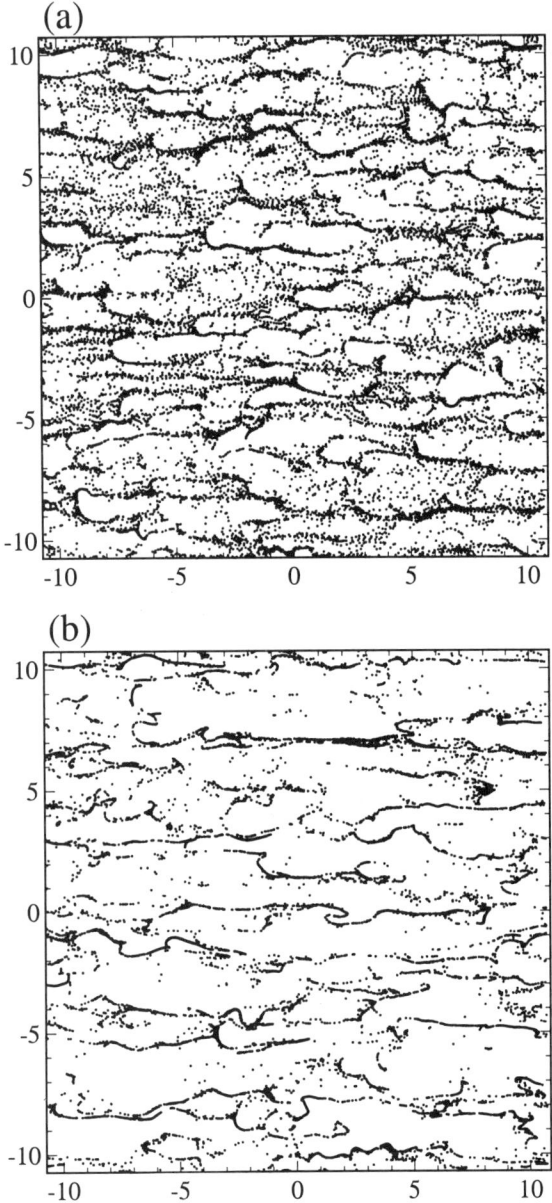

Figure 1. Instantaneous distribution of the surface floating Lagrangian particles at times $t=$ (a) 4.3 and (b) 6 s from the start of the simulation. 128^2 uniformly distributed particles are released at time $t=3.6$ s.

2. NUMERICAL SIMULATION

To elucidate the dynamic process within the wind-driven surface layer, turbulent shear flow induced by a wind stress acting at the water surface is simulated numerically in this study. The numerical simulation largely focuses on providing further insight into the hydrodynamic mechanism, particularly the coherent flow structures that govern the process of gas exchange across the air-sea interface. Here, the emphasis is on educing the surface features resulting from the interaction between a wind-sheared interface and the subsurface turbulence structures. The interaction dynamics of subsur-

down-wind much faster than that between the streams. The streaks are approximately a centimeter across and with a distance of the order of centimeters. While performing similar experiments using cork powder to dye the surface, Gemmrich and Hasse observed that lines or streaks are the most frequently appearing patterns of surface water movement at wind speeds of 2 or 3 m s−1 or higher. These predominantly line structures face turbulent shear flow with a shear-free interface were considered by the present author in *Tsai* [1998a], as well as with a surfactant-contaminated surface in *Tsai* [1996 and 1998b].

The approach adopted herein is a direct numerical simulation of the flow without using any turbulence parameterization to model the subgrid flow processes. Details of the mathematical formulation and the numerical implementation of the simulation resemble those in *Tsai* [1998a] except for the tangential stress condition on the interface. A constant stress is applied at the interface, which is balanced by the interfacial local tangential stress. The water surface is free to move according to the kinematic and dynamic free-surface conditions, which are satisfied at the mean water surface [*Tsai*, 1998a].

Numerical simulation of the flow is initiated with a mean streamwise velocity distribution (cm s^{-1}):

$$U(z) = t_0(1 + 2z^2)\,\text{erf}(-z) + \frac{2}{\pi^{1/2}}\,z\,\exp(z^2),$$

corresponding to the velocity profile of a plane laminar shear layer driven by a constant surface stress, which induces a linear increase in the surface velocity for a period of $t_0 = 20$ s. Accordingly, the simulated flow corresponds to the subsequent development of the flow of water which has been driven by a wind field accelerating constantly from rest to a final speed of 5 m s^{-1} in approximately 20 s as in the experiment of *Melville et al.* [1998].

The shear stress acting at the water surface, $\tau_0 = \nu_w(dU/dz)_{z=0}$ is assumed to be constant and invariant with time from the start of the simulation, and the surface velocity develops freely with the maximum turbulence production occurring near the water surface. Random but solenoid velocity fluctuations, which are homogeneous in the horizontal plane (x-y plane), are added to the initial mean flow.

The length and width of the computational domain are both 21.5 cm. Imposing a free-slip bottom condition at the depth of 5.4 cm to emulate the infinite depth closes the computational domain. There is no steady state for the problem in the absence of the no-slip bottom (as in the open-channel flow) since the total momentum in the water grows linearly with time in proportion to the applied wind stress. The thickness of the shear layer develops to about one third of the depth of the computational domain at the end of the simulation.

The simulated flow field, which is maintained by applying a constant shear stress at the water surface, may not be identical to the actual flow of water driven by the boundary forcing of a developing wind field. The focus of the simulation, however, is on the inception and

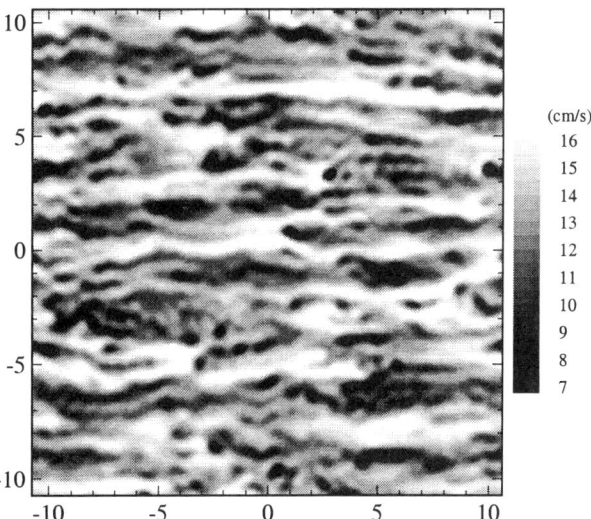

Figure 2. Contours of instantaneous streamwise velocity on the water surface showing predominant high-speed streaky structure (indicated by white streaks) and localized low-speed spots (in black) at time $t = 6$ s from the start of the simulation. The flow travels from left to right.

initial development of the coherent surface signatures, which occur within a few seconds [see, e.g., Figure 2 in *Melville et al.*, 1998] as well as the possible underlying generation mechanisms. For such purposes, the present free-surface flow driven by a constant surface shear should be sufficient. It, nevertheless, would be invalid to describe longer-term evolution of the actual wind-driven flow. In fact, as will be discussed later, the present numerical simulation results reproduce all the surface features observed in the experiment of *Melville et al.* [1998].

3. RESULTS AND DISCUSSION

Although the turbulent flow is initiated artificially, high-speed surface streaming, similar to that observed in the ocean and lake [*Woodcock*, 1941; *Gemmrich and Hasse*, 1992; *Kenney*, 1993] and in the laboratory experiment of *Melville et al.* [1998], emerges on the interface at time approximately 3 s from the start of the simulation (corresponding to 23 s in the experiment of *Melville et al.* [1998]). To visualize the structures and also the inception and evolution of the surface streaming, uniformly distributed floating Lagrangian particles are released at time $t = 3.6$ s and their trajectories are tracked.

Figure 1 depicts the distributions of the Lagrangian particles about 0.7 s (at $t = 4.3$ s) and 2.4 s (at $t = 6$ s) after the release of the particles. Immediately after the particles are released, they aggregate and form several

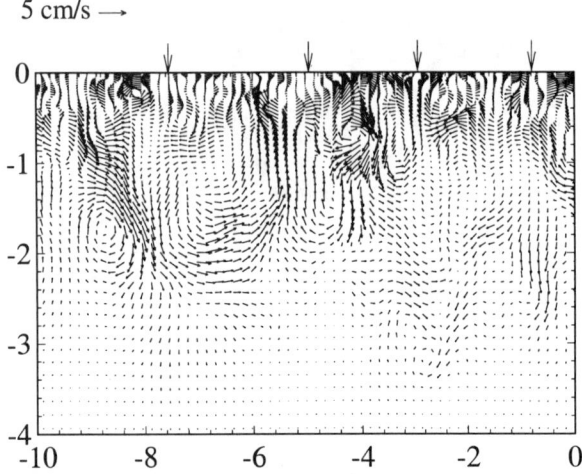

Figure 3. Velocity vectors on a vertical plane perpendicular to the wind direction showing the spanwise and vertical velocity components at time $t = 6\,\text{s}$ from the start of the simulation.

elongated streaky groups. The rapidity of the formation of this surface streaky pattern is the same as the observation of *Melville et al.* [1998] and as that described by *Woodcock* [1941]: "When the powder is dusted over the water it moves into lines so quickly that, even in moderate breeze, the eye can hardly detect the transition from random scattering to the linear pattern." Some particles of the individual streaks may diverge and join other streaks. Such a feature is identical to the streak dislocation or bifurcation observed by *Melville et al.* [1998].

The corresponding contours of instantaneous streamwise velocity on the water surface at $t = 6\,\text{s}$, showing predominant high-speed streak structure, are plotted in Figure 2. Comparing the distribution of particles in Figure 1 and the velocities in Figure 2, the Lagrangian particles travel along the high-speed streams (indicated in white in Figure 2). Velocity vectors of on a vertical plane perpendicular to the wind direction, showing the spanwise and vertical velocity components, are plotted in Figure 3. The arrows on the water surface indicate the spanwise positions of the surface streaks. It can be seen, from the distribution of the velocity vectors, that the underlying streamwise vortical structure more complex than just an array of counter-rotating vortices, a then-prevalent notion.

The bifurcation of streak occurs when it encounters a localized low-speed spot (indicated in black). Looking into the velocity field at the surface and underneath, it is found that a local upwelling flow brings up the submerged slowly moving fluids and induces divergence of the velocity field at the surface. An explanation of this three-dimensional flow is the existence and movement of coherent horseshoe vortices in the turbulent shear layer, as has also been identified in our previous studies involving the interaction of a shear-free interface with the underlying turbulent shear flow [*Tsai*, 1998a]. The underlying coherent horseshoe vortices move upwards and impinge the interface. During impingement, these vortices induce upwelling flows, thereby enhancing the renewal of water in contact with the interface.

While the present numerical simulations are not completely compatible with the conditions of the experiments of *Melville et al.* [1998], it is worth comparing the characteristic scales of the surface features with the measurements. The spacing of the surface streaks is estimated by examining the spanwise spectrum of the x-averaged streamwise surface velocity, $\int \hat{u}(x, n, 0) dx$, as shown in Figure 4 at various times, where n is the spanwise modal number. The simulation begins with a wide spectrum of streamwise velocity without any predominant spanwise wavelength. As time progresses, streamwise streaks appear and later on in the simulation, the streamwise velocities are dominated by the component with modal number six corresponding to a streak spacing of 3.58 cm (= width of computational domain 21.5 cm/6). This is close to the measurement of *Melville et al.* [1998].

In addition to the streak spacing, the wavelength of the generated surface waves is another characteristic length scale in the simulated flow process. The streamwise spectrum of the y-averaged surface elevation, $\int \hat{\eta}(m, y) dy$, is shown in Figure 5, where m is the streamwise modal number. The initial water surface is flat in the simulation. Surface waves are generated as

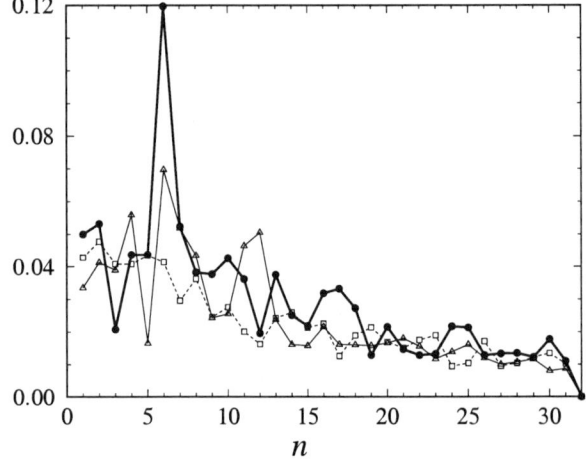

Figure 4. Spanwise spectrum of the x-averaged streamwise surface velocity, $\int \hat{u}(x, n, 0) dx$, at times $t = 1.6$ (rectangle), 5.4 (triangle) and 8 s (solid circle) from the start of the simulation. n is the spanwise modal number.

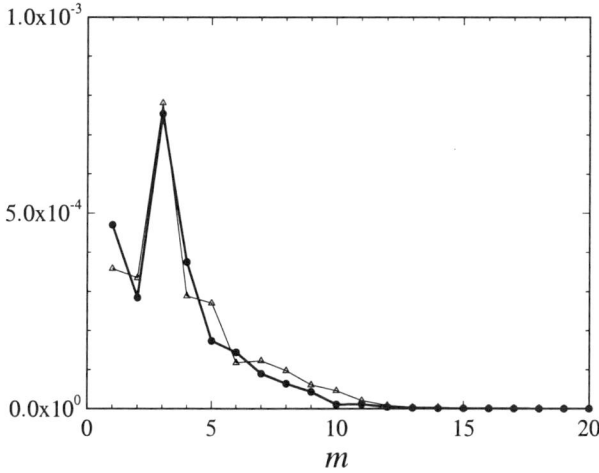

Figure 5. Streamwise spectrum of the y-averaged surface elevation, $\int \hat{\eta}(m,y)dy$, at times $t = 5.4$ (triangle) and $8\,\mathrm{s}$ (solid circle) from the start of the simulation. m is the streamwise modal number.

the underlying turbulent flow develops. The dominant surface wavelength of the generated waves is 7.2 cm (= length of computational domain 21.5 cm/3). This computed surface wavelength is about three times that measured by *Melville et al.* [1998], despite the resemblance in the features of surface streaming and bifurcation of streaks and also the good agreement in the streak spacing. The possible cause of such a discrepancy in the surface wavelength lies in the different generation mechanisms of surface waves. In our numerical simulation, the surface waves arise solely from the instability of the turbulent shear layer of water. However, in reality, the mechanism of surface wave generation is the instability of the coupled air-water shear flow, and in particular the growth of disturbance of the air flow. This, nevertheless, poses a question of the role of surface waves in the onset of coherent surface patterns. This is the issue we are currently investigating.

Acknowledgments. This research was supported by grants from the National Science Council of Taiwan under Contract No. NSC 88-2611-M-019-003 and 89-2611-M-009-002. Most of the computations were performed on the computers at the National Center for High-Performance Computing of Taiwan.

REFERENCES

Kenney, B. C. Observations of coherent bands of algae in a surface shear layer, *Limnol. Oceanogr.*, *38*, 1059-1067, 1993.

Gemmrich, J. and L. Hasse, Small-scale surface streaming under natural conditions as effective in air-sea gas exchange, *Tellus*, *44B*, 150-159, 1992.

Langmuir, I. Surface motion of water induced by wind, *Science*, *87*, 119-123, 1938.

Leibovich, S., The form and dynamics of Langmuir circulations, *Annu. Rev. Fluid Mech.*, *15*, 391-427, 1983.

Melville, K., R. Shear, and F. Veron, Laboratory measurements of the generation and evolution of Langmuir circulations, *J. Fluid Mech.*, *364*, 31-58, 1998.

Tsai, W.-T., Impact of a surfactant on a turbulent shear layer under the air-sea interface, *J. Geophys. Res.*, *101*, 28557-28568, 1996.

Tsai, W.-T., A numerical study of the evolution and structure of a turbulent shear layer under a free surface, *J. Fluid Mech.*, *354*, 239-276, 1998a.

Tsai, W.-T., Vortex dynamics beneath a surfactant contaminated ocean surface, *J. Geophys. Res.*, *103*, 27919-27930, 1998b.

Woodcock, A. H., Surface cooling and streaming in shallow fresh and salt waters, *J. Marine Res.*, *4*, 153-161, 1941.

W.-T. Tsai, Department of Civil Engineering, Chiao Tung University, Hsinchu 300, Taiwan. (e-mail: tsai@cc.nctu.edu.tw)

Direct Numerical Simulation of Turbulent Free Surface Flow with Carbon-Dioxide Gas Absorption

Tomoaki Kunugi

Department of Nuclear Engineering, Kyoto University, Kyoto, Japan

Shin-ichi Satake

Department of Mechanical and Intellectual Systems Engineering, Toyama University, Toyama, Japan

The turbulent behavior of liquid and gas flows at a free surface is not very well understood in comparison with that of the motions at the solid boundary. In the present study, the DNS for the turbulent free surface flow with a shear wind has been carried out by means of a coupled gas-liquid flow solution procedure, i.e., MARS (Multi-interface and Advection and Reconstruction Solver) developed by one of the authors. From the present DNS results, the turbulent statistics of the free surface characteristics, such as surface-shape, velocity fluctuations and the budget of turbulent kinetic energy, are obtained. The Henry's Law is applied to the evaluation of a saturated gas concentration at the free surface caused by the carbon-dioxide gas absorption. The mean concentration profile of carbon-dioxide and the root-mean-square of concentration fluctuation distribution throughout both the gas and liquid flow fields are obtained. According to the computational flow visualization, the carbon-dioxide gas is transferred by the low-speed streaks corresponding to low-pressure regions and also transported by the vortexes from the free surface to the bottom in the water region. It is found that this is one of the mechanisms of carbon-dioxide gas transport due to turbulence at the free surface. Moreover, an exchange coefficient of carbon-dioxide gas absorbed at the turbulent free surface is estimated.

1. INTRODUCTION

The turbulent behavior of liquid and gas flows at a free surface is not very well understood in comparison with that of the motions at the solid boundary. However, turbulent behaviors of both free surface and solid wall cases are equally important in the practical application. In addition, the mechanism of scalar transport into a turbulent liquid across the gas-liquid interfaces or free surfaces is great importance in the industrial devices, especially a chemical process like gas-absorption equipment and so on. Recently, the turbulent mass transport of the carbon-dioxide is very important for the global warming because of understanding the "Missing sink" due to a mass imbalance of the carbon production and consumption after the industrial revolution. There are many studies on the liquid-side turbulence near the free surface imposed with and without very low shear flow [*Rashidi and Banerjee*, 1988; *Komori et al.*, 1989]. Moreover, a few experimental investigations with shear wind have been performed. Thus, in case of the low wind velocity, most of previous studies have been considered the turbulence structure caused by the shear rate in the liquid-side flow rather than the coupling effect between the gas- and liquid-flows. However, in case of the capillary and capillary-gravity range wind velocity, i.e., around 2-7m/s, the turbulent characteristics near the free surface could be influenced by a deformation of the free surface due to the shear wind.

Direct numerical simulations (DNSs) and experimental investigations for the turbulent free surface flows have been less extensive because of the numerical and experimental difficulties for tracking and measuring the transient free surface deformation. Recently, the DNS with coupling between gas and liquid flows was performed by Lombardi, Angelis and Banerjee [1996]. However, they still assumed the interface kept to be flat, i.e., this assumption was similar to the previous study [*Lam and Banerjee*, 1992]. Angelis and Banerjee [1999] focused on the capillary and capillary-gravity range wind velocity and presented some correlation between mass flux and turbulent structure. Although their surface can be deformed by the shear wind, the interface momentum and continuity treatments are almost the same as Lombardi et al. [1996]. Another free surface treatment by Tsai [1998] was linearized by assuming a free surface deformation of $\varepsilon \sim O(Fr)$ and a free surface boundary layer thickness $\delta \sim O(Re^{-1/2})$, with $\delta^2 \ll \varepsilon \ll \delta \ll 1$. Here, Reynolds number is based on the shear layer thickness h, the free-stream velocity U_o and kinematic viscosity of bulk fluid ν and Froude number $Fr=U_o/(gh)^{1/2}$, where g is the gravitational acceleration. However, most of liquid surfaces are treated like a solid wall boundary condition if you solve the gas flow. These treatments have an essential difficulty when the liquid surface breaks due to a high shear flow.

On the other hand, in order to understand the gas-liquid interaction at the turbulent free surface up to higher velocity range, the authors have been developing a precise free surface tracking for two-phase flows, i.e., MARS (Multi-interfaces Advection and Reconstruction Solver) [*Kunugi*, 1997], and provided a high quality initial turbulent fields for both the gas and liquid flows via the DNS [*Satake and Kunugi*, 1998]. Moreover, we have been investigating the turbulent scalar transport from the gas to the sea due to the strong interaction between liquid and gas [*Kunugi, Satake and Ose*, 1999]. However, in our previous study the capability of the MARS to solve the multi-phase flows was only emphasized and the very rough and preliminary results were obtained.

In the present study, DNS for the turbulent free surface flow with the relatively low shear wind has been carried out by means of the coupled gas-liquid flow solution procedure (MARS). From the present DNS results, the turbulent statistics of turbulent free surface flow, the mean concentration gradient and velocity-concentration fluctuation distributions of carbon-dioxide throughout both flow fields are obtained. Finally, the gas exchange coefficient of carbon-dioxide absorbed by the water at the turbulent free surface is estimated.

2. GOVERNING EQUATIONS

Since the treatment of multiphase flows in the MARS [*Kunugi*, 1997] is based on a volume fraction function $F(t, x_i)$, the spatial distribution of each material $F(t, x_i)$ can be defined as:

$$\langle F \rangle = \sum F_m \quad (1)$$

here, F_m denotes a volume fraction of m-th fluid and $\langle \ \rangle$ represents a spatial mean.

In the present study, the following assumptions for deriving the governing equations are made:
1) Newtonian fluid and,
2) thermal properties are constant and,
3) thermal radiation and viscous dissipation are neglected.

The continuity equation for the multiphase flows of m fluids can be expressed as:

$$\frac{\partial F_m}{\partial t} + (U \cdot \nabla) F_m = \frac{\partial F_m}{\partial t} + \nabla \cdot (F_m U) - F_m \nabla U = 0 \quad (2)$$

The momentum equation with the CSF (Continuum Surface Force) model for a surface tension term by Brackbill [1991] can be written as:

$$\frac{\partial U}{\partial t} + \nabla(UU) = g - \frac{1}{\langle \rho \rangle} \nabla P - \nabla \cdot \tau + \frac{1}{\langle \rho \rangle} F_V \quad (3)$$

The CSF term, F_V at the certain surface position x_s, can be modeled as:

$$F_V(x_s) = [\sigma \kappa(x_s) + \tau_2(x_s)_{ii} - \tau_1(x_s)_{ii}] n(x_s) \langle \rho(x_s) \rangle / \overline{\rho} \quad (4)$$

where, σ is a surface tension coefficient, $\kappa(x_s)$ is a surface curvature at x_s and n is the normal vector to the surface at x_s. Other thermal properties, such as the mean density ρ, the viscosity μ, and the viscous stress τ are defined as follows:

$$\left. \begin{aligned} &\langle \rho \rangle = \sum (F_m \rho_m), \quad \overline{\rho} = (\rho_g + \rho_l)/2 \\ &\langle\langle \phi \rangle\rangle = \left[\phi_g \times \{sign(F_g - F_l) + 1\} + \phi_l \times \{1 - sign(F_g - F_l)\} \right]/2 \\ &\tau = -\left[\langle\langle \mu \rangle\rangle / \langle\langle \rho \rangle\rangle \right] \left[(\nabla U) + (\nabla U)^T \right] \end{aligned} \right\} \quad (5)$$

here, the suffices, g and l, denote the gas and the liquid phases, respectively. ϕ is a general variable such as ρ and μ. Another average of the property at the free surface region is defined as $<<\phi>>$, i.e., a kind of "up-winding" of property.

The momentum equation (3) can be solved by means of the well-known projection method [Chorin, 1968]. Once the velocity field can be obtained, it can be transferred the fluid volume flux by using Eq. (2) with the MARS. The mass transport equation for the carbon-dioxide gas can be expressed as follows:

$$\frac{\partial}{\partial t}\langle\rho\rangle C + \nabla\cdot(\langle\rho\rangle CU) = \nabla\cdot(D_m \nabla C) + S \quad (6)$$

where, C is a concentration, D_m is a molecular diffusivity and S is an absorbed mass flux at the free surface based on the following Henry's law;

$$C_s = P_s/H, \quad (7)$$

here C_s is a concentration at free surface corresponding to the saturation solubility, P_s is an instantaneous pressure at free surface and H is the Henry's constant corresponding to the pressure at the water surface.

As for the discretization of the governing equations on the Cartesian coordinate system, the second-order scheme for the spatial differencing terms is used on the staggered grid system and the Euler implicit scheme is used for the first stage of projection method to solve the momentum equation. The gas absorption term is treated as an internal boundary condition based on Eq. (7) in the computation program.

3. PBLEM DESCRIPTION

The physical problem treated here is the motion of two Newtonian incompressible fluids with the deformation of the interface between them. Both gas and liquid flows flow in parallel. The computational domain is 0.157m in width (x-direction), $2h$=0.05 m in height (y-direction) and 0.281 m in length (z-direction). The periodic boundary conditions in the spanwise (x) and streamwise (z) directions are imposed and the free-slip boundaries are applied to both upper and lower boundaries in y-direction. The initial free surface is located at the mid-plane of the total height, i.e., y=h=0.025m. The free-stream velocity U_o in gas phase is 2.1 m/s and a bulk Reynolds number based on the gas layer height (h) and U_o is 3380. The liquid mean velocity is assumed to be one-hundredth of the gas velocity in this study. The number of computational grids is 128 in x-direction, 300 in y-direction and 192 in z-direction, respectively. The

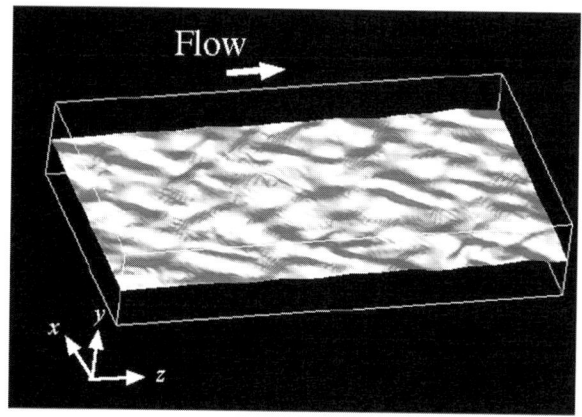

Figure 1. A snapshot of the instantaneous flow visualization of the free surface behavior

computational time step is $10^{-6} \sim 10^{-4}$ sec depending on the residual of mass conservation.

4. RESULTS AND DISCUSSIONS

Figure 1 shows a snapshot of the instantaneous flow visualization of the free surface behavior. The capillary-like wave motion can be observed and around five waves can be seen in the streamwise direction and around three in the spanwise direction. No wave breaking and droplet formation were observed. The friction velocity is calculated from the following equation,

$$\left.\begin{array}{l}\dfrac{\tau_0}{\rho} = \sqrt{\left(\langle\langle v\rangle\rangle\dfrac{\partial W}{\partial x}\right)^2 + \left(\langle\langle v\rangle\rangle\dfrac{\partial W}{\partial y}\right)^2 + \left(\langle\langle v\rangle\rangle\dfrac{\partial W}{\partial z}\right)^2}, \\ u_{\tau g} = \sqrt{\tau_0/\rho}: \; F_g > F_l, \; u_{\tau l} = \sqrt{\tau_0/\rho}: \; F_g < F_l\end{array}\right\} \quad (8)$$

here, W is the streamwise velocity and τ_0 is the boundary shear stress. The friction velocity of gas phase $u_{\tau g}$ is around 0.05m/s and for the liquid phase $u_{\tau l}$ is around 0.02m/s. These values are corresponding to $Re_{\tau g}(=u_{\tau g}h/\nu)$=80 and $Re_{\tau l}(=u_{\tau l}h/\nu)$=553. In the early stage of the computation, the turbulence was suppressed at certain time steps because the gas flow became a low $Re_{\tau g}$ due to the interaction with the water surface and then it recovered later due to producing the high shear in the water underneath the free surface. It seems that the turbulence energy is transferred from the liquid flow to the gas flow through the free surface.

Figure 2 shows the mean velocity distribution. The velocity profile shows the continuous distribution around the mean surface (y=0.025m). This means that the liquid flow near the free surface is driven by the gas flow.

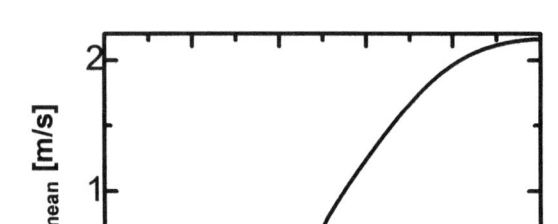

Figure 2. Mean velocity distribution in y-direction (y=0.025m is the mean surface location)

Figure 3 shows the root-mean-square (rms) values of the velocity fluctuation in each direction. It can be seen that the high peak rms values appear at the free surface region. Especially, the normal componet (v_{rms}) is comparable to the spanwise component (u_{rms}). This also suggests that the normal velocity fluctuation component is very important to generate the turbulence at the free surface.

In order to investigate the magnitude of the surface deformation, the mean free surface position (F_{mean}) and the rms-value distributions of F_{rms} are shown in Fig. 4. According to this result, the mean wave height at the surface is estimated around 1mm. This value is close to the wave height taken from experimental data [*Komori*, 1989].

Figure 5 shows the budget of turbulent kinetic energy. It can be seen that a large turbulent energy is generated in the free surface region. In the figure, a circle shows the production term, a triangle shows the dissipation term, a diamond shows the viscous diffusion term, a plus shows the pressure diffusion term, a square shows the turbulence diffusion term and an inverted triangle shows the turbulent convection term, respectively.

In order to investigate the correlation of the turbulence structure between the gas and water phases, Plate 1 shows the contour surface of low-speed regions from the view of water. In the plate, the green color shows the low gas velocity region corresponding to w'<-0.015 [m/s] and the blue color shows the low water velocity region corresponding to w'<-0.047 [m/s]. Both streaks are located almost the same position. It suggests that the turbulence structure in water is strongly correlated with that in gas. This fact supports the assumption of the previous DNS without surface deformation under low shear wind condition such as Lombardi, Angelis and Banerjee [1996] and Angelis and Banerjee [1999].

Plate 2 shows the contour surface of low-pressure region, the water surface and the low-speed regions from the view of water: the gray color shows the region of p'<-5 [Pa], the blue color shows the region of w'<-0.047 [m/s] and the white color shows the surface. The low-pressure regions are located at the tip of free surface in the water. It seems that the low-pressure regions are not directly related to the low-speed region in this figure. However, the location of low-speed regions are overlapped the low-pressure region.

Figure 6 shows the gradients of the mean concentration distribution in each direction. The gradient in normal direction to the free surface is four-order higher than that of other gradients and the peak value is the order of $10^2 m^{-1}$.

Figure 7 shows the mean velocity-concentration correlation distributions along y-axis. All correlations show the similar tendency, however, the magnitude of $\overline{v'c'}$ is five

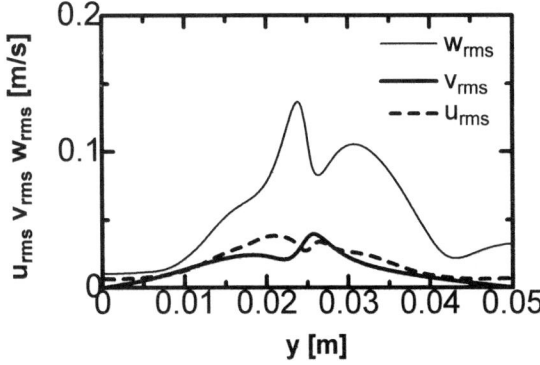

Figure 3. Velocity fluctuation distribution

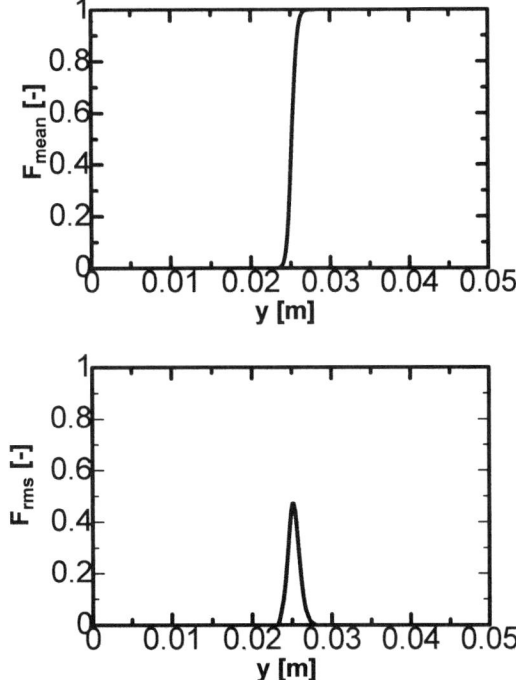

Figure 4. Mean free surface position, Fmean and the root mean square (rms) value distribution of free surface fluctuation, Frms

times larger than $\overline{u'c'}$ at the free surface region. As for the streamwise direction, $\overline{w'c'}$ is one order larger than $\overline{v'c'}$ at the free surface region.

From these results, the turbulent gas exchange coefficient can be estimated. At first, a turbulent diffusivity D_{turb} can be calculated by using a usual gradient diffusion model:

$$-\overline{v'c'} = D_{turb}\, \partial \overline{C}/\partial y \quad (9)$$

According to the present DNS, $D_{turb}=6\times10^{-9}\sim10^{-7}$ m^2/s.

The turbulent gas exchange coefficient in y-direction, k_L can be calculated by using the following the turbulent exchange molar flux, J equations:

$$J = -D_m \overline{\partial C/\partial y} \quad (10)$$

$$J = k_L \Delta C \quad (11)$$

Substituting Eq. (10) to Eq. (11), k_L can be obtained by

$$k_L = \left(-D_m \overline{\partial C/\partial y}\right)/\Delta C \quad (12)$$

here, ΔC is a gas concentration difference between the surface and the upper boundary, however, this value is almost unity. The turbulent gas exchange coefficient k_L can be estimated around 6×10^{-7}-10^{-5} m/s and is almost the same order of the existing measuring data [*O'Connor*, 1983; *Komori*, 1993].

Plate 3 shows the contour map of carbon-dioxide concentration on the x-y plane and the contour surface of low-pressure region, water surface and low-speed regions from the view of water: the red color to the blue color shows the concentration: 0.0<C<0.0006, the gray iso-surface shows the low-pressure region: p'<-5 [Pa], the blue iso-surface shows the low-speed region: w'<-0.047 [m/s] and the pink iso-surface shows the water surface. It can be seen that the higher concentration region is located at the low-speed region and the low-pressure region. The carbon-dioxide gas is absorbed at the free surface and transferred to the deeper water region by various turbulent motions such as the existence of low-pressure and the low-speed regions underneath the free surface of the water.

5. CONCLUSIONS

The DNS for the turbulent free surface flow with shear wind has been carried out by means of the coupled gas-liquid flow solution procedure, i.e., MARS. From the DNS results, the turbulent statistics for the free surface characteristics are obtained. It is found that the interaction between gas and liquid flows is very important for turbulent generation at the free surface due to shear wind, and the Henry's Law can be used for the evaluation of a gas concentration at the free surface due to carbon-dioxide gas absorption. From the flow visualization results, low-speed streaks and vortexes (i.e., low-pressure region) transport the carbon-

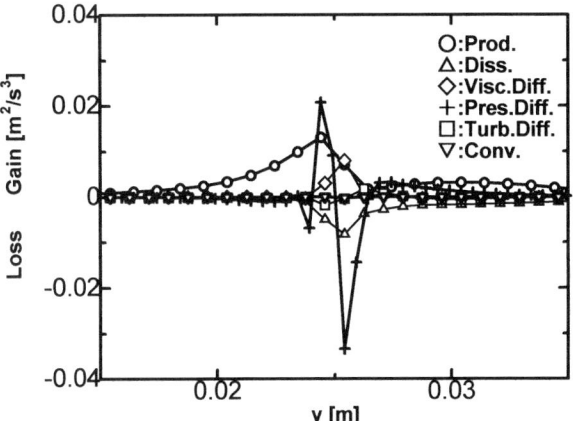

Figure 5. Budget of turbulent energy

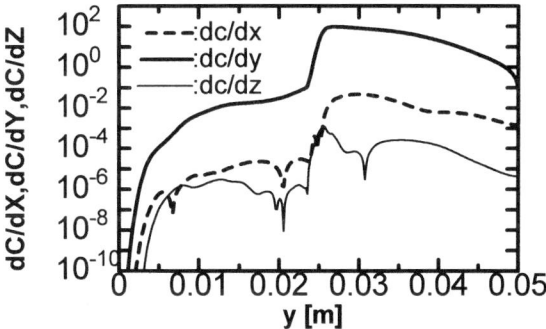

Figure 6. Distributions of mean concentration gradient

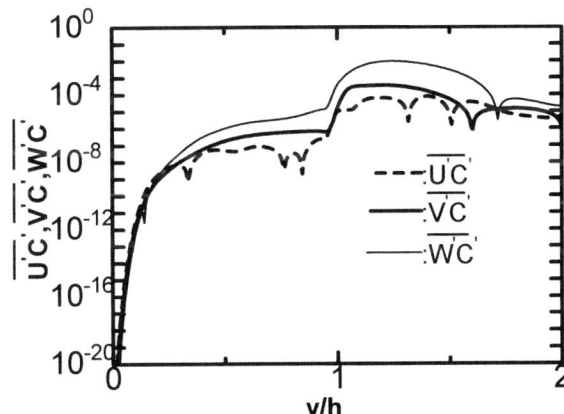

Figure 7. Mean velocity-concentration correlation distributions along y-axis

dioxide gas from the surface to the deeper region of the water. It is found that this is one of the mechanisms of carbon-dioxide gas transport due to turbulence at the free surface. Moreover, an exchange coefficient of carbon-dioxide gas is estimated and agrees well with the existing experimental and/or observation data. Therefore, it can be said that the low shear wind with the water surface deformation generates the turbulence underneath the water surface and this induced turbulent motion is a key mechanism of the scalar transport at the turbulence free surface.

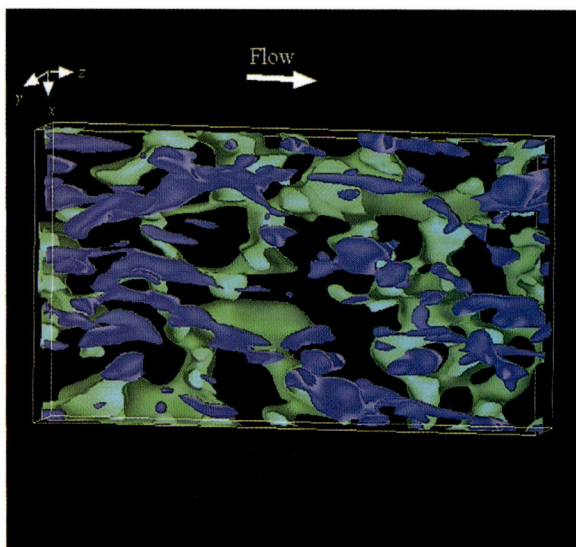

Plate 1. Contour surface of low-speed regions inside the water

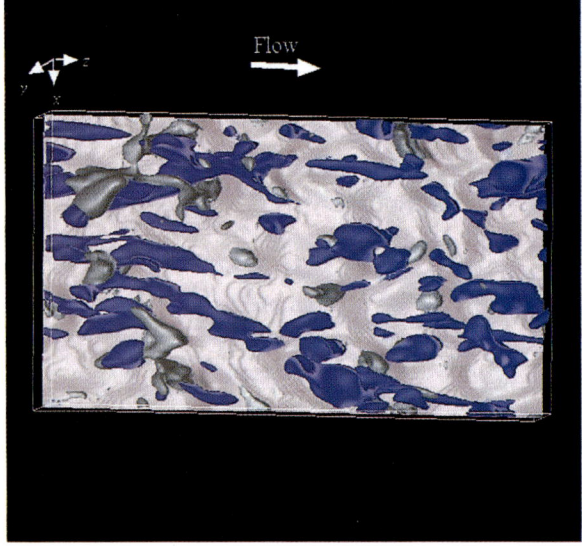

Plate 2. Contour surface of low-pressure, water surface and low-speed regions inside the water

Acknowledgments. The Japan Science and Technology Corporation supported this work. We also acknowledge the collaboration program of the division of computer and information center of RIKEN (The Institute of Physical and Chemical Research) and the Kansai Research Establishment of Japan Atomic Energy Research Institute for using the Fujitsu vector parallel supercomputers.

REFERENCES

Angelis, V. D. and Banerjee, S. Heat and mass transfer mechanisms at wavy gas-liquid interfaces, *Turbulence and Shear Flow Phenomena-1*, Edited by Banerjee S. and Eaton, pp.1249-1254, J. K., Begell House Inc., 1999.

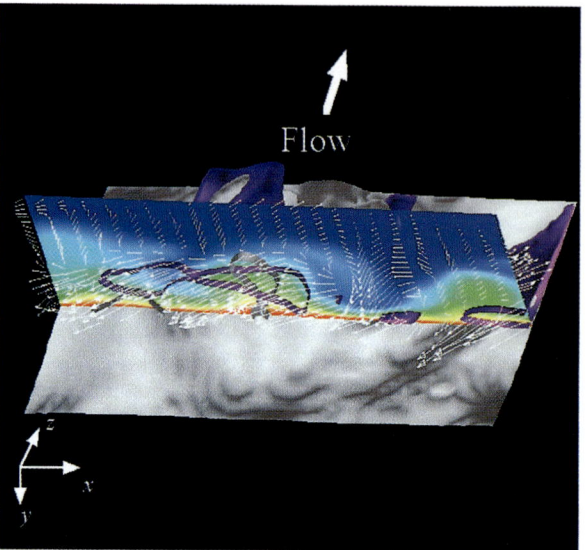

Plate 3. Contour map of carbon-dioxide concentration on the x-y plane and contour surface of low-pressure region, water surface and low-speed regions inside the water

Brackbill, J. U., Kothe, D. B. and Zemach, C., A continuum method for modeling surface tension, *J. Comput. Phys.*, 100, 335-354, 1992.

Chorin, A. J., Numerical Solution of the Navier-Stokes Equations, *Math. of Comput.*, 22, 745—762, 1968.

Komori, S., Murakami, Y. and Ueda, H., The relationship between surface-renewal and bursting motions in an open-channel flow, *J. Fluid Mech.*, 203, 103-123, 1989.

Komori, S., Nagaosa, R. and Murakami, Y., Turbulence structure and mass transfer across a sheared air-water interface in wind-driven turbulence, *J. Fluid Mech.*, 249, 161-183, 1993.

Kunugi, T., Sakate, S. and Ose, Y., Direct numerical simulation on wave formation and breaking of turbulent free surface flow, *Proc. 2nd International Symposium on Two-Phase Flow Modelling and Experimentation*, 2, 819-826, 1999.

Lam, K. and Banerjee, S., On the condition of stream formation in a bounded turbulent flow, *Phys. Fluid*, A 4, 306-320, 1992.

Lombardi, P., Angelis, V. D. and Banerjee, S., Direct numerical simulation of near-interface turbulence in coupled gas-liquid flow, *Phys. Fluids*, 8, 1643-1665, 1996.

O'Connor, D. J., Wind effects on gas-liquid transfer coefficients, *J. Eng.*, ASCE, 109, 731-752, 1983.

Rashidi, M. and Banerjee, S., Turbulence structure in free surface channel flows, *Phys. Fluids*, 31, 2491-2503, 1988.

Sakate, S. and Kunugi, T., Direct numerical simulation of an impinging jet into parallel disks, *Int. J. Numerical Methods for Heat and Fluid Flow*, 8, 768-780, 1998.

Tsai, W-T., A numerical study of the evolution and structure of a turbulent shear layer under a free surface, *J. Fluid Mech.*, 354, 239-276, 1998.

Tomoaki Kunugi, Yoshida, Sakyo, Kyoto, 606-8501 Japan
Shin-ichi Satake, 3190 Gofuku, Toyama, 930-8555 Japan

LIF Measurements of Oxygen Concentration Gradients Along Flat and Wavy Air-Water Interfaces

Philip T. Woodrow, Jr. and Steve R. Duke

Chemical Engineering Department, Auburn University, Auburn, Alabama

Instantaneous spatially-varying measurements of concentration gradients occurring during aeration for flat, stagnant air-water interfaces and for interfaces with mechanically-generated waves are presented. Measurements were obtained in a laboratory wave tank using a laser-induced fluorescence (LIF) technique that images planar oxygen concentration fields near air-water interfaces. Pulsed nitrogen laser light focused to a thin sheet induces the fluorescence of pyrene butyric acid (in micromolar concentration) in deoxygenated water. The PBA fluorescence is quenched by dissolved oxygen. A high-resolution CCD camera images in two dimensions the intensities of the fluorescence field, providing spatial measurements of oxygen concentration with magnification of 7 μm per pixel. The concentration fields, gradients, and boundary layer thicknesses along the flat and wavy air-water interfaces are quantified and compared to previous measurements associated with sheared gas-liquid interfaces and with wind-generated waves.

INTRODUCTION

For many environmental and industrial situations, the air-water interface takes the form of a flat or wavy interface. Oxygen transfer from air to water is controlled in a thin surface concentration boundary layer in the liquid near the surface where DO concentration gradients are present. The flux of oxygen across the interface, or the rate of aeration, is commonly modeled as the product of the mass transfer coefficient k_l and the difference between the oxygen concentration at the interface and in the bulk ($C_I - C_B$). An accurate model for k_l is needed to effectively predict oxygen exchange at these interfaces. Moog and Jirka [1999] show that, in most instances, present models yield very large errors. The difficulty is due to the inability to account for the wide range of factors influencing k_l.

The ultimate goal of present research is to determine how interfacial structure and dynamics can be related to k_l. The mass transfer coefficient can be estimated as.

$$k_l = D/\delta_c \quad (1)$$

where D is the molecular diffusivity of oxygen in water and δ_c is the surface concentration boundary layer thickness.

Measurements of δ_c can be obtained in a laboratory wave tank using a laser-induced fluorescence (LIF) technique that images planar oxygen concentration fields near air-water interfaces [*Woodrow* and *Duke*, 2001]. The LIF technique is similar to techniques presented at previous symposia on Gas Transfer at Water Surfaces [*Wolff* and *Hanratty*, 1990; *Duke* and *Hanratty*, 1995]. LIF techniques provide visualizations of the concentration boundary layer, yielding far more mechanistic information than time-averaged mass transfer experiments. LIF studies of sheared, wavy interfaces have shown large temporal variations in k_l and evidence that the enhancement in k_l is attrib-

84 LIF MEASUREMENTS OF O₂ CONCENTRATION GRADIENTS

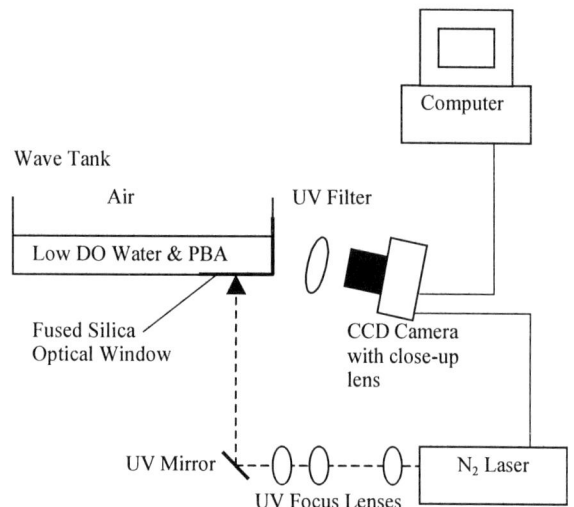

Figure 1. Experimental LIF setup.

uted to large-scale circulations in the liquid and surface renewal effects [*Duke*, 1996; and *Münsterer* and *Jähne*, 1996].

This paper presents LIF measurements of two cases in the absence of surface shear: a flat stagnant interface and a wavy interface.

LIF TECHNIQUE AND EXPERIMENTS

The LIF technique uses pyrene butyric acid (PBA) as a molecular indicator of dissolved oxygen. Vaughan and Weber [1970] demonstrated that PBA could be used as a molecular probe for oxygen. When illuminated by a nitrogen laser ($\lambda = 337.1$ nm), PBA molecules fluoresce with λ between 375 and 425 nm. Fluorescence of the PBA is perturbed by the presence of dissolved oxygen. In the absence of a quencher (oxygen), the molecules fluoresce with a fluorescence lifetime of 160 ns. With oxygen present, collisional quenching occurs and reduces the lifetime to 65 ns in air-saturated water. Fluorescence intensity, F, is proportional to fluorescence lifetime, τ, and is related to oxygen (quencher) concentration, C', by the Stern-Volmer equation:

$$\frac{F_0}{F} = \frac{\tau_0}{\tau} = 1 + k_{sv}C' \qquad (2)$$

where F_0 and τ_0 are values in the absence of the quencher and k_{sv} is a constant dependent on the interaction between the fluorescence molecule and the quencher.

Figure 1 shows a diagram of the experimental wave tank and LIF setup. The wave tank is constructed of Plexiglas and has dimensions 1.219 m × 0.305 m × 10.160 cm. The tank has a fused silica optical window, which allows transmittance of UV light. Waves are generated by an oscillating paddle powered by a variable speed DC motor. Wave properties can be controlled by adjusting the wave maker position settings and the motor speed. Foam beaches at both ends of the wave tank dampen wave reflections.

The tank is filled with 4×10^{-5} M PBA to a depth of 11 mm and deoxygenated through nitrogen stripping to a DO concentration ≤ 1.5 mg/L. At the air-water interface, the DO concentration, C_I, is estimated from Henry's Law.

A pulsed beam emitted from a nitrogen laser (PTI model GL-3300) is expanded and focused through a series of UV-grade lenses to form a thin vertical sheet. As the vertical light sheet passes through the wave tank (25 mm from the side wall), it causes the PBA molecules to fluoresce.

A Photometrics Sensys model CCD camera (1035 × 1317 pixels, 12-bit intensity resolution), with a Nikon 60 mm 1:1 focus lens and UV filter, images the intensities of the laser-induced fluorescence field in two dimensions with a magnification of 7 μm/pixel. The CCD chip has a UV-enhancing coating and the camera is tilted at 10° to avoid interference from the water surface in the foreground.

Figure 2 shows a raw image of the fluorescence field obtained for a flat interface. The dark band across the image indicates a region of higher oxygen concentration (fluorescence quenching), with the center corresponding to the air-water interface. The signal above the interface is the result of reflection.

LIF image sets were obtained for a flat, stagnant interface and a mechanically generated wavy interface. Table 1 lists experimental conditions for seven image sets. DO concentration of the water in the tank was measured with a YSI model 50B dissolved oxygen meter equipped with a field probe. To initiate a run, the DO was reduced by nitro-

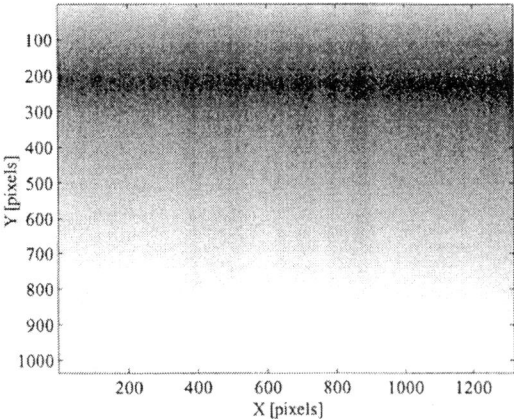

Figure 2. Raw fluorescence field for a flat interface.

Table 1. Experimental conditions.

Set	# Images	Interface	A (mm)	f (Hz)	λ (mm)	T_w (°C)	T_{air} (°C)	P (in Hg)	C_0 (mg/L)	C_I (mg/L)
1	33	Flat	-	-	-	21.1	22.0	29.65	1.60	8.90
2	35	Flat	-	-	-	20.6	23.0	28.41	1.38	8.55
3	35	Flat	-	-	-	21.5	23.5	28.30	1.20	8.44
4	35	Flat	-	-	-	21.5	23.0	28.29	1.11	8.44
5	35	Flat	-	-	-	20.8	23.0	28.29	1.34	8.50
6	30	Wavy	1.0	3.0	90.0	20.8	21.5	29.65	1.50	8.95
7	33	Wavy	2.3	2.8	120.0	21.3	23.0	29.23	0.81	8.83

gen stripping; $t = 0$ occurs just after nitrogen bubbling is ceased. LIF images were obtained within the first 5 minutes and probe measurements were recorded for at least 50 minutes. The absorption of oxygen into water is described by the expression

$$\ln \frac{C'_I - C'(t)}{C'_I - C'(0)} = -K_{abs}\, t \qquad (3)$$

where $C'(t)$ is the measured oxygen concentration at time t, $C'(0)$ is the measured oxygen concentration at $t = 0$, and K_{abs} is the experimentally observed absorption rate constant. Concentration-time data was fit to equation (3) to yield K_{abs}; K_{abs} is multiplied by the water depth to obtain the mass transfer coefficient from concentration-time measurements ($k_{l,t}$).

IMAGE PROCESSING AND ANALYSIS

A raw image (figure 2) from the LIF technique is a 1035 × 1317 matrix of fluorescence intensities. Image processing steps corrects the intensities for variations in fluorescence attributed to factors other than oxygen quenching. The steps include: (1) background subtraction, (2) use of a 2-D adaptive smoothing filter, (3) interface location, (4) removal of Lambert-Beers decay and intensity normalization, and (5) near interface correction for an optical blurring effect.

An adaptive smoothing filter was developed and used to remove noise from the raw image. Filtering was performed in the vertical and in the horizontal directions for twenty iterations. Examination of a single column profile illustrates the effectiveness of the adaptive filter (figure 3). The raw fluorescence intensity, F', is shown as a function of Y (vertical pixel coordinate). The adaptive smoothing filter preserves the integrity (particularly the sharp peak at the interface) of the signal while removing noise more effectively than a typical lowpass filter

In figure 3, the signal minimum around $Y = 220$ represents the general location of the interface. The precise location of the interface is determined by employing a maximum symmetry filter [*Duke*, 1996; and *Münsterer*, 1996]. The filter takes advantage of the symmetry created at the interface between the fluorescence signal and its reflection.

As the light sheet travels through the water (containing PBA), there is an exponential decrease in the fluorescence intensity (Lambert-Beers decay).

$$\frac{I}{I_0} = \exp(-\alpha C_{PBA} l) \qquad (4)$$

where I is the energy of the laser in the fluid, I_0 is the energy of the laser as it enters the fluid, α is a molecule-

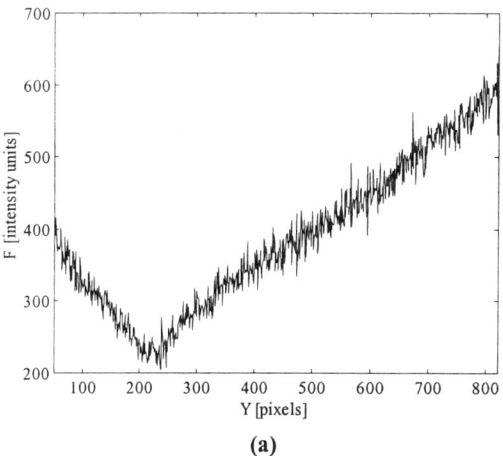

(a)

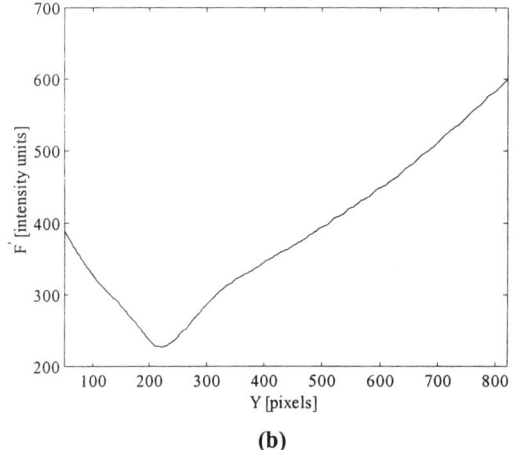

(b)

Figure 3. Column profile: (a) unfiltered (b) filtered.

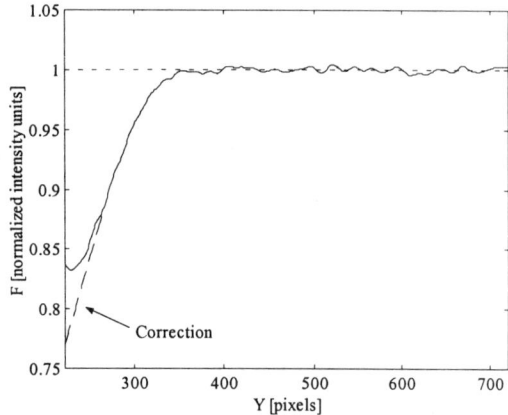

Figure 4. Near-interface optical blurring correction.

dependent constant, C_{PBA} is the concentration of PBA, and l is the distance through the fluid. Lambert-Beers decay for each column is determined from an exponential fit of intensity values in the region of non-varying oxygen concentration (bulk). Lambert-Beers decay and normalization is achieved by dividing intensity values by the exponential fit. Normalization diminishes the effects of the variability of the PBA concentration, the laser pulse energy, the CCD controller signal gain, and the room light level.

A near-interface optical blurring correction is applied to each column. The Lambert-Beers corrected column profiles exhibit a diminishing of the gradients near the interface, rather than the expected sharp and increasing gradients. The near-interface blurring has been noted in other LIF systems [*Duke*, 1996; and *Münsterer*, 1996]. The correction selects the largest gradient from the Lambert-Beers corrected profile and extends the profile to preserve the gradient from the point where it occurred to the interface. Figure 4 shows the near-interface optical blurring correction applied to the column of figure 3(b). The region of non-varying O_2 concentration (well-mixed bulk concentration) is at $Y > 350$. Variations in oxygen concentration occur in the region from $Y = 221$ (interface location) to $Y = 350$.

Image analysis steps transform corrected fluorescence intensities into measurements of dissolved oxygen concentration. Equation (2) relates O_2 concentration to fluorescence intensity. This is implemented by defining a normalized (dimensionless) concentration of the form:

$$C(Y) = \frac{C'(Y) - C'_B}{C'_I - C'_B} = \frac{F_I(F(Y) - F_B)}{F(Y)(F_I - F_B)} \qquad (5)$$

where C' are dimensional concentration values at a particular Y pixel, the interface, and the bulk. Dimensionless concentration varies from 1.0 at the interface to 0.0 in the bulk.

Prior normalization of fluorescence intensity sets F_B (bulk intensity) equal to 1.0; F_I is the average of interface intensities of the particular column and the ten nearest columns.

Implementation of equation (5) yields a vertical concentration profile. Using the magnification, pixels are converted to units of length (mm). Each image provides more than 1300 distinct vertical profiles that are combined to form a two-dimensional concentration field.

Further analysis of the final concentration values reveals properties of the concentration boundary and surface layers. The Photometrics V for Windows software package that accompanied the CCD camera was used for image acquisition and background subtraction. Image processing and analysis steps were carried out using MATLAB (version 5.3).

RESULTS

Figure 5 shows the vertical concentration profile resulting from figure 4. The concentration boundary layer thickness (δ_c) for each concentration profile is determined to be the depth (distance from the interface) at which the gradient crosses zero (bulk concentration value). The gradient is a linear fit based on the maximum slope of the concentration profile near the interface. The concentration gradient and δ_c are shown in figure 5.

The collection of concentration profiles for an image results in a two-dimensional concentration field. Representative two-dimensional concentration fields for a flat and wavy interface are displayed as contour plots in figure 6. Figure 6(a) is the concentration field that resulted from the fluorescence field shown in figure 2. The streamwise (horizontal) coordinate is shown as x and the vertical coordinate is y. The y values are relative to the mean interface location of each concentration field. Flat interfaces are characterized by larger δ_c than the wavy interfaces.

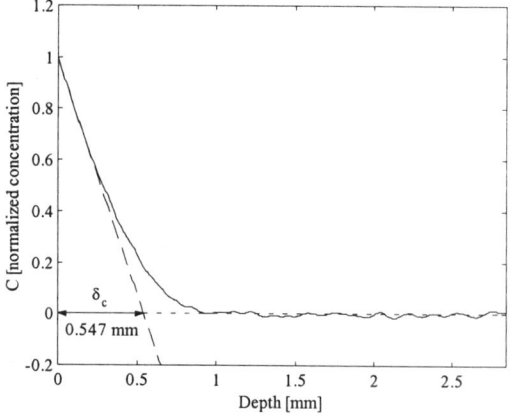

Figure 5. Vertical concentration profile

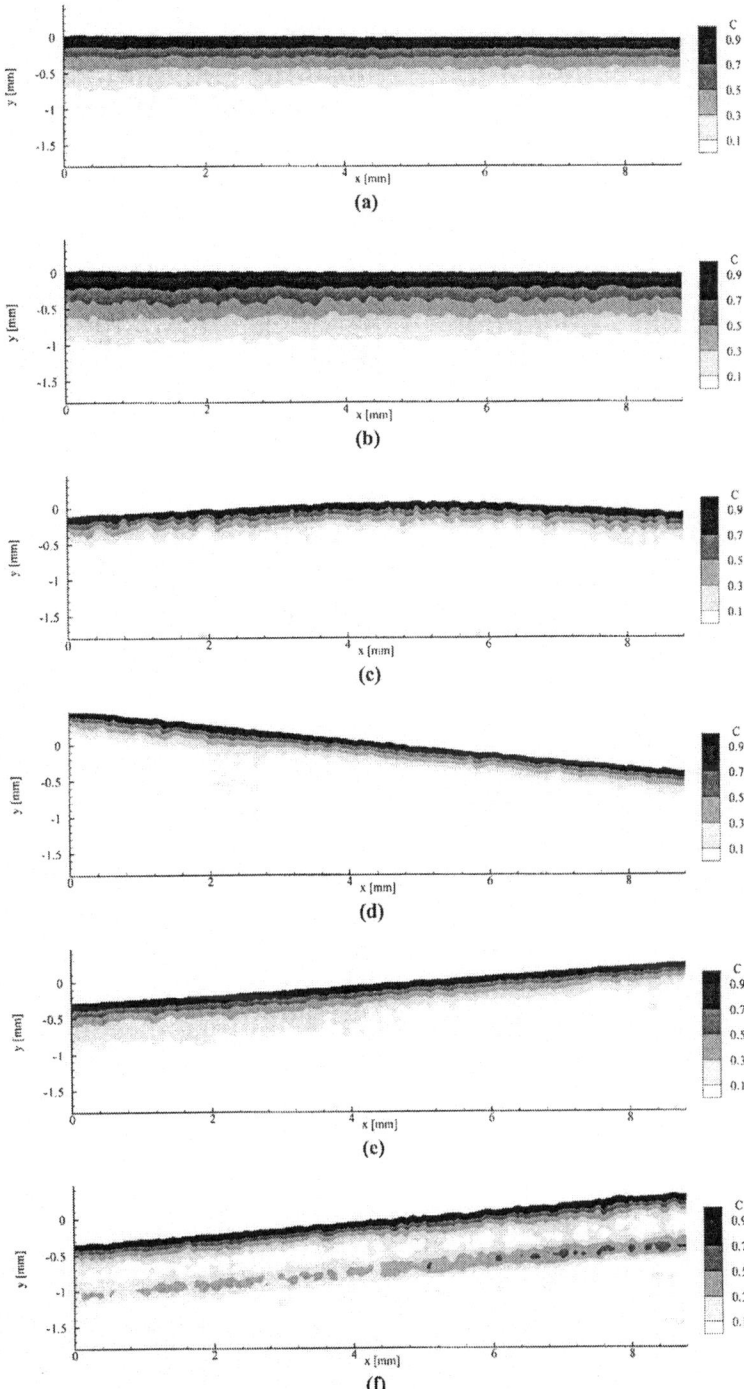

Figure 6. Two dimensional concentration fields: (a), (b) flat, (c) crest, (d) upslope, (e) downslope, and (f) detached layer.

A number of images exhibited detached concentration layers, a phenomena that has been observed for sheared interfaces [*Duke*, 1996; and *Münsterer*, 1996]. This is depicted in figure 6(f). A detached layer is a secondary region of high concentration located beneath the concentration boundary layer.

The mass transfer coefficient ($k_{l,i}$) can be estimated from equation (1) using δ_c obtained from LIF image measure-

Table 2. Results.

Set	Interface	D (cm²/s)	K_{abs} (min⁻¹)	$k_{l,t}$ (cm/s)	δ_c (mm)	$k_{l,i}$ (cm/s)
1	Flat	2.18E-05	0.0083	1.52E-04	0.581	3.76E-04
2	Flat	2.15E-05	0.0102	1.87E-04	0.871	2.47E-04
3	Flat	2.20E-05	0.0104	1.91E-04	0.692	3.19E-04
4	Flat	2.20E-05	0.0118	2.16E-04	0.474	4.65E-04
5	Flat	2.16E-05	0.0112	2.05E-04	1.026	2.11E-04
6	Wavy	2.16E-05	0.0478	8.76E-04	0.366	5.91E-04
7	Wavy	2.19E-05	0.0586	1.074E-03	0.298	7.35E-04

ments. Table 2 lists the results of the seven image sets in table 1.

For approximately the same C_I and C_B, both methods of determining k_l ($k_{l,t}$ from concentration-time measurements and $k_{l,i}$ from the direct measurement of δ_c obtained from the images) revealed that the presence of waves gives a higher mass transfer rate and a larger value of k_l.

For the flat interface sets, $k_{l,i}$ is larger than $k_{l,t}$. For the wavy interface sets, $k_{l,i}$ is smaller than $k_{l,t}$. The detached layers observed for wavy interfaces (figure 6(f)) produce secondary concentration gradients, which contribute to an increase in the mass transfer rate. These subsurface gradients are not accounted for when $k_{l,i}$ is estimated from the measured δ_c.

The LIF technique allows for evaluation of $k_{l,i}$ at each point along an interface, providing information that is not obtainable from traditional k_l measurements. Flat interfaces show small fluctuations in $k_{l,i}$. In comparison, wavy interfaces reveal significant fluctuations in $k_{l,i}$, which are related to streamwise variations in δ_c. Other investigators [*Duke*, 1996; *Wolff* and *Hanratty*, 1994; and *Münsterer* and *Jähne*, 1998] have made this observation for wind waves.

DISCUSSION

LIF measurement of the sharp, near-interface concentration gradients has been enhanced by employing a two dimensional adaptive smoothing filter and a near-interface correction for optical blurring.

Table 2 compares mean values of δ_c, $k_{l,i}$, and $k_{l,t}$ for the flat and wavy data sets. Mass transfer is not totally diffusion controlled. Convection from vibration, surface tension, or temperature driven flows is likely aiding the mass transfer and impacting δ_c. The presence of waves even in the absence of surface shear (wind) produced a several fold increase in oxygen transfer, which has been noted previously [*Downing* and *Truesdale*, 1955; *Daniil* and *Gulliver*, 1991; *Hosoi et al.*, 1977; and *Kakuno et al.*, 1995]. The increase for wavy interfaces is attributed to a thinner δ_c (steeper concentration gradients) and the presence of a detached concentration layers. Similar mechanisms were observed for wind waves [*Duke*, 1996; *Wolff* and *Hanratty*, 1994; and *Münsterer* and *Jähne*, 1998], thus implying that wave motions in the absence of surface shear contribute significantly to high transfer rates in environmental systems.

Acknowledgements. This work was supported by a U.S. Geological Survey Water Resources Research Grant and by the Auburn University Competitive Research Grants.

REFERENCES

Daniil, E., Gulliver, J. S. Influence of Waves on Air-Water Gas Transfer. *J. Envir. Engrg.*, 1991, *117(5)*, 522-539.

Downing, A. L. Truesdale, G. A. Some Factors Affecting the Rate of Solution of Oxygen in Water. *J. Appl. Chem.*, 1955, *5*, 570-581.

Duke, S. R., Hanratty, T.J. Measurement of the Concentration Field Resulting from Oxygen Absorption at a Wavy Air-Water Interface. *Air-Water Gas Transfer* (ed. Jähne, B.; Monahan, E. C.), 627-635, AEON Verlag, Hanau, Germany, 1995.

Duke, S. R. Air-Water Transfer at Wavy Interfaces. Ph.D. Thesis, Univ. of Illinois, 1996.

Hosoi, M., Ishida, A., Imoto, K. Study on Reaeration by Waves. *Coastal Eng. Japan*, 1977, *20*, 121-127.

Kakuno, S., Saitoh, M., Nakata, Y., Oda, K. The Air-Water Oxygen Transfer Coefficients with Waves Determined by Using a Modified Method. *Air-Water Gas Transfer* (ed. Jahne, B.; Monahan, E. C.), 577-587, AEON Verlag, Hanau, Germany, 1995.

Moog, D. B., Jirka, G. H. Analysis of Rearation Equations Using Mean Multiplicative Error. *J. Envir. Engrg.*, 1998, *124(2)*, 104-110.

Münsterer, T. LIF Investigation of the Mechanisms Controlling Air-Water Mass Transfer at a Free Interface, Ph.D. Dissertation, Univ. of Heidelberg, Germany, 1996.

Münsterer, T., Jähne, B. LIF Measurements of Concentration Profiles in the Aqueous Mass Boundary Layer. *Exp. Fluids*, 1998, *25*, 190-196.

Vaughan, W. M., Weber, G. Oxygen Quenching of Pyrenebutyric Acid Fluorescence in Water. A Dynamic of the Microenvironment. *Biochemistry*, 1970, *9*, 464-473.

Wolff, L. M., Hanratty, T. J. Instantaneous Concentration Profiles of Oxygen Accompanying Absorption in Stratified Flow. *Exp. Fluids*, 1994, *16*, 385-392.

Woodrow, P. T., Jr., Duke, S. R. Laser-Induced Fluorescence Studies of Oxygen Transfer Across Unsheared Flat and Wavy Air-Water Interfaces. *Ind. and Eng. Chem. Res.*, 2001, *8(40)*, 1985-1995.

A Direct Visualization Method of CO_2 Gas Transfer at Water Surface Driven by Wind Waves

Kohsei Takehara and Goji T. Etoh

Department of Civil Engineering, Kinki University, Higashi-Osaka, Japan

A visualization technique for CO_2 gas transfer at water surfaces under wind wave condition has been developed by using the laser induced fluorescence (LIF) technique. Water-soluble fluorescent dye was used for the CO_2 indicator. The above technique was applied for the visualization of CO_2 gas transfer at water surfaces under wind wave conditions. CO_2 gas transfer phenomena at water surfaces are successfully visualized

1. INTRODUCTION

In the ocean, one of the primary driving forces of the gas transfer at water surfaces is wind stress, which acts on the sea surfaces and enhances the gas transfer between air and sea. In general, winds over seas generate wind waves and drift currents at the sea surfaces. Under the strong wind condition, the wind waves often break and yield whitecaps. These phenomena considerably enhance the gas transfer through the water surfaces.

In the evaluation of the gas transfer at water surfaces, it is necessary to measure the concentration of the gases and velocity fields in air and water very close to the surface. In most of previous studies, bulk concentrations of gases in water and air velocity far from the water surface have been measured, since it was difficult to measure the information very close to the water surface. The transfer velocities of gases into water were related to the measured bulk concentration and air velocity.

Recently, *Asher and Pankow* [1989] applied the LIF (Laser Induced Fluorescence) technique for measurements of CO_2 concentration fluctuation in water at the surface. They measured the dependence of surfactant on the surface renewal phenomena in detail. *Jähne and Haußecker* [1998] and *Müsterer and Jähne* [1998] also applied LIF technique for the visualization of HCl gas transfer through the water surface under wind wave conditions. *Wolff et al.* [1990, 1994] and *Duke et al.* [1995] applied oxygen quenching of fluorescence technique in order to measure the O_2 concentration fluctuation by wind waves.

In this study, the direct visualization technique of CO_2 gas transfer at water surface has been developed in order to reveal the mechanism of the transfer at water surface.

2. EXPERIMENTS

2.1 Visualization of pH with Fluorescence

It is known that the fluorescent characteristics of many substances vary with pH and temperature. As indicated by *Asher and Pankow* [1989], the pH value of water depends of the concentration of CO_2. Some fluorescent substances can be used as indicators of the concentration of CO_2.

In this experiment, water-soluble fluorescein was used. The fluorescent intensity is maximum at an excitation wavelength of 494 nm and pH of 7.1, and the fluorescent spectrum is maximum at a wavelength of 518 nm. The light sheet from a 488- and 514-nm argon-ion laser was employed as the excitation source to visualize the process of the dissolution of carbon dioxide. Fluorescein was dissolved in tap water to a concentration of 2.5×10^{-6} mol/l.

The fluorescent spectrum of fluorescein is broader on the longer wavelength side than the spectrum peaks of the argon-ion laser. An optical filter that cut wavelengths below

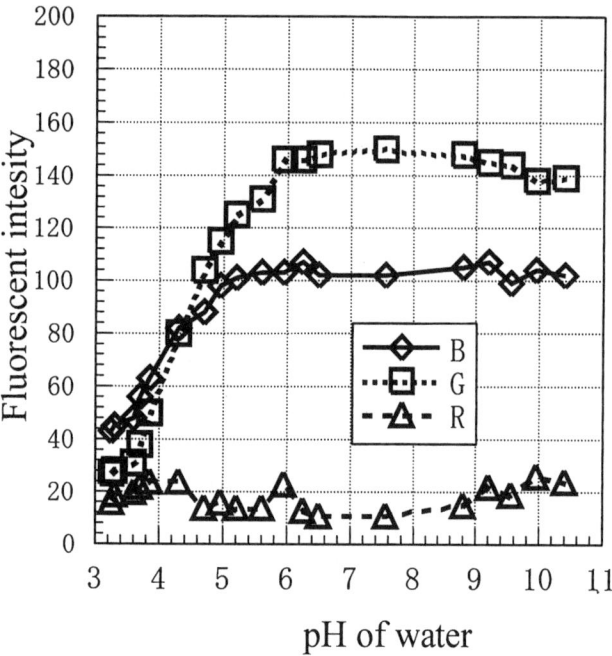

Figure 1. Dependence of fluorescent intensity on pH of water.

520 nm was installed in front of the lens on the video camera to record only images due to fluorescence.

First, the relationship between the pH of the fluorescein solution and the fluorescent intensity was studied (see Figure 1). In this measurement, the optical filter described above and Toshiba ½ inch CCD video camera were used. RGB images recorded on a video tape were processed with an image processor into digital images having 8 bit (256) colors resolution for R, G, and B and 512 × 480 pixels. A vessel, which is 10cm wide, 10 cm long, and 15 cm depth, was used for the test. The vessel is made of transparent plastic plates.

The recording conditions were kept constant except pH; the water temperature was 17.7 C. The light intensity was averaged within a fixed area of 50 × 50 pixels for all the images recorded. In addition, three images were recorded at the same pH and the average values were employed. The pH was adjusted with 0.1 N HCl and 0.1 N NaOH.

It was not possible to check accurately which wavelengths corresponded with the R, G, and B output of the video camera because detailed information on the color filters and the color response of the CCD video camera was not available. However, the following could be confirmed using Figure 1.

1. Fluorescent intensity for blue: Weak. Little change within the tested pH range. The blue light is cut by optical filter, so all the light with wavelengths below 520 nm is eliminated.
2. Fluorescent intensity for green: With a pH of 6 or more, the fluorescent intensity for green is the highest of the three colors and does not depend on pH. With a pH of less than 6, however, the fluorescent intensity decreases rapidly with the drop in pH. This indicates that the fluorescent intensity for green depends most strongly on pH when it is less than 6.
3. Fluorescent intensity for red: With a pH of 5 or more, the fluorescent intensity for red is the second highest next to green and barely changes with the increase in pH. With a pH of less than 5, the fluorescent intensity reduces in the same manner as that for green, though the rate of decrease is lower.

2.2 Wind-Wave Tank

The wind-wave tank is 50 cm height, 50 cm width, and 16 m long. Wind blows into the wind-wave tank at 8.4 m from the test section of the tank and blow out into an air circulation pipe at the end of the tank. The both sidewalls and the top cover of the tank are made of glass plates in order to visualize the gas transfer phenomena at water surface. In the test section of the wind-wave tank, the bottom wall is also made of transparent plate.

The air inside the wind-wave tank is separated from outside in order to control the concentration of CO_2 gas. Therefore, the wind-wave tank is an air-tight structure. The air inside the wind-wave tank circulates through the circulation pipe. The fan driving wind is placed in the middle of the circulation pipe.

The concentration of CO_2 gas in the wind wave tank is increased up to about 20 % by adding the CO_2 gas from a gasometer. The gasometer can supply pure CO_2 gas from the liquid CO_2. The pure CO_2 gas was supplied into the wind wave tank at the end of the tank.

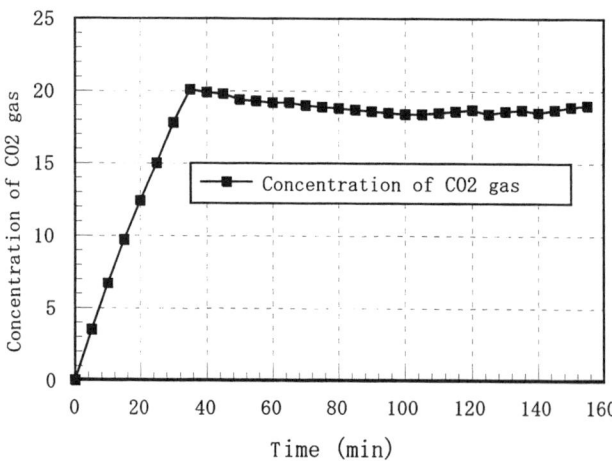

Figure 2. Concentration change of CO_2 gas in the experiment.

The CO_2 gas, however, dissolves into water and leaks from the wind-wave tank, so that the concentration of CO_2 gas reduces gradually. The concentration of CO_2 gas inside the wind-wave tank is maintained by adding the loss of CO_2 gas from the gasometer, of which value is previously measured. The concentration of CO_2 gas in airside was measured by a non-dispersive infra-red CO_2 analyzer.

The performance of the control method of the CO_2 concentration is shown in Figure 2. During the first 40 minutes, CO_2 gas is added to the wind-wave tank at a rate of 20 l/min. After the first 40 minutes, CO_2 gas is supplied at a rate of 6 l/min. The average concentration of CO_2 gas in the wind-wave tank is 18.9 % and the standard deviation of the fluctuation of the concentration is less than 1.0 %.

2.3 Experimental Method

The above mentioned method was applied for visualization of CO_2 gas transfer phenomena at water surfaces under wind wave conditions. An argon-ion laser light sheet is used as the excitation light of the fluorescein.

The still water depth and average wind speed were set to 32 cm and about 2.62 m/sec, respectively. The concentration of the fluorescein in water is 2.5×10^{-6} mol /l. The images were taken at a point 8.4 m from the outlet of the wind blower. The visualized images were captured by ½ inch CCD color video camera (Toshiba IK-C 40 MF).

After the wind waves were fully developed, CO_2 gas is supplied from a gasometer. In the first 30 minutes, the flow rate of CO_2 gas was 20 l/min. After that, the flow rate of CO_2 gas reduced to 6 l/min.

The CO_2 gas transfer phenomena are visualized on the following three cross sections in order to understand the three dimensional structure.
1. Vertical plane parallel to the wind direction (Case 1, Figure 3 (a))
2. Horizontal cross section close to the water surface (Case 2, Figure 3 (b))
3. Vertical plane perpendicular to the wind direction (Case 3, Figure 3 (c))

3. RESULTS

3.1 Vertical Plane Parallel to the Wind Direction (Case-1)

When the concentration of CO_2 gas reached to 7 % in the airside of the wind-wave tank, a dark film at the water surface intermittently appeared. It was caused by dissolution of CO_2 gas into water. If the concentration of CO_2 gas is higher than 15 %, the gas transfer phenomena are clearly visualized. Examples of the visualized images are shown in Plate 1(a) and (b) taken in Case 1. As the images were taken slightly below the water surface, fluorescent light reflected at water surface. Water surfaces on images are displayed as dark lines indicated by arrows.

The results from the visualization of Case 1 are as follows:
1. The dark film at the water surface is visualized by dissolution of CO_2 gas into the water.
2. Thicker and thinner parts appeared in the dark line.
3. The thicker part is entrained into bulk water intermittently. The tail of the entrained darker part is stretched from the water surface to the entrained front.
4. Sometimes, the thicker part of the dark film is entrained by small a eddy motion into bulk water.

3.2 Horizontal Plane Close to the Water Surface (Case 2)

The representatives of visualized images just below the water surface are shown in Plate 2 (a) to (c). The laser light sheet illuminated 3 mm below the still water surface. On these images, wind blows from right to left. At the bottom of the images, the sidewall of the wind-wave tank is appeared as dark lines indicated by arrows. The top of the images is located at about 20 cm from the sidewall.
1. A few narrow dark lines parallel to the wind direction appear and move downward with slightly meandering (Plate 2(a)).
2. The narrow dark lines spreads and often breaks up (Plate 2(b)).
3. Sometimes, the dark lines deforms into a large vortex motion (Plate 2(c)).

3.3 Vertical Plane Perpendicular to the Wind Direction (Case-3)

The vertical plane perpendicular to the wind direction is visualized. The successive images are shown in Plate 3 (a) to (d). At the left-hand side of the images, the side wall of the wind-wave tank is indicated by arrows. The water surfaces are displayed as a dark line in the middle of the images. Reflections of fluorescent light at water surface appear on the images.

The results from the visualization of Case 3 are as follows:
1. Thicker parts of the dark film, indicated by the arrow A and B in Plate 3(a), appear in the thin dark film at water surface.
2. The thicker dark parts are entrained into bulk water by an eddy motion (arrow A), and the tail of the entrained part remained at water surface (arrow B), as shown in Plate 3(b).
3. The entrained part moves downward with time (arrow

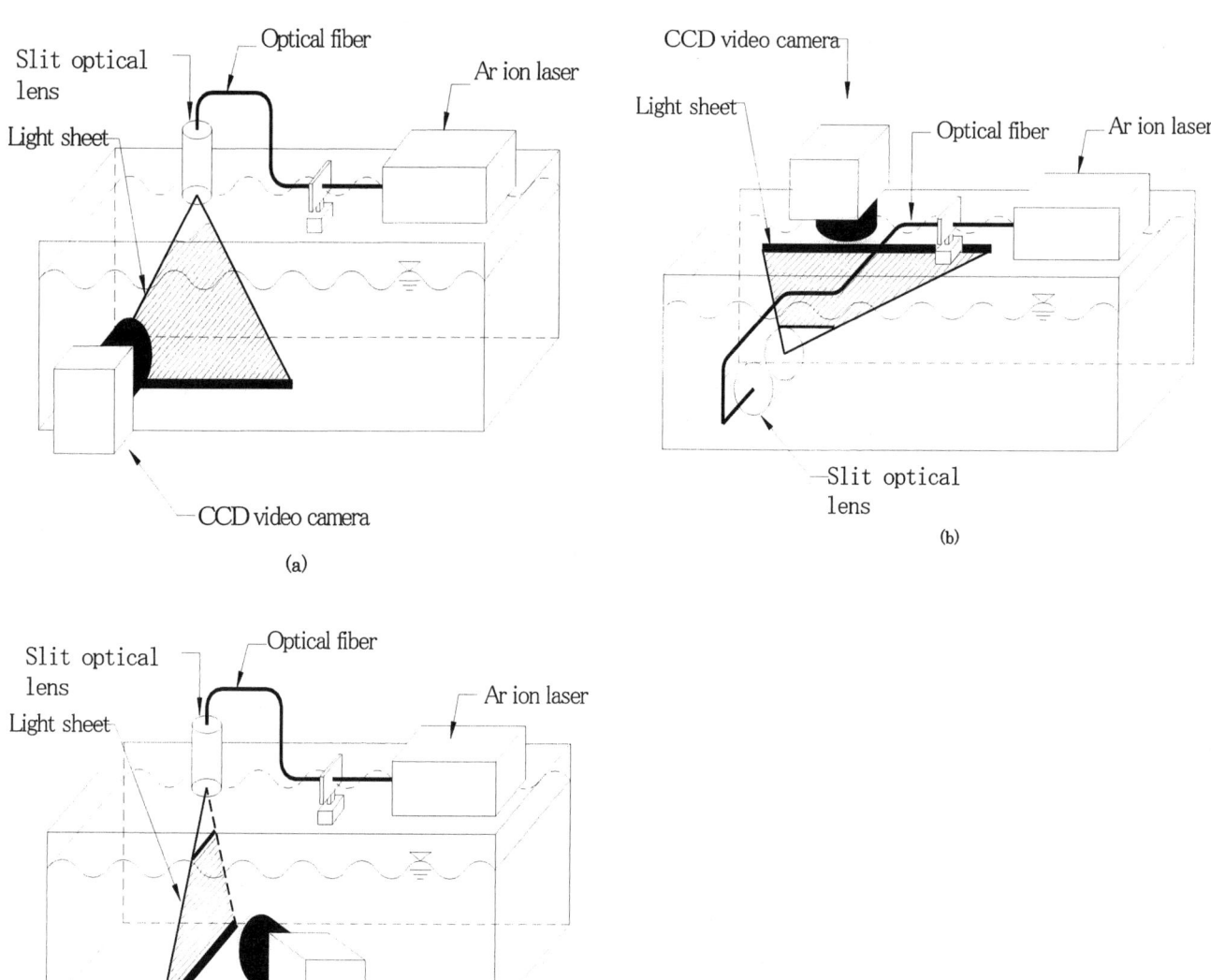

Figure 3. Visualized plane by the light sheet of argon-ion laser. (a) Vertical plane parallel to the wind direction (Case-1); (b) Horizontal plane just below the water surface (Case-2); 3 mm from the still water surface; (c) Vertical plane perpendicular to the wind direction (Case-3).

A) and the tail part still contacts the water surface (arrow B) as shown in Plate 3(c).

4. In Plate 3(d), the entrained part still moves downwards and the tail part is separated from the water surface.

4. CONCLUSION

A direct visualization method of CO_2 gas transfer at water surface under wind wave conditions is developed by using the LIF (Laser Induced Fluorescence) technique. The fluorescent intensity of water-soluble fluorescent dye, Fluorescein, depends on the pH value of water. The fluorescent intensity of high concentrated parts of CO_2 gas in water is low. The CO_2 gas transfer phenomena at water surface was visualized under high concentration CO_2 air blowing over the water surface by using the technique.

The results from the experiments indicate that the mechanism of the gas transfer at water surface under wind

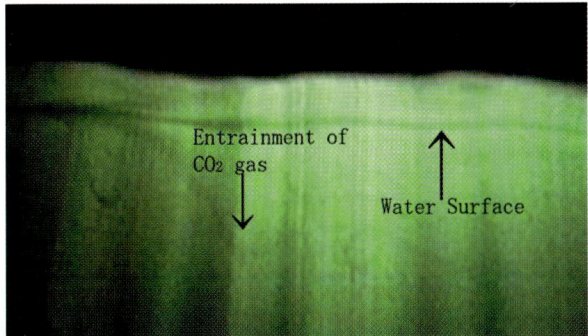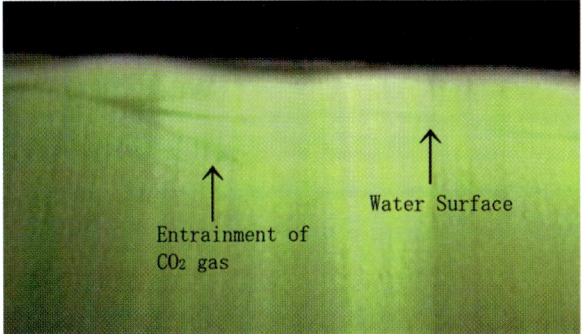

Plate 1 CO_2 gas transfer at water surface on the vertical plane parallel to the wind direction (Case-1); The wind blows from right to left. (a) Entrainment of CO_2 gas by large motion; (b) Entrainment of CO_2 gas by small eddy

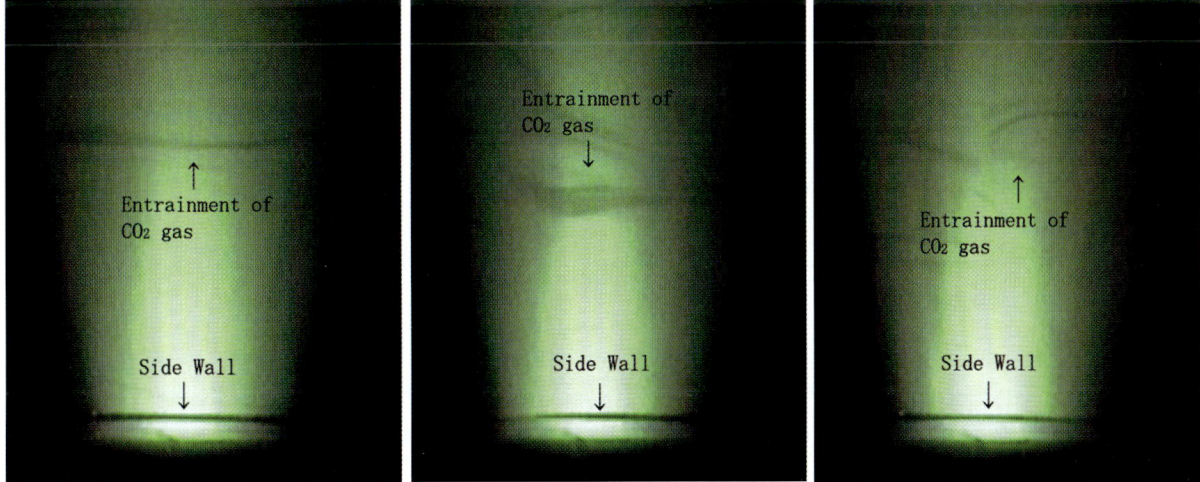

Plate 2 CO_2 gas transfer at water surface on the horizontal plane just below the water surface (Case-2); The wind blows from right to left.(a) A streak of entrainment; (b) Breakdown of the streak; (c) Vortex pattern entrainment

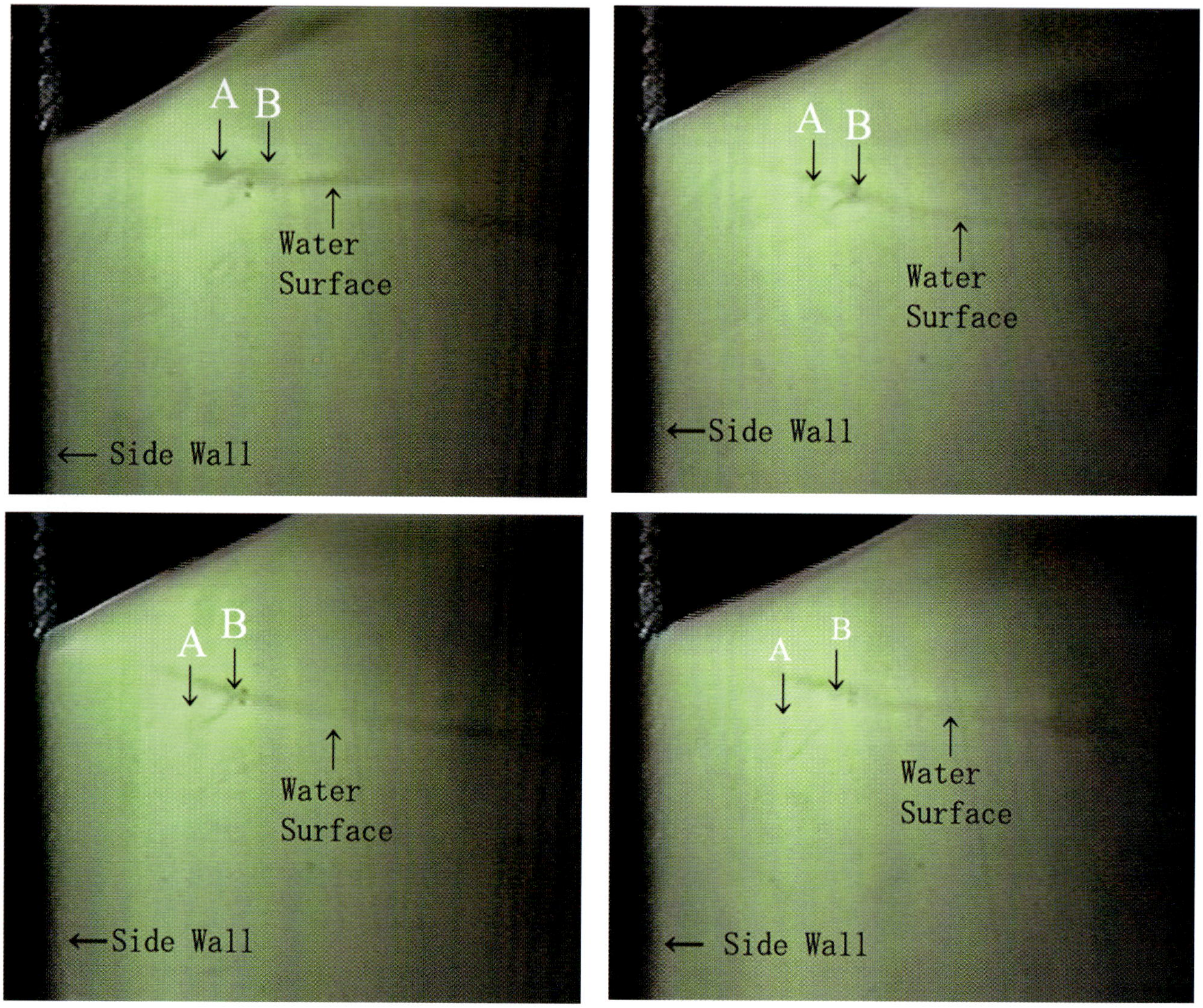

Plate 3 CO_2 gas transfer at water surface on the vertical plane perpendicular to the wind direction (Case-3). (a) t = 0 sec; (b) t = 0.100 sec; (c) t = 0.233 sec; (d) t = 0.500 sec.

wave condition is dominated by three-dimensional motions of water very close to the water surface. The longitudinal fluid motion close to the water surface exists and is strongly related to the gas transfer at water surface.

REFERENCES

Asher, W.E. and Pankow, J.F.: Direct observation of concentration fluctuations close to a gas-liquid interface, *Chem. Eng. Sci.*, 44, pp. 1451 – 1455, 1989.

Jähne, B. and Haußecker, H.: Air-water gas exchange, *Annu. Rev. Fluid Mech.*, pp. 443-468, 1998.

Münsterer, T. and Jähne, B.: LIF measurement of concetnration profile in the aqueous mass boundary layer, *Exp. Fluids*, 25, pp.190-196, 1998.

Wolff, L.M., Z-C Liu and Hanratty, T.J.: A fluorescence technique to measure concentration gradients near an interface, *Air Water Mass Transfer,* Wilhelms, S.C. and Gulliver J.S. (eds.), ASCE, pp.210 – 218, 1990.

Wolff, L.M. and Hanratty, T.J.: Instantaneous concentration profiles of oxygen accompanying absorption in stratified flow, *Exp. Fluids*, 16, pp. 385-392, 1994.

Duke, S.R. and Hanratty, T.J.: Measurement of the concentration field resulting from oxygen absorption at a wavy air-water interface, *Air-Water Gas Transfer,* Jähne, B. and Monahan, E.(eds.), AEON Verlag, pp.627-635, 1995

3-4-1 Kowakae, Higashi-Osaka, 577-8502 Japan

Measurements of Wind-Driven Current Near the Water Surface by the Use of the PTV Method

Hajime Kato and Hisamichi Nobuoka

Department of Urban and Civil Eng., Faculty of Eng., Ibaraki University, Hitachi, Japan

Naoki Ooshima

Toyo Construction Co., LTD., Tokyo, Japan

Using a wind-wave channel, the measurements of velocity field near the water surface as well as the surface velocity under the wind-waves were made by means of particle tracking velocimetry (PTV). We removed the orbital velocity by using the stream function method (Thais and Magnaudiet, 1995), though we neglected the rotational part. Vertical profiles of the net wind-driven velocities were obtained in a substantial layer below the wind-waves as a function of the wave phase angles with taking the vertical coordinate z from the temporal free surface. The surface wind drift speeds were also obtained as a function of the phase angles. The results show the vertical profiles of the drift currents vary as a function of the wave phases. The mean profile near the water surface was found to be expressed by a logarithmic distribution with z_0 =0.10cm. In this work the resulting scatter of measured points both at various phases and depths might correspond to turbulence that dominates the gas transfer rate at the air-water interface. The relative magnitude of turbulence in the aqueous layer near the water surface at various wave phases can be seen qualitatively.

1. INTRODUCTION

A wind-driven current is of importance in many aspects for experimental studies of wind-waves. Since it has large vertical gradient just below the water surface, it generates intense turbulence, which dominates the gas transfer rate at the air-water interface. On the other hand the vertical profile of the drift current is required for estimating the wave dispersion relation correctly. The profiles have long been considered to be logarithmic near the water surface. Recently, however, some questions have been raised of whether the profiles are equally logarithmic when the amplitude of wind-waves are not small (e.g. Cheung and Street, 1988). Banner and Peirson (1998) have measured the vertical gradients of current speed within the surface laminar layer of order 1mm using the PIV technique and showed that the tangential stress varies at the wave phase. They also discussed the fraction of the tangential stress in the total wind stress. However, the velocity field under developing wind-waves consists of the drift current, orbital velocities and turbulence. To our knowledge, there have been no measurements of the net drift current including turbulence above the trough level of wind-waves. In this paper we describe the results of laboratory measurements of the wind drift current just below as well as at the water surface by applying the PTV technique.

Thais and Magnaudiet (1995) developed the stream

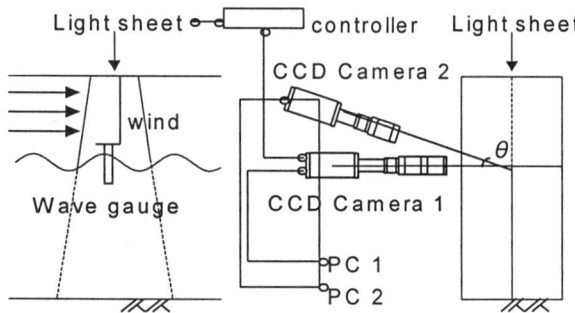

Figure 1. The general arrangement of the PTV system with two CCD cameras and a wave gauge.

function method (abbreviated as SFM hereafter) to wind waves that was originally proposed by Dean (1965) for non-linear regular waves. Kato et al. (1999) used the method of Thais and Magnaudiet (1995) to obtain the turbulence below the wind waves under stratified wind conditions, where the velocity measurements were made with a hot-film towed with a constant speed. In this study we use the SFM to remove the orbital motion from velocity measurements, following which the scatter of the resulting velocities would indicate the degree of turbulence.

2. EXPERIMENTAL APPARATUS AND PROCEDURE

The wind-wave tunnel used is 30cm wide and 80cm high (45 cm of water depth) and the effective fetch is 9m. We used a wave gauge and two CCD cameras connected to personal computers (Fig.1). One camera was used to record the particle images from a close-up viewing angle as shown in Fig.5 (b) and the other to record temporal wave forms as in Fig.5 (a) from which the wave phases of the measurement points were evaluated using wave gauge records. A vertical light sheet was produced along the centerline of the wave tank with two strobe lights synchronized with the CCD cameras. From three consecutive frames of tracer particle images captured every 1/60 second by the CCD camera 1, the spatial vector map was calculated using the "Triple Pattern Matching Algorithm" (Nishino & Torii, 1993). The tracers used were high porous polymer particles with diameters from 37 to 75 μm and a relative density of 1.01. The measurements were made at a fetch of $F=5.5$m with a center line wind speed of $U_a \cong 5.3$ m/s.

3. DECOMPOSITION OF ORBITAL VELOCITY FROM THE TOTAL VELOCITY MEASURED

The velocities $U(t)$ under the wind-waves are written as:

$$U(t) = \bar{u} + u(t) + u'(t) \qquad (1)$$

where the vertical z-dependence is omitted implicitly, $\bar{u}$ is the drift current, $u(t)$ the orbital velocity and $u'(t)$ turbulent velocity. Thais and Magnaudiet (1995) indicated the existence of orbital rotational motions below waves when wind is present and expressed the orbital velocity as

$$u(t) = u_p(t) + u_R(t) \qquad (2)$$

where u_p is the potential flow orbital velocity and u_R the rotational one. They calculated the potential components u_p by the SFM and obtained the turbulent components $u'(t)$ in the frequency domain by using a linear filtration technique (Benilov et al., 1974) for the decomposition of the rotational orbital velocities u_R.

The stream function of the mean and orbital potential motions $\overline{\Psi} + \Psi_p$ and water surface displacement η_p are written after Thais and Magnaudiet (1995) as follows:

$$\overline{\Psi} + \Psi_P = (c - U_0)z - U_0'\frac{1}{2}z^2 + C$$
$$+ \sum_{n=2,4,6,...}^{N-1} \exp\frac{n\pi}{L}z \cdot \left[X(n)\cos\frac{n\pi}{L}x + X(n+1)\sin\frac{n\pi}{L}x \right] \qquad (3)$$

$$\eta_{pi} = \frac{1}{c - U_0}X(1) + \frac{1}{c - U_0}\left[\frac{1}{2}U_0'\eta_{pi}^2 - C\right] - \frac{1}{c - U_0}$$
$$\sum_{n=2,4,6,...}^{N-1} \exp\frac{n\pi}{L} \cdot \eta_{pi}\left[X(n)\cos\frac{n\pi}{L}x + X(n+1)\sin\frac{n\pi}{L}x \right] \qquad (4)$$

where L is the wavelength of the carrier wave, c the dominant wave speed, U_0 and U_0' are the surface drift speed and its vertical gradient there, and $X(n)$ are coefficients to be

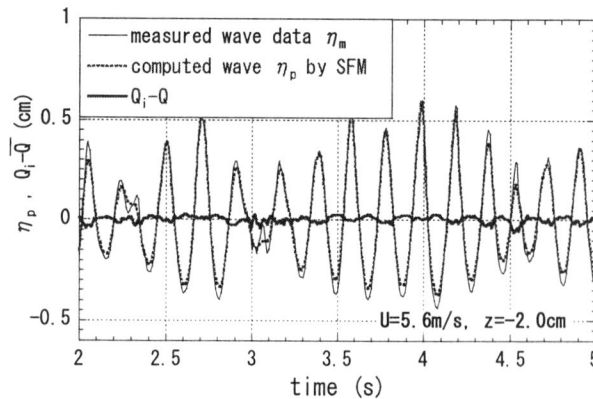

Figure 2. An example of the convergence of the stream function method used, from Kato et al. (1999).

calculated by the least square mean method. N is the number of Fourier mode assumed, and $N=251$ was used.

Thais & Magnaudiet assumed a drift current with a linear vertical profile because the vorticity of the mean flow $\bar{\omega}$ must be zero so that the stream function Ψ_p exists and $\bar{\Psi}$ is involved only in the surface boundary conditions.

Kato et al. (1999) used the SFM of Thais & Magnaudiet to obtain the turbulence of the velocity field under wind waves. Some results of calculation by the SFM are shown in Fig.2 to Fi.g.4. Figure 2 shows an example of the convergence for Bernoulli constants and surface displacements. Measured total velocities and computed potential orbital velocities with linear drift current are compared in Fig.3.

Figure 4 shows the spectra of measured velocity u and calculated components u_p, u_R and u' and it is seen that the rotational orbital component u_R is much smaller than the potential component u_p in the dominant wave frequency range. Considering this fact we neglected the rotational components in the following procedure. Then the orbital potential velocity is given by

$$u_P = -\frac{\partial \Psi_p}{\partial z} = -c - \sum_{n=2,4,6,\ldots}^{N-1} \frac{n\pi}{L} \exp\frac{n\pi}{L} z$$
$$\cdot \left[X(n)\cos\frac{n\pi}{L}x + X(n+1)\sin\frac{n\pi}{L}x \right] \quad (5)$$

4. COMPARISON OF THE RESULTS MEASURED BY PTV WITH THOSE CALCULATED BY THE STREAM FUNCTION METHOD

In order to check the movability of the small tracer particles added in water, the calibration of the PTV method used

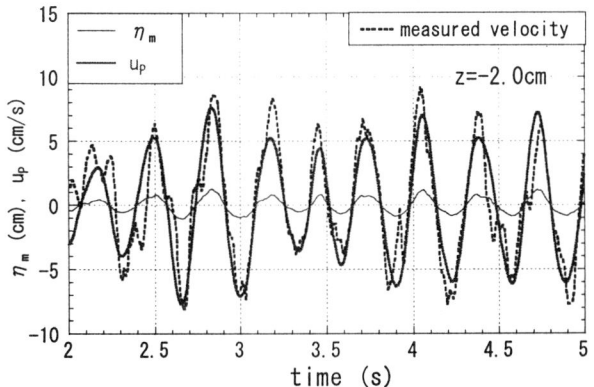

Figure 3. Measured velocity u_m is compared with the total velocity (potential orbital velocity u_p and linear drift current) computed by the stream function method for laboratory wind waves. Velocity measurements were made with a hot-film (Kato et al., 1999).

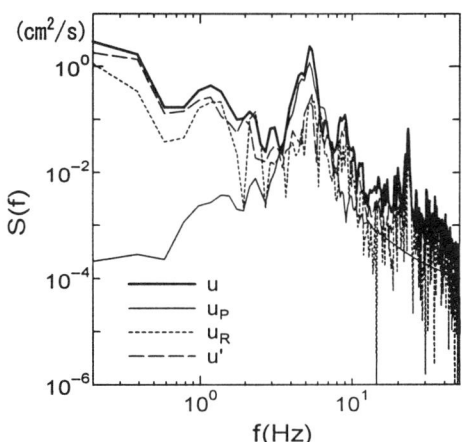

Figure 4. Examples of spectral decomposition of potential and rotational power of drift currents (Kato et al., 1999).

in this study was made by using the regular waves generated by the flap-type wave maker attached to the wind-wave channel. An example of the captured images by two CCD cameras is shown in Fig.5.

Since in this case there is no wind, the wave-induced motions should be only potential orbital velocities. The PTV results shown in Fig.6 are in good agreement with the calculated velocities by the SFM and this certifies the validity of the present PTV procedures.

5. THE MEASUREMENTS OF DRIFT CURRENT

5.1. Vertical Profiles of Drift Current Below the Water Surface

Spatial and temporal measurements of drift current profiles were made with the same method as stated in the preceding sections and the velocities $u_T(z,t)$ were obtained as functions of wave phases. In order to synchronize the two kind of images captured by the two CCD camera and the record of the water surface, the two cameras were started simultaneously first; then the start buttons of both strobes and wave recorder were turned on. The resulting images we analyzed are similar to but more complicated than those shown in figure 5.

The total velocities below the surface $u_T(z,t)$ were obtained by PTV as described in section 2. The potential parts of the orbital velocities $u_p(z,t)$ were evaluated from the wave gauge data using the SFM. Neglecting the rotational part of them as stated earlier the net wind-driven velocities $u_T - u_p$ were calculated as a function of the wave phase positions (-180 to +180 degrees) taking the vertical coordinate z both from the still water level and from the temporally changing free surface.

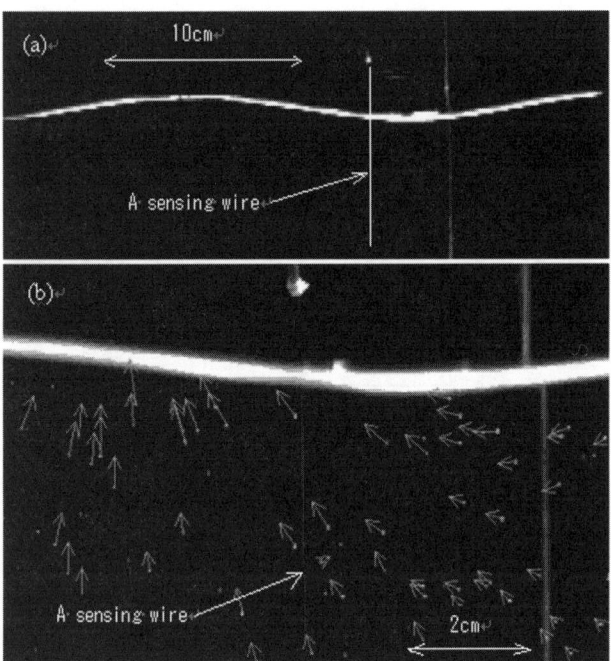

Figure 5. Example of the simultaneously captured images of regular waves: (a) A wide images by CCD camera 2; (b) a close-up images for PTV calculation by CCD camera 1.

The results of velocity measurements at 4 different wave phases are shown in Fig.7 in which the individual velocities u measured by PTV and the corresponding net drift speeds $(u - u_p)$ are plotted. The points of $(u - u_p)$ show considerable scatter and this must be due to turbulence as well as to measurement errors. We can assume that the error contributions exist uniformly at every phase point. Therefore the scatter of the points may be considered to represent the relative degree of turbulence near the surface. Then Fig. 7 indicates that turbulence just below the surface is larger at the phase of near the wave crest (no. 2 & 3 in Fig.7) and smaller at the leeward trough (no.4), and that the profiles of drift current vary appreciably with the wave phases.

These data points were gathered in small vertical increments, although the points are rather few very close to the surface, and mean values were evaluated for each wave phase. The vertical profiles for different phases obtained in this way are shown in figure 8. From this figure it is noted that at the phases from the wave crest to the leeward surface, such as no.5 to 8 designated in the sub-figure in Fig. 8, there exist appreciable vertical velocity gradients. On the other hand at the phases from upwind surface to the crest, such as no.2 to 4, there are only small gradients.

It is noted here that the wave phase showing smaller turbulence mentioned above correspond to the phase with larger drift current gradient. Although the reason for this is not clear for the time being, it will be mentioned again in the next section.

5.2. Measurements of surface drift velocity by PTV

For the measurement of surface drift velocity with respect to the wave phases we adopted the camera and light source positions as shown in Fig.9, which is different from the ones shown in Fig.1. CCD camera 1 was used from the air side to capture the tracers on the water surface. CCD camera 2 was set at nearly the same position as in Fig.1 to observe the whole waveforms. A strobe light sheet was projected horizontally near the water surface. The tracers used were very thin disk of Polypropylene about 5mm diameter and 0.1mm thick. Using similar PTV procedures as in Fig.1, the speeds of the tracers were measured with the wave phase data. The results are shown in Fig.10. The mean speeds were calculated for each wave phase and they have the maximum values near at the wave crest (note the waveform is slightly skewed to the forward direction). The measured (total) mean speeds calculated for each wave phase change appreciably with the phase from about 4 cm/s to 29 cm/s. However, the net wind-driven speeds $(u_0 - u_p)$ change only slightly with the phase.

By comparing the results of Fig.10 with those of Fig.8, one conjecture emerges, viz: the wind-driven current has a very large vertical gradient just below the surface at the phases near the wave crest. This coincides with the variation of the turbulence with wave phases inferred from the scatter of $(u - u_p)$ points in Fig.7. Indeed Okuda *et al.* (1974) and Banner and Peirson (1998) showed the skin friction stress has the maximum values at the wind wave crest.

6. THE MEAN VERTICAL PROFILE OF THE DRIFT CURRENT

Many investigators have studied the vertical profiles of the drift current. However when the wave heights were large, the real profile was difficult to obtain. From the data of drift current measured when the wind waves were small or suppressed with detergent Kato (1975) proposed the logarithmic distribution (7):

$$\bar{u}(z) = u_0 - U_1 \ln[(z_0 - z)/z] - bz \qquad (7)$$

where $\bar{u}(z)$ is the drift current, u_0 the surface drift speed and U_1, z_0, and b are constants. In the case of no wind waves Kato (1975) showed $z_0 = 0.01$cm, and he calculated the wave speed c for the drift current expressed by eq. (7). Since on that occasion the speed c of young wind waves naturally depend on the constant z_0, the correct value of z_0 is very important for the dispersion relation for laboratory wind waves. Figure 11 shows the mean profile of drift

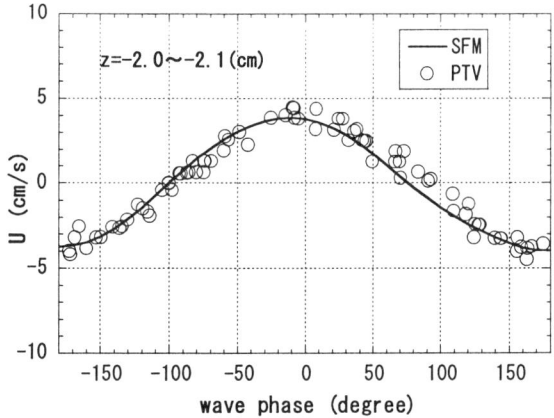

Figure 6. Comparison of particle velocities measured by PTV and calculated by the SFM for regular waves (L=17.5cm, H=1.0cm, and T=0.335s). The depths are from the instantaneous water surface.

current obtained by averaging all phase profiles and the best fit curve of eq.(7) yields z_0=0.10 cm. Just for reference the mean drift current in the spatially fixed coordinate, that is the mean values for the period when the points are in water, is also shown with a dotted line. The re-

sulting value (z_0=0.01cm) is equivalent to the results measured with a current meter when it is in water.

7. CONCLUSIONS

We measured the vertical profiles of wind-driven current near the surface and above the trough level by using Particle Tracking Velocimetry (PTV) as a function of the wave phases. The main results obtained are as follows:

(1) We removed the orbital velocities by using the stream function method. The net drift current profile obtained was found to vary greatly with the wave phases. On the other hand the surface speeds of the drift current do not change very much with the phase.

(2) The total mean profiles averaged over whole phases, which were expressed with the depth from the changing water surface, agreed with the logarithmic distribution (7), which had been proposed by Kato (1974) and the important parameter z_0 in eq.(7) was found to be 0.1cm.

(3) It was found that there existed considerable gradients of the drift current close to the surface at the phases near the wave trough, while the turbulence in aqueous layer was inferred to be relatively small at those phases. The variation of tur-

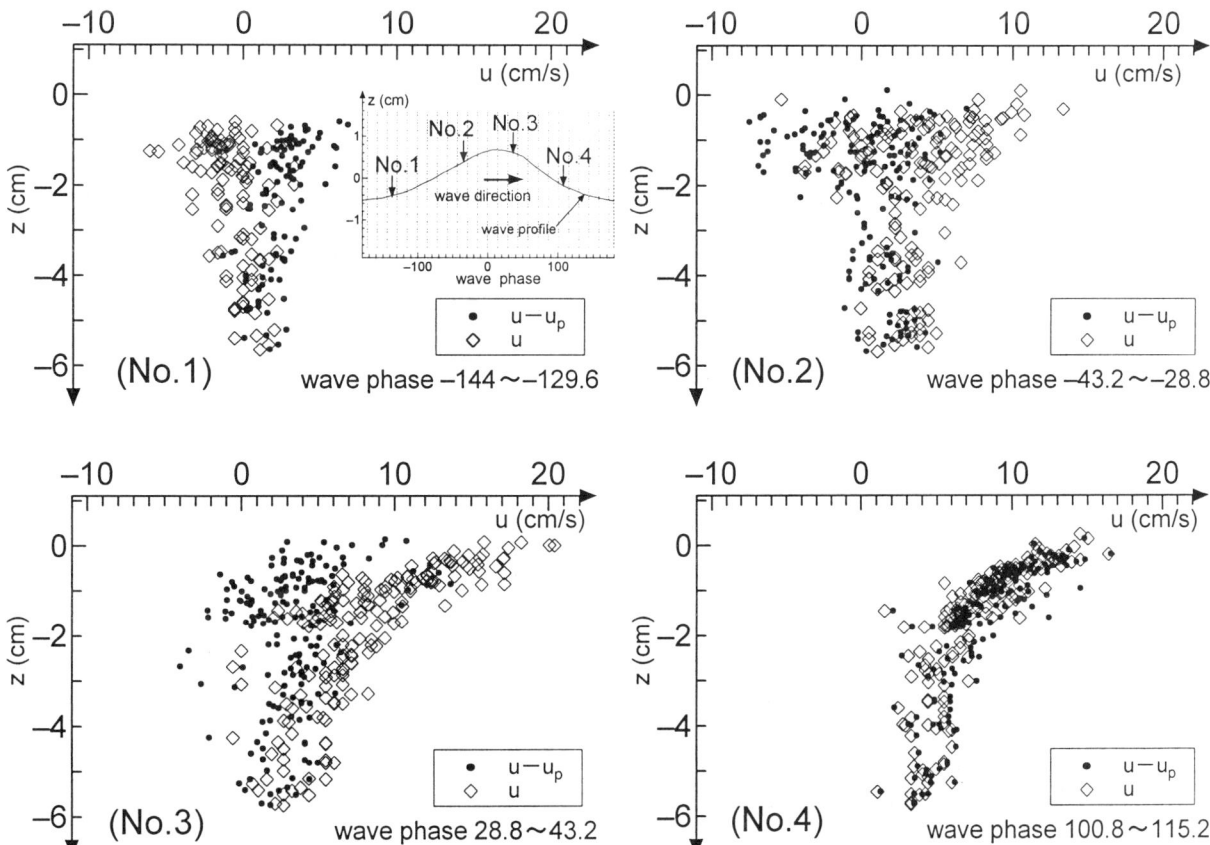

Figure 7. Examples of vertical profiles of drift current at various wave phases. ($U \cong 5.5$ m/s, $H_{1/3}$=1.47 cm, $T_{1/3}$=0.30s)

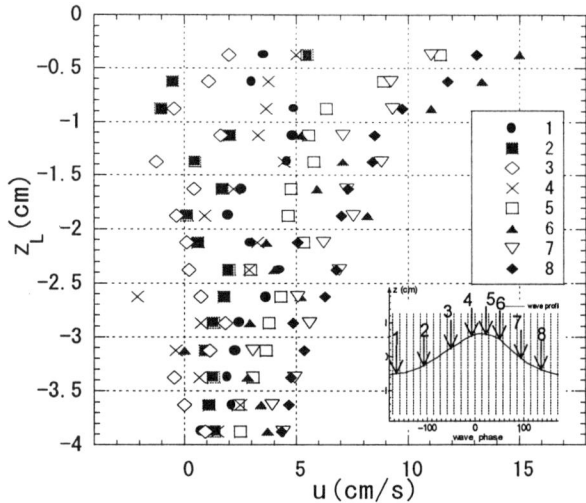

Figure 8. Vertical profiles of the drift current measured at different wave phases.

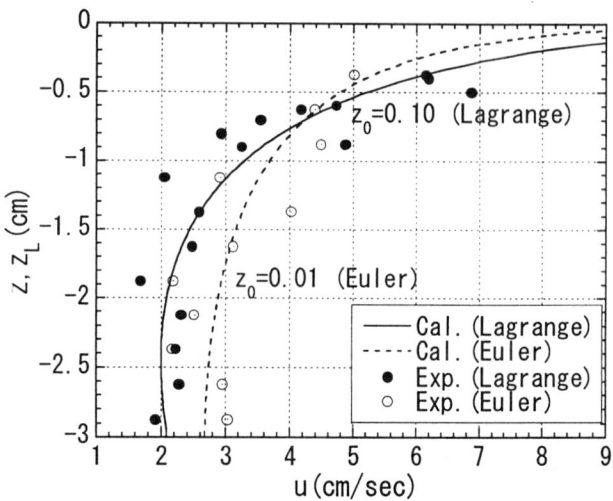

Figure 11. Application to the logarithmic profile of Eq. (7).

Acknowledgments. The authors thank the anonymous reviewers for their valuable comments, which were helpful in revising this paper.

REFERENCES

Banner, M. L. and W. L. Peirson (1998): Tangential stress beneath wind-driven air-water interfaces, *J. Fluid Mech.*, Vol.364, pp.115-145.

Benilov, A.Yu., O.A. Kouznetsov, and G.N.Panin (1974): On the analysis of wind wave-induced disturbances in the atmospheric turbulent surface layer, *Boundary Layer Meteorol.*, Vol.6, pp.269-285.

Cheung, T. K. and R. L. Street (1988): Turbulent layers in the water at an air-water interface, *J. Fluid Mech.*, Vol. 194, pp.133-151.

Dean, R.G. (1965): Stream function representation of nonlinear ocean waves, *J.Geophys.Res.*, Vol.70, no.18, pp.4,561-4,572.

Kato, H. (1974): Calculation of the wave speed for a logarithmic drift current, *Rep. Port & Harbour Res. Inst.*, Vol.13, No.4, pp.3-32.

Kato, H., M. Mori, H. Nobuoka and J. Ooyama (1999): Study of the temperature stratification on turbulence in water under the wind waves, *Proc. Coastal Eng., JSCE* (in Japanese), Vol.46, pp.91-95.

Nishino, K. and K. Torii (1993): A Fluid-Dynamically Optimum Particle Tracking Method for 2-D PTV: Triple Pattern Matching Algorithm, *Transport Phenomena in Thermal Engineering*, Eds. Lee, J. S., S. H. Chung and K. H. Kim, Begell House, Vol. 2, pp. 1411-1416.

Okuda, K., S. Kawai and Y. Toba (1974): Measurements of skin friction distribution along the surface of wind waves, *J. Oceanogr. Soc. Japan*, Vol.30, pp.190-198.

Thais, L. and J. Magnaudet (1995): A triple decomposition of the fluctuating motion below laboratory wind water waves, *J. Geophys. Res.*, Vol. 100, no.C1, pp.741-755.

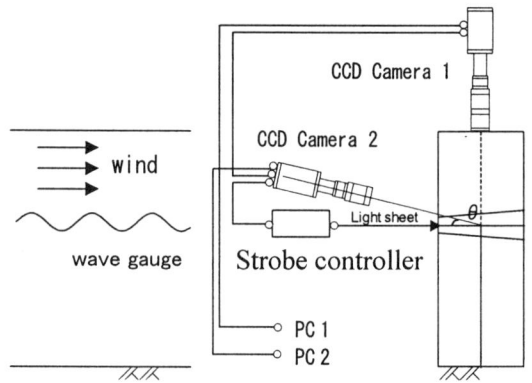

Figure 9. Setup for measurement of surface velocity u_0.

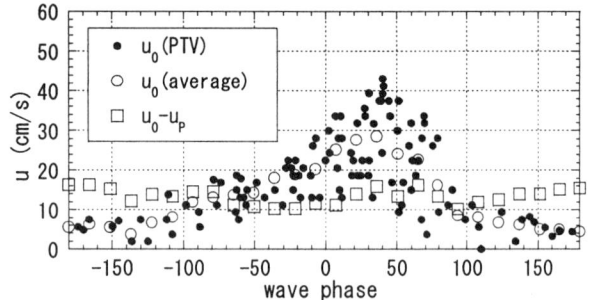

Figure 10. The results of the surface drift speed u_0 (Polypropylene tracer)

bulence with phase is similar to the variation of the tangential stress measured by Banner and Peirson (1998).

(4) With respect to the air-water gas transfer problem, we showed qualitatively that the turbulence just below the surface varied with wave phase. The procedures showed in this paper could be used for quantitative estimation of turbulence very close to the surface with much more measurements.

H. Kato and H. Nobuoka, Department of Urban & Civil Eng., Faculty of Eng., Ibaraki University, 4-12-1 Nakanarusawa, Hitachi, 316-8511, Japan.

N. Ooshima, Civil Eng. Div., Toyo Construction Co., LTD., 8-7, 3-Banchou, Tokyo, 102-0075, Japan.

Physics from IR Image Sequences: Quantitative Analysis of Transport Models and Parameters of Air-Sea Gas Transfer

Horst W. Haussecker

Xerox Palo Alto Research Center, Palo Alto, California

Uwe Schimpf, Christoph S. Garbe, and Bernd Jähne

Interdisciplinary Center for Scientific Computing, University of Heidelberg, Germany

A number of infrared techniques have been developed to investigate the physical transport processes across the air-water interface. Exploiting the similarity between the transport of heat, and gas tracers, these techniques allow visualizing the transport processes, and to remotely measuring the transport parameters. Among the relevant parameters are the transfer velocities of heat and gases across the air-sea interface, the water surface velocity field, and the heat flux density across the interface. This paper gives an overview of how a combination of physical modeling and quantitative signal processing can be used to identify transport models, verify model assumptions, and to measure the parameters of air-sea gas transfer.

1. INTRODUCTION

Quantitative measurements of parameters influencing air-sea interaction are extremely difficult to obtain due to the complex dynamic processes within a microscopic layer on top of the wavy ocean surface. One task is to get a fast and reliable estimate of the mean transport velocity of gases across the interface. Additionally, it is of equal importance to identify and parameterize the underlying physical transport processes. Both tasks can be achieved using non-invasive techniques to measure the relevant transport parameters. This paper surveys new techniques that are exploiting the similarity between the transport of heat, and gas tracers, across the air-sea interface. Given the transfer velocity of heat, k_h, the transfer velocity of a gas tracer, k_g, is given by:

$$k_g = k_h \, (\text{Sc}_h/\text{Sc}_g)^{-n}, \qquad (1)$$

with the *Schmidt numbers* Sc_h (heat) and Sc_g (gas), and the *Schmidt number exponent* n, respectively [*Jähne et al.*, 1989; *Jähne and Haussecker*, 1998]. The heat transfer velocity, k_h, relates the heat flux density, j_h, across the air-water interface to a temperature difference, ΔT, that establishes between the water surface and deeper water layers:

$$k_h = \frac{j_h}{\rho \, c_p \, \Delta T} = \sqrt{\frac{D_h}{t_*}}. \qquad (2)$$

D_h denotes the molecular diffusion coefficient for heat in water, t_* is the mean time constant for the transport process, and ρ and c_p are the density and specific heat of water, respectively. Hence, measuring j_h together

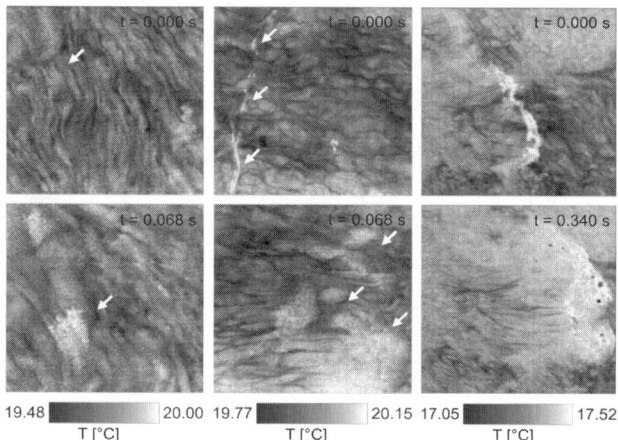

Figure 1. SR events at different scales visible in IR sequences (size 1×1 m) of the ocean surface. The arrows indicate positions of SR events in the moving coordinate frame.

with ΔT, or, alternatively, measuring t_*, allows one to compute k_h from (2), and k_g, using (1).

Infrared techniques that are aiming at measuring these parameters have been developed over the past few years [*Haussecker et al.*, 1995; *Haussecker*, 1996] and found widespread acceptance [*Jessup et al.*, 1996; *Jessup and Hesany*, 1996; *Jessup et al.*, 1997]. Part of this success is due to the accessibility of sensitive IR cameras that allow to remotely measure the dynamic, infrared signature of the sea surface temperature (SST).

The dynamics of water surface heat patterns provides important information about the hydrodynamic processes underneath the water surface. If the transport process is appropriately parameterized, the 3D transport parameters can be inferred from the estimated image parameters [*Haussecker*, 2000].

2. SYSTEM SETUP

The main components of the system setup are a focal plane array infrared (IR) camera, a calibration device, a CO_2 infrared laser, and electronics for remote control of all components. The instrument was designed for both, laboratory, and field experiments. The first realization of the field instrument was successfully used in 1995, and it has been continuously improved in different implementations [*Haussecker et al.*, 1995; *Haussecker*, 1996; *Schimpf*, 2000].

3. TRANSPORT MODEL

3.1. Surface Renewal Model

At the ocean surface the net heat flux, j_h, is composed of latent, j_l, sensible, j_s, and long-wave radiative, j_r, fluxes: $j_h = j_l + j_s + j_r$. This causes a temperature difference across the interface of $\Delta T = j_h/(\rho c_p k_h)$ according to (2) (*cool skin* of the ocean [*Saunders*, 1967]). The temperature difference under natural flux conditions is in the order of some $1/10$ K. Any process that causes mixing of water across this thermal boundary layer causes temperature patterns on the water surface that are visible in images of high-resolution IR cameras (Fig. 1). An appropriate model of the transport processes needs to accommodate both, molecular diffusion, as well as turbulent transport. Among many possible models, a simple surface renewal model with depth-independent renewal rate [*Danckwerts*, 1970] turned out to predict both, the measured statistical surface temperature distribution, as well as its dynamic changes. The choice of this model was motivated by the fact, that surface renewal events are visible in IR sequences at all wind speeds and at different scales (Fig. 1).

The surface renewal theory considers fluid elements to be cyclically exposed to the water surface where molecular diffusion governs the exchange (Fig. 2). The assumption of a constant surface heat flux will lead to a temporal change of the temperature difference across the molecular boundary layer according to [*Liu and Businger*, 1975; *Soloview and Schlüssel*, 1994]:

$$T_s(t) = T_b + \alpha j_h \sqrt{t'}, \quad \text{with} \quad \alpha = 2(\rho c_p \sqrt{\pi D_h})^{-1}, \quad (3)$$

where $t' = t - t_0$ is the time elapsed since the previous surface renewal event. The heat flux j_h is measured positive downwards.

3.2. Statistical Analysis of the SST

In order to find a generative model of the statistical SST distribution, we separate the transport process into two parts, namely, molecular diffusion, and statistical renewal. Within the time between two renewal events the SST is changing according to (3). Given a fixed

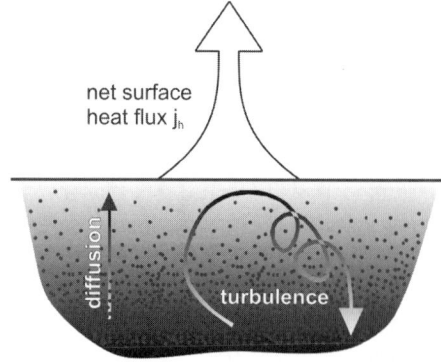

Figure 2. Surface renewal transport model.

renewal time interval, τ, and the bulk temperature, T_b, the normalized likelihood to measure the surface temperature T_s is given by [*Haussecker*, 1996]

$$p(T_s|\tau) = 2/\left(\tau\left(\alpha j_h\right)^2\right)(T_s - T_b). \tag{4}$$

The temporally averaged temperature distribution can be derived from (4) by weighing $p(T_s|\tau)$ by the likelihood of observing τ, and integrating out the nuisance parameter τ, which is not directly observable:

$$h(T_s) = \int_{\left(\frac{T_s-T_b}{\alpha j_h}\right)^2}^{\infty} p(T_s|\tau)\, p(\tau)\, d\tau, \tag{5}$$

with the - a priori unknown - probability distribution $p(\tau)$ for the time in between two consecutive surface renewal events. The lower integration limit is given by the minimum renewal time required to reach the temperature T_s. Early applications of the surface renewal theory [*Brutsaert*, 1965] assumed an exponential distribution for $p(\tau)$. This, however, predicts a maximum for instantaneous renewals with very few longer periods, which hardly matches reality. A more realistic probability density function is the log-normal distribution

$$p(\tau) = \pi^{-0.5}(s\tau)^{-1} \exp\left(-\frac{(\ln \tau - m)^2}{s^2}\right), \tag{6}$$

which was found for statistical fluctuations in a turbulent air-flow [*Rao et al.*, 1971]. The parameters s and m of the log-normal distribution are directly related to the time constant t_* by

$$t_* = \int \tau\, p(\tau)\, d\tau = \exp\left[m + s^2/4\right], \tag{7}$$

which allows estimating k_h by equation (2).

Using (6) the integration in (5) can be carried out analytically, resulting in the following theoretical distribution $h(T_s)$:

$$h(T_s) = \begin{cases} \Theta(T_s - T_b) H(T_s) & \text{if } j_h > 0 \\ -\Theta(T_b - T_s) H(T_s) & \text{if } j_h < 0 \\ \delta(T_s - T_b) & \text{if } j_h = 0 \end{cases}, \tag{8}$$

where

$$H(T_s) = \frac{(T_s - T_b)}{(\alpha j_h)^2} \exp\left[\frac{s^2}{4} - m\right] Q, \tag{9}$$

with

$$Q = \left(1 - \operatorname{erf}\left[\frac{s}{2} - \frac{m}{s} + \frac{1}{s} \ln\left(\frac{T_s - T_b}{\alpha j_h}\right)^2\right]\right).$$

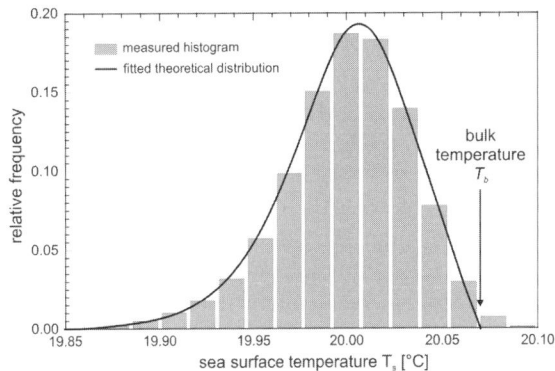

Figure 3. Measured histogram together with the fitted theoretical surface temperature distribution.

$\Theta(T)$ and $\delta(T)$ denote the binary step function, and Dirac's delta distribution, respectively:

$$\Theta(T) = \begin{cases} 1, & T > 0 \\ 0, & T \leq 0 \end{cases}, \quad \delta(T) = \begin{cases} 1, & T = 0 \\ 0, & \text{else} \end{cases}.$$

Fitting the theoretical temperature distribution (8) to the measured data allows to estimate the model parameters, T_b, s, and m, together with accuracy bounds. Furthermore, the goodness of fit allows validating the model assumptions. Figure 3 shows that this distribution can be fitted to measured SST distributions with high accuracy, which supports the surface renewal model. Although the model fits the measurements, an analysis of (8) shows that the parameters s and m, and hence t_*, cannot be estimated independently from the heat flux. To circumvent this problem a new technique has been developed that allows to directly estimate j_h from IR sequences (Sect. 3.3, [*Garbe et al.*, 2000]).

The *temperature difference* ΔT across the interface can be inferred from the estimated bulk temperature T_b and the fitted temperature distribution $h^f(T_s)$ by

$$\Delta T = \left[\int T_s\, h^f(T_s)\, dT_s\right] - T_b = \langle T_s \rangle - T_b, \tag{10}$$

where $\langle T_s \rangle$ is the expectation value of the SST. With ΔT, an independent estimate of the heat transfer velocity, k_h, is derived from (2).

Results using this technique during the 1997 CoOP North Atlantic experiment and more detailed model verifications from laboratory measurements can be found in [*Schimpf et al.*, 2000].

3.3. Heat Flux from IR Image Sequences

Within two renewal events the water surface temperature changes with the square-root of time, proportional to the heat flux, j_h (3). Tracking water surface elements and fitting the temporal variation should allow to estimate j_h. However, even for large heat fluxes the temperature change within two consecutive images is below the signal-to-noise-ratio. Furthermore, tracking objects of changing brightness inevitably leads to biased estimates of both the object motion as well as the temperature change. To solve these problems, new image sequence processing techniques had to be developed that incorporate physical models of the expected brightness changes into the motion analysis (Sect. 4).

In the coordinate frame of reference moving with the water surface motion, the total temporal derivative of the surface temperature (3) is given by:

$$\frac{dT_s}{dt} = \frac{\alpha\, j_h}{2\sqrt{t-t_0}} = j_h^2 \frac{\alpha^2}{2\,(T_s - T_b)}, \quad (11)$$

where $(t - t_0)$ is the time since the previous surface renewal event. If the bulk temperature T_b is known (which can be estimated from the statistical technique Sect. 3.2), the total temporal derivative is proportional to the parameter j_h^2. Hence, computing dT_s/dt in the coordinate frame moving with the image structures allows to estimate the heat flux density, j_h (Sect. 4). Variations of this technique, including a statistical analysis of temperature fluctuations, and results from laboratory measurements can be found in [Garbe et al., 2000].

3.4. Estimation of the Renewal Distribution

So far, the statistical temperature distribution allowed to retrieve the parameters s, and m, of the log-normal distribution. However, the distribution itself was not directly accessible. Analyzing the dynamical behavior of the SST allows to further estimating the probability distribution, $p(\tau)$, of surface renewal. Combining (11) and (3) yields the following relationship between the surface temperature, T_s, the bulk temperature, T_b, and the total temporal derivative of the SST at time t:

$$(t - t_0) = \frac{1}{2} \frac{(T_s(t) - T_b)}{(dT_s(t)/dt)}. \quad (12)$$

In a statistical average, the distribution of all time intervals $(t-t_0)$ should converge to the probability distribution of the surface renewal events, $p(\tau)$. Hence, measuring $(t-t_0)$ for an ensemble of points and images provides an estimate of $p(\tau)$. Again, it is necessary to analyze the temporal derivatives in the coordinate frame of reference moving with the water surface (Sect. 4). Further details on the dynamic temperature change technique, and model verification can be found in [Garbe et al., 2000].

3.5. Active Thermography (CFT)

The *Controlled Flux Technique*, CFT, uses an IR laser to locally heat the water surface and estimates the heat transfer rate from the temporal decay of the heat spots [Jähne et al., 1989]. As shown by *Haussecker et al.*, [1995], the temporal decrease of the surface temperature within the heated area is given by

$$T_s(t) = T_0 \frac{h}{\sqrt{h^2 + 4 D_h t}} e^{-\lambda t}, \quad (13)$$

where T_0 is the surface temperature at time $t=0$ (right after the laser is turned off), and $\lambda = t_*^{-1}$ is the mean surface renewal rate. The solution (13) is derived by approximating the vertical temperature profile of the heated area at time $t=0$ by a Gaussian distribution

$$T(z,0) = T_0\, e^{-\frac{z^2}{h^2}}, \quad (14)$$

where h is related to the laser penetration depth η by $\eta = h/2$. Given (13) the temporal derivative of the heat spot temperature can be derived as

$$\begin{aligned}\frac{dT_s(t)}{dt} &= -\lambda T_s(t) - \frac{2 D_h}{(h^2 + 4 D_h t)} T_s(t) \quad (15)\\ &= -\kappa T_s(t), \quad \text{with} \quad \kappa = \lambda + \frac{2 D_h}{(h^2 + 4 D_h t)}.\end{aligned}$$

For IR radiation emitted by a CO_2 laser at a wavelength of $10.6\,\mu m$, the penetration depth is $\eta = 11\,\mu m$. The minimal observable time difference is $t_{min} > 0.016\,\text{s}$, which is imposed by the frame rate of the IR camera. Hence, $h^2 \ll 4 D_h t$, for all $t > t_{min}$, given the diffusion coefficient of heat in water, $D_h = 0.0014\,\text{cm}^2\,\text{s}^{-1}$. This allows using the approximation $(h^2 + 4 D_h t) \approx 4 D_h t$, and (15) reduces to

$$\frac{dT_s(t)}{dt} = -\kappa(t)\, T_s(t), \quad \text{with} \quad \kappa(t) = \lambda + (2\,t)^{-1}, \quad (16)$$

i. e., the temperature change follows an exponential decay law with time varying decay rate. Tracking heated spots on the ocean surface should allow to estimate the decay rate, κ, from the temporal derivative in the coordinate frame of reference fixed on the water surface. However, tracking the rapidly decaying heat spots requires image sequence analysis techniques that incorporate the decay process into motion estimation. Otherwise both, the resulting motion, as well as the estimated decay rates, will be severely biased, as outlined in Sect. 4.

4. IMAGE SEQUENCE ANALYSIS

4.1. Brightness Constancy Assumption

Common techniques that aim at estimating motion from image sequences assume brightness constancy of objects. For temperature images $T(\mathbf{x}, t)$, where $\mathbf{x} = (x, y)$, the tracking of points of constant temperature amounts to finding a path $\mathbf{x}(t)$ along which the imaged temperature is constant. However, the physical heat transport processes are violating the assumption of constant temperature (Sec. 3). Without an appropriate model of the combined heat transport and object motion, motion estimates and object tracking in IR sequences can be severely biased [Haussecker and Fleet, 2000; Haussecker, 2000].

4.2. Physics-Based Brightness Variation

In order to incorporate temporal temperature variation into motion analysis, we define a trajectory $\mathbf{x}(t)$ through the IR sequence along which the temperature can change according to a parameterized function, h:

$$T(\mathbf{x}(t), t) = h(T_0, t, \mathbf{a}), \quad (17)$$

where $T_0 = T(\mathbf{x}(t_0), t_0)$ denotes the temperature at time 0, and $\mathbf{a} = [a_1, \ldots, a_Q]^T$ denotes a Q-dimensional parameter vector for the temperature change model.

Taking the total derivative of both sides of (17) provides us with a *temperature change constraint equation*

$$(\nabla T)^T \mathbf{v} + T_t = f(T_0, t, \mathbf{a}), \quad (18)$$

where f is defined as

$$f(T_0, t, \mathbf{a}) = \frac{d}{dt}[h(T_0, t, \mathbf{a})]. \quad (19)$$

Given constraints like that in (18), our goal is to estimate the parameters of the motion field $\mathbf{v}$, and the parameters $\mathbf{a}$ of the physical model f. This will be illustrated below for the examples of parameterized brightness changes derived in Sec. 3.

4.2.1. Instantaneous temperature derivative.
If we are not interested in the analytical shape of the temporal temperature variation, h can be approximated as piecewise linear brightness change within a small temporal window, i. e.,

$$h(T_0, t, a) = T_0 + a t. \quad (20)$$

In this case (18) reduces to the simple linear constraint on the parameters $\mathbf{v}$, and a:

$$(\nabla T)^T \mathbf{v} + T_t - a = 0, \quad (21)$$

where the scalar parameter $a = dT/dt$ is the instantaneous temperature derivative in the coordinate frame of reference moving with the water surface, which is required to compute the probability distribution of surface renewal given by (12).

4.2.2. Direct heat flux estimates.
In Sect. 3.3 we derived that the total temporal derivative of the surface temperature, T, exposed to a net heat flux, j_h, is given by (11). As h depicts the temporal variation of the surface temperature in the coordinate frame of reference moving with the water surface, f can be identified with dT/dt, and, using (11), the combined motion and temperature change constraint can be written as

$$(\nabla T)^T \mathbf{v} + T_t - j_h^2 \frac{\alpha^2}{2(T - T_b)} = 0. \quad (22)$$

If the bulk temperature, T_b, is estimated independently by the statistical technique described in Sect. 3.2, (22) constitutes a linear constraint relating the unknown parameters, $\mathbf{v}$, and j_h^2, to the observable quantities, ∇T, T_t, and T.

4.2.3. Temperature decay.
For the active technique in Sect. 3.5 the increased surface temperature is asymptotically returning to its equilibrium state according to an exponential decay law. Again, f can be identified with dT/dt, and (18) yields the following linear constraint on the unknown parameters $\mathbf{v}$, and λ:

$$(\nabla T)^T \mathbf{v} + (T_t + (2t)^{-1} T) + \lambda T = 0. \quad (23)$$

4.3. Computational Framework

Each of the formulations in Secs. 4.2.1 - 4.2.3 yields linear constraints that relate the variables of interest and noisy measurements. As they are single constraints in multiple unknowns, the problem is ill-posed and cannot be solved for individual points. To further constrain the solution we assume that the parameters are constant within a local space-time region. We then use a collection of N such constraints at neighboring pixels in the region to obtain a linear equation system. Assuming IID Gaussian noise in the measurements, a maximum likelihood estimate of the parameters is given by using linear techniques. Details on the numerical techniques can be found in [Haussecker and Fleet, 2000].

4.4. Example: Exponential Temperature Decay

Figure 4 shows an exponentially decaying heat spot on the water surface in a wind/wave tank. The size of the image is 5×5 cm. In addition to the exponential decay the pattern is subject to deformation accord-

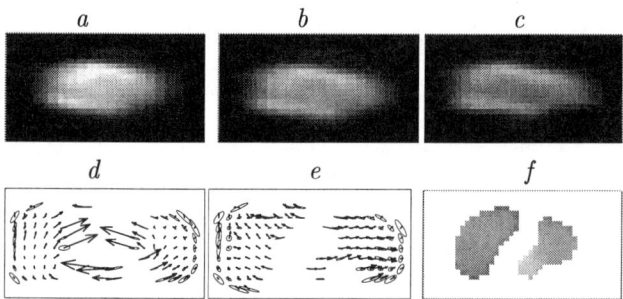

Figure 4. Decaying heat spot on a wavy water surface (*a - c*). Motion field **v** and uncertainty ellipses estimated with the constant brightness assumption (*d*), and an exponential decay model (*e*). (*f*) decay rate λ.

ing to the underlying turbulent flow field. If the temperature is assumed to remain constant (Figure 4 d), the estimated motion, especially the convergent motion field in the center, is unrealistic. In fact the heat spot is sheared and elongated from one image to the next. Using an exponential decay model for the temperature change (Figure 4 e), the motion field can be accurately estimated together with the decay rate λ (Figure 4 f).

5. CONCLUSIONS

This paper presents a comprehensive overview of a set of techniques to estimate the parameters of air-sea gas transfer using infrared thermography. It shows how an interaction of physical modeling and quantitative image sequence analysis allows to validate model assumptions on the transport processes, and to cross-verify independently obtained parameter estimates. Among the estimated parameters are the time constants and transfer rates of heat, and gases, across the air-water interface, the bulk water temperature, and the net heat flux density across the interface. Results from field experiments are consistent with gas transfer measurements obtained by independent techniques.

Acknowledgments. Portions of this work were supported by the 'Deutsche Forschungsgemeinschaft', DFG, by the Office of Naval Research, ONR, and by the National Science Foundation, NSF.

REFERENCES

Brutsaert, W., A model for evaporation as a molecular diffusion process into a turbulent atmosphere, *J. Geophys. Res., 70*, 5017-5024, 1965.

Danckwerts, P. V., *Gas-liquid reactions*, MacGraw-Hill, New York, 1970.

Garbe, C., H. Haussecker, and B. Jähne, Measuring the surface heat flux and probability distribution of surface renewal events, in Gas Transfer at Water Surfaces, edited by M.A. Donelan, W.M. Drennan, E.S. Saltzman and R. Wanninkhof, AGU, this volume, 2001.

Haussecker, H., S. Reinelt, and B. Jähne, Heat as a proxy tracer for gas exchange measurements in the field: principle and technical realization, in *Air-Water gas Transfer*, edited by B. Jähne, and E. C. Monahan, 405-413, Aeon, Hanau, 1995.

Haussecker, H., Messung und Simulation von kleinskaligen Austauschvorgängen an der Ozeanoberfläche mittels Thermographie, Dissertation thesis, Heidelberg University, Germany, May 1996.

Haussecker, H., and D. Fleet, Computing optical flow with physical models of brightness variation, *Proc. Computer Vision and Pattern Recognition (CVPR) 2000*, Hilton Head, SC, Vol. 2, 760-767, 2000.

Haussecker, H., Simultaneous estimation of optical flow and heat transport in infrared image sequences, *Proc. IEEE Workshop on Computer Vision Beyond the Visible Spectrum*, Hilton Head, SC, 85-93, 2000.

Jähne, B., P. Libner, R. Fischer, T. Billen, and E. Plate, Investigating the transfer processes across the free aqueous viscous boundary layer be the controlled flux method, Tellus *41*, 177-195, 1989.

Jähne, B., and H. Haussecker, Air-Water Gas Exchange, Ann. Rev. Fluid Mech. *30*, 443-468, 1998.

Jessup, A. T., C. J. Zappa, M. R. Loewen, and V. Hesany, Infrared remote sensing of breaking waves, Nature *385*, 52-55, 1996.

Jessup, A. T., and V. Hesany, Modulation of ocean skin temperature by swell waves, *J. Geophys. Res., 101* 6501-6511, 1996.

Jessup, A. T., C. J. Zappa, and H. H. Yeh, Defining and quantifying microscale wave breaking with infrared imagery, *J. Geophys. Res., 102*, 23145-23153, 1997.

Liu, W. T., J. A. Businger, Temperature profile in a molecular sublayer near the interface of a fluid in turbulent motion, Geophys. Res. Letters. *2*, 403-404, 1975.

Rao, K. N., R. Narasimha, and B. Narayanan, The bursting phenomenon in a turbulent boundary layer, J. Fluid Mech. *48*, 339-352, 1971.

Saunders, P. M., Aerial measurements of the sea surface temperature in the infrared, *J. Geophys. Res., 72*, 4109-4117, 1967.

Schimpf, U., H. Haussecker, and B. Jähne, On the investigations of air-water gas exchange and surface micro turbulence using thermography, in Gas Transfer at Water Surfaces, edited by M.A. Donelan, W.M. Drennan, E.S. Saltzman and R. Wanninkhof, AGU, this volume, 2001.

Schimpf, U., Untersuchung des Gasaustausches und der Mikroturbulenz an der Meeresoberfläche mittels Thermographie, Dissertation thesis, Heidelberg University, Germany, February 2000.

Soloview, A., and P. Schlüssel, Parameterization of the cool skin of the ocean and of the air-ocean gas transfer on the basis of modeling surface renewal, J. Phys. Ocean. *24*, 1339-1364, 1994.

H. Haussecker, Xerox Palo Alto Research Center, Palo Alto, CA 94304. (e-mail: hhaussec@parc.xerox.com)

U. Schimpf, C. Garbe, B. Jähne, Interdisciplinary Center for Scientific Computing, Im Neuenheimer Feld 368, 69120 Heidelberg, Germany. (e-mail: {uwe.schimpf,bernd.jaehne, christoph.garbe}@iwr.uni-heidelberg.de)

Measuring the Sea Surface Heat Flux and Probability Distribution of Surface Renewal Events

Christoph S. Garbe[1,2] and Bernd Jähne[1,2]

Interdisciplinary Center for Scientific Computing University of Heidelberg, Germany

Horst Haußecker

Xerox Palo Alto Research Center (PARC), Palo Alto, California

The net sea surface heat flux is a crucial parameter for quantitative measurements of air-sea gas exchange rates, as well as for climate models and simulations. Current techniques used to measure heat fluxes are based on direct radiative schemes and micro-meteorological parameterizations. The long time averages required by the meteorological estimates tend to obscure short time and space scale events. Our technique uses a single infrared camera in order to quantitatively estimate the parameters of a surface renewal model of heat transfer. Through the use of spatio-temporal image processing techniques the material derivative of the sea surface temperature with respect to time can be used to compute both the heat flux density, as well as the probability density function of the underlying surface renewal model. First results of our technique obtained at the Heidelberg Aeolotron showed excellent agreement to ground truth data.

1. INTRODUCTION

A temperature difference across the aqueous boundary layer gives rise to a transport of heat, the rate of which is given by

$$k_h = \frac{j}{\rho c_p \Delta T}, \qquad (1)$$

with the temperature difference ΔT, the heat flux density j, the density ρ and specific heat c_p of sea water. Affixed to any transport process of heat is a characteristic time constant t_*, the relation of which to k_h can be expressed as

$$t_* = \frac{D}{k_h^2} \qquad (2)$$

where the coefficient of thermal diffusivity D is known. The heat flux j for a substance of mass M is defined as the timely change of its quantity of heat Q across the surface A, or

$$j = \frac{dQ/dt}{A} = \frac{Mc_p}{A}\frac{dT}{dt} \qquad (3)$$

[1] Also at Institut für Umweltphysik, University of Heidelberg, Germany.
[2] Also at Scripps Institution of Oceanography, UCSD, California.

Gas Transfer at Water Surfaces
Geophysical Monograph 127
Copyright 2002 by the American Geophysical Union

Several models of surface renewal have been postulated ([*Danckwerts*, 1970], [*Rao et al.*, 1971]). In our work we present strong experimental evidence of a surface renewal model, where the probability density function (PDF) $p(t)$ of times in between consecutive renewal events can be described by a logarithmic normal distribution:

$$p(t) = \frac{1}{\sqrt{\pi}\sigma t/t'} e^{-\frac{(\ln t/t' - m)^2}{\sigma^2}}, \quad t > 0 \qquad (4)$$

where m is the mean value of $\ln t/t'$ and σ^2 the variance for the logarithm of the scaled random variable t. t' is a unit scaling factor. This type of model was proposed by [*Kolmogorov*, 1962] and [*Soloviev and Schlüssel*, 1994] and indicated by measurements of [*Rao et al.*, 1971].

The mean time between burst t_* is the expectancy value of this distribution, given by

$$t_* = \int_0^\infty p(t)\, t/t'\, dt = t' \cdot e^{\frac{\sigma^2}{4} + m} \qquad (5)$$

Following [*Soloviev and Schlüssel*, 1994] a fluid element adjacent to the sea surface is considered. Prior to its exposure to the surface, its temperature was that of the bulk water. Once exposed to the surface, temperature differences are governed by the applicable molecular diffusion laws. The temperature difference ΔT across the thermal sub layer is thus given by

$$\Delta T(t) = \frac{2j}{\sqrt{\pi D} c_p \rho} \sqrt{t - t_0}, \quad t \geq t_0 \qquad (6)$$

where j is the vertical heat flux density just below the sea surface and t_0 the time at which the surface renewal event occurred. For ease of notation, the temperature difference shall be denoted by T, that is $T = \Delta T = T_{\text{surface}} - T_{\text{bulk}}$, where T_{surface} is the temperature of surface water and T_{bulk} that of bulk water, respectively. In the following sections two techniques for deriving the total heat flux density j will be presented as will be experimental evidence of the PDF of surface renewal.

In common to all these techniques is, that they rely on calculations of the material derivative of the sea surface temperature (SST) with respect to time. This can be achieved with a digital image processing technique, described in [*Haußecker et al.*, 1999]. Through this technique simultaneous estimates of the surface velocity field and parameters of the temporal temperature change are given. The statistical significance can further be raised by introducing a global optimization technique indicated by [*Spies et al.*, 2000].

2. THE MATERIAL DERIVATIVE FROM IMAGE SEQUENCES

When trying to determine the heat flux at the air sea interface, we need to examine an individual fluid parcel as it moves about over a period of time. At the sea surface the fluid element will cool down or heat up, depending on the predominant direction of the total heat flux according to the laws of molecular diffusion. This change of temperature takes place in a Lagrangian frame of reference, with the measurements being taken in an Eulerian frame of reference. The task is therefore to represent a concept essentially Lagrangian in nature in Eulerian language. The change of temperature with respect to time dT/dt in the Eulerian frame of reference has of course to take the movement of the parcel into account. This is readily done by the material derivative, which is denoted by DT/Dt:

$$\frac{DT}{Dt} = \frac{\partial T}{\partial t} + \left(\vec{u}\vec{\nabla}\right)T, \qquad (7)$$

where $\vec{u} = (u_1, u_2, u_3)^\top$ is the three dimensional velocity of the fluid parcel. In the following superscript T denotes the transpose.

To determine the material derivative in equation (7) of a fluid parcel by digital image processing, it is not sufficient to extract the change of temperature $\partial T/\partial t$ at a fixed position of the image, which is quite trivial. Moreover an exact estimate of the velocity $\vec{u}$ has to be found. In the following sections a means for deriving the material derivative directly from an image sequence will be presented.

2.1. The Brightness Model

In digital image processing motion cannot be detected directly. Only through temporal variation in the image brightness can the real motion be deduced. Some constraints must of course be applied. A very common assumption is the brightness change constraint (BCCE). It states that the image brightness $g(\vec{x}, t)$ at the location $\vec{x} = (x_1, x_2)^\top$ should change only due to motion, that is the total derivative has to equal zero:

$$\frac{dg}{dt} = \frac{\partial g}{\partial t} + \frac{\partial g}{\partial x}\frac{dx}{dt} + \frac{\partial g}{\partial y}\frac{dy}{dt} = g_t + (\vec{f}\vec{\nabla})g = 0, \quad (8)$$

with the optical flow $\vec{f} = (dx/dt, dy/dt)^\top$, the spacial gradient $\vec{\nabla} g$ and the partial time derivative $g_t = \partial g/\partial t$.

When using an infrared camera for acquiring image sequences, the temperature T of a three dimensional object is mapped as a grey value g onto the image plane. When comparing the two equations (7) and (8) it is evident that the brightness change constraint does not hold in the context of this work. In order to satisfy this constraint, the temperature change we seek to measure would have to be equal to zero. To account for this fact, a linear term is introduced in the BCCE, that is

$$g_t + (\vec{f}\vec{\nabla}g) = c, \quad (9)$$

where c is a constant, which is proportional to DT/Dt. The equation (9) poses an underdetermined system of equations, as there is only one constraint with the three unknowns $\vec{f}$ and c. A common method for finding a solution to this ill posed problem is to introduce the assumption that all parameters (the optical flow $\vec{f}$ and c) are constant over a small spatio-temporal neighborhood surrounding the location of interest. If the neighborhood comprises of N pixels, the problem consists of N equations of the form of equation (9). This overdetermined system of equation can be solved in a total least squares sense [Haußecker and Spies, 1999]. The result of this calculation is shown in Figure 1.

2.2. Flow Regularization

In some cases the image sequences may be rather short and the location of computed material derivatives sparse. In order to raise the statistical significance of subsequent calculations a regularization scheme is used [Spies et al., 2000]. This is done by applying a simple membrane model of the form

$$\sum_{i=1}^{3} (\nabla v_i)^2 \longrightarrow \min, \quad (10)$$

with the dense and smooth flow field $\vec{v} = [f_1, f_2, c]$. A weighting factor is introduced reflecting the confidence measure of the prior calculated estimates from section 2.1.

3. THE PDF OF SURFACE RENEWAL

The probability density function (PDF) of the times in between consecutive surface renewal events $\Delta t = t - t_0$ is of great importance. It gives rise to speculations regarding the exact processes involved in the renewal events and justifies some of the techniques in deriving the heat flux described later on.

In order to be able to make a quantitative statement for this PDF, a value for j has to be found in equation (6). This can be achieved by differentiating the equation with respect to time t:

$$\dot{T}(t) = \frac{DT(t)}{Dt} = \frac{j}{\sqrt{\pi D}c_p\rho}(t-t_0)^{-1/2} \quad (11)$$

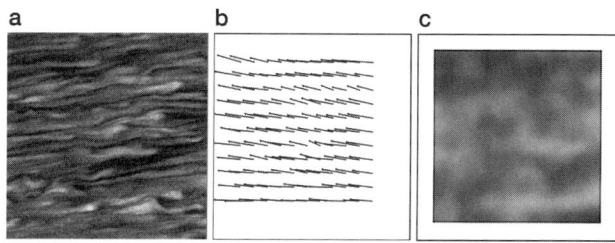

Figure 1. (a) original scaled infrared image g (b) flow field $\vec{f}$ derived from equation (9) (c) material derivative $DT/dt = c$ also derived from equation (9)

The two equations (6) and (11) thus lead to

$$(t - t_0) = \frac{1}{2}\frac{T(t)}{\dot{T}(t)} \quad (12)$$

In equation (12) the value for T is directly obtained from the measured infrared image sequence. $\dot{T}(t)$ can be computed from the sequence with the image processing technique mentioned in section 2.

From the Taylor hypothesis [Taylor, 1938] the temporal statistics for $(t-t_0)$ can be extended on the spatial domain. A value is thus computed for every image point for which a valid $\dot{T}$ could be found. To increase the statistical significance this step is repeated for each image in the sequence and the number of occurrences can be plotted in a histogram.

Given the theoretical PDF from equation (4) the values for σ and m can be computed together with error estimates for the individual parameters. This is achieved by fitting the PDF to the histogram by means of least squares. From the Parameters σ and m the mean time between surface renewals t_* can be computed from equation (5).

4. DETERMINATION OF THE HEAT FLUX

Through the use of infrared cameras in conjunction with modern digital image processing algorithms, new means of analyzing the Air-Sea gas transfer as well as measuring the heat flux j directly are readily available [Haußecker et al., 2001].

In this work two algorithms are presented. They both rely on the consistent estimation of the change of surface temperature with respect to time, as described in section 2. The two methods of calculating the heat flux are:

- The square root method. Here the heat flux can be calculated directly from the material derivative

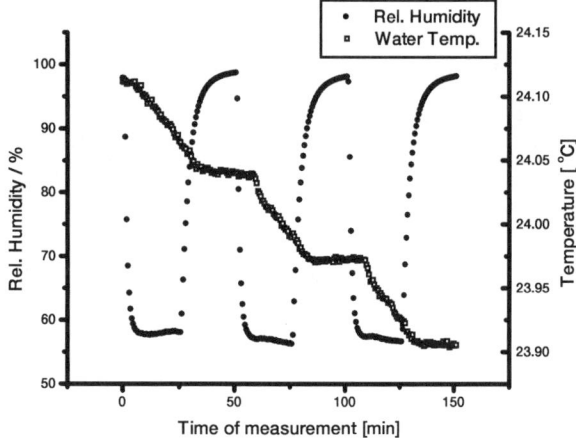

Figure 2. A plot of the water temperature and the relative humidity as functions of time since start of measurement.

DT/Dt and the temperature difference across the aqueous boundary layer ΔT.

- The mean heat flux over the whole sequence. For this technique to work, some assumptions have to be made on the statistics between surface renewal events. This method will be referred to as the PDF method, as the probability density function of the time between surface renewal events is used to evaluate the flux.

In the following section these different methods shall be further scrutinized.

4.1. Square Root Method

This method for calculating the heat flux is the most promising one. It relies on no further assumptions concerning the statistics of surface renewal events. The only physical model to enter the evaluation is the equilibration of a fluid parcel adjacent to the surface by the equations of molecular diffusion.

In section 1 a fluid element was pictured over the course of time. From the assumption alone that the heat transfer at the sea surface is governed by molecular diffusion, equation (6) was derived. This equation can be solved for the heat flux j, which leads to

$$j = \frac{T(t)}{\alpha\sqrt{t-t_0}}, \quad t \geq t_0, \quad \text{with} \quad \alpha = \frac{2}{\sqrt{\pi D c_p \rho}}. \quad (13)$$

Of course, the exact measurement of $\Delta t = t - t_0$ poses a very difficult problem. This can be circumvented by taking the material derivative of equation (6). In the Lagrangian description this leads to

$$\frac{D}{Dt}T(t) = \frac{1}{2}\alpha j \frac{1}{\sqrt{t-t_0}}, \quad t \geq t_0 \quad (14)$$

where DT/Dt denotes the material derivative. Equation (14) can be solved for Δt and the result introduced into equation (13). This leads to the expression

$$|j| = \sqrt{\frac{\pi D}{2}} c_p \rho \sqrt{T(t) \frac{D}{Dt} T(t)} \quad (15)$$

The sign of the heat flux j can of course be deducted from DT/Dt following equation (14).

Through the use of equation (15) it becomes feasible to determine the heat flux j from measurements with a single infrared camera. T can be assessed directly from the infrared imagery with the aid of a technique described by [*Haußecker et al.*, 2001]. Here an analytic function is fitted to the histogram of the temperature distribution of an IR sequence. T_{bulk} is one of the parameters of this analytic function. It can thus be gained by fitting the function to experimental data. An evaluation of this technique on laboratory data was presented by [*Schimpf et al.*, 1999].

The material derivative of T can be computed with the digital image processing technique as described in section 2. All the other dimensions are material constants and well known for sea water.

4.2. The PDF Method

This method can be employed if the mean flux is of interest. It makes a few assumptions regarding the processes involved in the sea surface boundary layer. These assumptions are mainly the log-normal probability density function (PDF) of the statistic of surface renewal and the thermal equilibration of a water element adjacent to the sea surface by molecular diffusion, experimental evidence of which will be presented in section 5.

The mean temperature difference across the aquatic boundary layer was given by equation (6) as

$$\Delta T(t) = \frac{2j}{\sqrt{\pi D c_p \rho}} \sqrt{t-t_0}, \quad t \geq t_0 \quad (16)$$

The average temperature difference across the cool skin of the ocean is given by [*Soloviev and Schlüssel*, 1994] as follows

$$\Delta \bar{T} = \int_0^\infty p(t) t^{-1} \left(\int_0^t \Delta T(t') dt' \right) dt \quad (17)$$

The integration of this equation with the PDF from equation(4) yields

$$\Delta \bar{T} = \frac{4j}{3\sqrt{\pi D} c_p \rho} \sqrt{t_*} \exp(-\sigma^2/16) \quad (18)$$

This expression can be solved for the heat flux j, which together with equation (5) leads to

$$j = \frac{3}{4}\sqrt{\pi D} c_p \rho \Delta \bar{T} \exp\left[-\left(\frac{\sigma^2}{16} + \frac{m}{2}\right)\right] \quad (19)$$

Both the parameters σ and m can be calculated from a fit of the log normal distribution from equation (4) against the histogram of DT/Dt as outlined in section 3.

5. EXPERIMENTAL RESULTS

In order to validate the presented algorithms and gain insight in the probability density function of surface renewal events, measurements on the Heidelberg Aeolotron were conducted. An experiment was carried out in three different wind regimes with different net heat fluxes. Great care was taken in ensuring that water and air temperatures were equal. This was only possible due to the excellent thermal isolation of the facility. Due to these measures only the relative humidity of the air gave rise to the heat flux.

In Figure 2 the data collected from analog sources is shown. When the humidity reaches 100% the water temperature stays constant, which underlines the good thermal properties of the Aeolotron. The heat flux was calculated from this analog data from equation (3).

Measurements were conducted at wind speeds of 2.0 m/s, 4.2 m/s and 8.0 m/s. During all the measurements a surface film was present. Due to this film the temperature of the bulk water T_{bulk} could not be computed

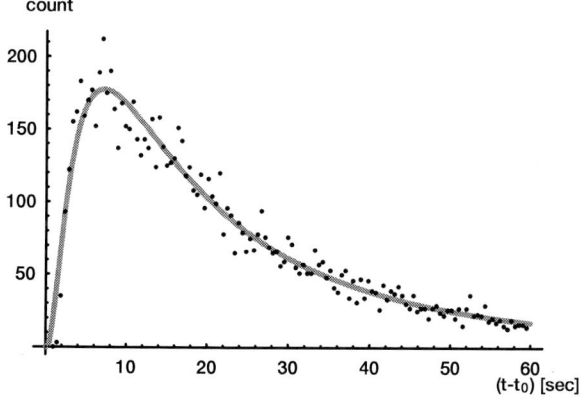

Figure 3. The log-normal distribution fitted to the histogram of $t - t_0$ at a wind speed of 2 m/s.

Table 1. Results for computation of t_*

Wind speed	m	σ	t_* [s]
2.0 m/s	2.934 ± 0.026	1.386 ± 0.026	30.38 ± 0.88
4.2 m/s	1.021 ± 0.011	0.812 ± 0.014	3.27 ± 0.04
8.0 m/s	0.277 ± 0.009	0.652 ± 0.012	1.47 ± 0.01

using the technique described in [Schimpf et al., 1999]. It was derived instead from the difference in mean temperature on sequences with to those without heat flux. The infrared camera was calibrated with no black body but by looking at the water directly. This was achieved by raising the air and water in the Aeolotron to different temperatures and waiting for equilibrium where no heat flux was present. The reason this technique was chosen is that it avoids the drawback of black bodies that have a different emissivity than that of water and thus makes for a more accurate mapping of temperature to grey value.

5.1. PDF of Surface Renewal

The times in between consecutive surface renewal events $t - t_0$ was calculated as described in section 3. An example of the histogram of this data from an sequence of 100 images for the experiment at 2 m/s is shown in Figure 3. The fitted log normal distribution seems to be in good agreement with the data. This presents strong experimental evidence of a surface renewal model in which the probability density function of surface renewal is log normal in nature, as outlined earlier. From the parameters σ and m of the log normal distribution in equation (4), the characteristic time constant t_* can be calculated, following equation (5). The results for all measurements are presented in Table 1.

5.2. Heat Flux Measurements

From the values for σ and m calculated from the data in section 5.1 the heat flux j can be calculated as outlined in section 4.2. This approach makes the assumption that the PDF of the times between surface renewal events is characterized by a log normal distribution, which seems to be valid as shown in the previous section. The results of this calculation can be found as j_{PDF} in Table 2. The predominant factor in the error of j_{PDF} is the noise equivalent temperature difference ($NE\Delta T$) of the infrared camera. Is is in the regime of $NE\Delta T = 26$ mK. The algorithm would thus benefit immensely from a camera with a lower noise ratio.

In Figure 4 the values for the measurements of j are shown. The algorithm used was that presented in section 4.1. The mean values over a period of 1.4 seconds is given in Table 2. Fluctuations in the measurements

Table 2. Results for calculations of j

Wind speed	j_{true} [W/m^2]	j_{PDF} [W/m^2]	j_{sqrt} [W/m^2]
2.0 m/s	-111 ± 3	-100 ± 11	-137 ± 13
4.2 m/s	-161 ± 2	-149 ± 31	-188 ± 12
8.0 m/s	-304 ± 3	-273 ± 43	-280 ± 49

are not only due to errors but are correlated to waves motion. This correlation was seen for the first time with this new technique and will be investigated further in the near future.

6. CONCLUSION

This paper presents a novel approach to estimating the heat flux j from infrared image sequences. This is made possible by recent advances in calculating the material derivative of sea surface temperature with respect to time. From this derivative the probability density function of times in between surface renewal events was obtained. Strong experimental evidence was given that this distribution is best approximated by a logarithmic normal distribution.

Apart from the probability density function the net heat flux across the aqueous boundary layer was computed. Two methods of calculating heat flux were presented. Measurements at the Heidelberg Aeolotron were conducted were the heat flux could be measured to a high accuracy by other means. This made it possible to compare the results of the new technique with ground truth. Slight variations in the new techniques from the real value for the heat flux can be explained by a surface film that was present during the measurements. Its influence on the data and possible adaptations on the algorithms shall be investigated in the future. Also, a modulation of the heat flux due to waves was seen for the first time.

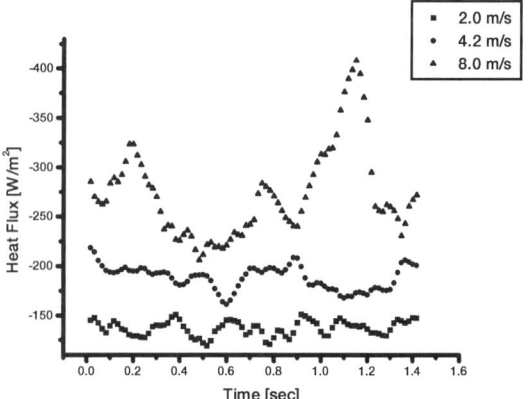

Figure 4. The heat flux calculated over a time period of 1.4 seconds for three different wind speeds.

Acknowledgments. We gratefully acknowledge financial support of this research by the German Science Foundation (DFG) through the research unit "Image Sequence Processing to Investigate Dynamic Processes", the Office of Naval Research, ONR, as well as the National Science Foundation, NSF.

REFERENCES

Danckwerts, P.V., *Gas-liquid reactions*, MacGraw-Hill, New York, 1970.

Haußecker, H., U. Schimpf, C.S. Garbe, and B. Jähne, Physics from ir image sequences: Quantitative analysis of transport models and parameters of air-sea gas transfer, in *Geophysical Monograph, this volume*, edited by M.A. Donelan, W.M. Drennan, E.S. Saltzman and R. Wanninkhof, AGU, 2001.

Haußecker, H., and H. Spies, *Handbook of Computer Vision and Applications*, vol. 2, chap. 13. Motion, pp. 309–396, Academic Press, 1999.

Haußecker, H., C. Garbe, H. Spies, and B. Jähne, A total least squares framework for low-level analysis of dynamic scenes and processes, in *21.Symposium für Mustererkennung DAGM 1999*, pp. 240–249, 1999.

Kolmogorov, A., A refinement of previous hypotheses concerning the local structure of turbulence in a viscous incompressible fluid at high reynolds number, *Journal of Fluid Mechanics*, *13*, 82–85, 1962.

Rao, K. N., R. Narasimah, and M. B. Narayanan, The 'bursting' phenomenon in a turbulent boundary layer, *Journal of Fluid Mechanic*, *48*, 339–352, 1971.

Schimpf, U., H. Haußecker, and B. Jähne, Studies of air-sea gas transfer and micro turbulence at the ocean surface using passive thermography, in *The Wind-Driven Air-Sea Interface: Electromagnetic and Acoustic Sensing, Wave Dynamics and Turbulent Fluxes*, edited by M. L. Banner, pp. 345-352, School of Mathematics, University of New South Wales, Sydney, Australia, 1999.

Soloviev, A. V., and P. Schlüssel, Parameterization of the cool skin of the ocean and of the air-ocean gas transfer on the basis of modeling surface renewal, *Journal of Physical Oceanography*, *24*, 1339–1346, 1994.

Spies, H., B. Jähne, and J. Barron, Regularised range flow, *Proc. of ECCV 2000*, Dublin, Ireland, 2000.

Taylor, G., The spectrum of turbulence, *Proc. R. Soc.*, *102*, 817–822, 1938.

C. S. Garbe and B. Jähne, Interdisciplinary Center for Scientific Computing, Im Neuenheimer Feld 368, 69120 Heidelberg, Germany. (e-mail: Christoph.Garbe@iwr.uni-heidelberg.de; bernd.jaehne@iwr.uni-heidelberg.de)

H. Haußecker, Xerox Palo Alto Research Center(PARC), 3333 Coyote Hill Road, Palo Alto, CA 94304, USA (e-mail: hhaussec@parc.xerox.com)

Effects of Counter-Swell on Both the Mean Current and Turbulent Structure Below the Wind-Waves

Shinjiro Mizuno

Department of Engineering, Hiroshima Institute of Technology

An interaction between wind waves and mechanically generated waves propagating against the wind suppressed the secondary flow in the tank, leading to a marked enhancement of the wind-driven current. This suppression effect of the secondary flow is consistent with the CL2 mechanism of Langmuir circulation.

1. INTRODUCTION

The interaction between wind-waves and swell is an important but unknown mechanism that controls various physical processes such as momentum transfer and gas exchange in air-sea interaction.

Cheng & Mitsuyasu (1992, referred to as CM) studied the effect of mechanically generated swells on the surface drift current. For the case of counter-swell they found the surface drift current to increase by about 1.5 times that of the pure wind-wave case, though such a marked increase of the surface drift was not obtained for the case of aligned swell. The primary purpose of the present current measurement is to examine a physical idea that a marked increase of the surface drift current obtained by CM may be due to modification of the wind-driven current caused by mechanically generated counter-swell rather than to an increase of the wind stress.

Recently Mizuno, Noguchi & Kimura (1998) have made systematic measurements on the subsurface current to extract organized convective motions in a vertical cross-section of a few wind-wave tanks. They found a steady secondary flow in the tanks. This secondary flow is called a two-dimensional Langmuir circulation (2D-LC) in this study because its lateral length scale remains fixed by the width of a tank in the wind direction. The CL2 mechanism of LC (Craik & Leibovich, 1976), that is an instability theory on the generation of LC based on the interaction between wind-driven lateral shear and Stokes wave drift, is incorporated into recent numerical experiments as a possible generation mechanism of LC observed in oceans and lakes (Skylling-stad & Denbo 1995; D'Alessio, Abdella & McFarlane 1998). Though the LC is well known to be efficient for the transfer of gas bubbles, the CL2 model itself still remains unconfirmed in both laboratory experiments and field observations. If waves propagate against the wind, then the CL2 mechanism tends to suppress the LC, as pointed out by Leibovich (1983). Hence, if mechanically generated counter-swell interacts with the 2D-LC in our tank, the CL2 mechanism will suppress its evolution. Since the decay of the 2D-LC tends to strengthen the wind-driven current shear, it is possible that this interaction will cause the surface drift current to increase markedly.

2. EXPERIMENT AND PROCEDURES

The experiments were carried out in a small wind-wave tank at Hiroshima Institute of Technology. The tank is 13 m long, B = 29 cm wide, 55 cm high. The experiments were made at a fetch of 5.5 m. Through the experiments the water depth D was set to 25 cm, and the height of wind-tunnel section H to the remaining 30 cm. A centrifugal fan for blowing wind through the tank and a transition plate for thickening the air boundary layer were situated on the upwind side of the test section. At the end of the downwind side a plunger-type wave generator was installed for generating regular oscillatory counter-swell waves

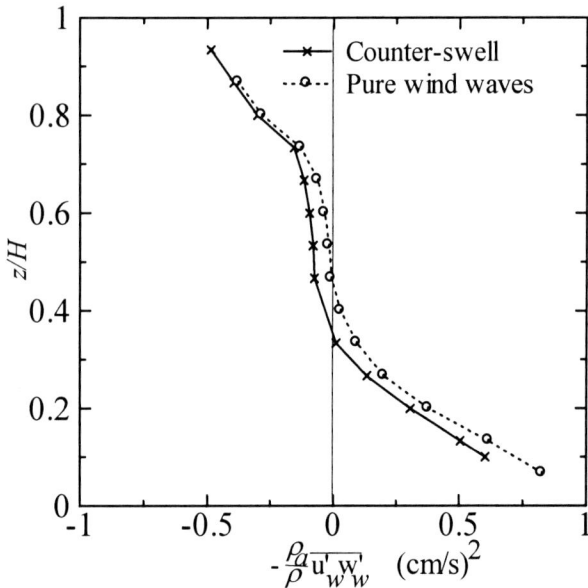

Figure 1. Vertical profiles of the Reynolds stress of wind turbulence at y = 0, $-\overline{u'_w w'_w}$, where the over bar denotes time average, and ρ_a and ρ denote air and water densities.

opposing to the wind. A beach for wave filter was also installed for absorbing the swell at the upwind end.

Two experiments were carried out at an air free-stream velocity of 7m/s:

Case 1: Case of pure wind waves;

Case 2: Coexisting case of wind waves and mechanically generated counter-swell against the wind (hereafter referred to as counter-swell case).

The frequency and height of swell were selected to f = 1 Hz and Ho = 5 cm, to make the swell steepness equal to that of CM (Ho/L =0.04), where Ho denotes the height of swell at the mechanical wave machine and L the wave length.

The x-, y-, and z- axes of the co-ordinate system are taken to the downwind direction from the inlet of the test section, to the left hand side of wind direction from the centerline of the tank (y = 0), and vertically upwards from the still water surface, respectively. A TSI two-component laser-Doppler anemometer (LDA) system was used in the backward scatter mode to measure downwind and vertical current components, u and w, where

$$u = U + \tilde{u} + u', \qquad w = W + \tilde{w} + w'. \qquad (1)$$

Here U and W denote mean current, $\tilde{u}$ and $\tilde{w}$ swell-induced components, and u' and w' consist of all the turbulent components in the higher and lower frequency range than the swell frequency, except for the swell frequency of 1 Hz. Measurement was made about 1 hour after the onset of wind to establish a steady state. The current was measured at 135 points (15 lateral and 9 vertical points at an interval of 2 cm) in a vertical cross-section at a fetch of 5.5 m. The record length of the data was 250 s, and the sampling frequency was 10 Hz.

3. EXPERIMENTAL RESULTS

Figure 1 shows two sets of vertical profiles of Reynolds shear stress above the surface at a fetch of 3.5 m at a free-stream wind speed of 7 m/s for the cases of pure wind waves and mechanically generated counter-swell. The Reynolds stress decreased a little for the case of counter-swell. Hence, it is not likely that the wind stress is the cause of an abnormally large increase of the surface drift current obtained by CM for the case of counter-swell.

Plate 1 shows effects of the counter-swell on the vertical cross-sectional distributions of U and W at a fetch of 5.5 m. It is evident that propagation of the counter-swell markedly intensified both the wind-driven current in the upper layer and its return flow in the lower layer. For the case of mechanically generated counter-swell, the transport of the primary circulation estimated from current measurements at 135 points increased by about 1.6 times that of the pure wind wave case at the same wind speed. From the above two facts, it follows that most of the increase of the surface drift current obtained by CM may be explained by intensification of the wind-driven current below the surface.

In the pure wind wave case there was a pair of strong secondary circulation, i.e., 2D-LC, as shown in the left panel of Plate 1(b). It should be stressed that this secondary circulation satisfies all of the basic features for being regarded as LC given by Craik & Leibovich (1976), namely, wind-driven vortices parallel to the wind direction, asymmetric structure with downwelling current larger than upwelling current, the strongest U in the downwelling zone, with the strongest W comparable to U. In addition, as a special feature of the tank flow the interface between the wind-driven current and its return current in the central zone of the tank (i.e., the boundary line of U = 0 in Plate 1(a)) shifted from the near-surface in the pure wind wave case to a greater depth in the case of counter-swell. This shift of the interface is due to the presence of upwelling current in the central zone and its disappearance.

It is certain from Plate 1 that propagation of the counter-swell suppressed the 2D-LCs, W, and simultaneously strengthened the primary circulation, U. As mentioned in Introduction, this fact is quite consistent with the CL2 mechanism (Leibovich, 1983). Thus, the primary

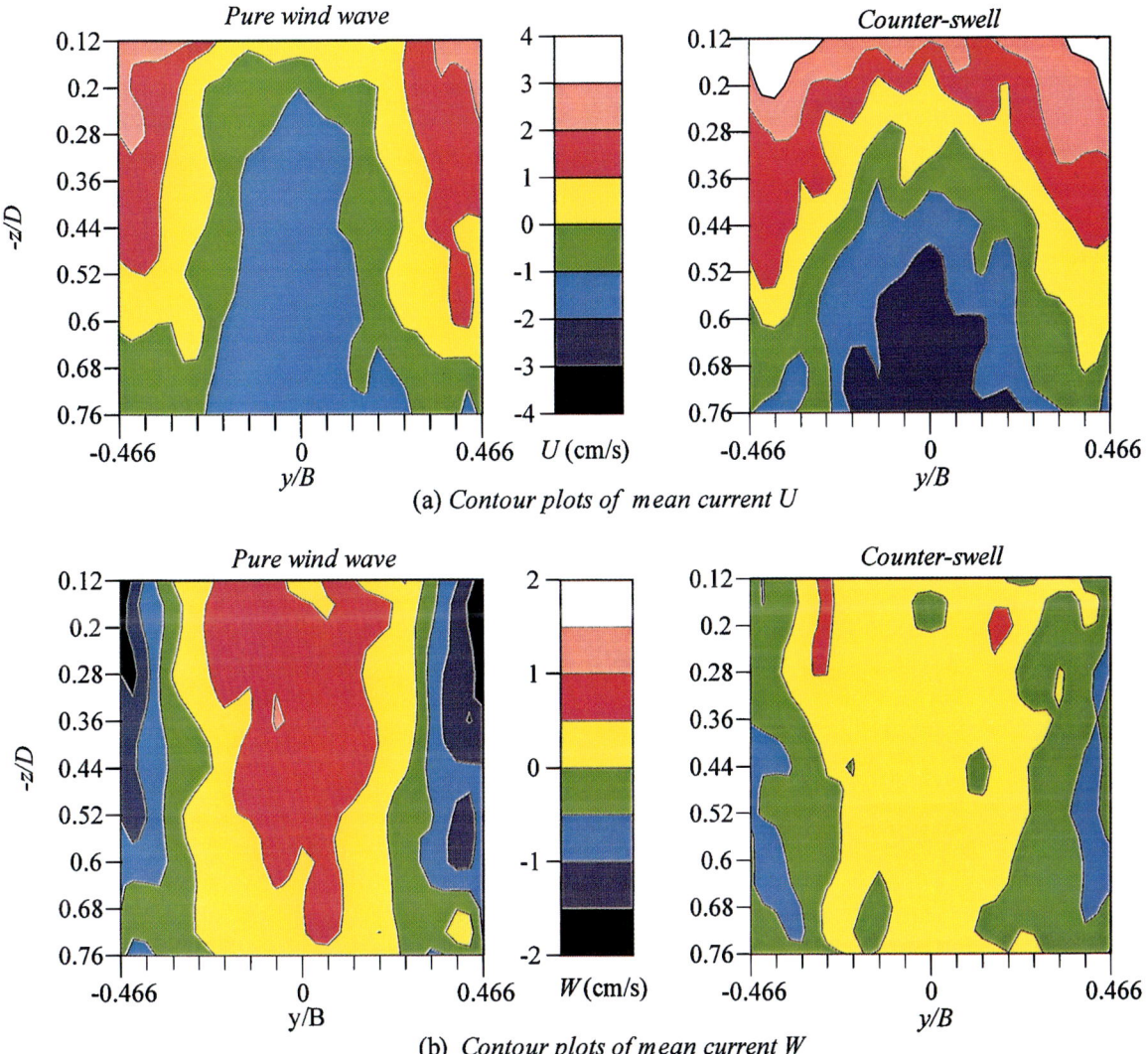

Plate 1. Contour plots of mean current U and W, obtained at a wind speed of 7 m/s for the two cases of pure wind wave and mechanically generated counter-swell with initial steepness of 0.04.

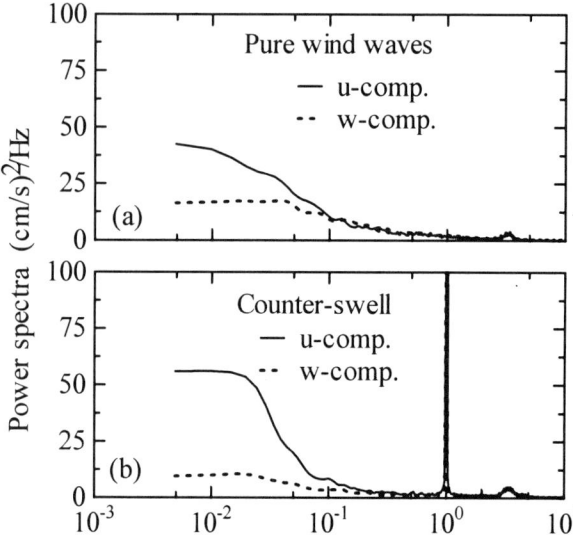

Figure 2. Power spectra of $(\tilde{u}+u')$ and $(\tilde{w}+w')$, which are averaged laterally over 15 points at a depth of $z = -3$ cm.

$$\frac{\partial <U>}{\partial t} + \frac{\partial <U^2 + \overline{\tilde{u}^2} + \overline{u'^2}>}{\partial x} = -g\frac{\partial \overline{\eta}}{\partial x}$$

$$-\frac{\partial}{\partial z}\{<UW> + <\overline{\tilde{u}\tilde{w}}> + <\overline{u'w'}>\} \quad (2)$$

where $\overline{\eta}$ denotes the time averaged water level. The two terms of the left hand side are omitted in this study because the first term vanishes for steady flow, and no current data were obtained at two stations along the x-direction.

The Reynolds shear stress consists of 3 different terms. For the pure wind wave case, there is no swell-induced Reynolds stress, $-<\tilde{u}\tilde{w}>$, because of the absence of swell in this case, while for the case of counter-swell, surprising enough, the largest Reynolds stress among the three terms was $-<\tilde{u}\tilde{w}>$ as shown in Fig. 5, particularly near the surface and decreased rapidly with depth.

Compare Figs. 4 and 5, and then it will be found that for the pure wind wave case the Reynolds stress associated with the 2D-LC, $-<UW>$, is of about one order of magnitude larger than for the case of counter-swell. According to Pollard (1977), this term plays an important role in redistributing momentum through the mixed layer below the surface. In the present tank flow, $-<UW>$ for the pure wind wave case was very effective at the momentum exchange between the wind-driven current in the upper layer and its return flow. For example, for the pure wind-wave case the mean velocities of U = 3 and W = -2 cm/s were obtained at (y/B = ± 0.467, z/D = -0.12) in the downwelling zone along the side-walls, giving the

circulation in the tank is not only driven directly by the wind stress, but also its strength is associated closely with the growth or decay of the secondary flow below the wind waves.

Figure 2 shows an example of power spectra of u and w for both cases of the pure wind wave and counter-swell. For the turbulence energy in the low-frequency range less than 1Hz, propagation of counter-swell increased slightly the power spectrum of u, but decreased slightly that of w (compare Figs. 2(a) with 2(b)), indicating that this anisotropic tendency observed in the power spectra of the two turbulent velocity components is quite similar to the intensification of U and suppression of W obtained for the case of counter-swell. From this similarity, we think that the source of high low-frequency turbulent energy observed in a wind-wave tank originates from unsteady motions of both the primary and secondary flow rather than from the energy of wind-waves, which corresponds to a small spectral peak in the frequency range of 2-4 Hz in Figs. 2(a) and 2(b). As is clear from a sharp spectral line at 1 Hz in Fig. 3, the counter-swell generated very strong positive swell-induced Reynolds shear stress, $-<\tilde{u}\tilde{w}>$, where the symbol $< >$ denotes a laterally averaged quantity of 15 points at a depth.

Next consider the acceleration mechanisms of the spanwise mean wind-driven current, $<U>$. The momentum equation for $<U>$ yields:

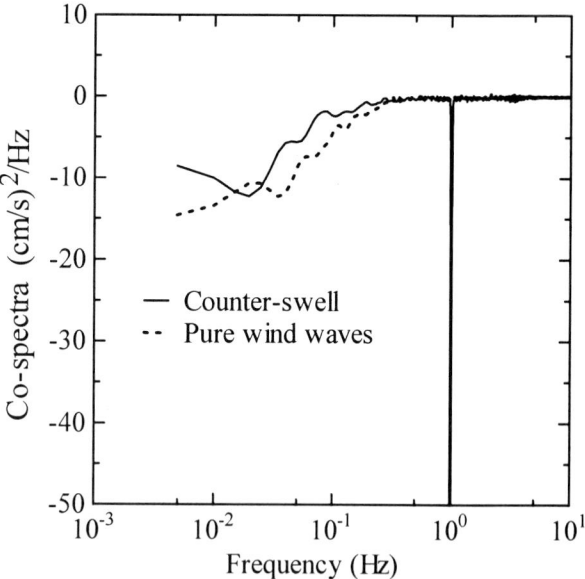

Figure 3. Co-spectra between $(\tilde{u}+u')$ and $(\tilde{w}+w')$ of the same current fluctuations as used in Figure 2.

product $-UW = 6$ (cm/s)^2, which is very large, i.e., about 6 times of the Reynolds stress of air turbulence near the surface (see Fig. 1 and Plate 1), while U = 3.6 cm/s and $W \approx 0$, i.e., $-UW \approx 0$, at the same position for the case of counter-swell. As shown in Plate 1, it is certain that this significant difference of $-UW$ between the two cases is due to the presence of nearly steady and active, organized convective motions in the former case.

Thus, for the pure wind-wave case the wind-driven current is not only accelerated by the wind stress, but also retarded by the large-scale momentum exchange associated with the 2D-LC between the downwind and upwind current, leading to slowdown of the primary circulation, whereas for the case of counter-swell, since the secondary flow is very weak, the wind-driven current is only accelerated by the wind stress, resulting in intensification of <U>. This is the primary reason why the primary circulation was intensified markedly for the latter case.

Finally, though the acceleration resulting from the slope of mean water level in Eq. (2) increased by 6% by adding the counter-swell, the change of the slope in the two cases was negligibly small in comparison with $-\partial/\partial z <\widetilde{u}\widetilde{w}>$ near the surface, i.e., about 0.4 cm/s^2, (see Fig. 5).

4. CONCLUSION

It was found that mechanically generated counter-swell suppresses evolution of the 2D-LC in the tank and simultaneously strengthens the transport of the primary

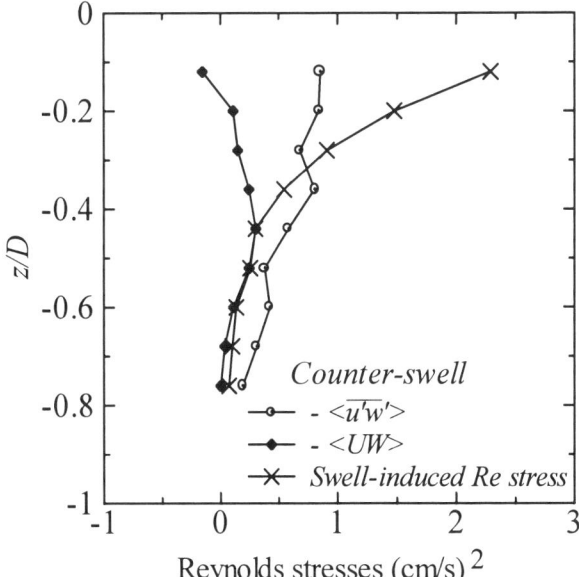

Figure 5. Vertical profiles of Reynolds stresses for the case of counter-swell.

circulation by about 60 % as compared with the pure wind wave case. Thus, we conclude that a greater part of the rate of increase of about 50 % of the surface drift current found by CM for the case of counter-swell arises from intensification of the primary circulation of the subsurface current rather than the air-water interaction between the wind or/and wind-wave and the swell.

We would like to stress that the present experimental results on response of the 2D-LC to the counter-swell qualitatively support the CL2 mechanism. Without the knowledge of the CL2 mechanism on waves propagating against the wind, it would have been difficult for us to try to make the present experiments.

The swell-induced Reynolds stress of the current was unexpectedly large and consistently positive. It may be the secondary important mechanism that accelerates the near-surface wind-driven current. However, further study will be required to understand its role, including the decay of the swell.

Acknowledgements. The author thanks Prof. H. Mitsuyasu for many discussions about the interaction between wind-waves and swells, valuable comments and hearty encouragement. He also thanks two reviewers for helping improve the manuscript. Thanks are extended to two graduate students Messrs. S. Miura and M. Tokuichi for helping to do the experiments. This work was supported financially by The First High Technology Research Project of Hiroshima Institute of Technology.

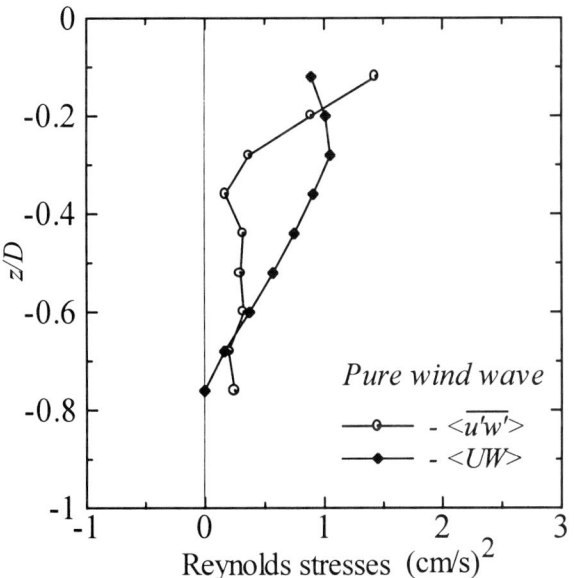

Figure 4. Vertical profiles of Reynolds stresses for the pure wind-wave case.

REFERENCES

Cheng, Z & H. Mitsuyasu (1992): Laboratory studies on the surface drift current induced by wind and swell. *J. Fluid Mech.*, 243, 247-259.

Craik, A. D. D. & S. Leibovich (1976): A rational model for Langmuir circulations, *J. Fluid Mech.*, 73, 401-426.

D'Alessio, S. J. D, K. Abdella & N.A. McFarlane (1998): A new second-order turbulence closure scheme for modeling the oceanic mixed layer, *J. Phys. Oceanogr.* 28, 1624-1641.

Leibovich S. (1983): The form and dynamics of Langmuir circulations. *Annu. Rev. Fluid Mech.*, 15, 391-427.

Mizuno, S., H. Noguchi & Y. Kimura (1998) :A pair of Langmuir cells in two laboratory tanks, (II) on generation mechanism, *J. Oceanogr.*, 54, 77-100.

Pollard, R.T. (1977): Observations and theories of Langmuir circulations and their role in near surface mixing. *In A Voyage of Discovery*, Pergamon Press, Oxford, 235-251.

Skyllingstad, E. D. & D. W. Denbo (1995):An Ocean large eddy simulation of Langmuir circulations and convection in the surface mixed layer. *J. Geophys. Res.*, 100, 8501-8522.

Department of Engineering, Hiroshima Institute of Technology, Miyake 2-1-1, Saeki-ku, Hiroshima 731-5193 JAPAN

A Closer Look at Short Waves Generated by Wave Interactions With Adverse Currents

Steven R. Long and Jochen Klinke

NASA GSFC / WFF, Wallops Island, VA and Scripps Institution of Oceanography, La Jolla, CA

Adverse currents over varying bottom topography can produce current gradients capable of blocking and trapping surface waves in the absence of wind. The blocking/trapping mechanism causes a shift in the longer incoming wave length until short waves in the capillary range are produced at the blocking point. These blocking areas effectively absorb the incoming wave energy, converting it to other forms, possibly turbulent mixing, which could effect the flux of gases across the interface. The short wave lengths produced are in the range that interacts with the present spacecraft radar instruments, raising the possibility of measurements over wide areas of gas exchange across the air-sea interface. The conditions reported here were produced at NASIRF, the NASA Air-Sea Interaction Research Facility at Wallops Island, VA. Using a novel digital imaging technique for measuring surface slope, these interactions were studied in detail, using the new Hilbert-Huang component analysis techniques for nonlinear, non-steady data. The results show that wave number can vary substantially during short periods of time, a phenomena not observed with the earlier Fourier and wavelet techniques, due to inherent limitations of these previous methods.

1. INTRODUCTION

1.1. Adverse Currents and Blocking: An Overview

Smith (1975) studied wave interaction with adverse current, and derived a uniform, asymptotic solution which indicated a reflection of waves instead of wave breaking. Basovich and Talanov (1977) independently confirmed this result. Later, other factors were also included, such as nonlinear and finite amplitude effects, as found in the work of Peregrine and Smith (1979) as well as Peregrine and Thomas (1979). Blocking and reflection were observed in the laboratory study of Badulin *et al*. (1983). Shyu and Phillips (1990) extended the earlier work to include a uniform asymptotic solution of blocking for capillary-gravity waves. Wave-current interaction has also been studied in the laboratory work of Lai *et al*. (1989). Evidence of double reflection and trapping of waves was found in the study of Long *et al*. (1993).

Theoretically, wave motions are governed by the kinematic and dynamic conservation laws given by

$$\partial \boldsymbol{k}/\partial t + \nabla n = 0 \quad (1)$$

and

$$\partial A/\partial t + \nabla \cdot [(\boldsymbol{C}_g + \boldsymbol{U})A] = 0. \quad (2)$$

Here $\boldsymbol{k}$ is the wave number, n is the apparent wave frequency, A is the wave action defined as E/σ with E as the wave energy density and σ as the intrinsic wave frequency, $\boldsymbol{C}_g$ is velocity, and $\boldsymbol{U}$ is the ambient current. Equation (1) is exact, while Equation (2) is the Wentzel-Kramers-Brillouin method approximation of the action conservation law. Equation (2) has neglected wave reflection, so it is not a uniformly valid solution for wave blocking or trapping. Using the steady-state assumption, Equations (1) and (2) will reduce to

---

Gas Transfer at Water Surfaces
Geophysical Monograph 127
Copyright 2002 by the American Geophysical Union

$$n = \sigma + \boldsymbol{k} \cdot \boldsymbol{U} = \text{constant} \qquad (3)$$

and
$$(C_g + \boldsymbol{U})A = \text{constant} \qquad (4)$$

As discussed by Long *et al*. (1993), Equation (3) can be rewritten as

$$n = (gk + \gamma k^3)^{1/2} + \boldsymbol{k} \cdot \boldsymbol{U}. \qquad (5)$$

Here the intrinsic frequency, σ, satisfies the dispersion relationship (where γ is the surface tension)

$$\sigma = (gk + \gamma k^3)^{1/2}. \qquad (5)$$

The results of Long *et al*. (1993) showed the existence of blocking conditions, as well as wave trapping. Certainly, as these conditions are approached, the resulting interactions between waves and adverse current are not fully understood. The present study was designed to examine these interesting and important phenomena closer.

1.2. Adverse Currents and Blocking: This Study

The measurements of this study were carried out at the NASA Air-Sea Interaction Research Facility (NASIRF) at Wallops Island, VA. The details of the facility and its capabilities can be found in Long (1992), Lai *et al*. (1989), and Long *et al*. (1993). The facility can also be visited on-line at http://airsea.wff.nasa.gov to see the latest projects completed and those in progress.

For these measurements, an SMD (Silicon Mountain Design) 1M-60 digital camera was used to obtain the 512 x 512 pixel images. The camera is capable of 60 images/sec at 12-bit resolution. The water surface was illuminated from below as in Plate 1. This is a novel approach using Schnell's law of refraction to encode the downstream surface slope at each cross channel pixel column by means of light intensity. Waves of different slope will refract light of different intensity originating from different areas of the intensity gradient of the source. Each cross channel column of pixels (512 columns) needs a complete lookup table of slope vs. intensity, in order to calibrate each image. Thus a set of 512 lookup tables is needed.

The concept of measuring surface slope in this manner has been developed over recent years by Bernd Jähne of the University of Heidelberg and Jochen Klinke of the Scripps Institution of Oceanography, and is described in detail in Klinke (1996). It is modified here to use an array of fluorescent tubes with an electronic ballast as a light source, and a thin film to provide the intensity variation needed.

2. THE INTERACTIONS

Waves were generated hydraulically and propagated into an adverse current. As the imaging area was approached by the waves, the current increased with fetch due to flow over a false bottom as shown in Plate 1. Thus the waves met a current that was increasing linearly with propagation distance of the waves. The wave frequency and amplitude, along with current speed, was adjusted to bring the progressive wave train to a stop within the imaging area, where the resulting interactions could be captured in digital image sequences. The following images show a sequence of interactions that will be studied further here. We start at image 127 of a set of 600 taken over 10 sec., as shown in Plate 2.

We followed these changes through several images, each 1/60 second apart, arriving at image number 133, which is 1/10 second after image 127. In these Plates, the waves are traveling from right to left against an adverse current flowing left to right. As the waves propagate left, the adverse current speed is linearly increasing. In Plates 4 and 5, the short wave length waves created at the maximum slope of the incoming longer wave have exploded towards the increasing current, in the manner of ring waves around a stone dropped through the surface. In 1/30 of a second more, Plates 6 and 7 show the rapid development of this feature. The slice is taken horizontally along a line at 17.88 cm along the y axis.

Plate 6 shows the expanding set of higher frequency waves near the 17.88 cm position in the image. From the slice shown in Plate 7, we note that this rapid change in wave number is happening near maximum slope just ahead of the crest of the incoming longer wave found at slope=0 on the plot. The end point in the sequence we present here is at image 133, shown in Plates 8 and 9.

3. ANALYSIS

To get an even closer look at these interactions, we employed EMD/HHT (Empirical Mode Decomposition/ Hilbert-Huang Technique) methods described in detail in Huang *et al*. (1998, 1999). This is a new, robust method just patented recently by NASA for analyzing nonlinear and non-stationary data. The method decomposes complicated data into a finite and often small number of 'intrinsic mode functions' that admit well-behaved Hilbert transforms. The decomposition portion of the technique is adaptive, and highly efficient. The conceptual innovations are centered around the introduction of 'intrinsic mode functions' based on local properties of the signal, which makes the instantaneous frequency meaningful, eliminating the need for spurious harmonics to represent nonlinear and non-stationary data, as introduced by the older Fourier and wavelet techniques.

LONG AND KLINKE 123

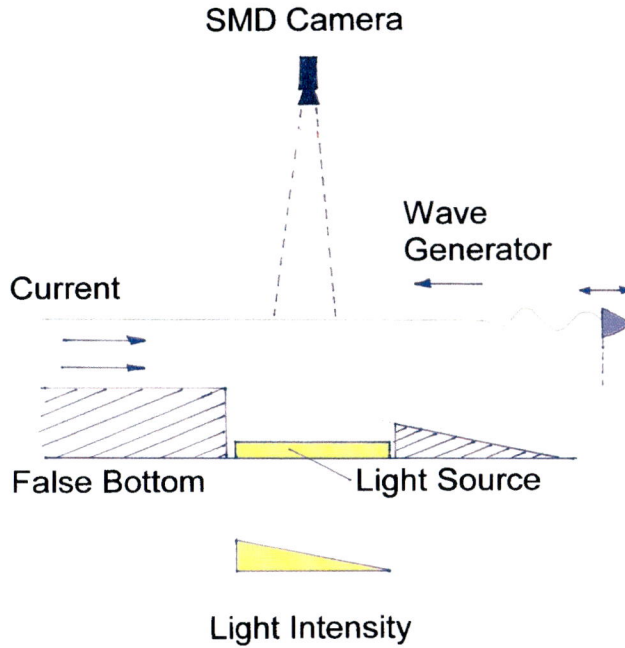

Plate 1. Slope images via intensity

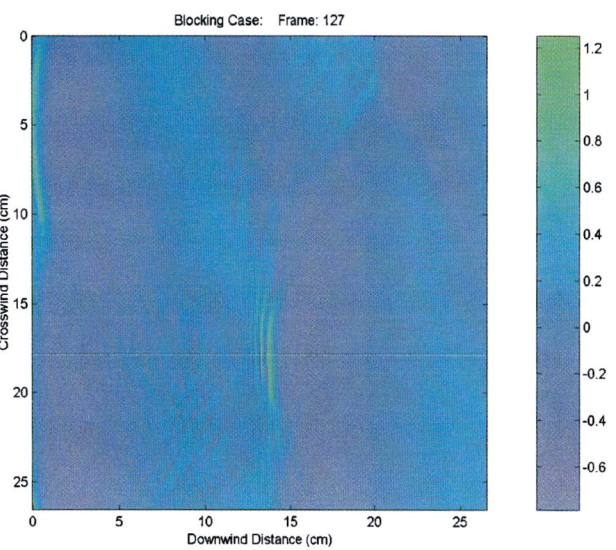

Plate 2. Start of digital slope sequence, image 127

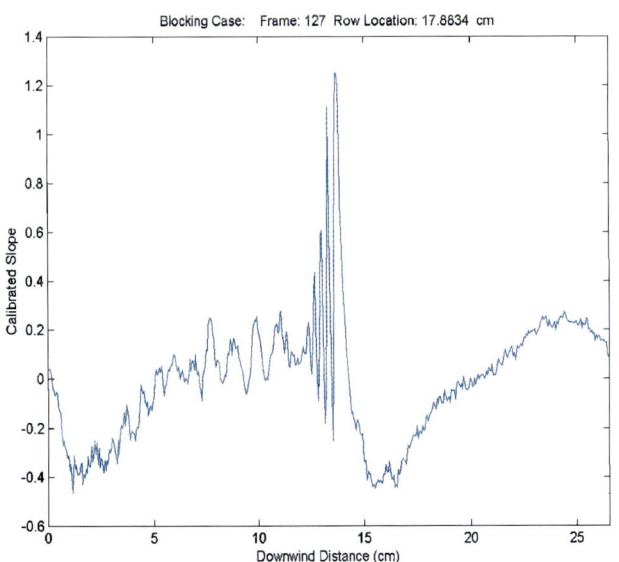

Plate 3. Horizontal slice of Plate 2

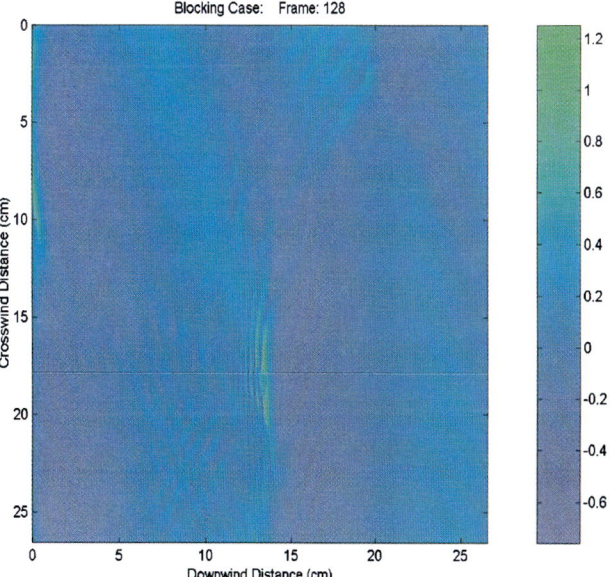

Plate 4. Sudden wave number change, image 128

124 A CLOSER LOOK AT SHORT WAVES

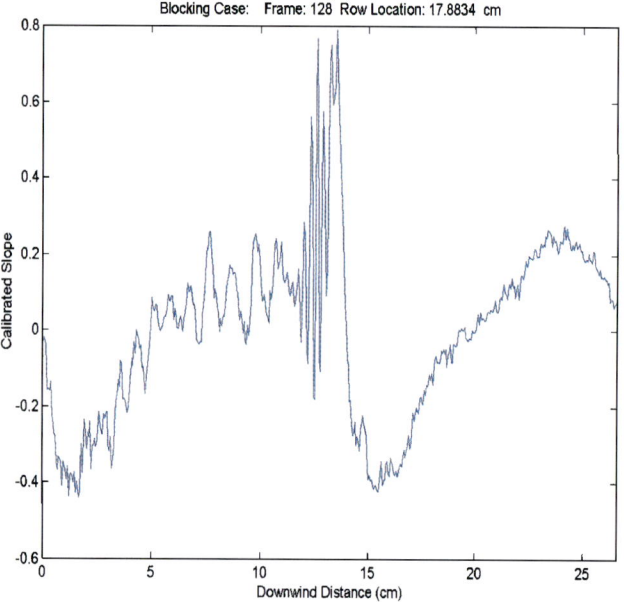

Plate 5. Slice through change in Plate 4

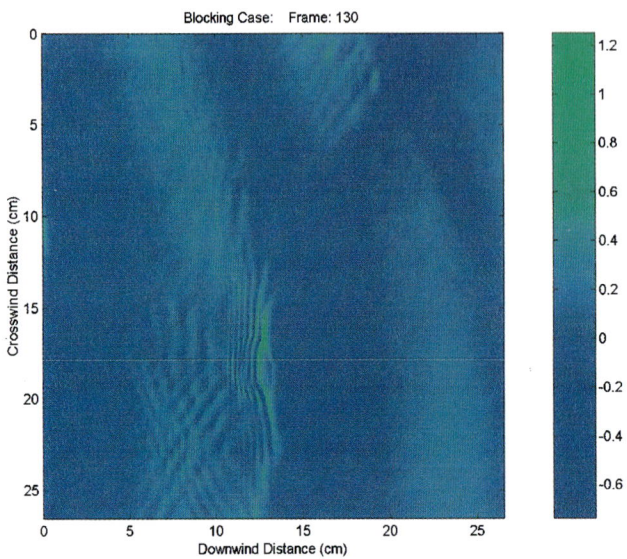

Plate 6. Rapid change, image 130

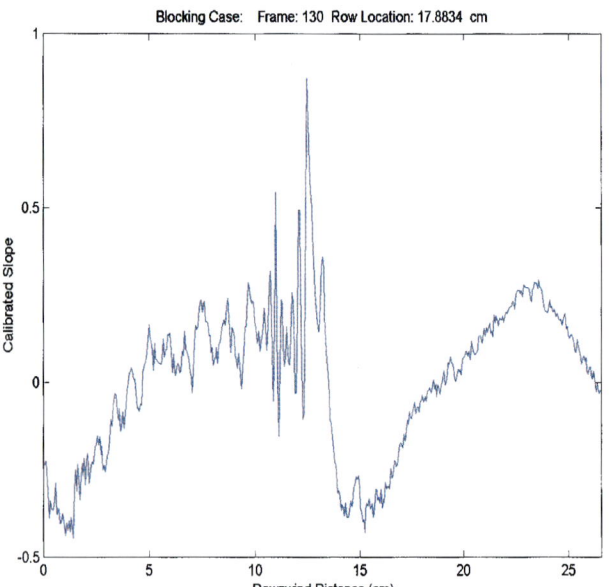

Plate 7. Slice through feature in Plate 6

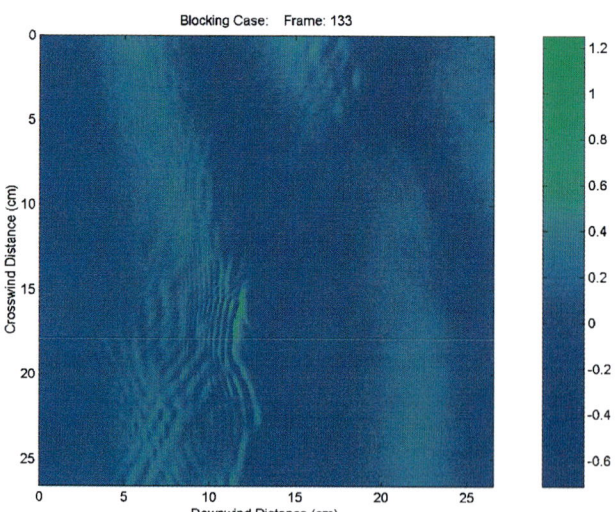

Plate 8. Final image in sequence, image 133

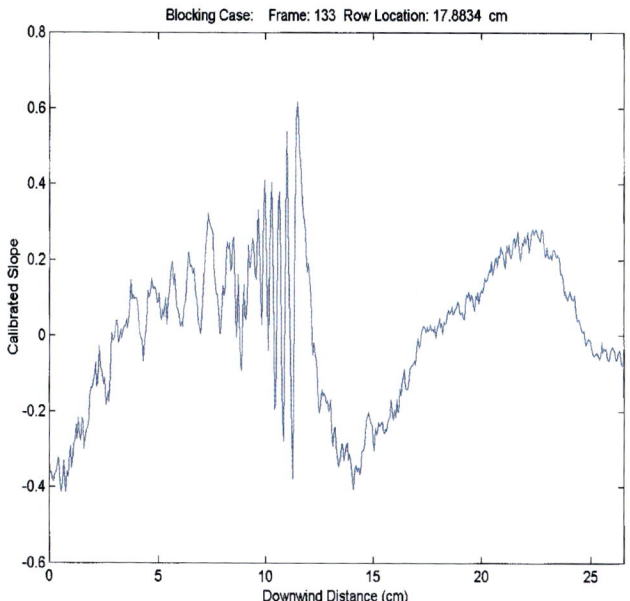

Plate 9. Slice through feature in Plate 8

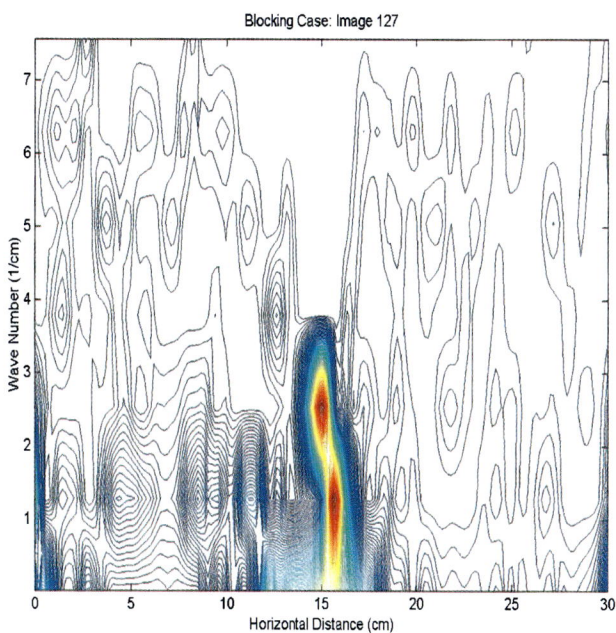

Plate 10. Wave numbers in image 127, the start

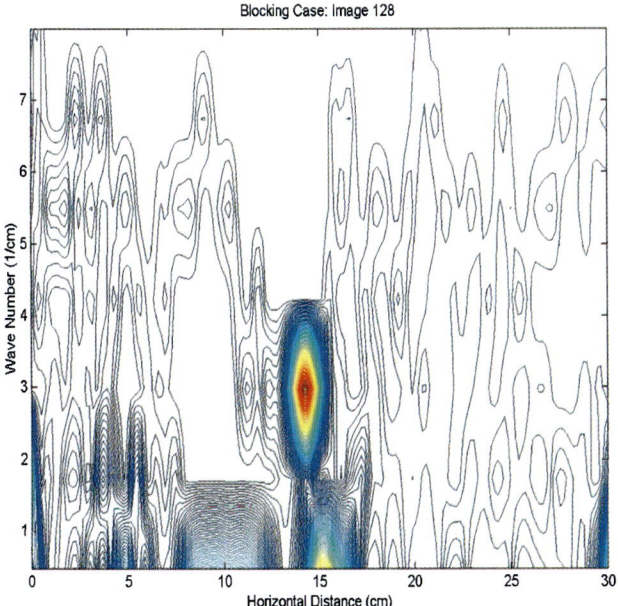

Plate 11. Image 128, 1/60 sec later

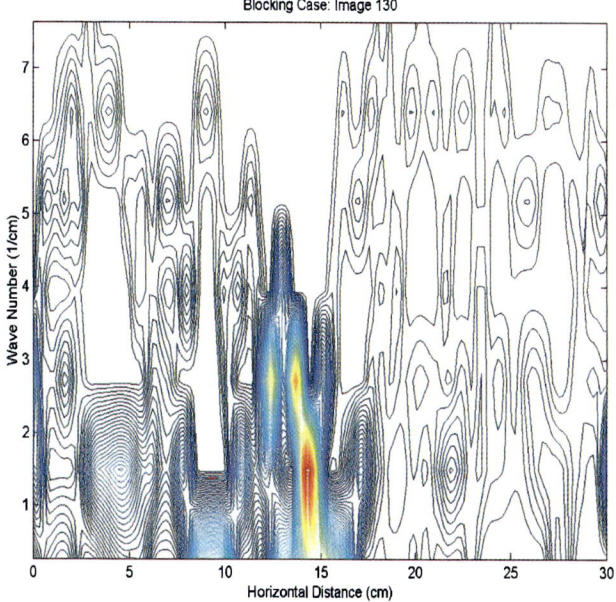

Plate 12. Image 130, 1/20 sec from start

126 A CLOSER LOOK AT SHORT WAVES

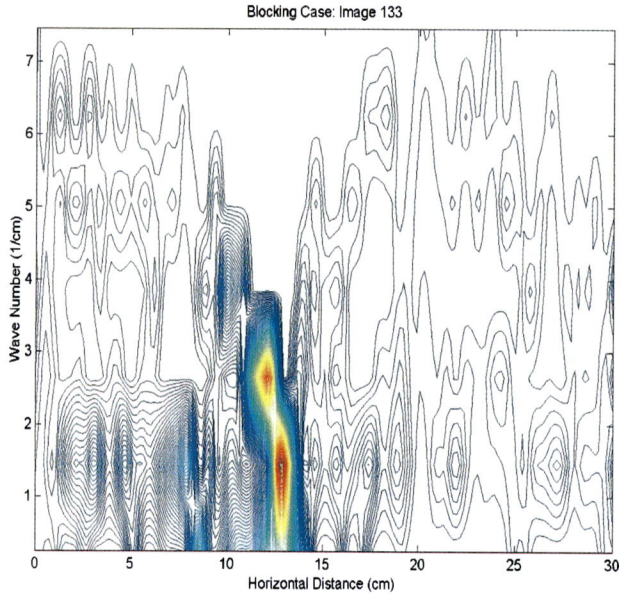

Plate 13. Image 133, 1/10 sec from start

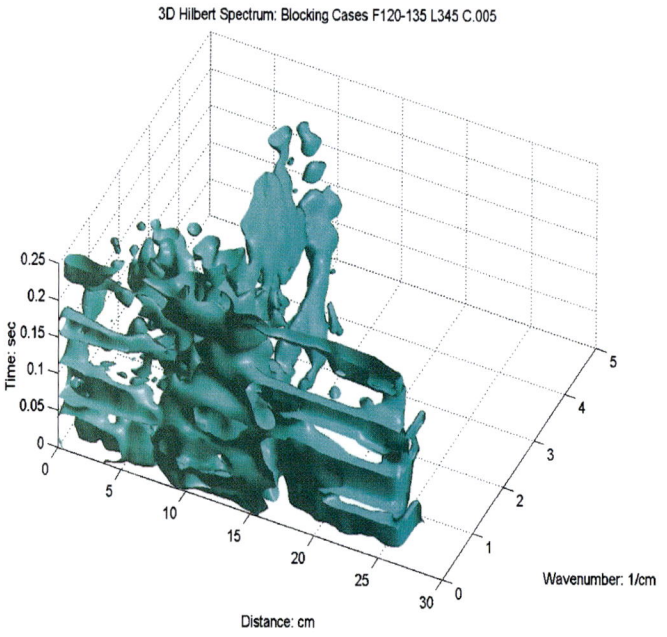

Plate 14. The 3-D Hilbert spectrum

The data shown in Plates 3, 5, 7, and 9 were analyzed using this new EMD/HHT approach. The results are as follows.

Plates 10 through 13 demonstrate the rapid change in wave number as the high frequency waves burst out of the maximum slope area of the incoming longer wave. The contour levels shown in color represent the slope existing at these wave numbers. Although theses processes occur at small scales, the sum of such breaking events has a large impact on the amount or rate of gas exchange. Zappa *et al*. (2001) has shown that the fractional area coverage of the surface affected by these features is significant, and was found to be linearly correlated with the gas transfer velocity, insensitive to surfactants, and independent of fetch. Micro-scale breaking was found by Zappa *et al*. (2001) to account for up to 75% of the gas transfer. Although wind was not used in our measurements, the events during micro-scale breaking as generated here were also seen to develop and progress rapidly. The non-steady nature of the data thus requires more than the earlier techniques can deliver.

As a final look into this rapid progression, we use the results from horizontal slices taken at 17.88 cm in the images from image 120 to 135, or covering a time period of 1/4 sec. Plate 14 shows the changes during this brief time period, using the EMD/HHT techniques. The stick-like structure starting around 15 cm (in the horizontal distance as shown) moves rapidly into larger wave numbers while also increasing along the vertical time axis. This represents the sudden change in wave length and wave number during the explosive event studied here. Plate 14 thus demonstrates the underlying nonlinear and non-stationary nature of these important processes. Earlier methods have revealed average measures, but failed to reveal the true workings of these processes so important in gas exchange and remote sensing studies.

4. CONCLUSION

Micro-breaking events occur when the local slope attains a critical steepness. This can occur at almost any wind speed, from light to moderate winds as observed by Zappa *et al.* (2001), to higher values as well. Currents can also raise the local slope to the critical values even in the absence of wind, as shown here. Throughout the world's oceans, these micro-breaking processes occur as a major source of gas transfer, whether wind-driven, current driven, or in combination. Zappa *et al* (2001) clearly demonstrated this in extensive measurements. Because they occur at the wave lengths sensed by active remote sensors, the feasibility of surveying the world's oceans for fractional coverage by micro-scale breaking by satellite instruments is now an ongoing possibility, as shown in the work of Gregg *et al.* (2000). Thus the nonlinear and non-steady nature of this micro-scale breaking will need to be better understood. This can be done in future studies in applying the new methods of EMD/HHT to this problem, and by other approaches yet to be developed.

Acknowledgments. The authors would like to thank Dr. Norden E. Huang of NASA/GSFC/Code 971 for his many helpful suggestions in the preparation of this manuscript. SRL gratefully acknowledges the support of NASA through Headquarters Code Y and its Physical Oceanography Program administered by Dr. Eric Lindstrom, as well as help from Al Waller and Richard Mitchell of EG&G in setting up the computing capability for this study. JK gratefully acknowledges the support of Scripps Institution of Oceanography and NASA Grant No. NAG5-5217.

REFERENCES

Badulin, S. T., Pokazeyev, K. V., and Rozenberg, A. D., A laboratory study of the transformation of regular gravity-capillary waves on inhomogeneous flows, *Izv. Atmos. Ocean. Phys.*, 19, pp. 782-787, 1983.

Basovich, A. Y. and Talanov, V. L., Transformation of short surface waves on inhomogeneous currents, *Izv. Atmos. Ocean. Phys.*, 13, pp. 514-518, 1977.

Gregg, W. W., Behrenfield, M. J., Hoge, F. E., Esaias, W. E., Huang, N. E., Long, S. R., and McCl;ain, C. R., NASA/GSFC Research Activities for the Global Ocean Carbon Cycle: A Prospectus for the 21st Century, NASA/TM No. 2000-209882, 22 pp., 2000.

Huang, N. E., Shen, Z., Long, S. R., Wu, M. C., Shih, H. H., Zheng, Q., Yen, N. C., Tung, C. C., and Liu, H. H., The empirical mode decomposition and the Hilbert spectrum for nonlinear and non-steady time series analysis, Proc. R. Soc. London, Ser. A, 454, pp. 903-995, 1998.

Huang, N. E.,Shen, Z., and Long, S. R., A new view of nonlinear water waves: the Hilbert spectrum, *Annu. Rev. Fluid Mech.*, 31, pp. 417-457, 1999.

Klinke, J., Optical measurements of small-scale wind-generated water surface waves in the laboratory and the field, University of Heidelburg, doctoral dissertation, 103 pp., 1996.

Lai, R. J., Long, S. R., and Huang, N. E., Laboratory studies of wave-current interactions: kinematics of the strong interaction, *J. Geophys. Res.*, 94, pp. 16201-16214, 1989.

Long, S. R., NASA Wallops Flight Facility: Air-Sea Interaction Research Facility, NASA Tech. Ref. Publ. RP-1277, 36 pp., 1992.

Long, S. R., Lai, R. J., Huang, N. E., and Spedding, G. R., Blocking and trapping of waves ini an inhomogeneous flow, *Dyn. Atmos. Oceans*, 20, pp. 79-106, 1993.

Peregrine, D. H. and Smith, R., Nonlinear effects upon near cautics, Philos. *Trans. R. Soc. London, Ser. A*, 293, pp. 341-370, 1979.

Peregrine, D. H. and Thomas, G. P., Finite-amplitude deep-water waves on currents, *Philos. Trans. R. Soc. London, Ser. A*, 293, pp. 371-390, 1979.

Shyu, J. H. and Phillips, O. M., The blockage of gravity and capillary waves by longer waves and currents, *J. Fluid Mech.*, 217, pp. 115-141, 1990.

Smith, R., The reflection of short gravity waves on a non-uniform current, Proc. Cambridge Philos. Soc., 78, pp. 517-525, 1975.

Zappa, C. J., Asher, W. E., Jessup, A. T., Klinke, J., and Long, S. R., Effect of microscale wave breaking on air-water gas transfer, Geophysical Monograph series (this volume), edited by M. A. Donelan, W. M. Drennan, E. S. Saltzman and R. Wanninkhof, American Geophysical Union, Washington, DC, 2000.

Jochen Klinke, Scripps Institution of Oceanography, Physical Oceanography Research Division 0230, La Jolla, CA 92093-0230. jklinke@ucsd.edu

Steven R. Long, NASA Goddard Space Flight Center, Wallops Flight Facility, Laboratory for Hydrospheric Processes, Observational Science Branch, Code 972, Bldg. N-159, Wallops Island, VA 23337. steve@airsea.wff.nasa.gov

Surface Divergence and Air–Water Gas Transfer

Sean P. McKenna and Wade R. McGillis

Department of Applied Ocean Physics and Engineering, Woods Hole Oceanographic Institution, Woods Hole, Massachusetts

Recent laboratory measurements of bulk gas-transfer velocity and free-surface hydrodynamics were made for the case of oscillating grid-stirred turbulence. Air–water gas-transfer velocities were determined from the waterside dissolved oxygen mass balance, and an innovative digital particle image velocimetry technique was used to measure the two-dimensional free-surface flow field. Bulk turbulence was found unable to provide a unique relationship for the gas-transfer rates realized, owing predominantly to surface contamination effects (adventitious and/or purposely introduced films). However, it was found that the surface divergence computed from the free-surface velocity field information was capable of reconciling the gas-transfer data in a physically meaningful way. This appears to be evidence confirming that the free-surface divergence is an important process involved in interfacial gas transport.

1. INTRODUCTION

Gas transfer at air–water interfaces ultimately occurs on a molecular level through a process of diffusion. This is true regardless of the nature of the flow regimes on either side of the interface, be they quiescent or turbulent. For the case of sparingly soluble gases, the transfer rate is controlled by the waterside resistance and the airside resistance can be neglected. Under turbulent conditions, the diffusional process results in a very thin aqueous mass boundary layer at the interface. The concentration gradient across this layer determines the flux of gas through the interface. Waterside mixing processes such as free-surface waves and near-surface turbulence act to thin this boundary layer, steepening the concentration gradient, and enhancing gas exchange.

Turbulent eddies near the free surface are important to air–water gas exchange because these eddies bring fresh fluid near the interface for diffusion. Frequent vertical transport of fresh fluid toward the surface results in a reduction of the mass boundary layer thickness. In this way, the vertical velocity fluctuations very near the interface are considered vital to gas-transfer enhancement. By continuity, and using Taylor expansion, these fluctuations, $w(z)$, can be expressed as

$$w(z) = -(\nabla_h \cdot \mathbf{v})_0 z - \left[\frac{\partial}{\partial z}(\nabla_h \cdot \mathbf{v})\right]_0 \frac{z^2}{2} + \ldots \quad (1)$$

near the interface, where ∇_h is the horizontal divergence operator and the subscript 'o' indicates evaluation at the free surface, $z = 0$. For small z, higher order terms can be neglected and (1) reduces to

$$w(z) \approx -\left(\frac{\partial u}{\partial x} + \frac{\partial v}{\partial y}\right)_0 z. \quad (2)$$

Thus, the velocity fluctuations responsible for thinning the mass boundary layer and increasing the interfacial gas flux are closely related to the divergence of the velocity field at the surface. The gas-transfer models of *Brumley and Jirka* [1988], *Chan and Scriven* [1970], and

Table 1. Grid Forcing Conditions

S (cm)	f (Hz)	$u_{HT}(z_s)$ (cm/s)	Re_{HT}
6.35	1.40	0.56	282
8.89	1.15	0.76	384
10.16	1.15	0.92	469
7.62	2.25	1.17	596
11.43	1.50	1.44	730
8.89	2.25	1.48	751
10.16	2.20	1.77	898
11.43	2.00	1.92	974

Csanady [1990], for example, incorporate the surface divergence as a key mechanism for gas-transfer enhancement. Such approaches may provide improved insight into the mechanisms of gas exchange.

The hydrodynamics at a free surface are inherently complex and are made even more so due to surfactant effects. Through their rheological behavior, surfactants act to damp flow at the free surface by resisting surface compression and dilation. By introducing an interfacial stress (the tangential stress at a perfectly clean air–water surface is essentially zero) in the plane of the interface, surfactants create a highly dissipative viscous boundary layer where flow damping takes place. Such flow damping acts to thicken the aqueous mass boundary layer, and can retard gas exchange significantly. Surfactants are ever-present at the air–sea interface, and are arguably present in all laboratory experiments. Therefore, careful consideration of the role surfactants play in interfacial studies is requisite. In the present work, fundamental laboratory experiments are performed that add to our current understanding of surface divergence, surfactants, and gas transfer.

2. EXPERIMENTAL APPROACH

2.1. Oscillating Grid-Stirred Turbulence

A recently constructed oscillating grid-stirred tank was used to explore the relationship between free-surface dynamics and air–water gas exchange. For such a study, a grid-stirred tank has the desirable benefits of generating repeatable levels of turbulence with very little mean flow in a closed, gas-tight system that can be successfully managed for chemical cleanliness. Furthermore, the turbulence produced due to an oscillating grid has been examined extensively and is discussed as being near-isotropic and quasi-homogeneous in the horizontal plane (e.g., *Hopfinger and Toly* [1976]; *Brumley and Jirka* [1987]). The present tank has dimensions 45.4 cm × 45.4 cm × 57.2 cm deep and was constructed entirely from polycarbonate, with all joints chemically bonded leaving no residual solvents. This avoided spurious contaminants from other types of materials and/or sealants. The system was made gas-tight using a neoprene gasket and tank lid. A 7 × 7 grid made of $d = 1.27$ cm square cross-section polycarbonate bars having a mesh size $M = 6.35$ cm was used for all experiments discussed. The grid was rigidly attached at its center vertex to a stainless steel shaft that passed through the tank floor with a watertight seal provided by a stack of Teflon V-rings. A gearmotor-driven reciprocating mechanism beneath the tank vertically oscillated the shaft and grid. Using this arrangement, the water surface was completely free of obstructions. Complete details of the grid tank and the generated turbulence can be found in *McKenna* [2000].

In all experiments, the water depth was 50.8 cm and the distance from the mean grid position to the free surface was $z_s = 25.4$ cm. Several different grid stoke, S, and frequency, f, combinations were explored. These are summarized in Table 1. Other parameters appearing in Table 1 were determined using the empirical expressions of *Hopfinger and Toly* [1976], who found that for a planar grid of square cross-section bars with $M/d = 5$, the RMS horizontal turbulent velocity could be related to the grid parameters as $u_{HT} = 0.25 M^{0.5} S^{1.5} f z^{-1}$, where z is the vertical distance from the mean grid position. Furthermore, the longitudinal integral lengthscale was found to increase linearly with distance from the grid: $L_{HT} = \alpha z$. The constant α has been shown to fall in the range 0.1–0.4; here, as in *Brumley and Jirka* [1987], our measurements indicate a choice of 0.1. The turbulent Reynolds number based on the Hopfinger and Toly values was defined as $Re_{HT} = 2 u_{HT} L_{HT}/\nu$ and is used to characterize the bulk turbulent mixing.

2.2. Free-Surface Particle Image Velocimetry

Particle image velocimetry (PIV) is a powerful flow measurement technique that, in its most common form, provides quantitative, whole-field fluid velocity information in a two-dimensional plane of interest. While PIV is becoming prevalent in experimental fluid mechanics, use of PIV at the free surface is still novel. *Gulliver and Tamburrino* [1995] used a quantitative imaging technique to investigate the flow field in a moving bed flume. A similar concept was implemented in this work using a PIV-based approach. Figure 1 illustrates the free-surface PIV technique, and a full description of the technique can be found in *McKenna* [2000]. An 8-bit digital CCD camera (1008 × 1018 resolution, maximum frame rate 30 Hz) recorded the flow, which was

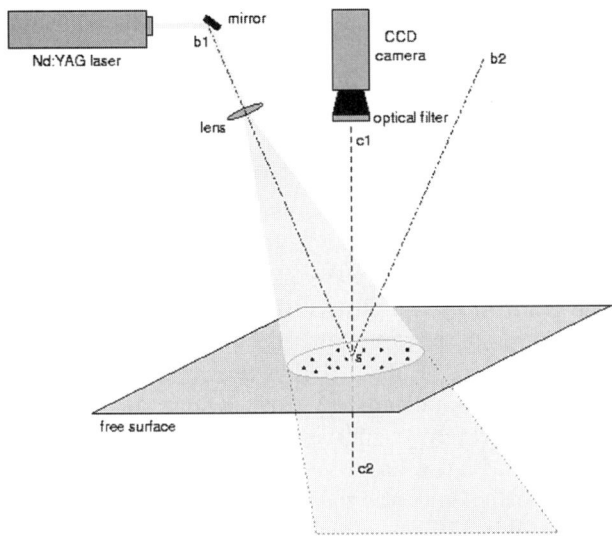

Figure 1. PIV surface flow measurement optical arrangement. The laser beam is redirected by a high energy mirror along beam axis b1–s. A spherical lens fans the beam into a three-dimensional, diverging light cone incident on the free surface. The camera optical axis is c1–s–c2. This axis lies in the plane of incidence formed by b1–s–b2.

seeded with fluorescent acrylic spheres (20–40 μm or 80–120 μm in diameter, specific gravity near 1.0) impregnated with Rhodamine 6G and dichloro-fluorescein dyes. To remove the unwanted laser light reflection from the water surface, a precision optical bandpass filter was used. By bandpass filtering the camera, the laser light is eliminated from the video image and only the PIV particles are detected. A potential concern with the approach depicted in Figure 1 is the fact that the cone of laser light penetrates the water column, illuminating particles at depth as well as at the surface. However, the details of the optical filtering and the particle fluorescence, along with the light scattering behavior of particles in water versus air, resulted in only surface particles being imaged by the camera. A complete explanation is given by *McKenna* [2000], where a complementary experimental validation of the free-surface PIV technique showed that any effect of imaging particles at depth is insignificant. An advantage of the specific implementation of this surface measurement technique is its potential extension to undulating free-surface flows exhibiting small slopes and small displacements from equilibrium.

2.3. Measurement of Gas Transfer

Measurements of O_2 evasion were used to determine the gas-transfer velocity. Bulk waterside dissolved O_2 concentration was measured at 0.1 Hz using a commercial water quality sensor located below the grid, 10 cm from the tank floor. Diagnostics indicated that the bulk fluid was well-mixed and the measurement of the transfer velocity was independent of sensor location. For each experiment, N_2 gas was used to continually flush the tank headspace to yield a known zero O_2 surface concentration, $C_s(t) \approx 0$ mg/l. The gas-transfer velocity k was obtained using the bulk dissolved O_2 time series data, $C_b(t)$, to solve the expression

$$C_b(t) = C_b(0)e^{-kt/H}, \qquad (3)$$

where H is the water depth and $C_s(t)$ has been taken to be zero.

3. RESULTS AND DISCUSSION

3.1. Bulk Flow Dynamics

Three-component fluid velocity measurements at 25 Hz using an acoustic Doppler velocimeter were used to assess the applicability of the Hopfinger–Toly (HT) relationships discussed in Section 2.1. A collection of measurements at nine different locations at a depth of 8 cm indicated that u_{HT} was an accurate measure of the RMS horizontal turbulent velocity and that the RMS vertical turbulent velocity was typically 10% greater than the horizontal. A separate set of experiments using PIV in a standard light sheet mode explored the turbulent flow field in a number of two-dimensional vertical planes that spanned the tank width. Data was collected for $Re_{HT} = 596$ and $Re_{HT} = 730$, and provided additional evidence of the applicability of both HT relations in the bulk flow (see Figure 2). These PIV measurements also revealed the influence of the free surface on both the turbulent velocities and the turbulent integral lengthscales. The presence of the free surface was found to slightly increase the magnitude of the near-surface RMS horizontal turbulent velocity, while reducing the RMS vertical turbulent velocity toward zero as the interface was approached. Furthermore, the horizontally measured longitudinal integral lengthscale was observed to undergo a reduction near the surface, and the horizontally measured transverse integral lengthscale was found to tend to zero at the free surface. Each of these effects was well-predicted by combined relationships using the source theory of *Hunt and Graham* [1978] and the HT expressions [*Brumley and Jirka*, 1987].

3.2. Surface Dynamics and Air–Water Gas Transfer

The surface PIV technique was used to quantify the free-surface flow field at four different surface locations under a variety of forcing conditions and degrees of sur-

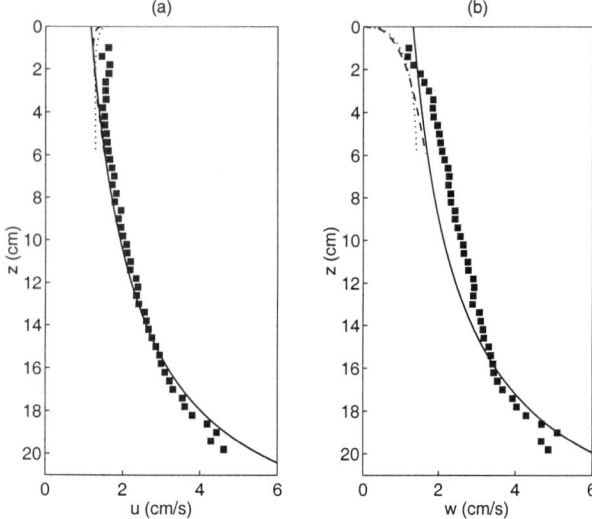

Figure 2. Vertical profiles of spatially averaged RMS (a) horizontal and (b) vertical turbulent velocity fluctuations for $Re_{HT} = 596$. Shown are the PIV data (squares), the empirical predictions u_{HT} and $w_{HT} = 1.1 u_{HT}$ (solid lines), the Hunt–Graham profiles (dotted lines), and the combined profiles (dashed lines). The free surface is located at $z = 0$ and the mean grid position is at $z = 25.4$ cm.

face cleanliness. Each spatial measurement domain was roughly 6 cm square and the physical dimensions of the PIV interrogation windows were either 2.9 mm or 5.8 mm square. The majority of surface velocity data was collected with the image acquisition arranged such that a pair of PIV images were collected (producing a single velocity field) every 5 seconds for 16 minutes. This was done for four surface measurement locations. In this manner, spatially intensive ensemble statistics could be computed. A second imaging setup involved a very small PIV image of the surface that yielded a 3×3 velocity field matrix about a single point. These images were collected continuously at 30 Hz for 10 minutes. Data from these sequences were used to compute time series of turbulent velocities and surface divergence. Concurrently, dissolved O_2 measurements were made to infer the gas-transfer velocity. Measurements of the surface tension using the Wilhelmy plate technique before and after each experimental run were used to qualify the interfacial contamination. Reported here is the surface pressure, which is defined as $\pi = \sigma_o - \sigma$, where σ_o is the reference surface tension and σ is the measured tension.

Figure 3 shows a collection of gas-transfer measurements as a function of the bulk turbulent mixing, parameterized through Re_{HT}. The data include two groups of runs where the tank and surface were kept as clean as possible, a group where a surfactant was added to reduce apparent surface bursting events, and a number of runs with spread oleyl alcohol films of varying concentrations present. Easily apparent from this figure is the inability of the bulk turbulent mixing to uniquely account for the observed transfer rates. The presence of surface-active material is seen to reduce the transfer velocity by a factor of four in certain cases. This is not a surprising result; air–water transport is a process that is determined by the physics very near the free surface, which is where the dominant effects of surfactants manifest themselves. Bulk turbulence away from the interface is largely unaffected by surface films and therefore cannot accurately predict the levels of gas transport.

Shown in Figure 4 are the same gas-transfer measurements plotted against a measure of the free-surface turbulent mixing, $<u>$, which reflects the temporally or ensemble averaged surface RMS velocity fluctuations. Each $<u>$ value derives from a measurement record that is coincident with the measurement of the gas-transfer velocity. Also included in Figure 4 are data collected from another experiment that investigated the ef-

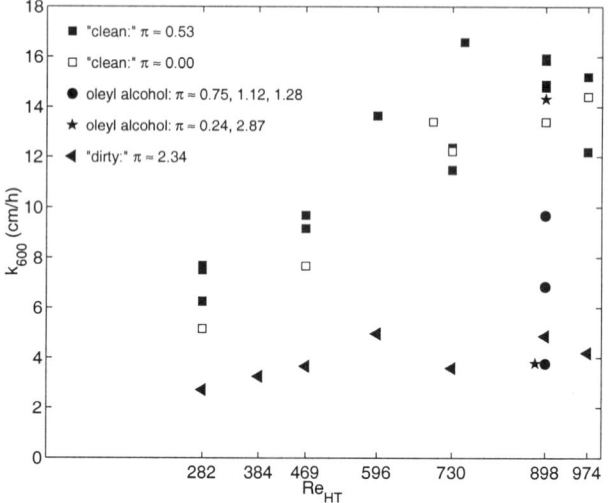

Figure 3. Dependence of the gas-transfer velocity on bulk mixing Reynolds number for grid-stirred turbulence. Transfer velocity has been normalized to that of CO_2 ($Sc = 600$ at $20°C$) with $n = 1/2$. Values of surface pressure π are in units of mN/m and have 95% confidence intervals of ± 0.27 mN/m. "Clean" data refers to cases where the water surface was aspirated and surface films were not introduced. "Dirty" refers to cases where a surfactant was added arbitrarily to minimize free-surface bursting events. Surface pressure values for the "clean" and "dirty" cases are averaged over all runs. The oleyl alcohol data are for spread films of increasing concentrations with individual surface pressures noted. Increasing oleyl alcohol π corresponds to decreasing k_{600} for all data points.

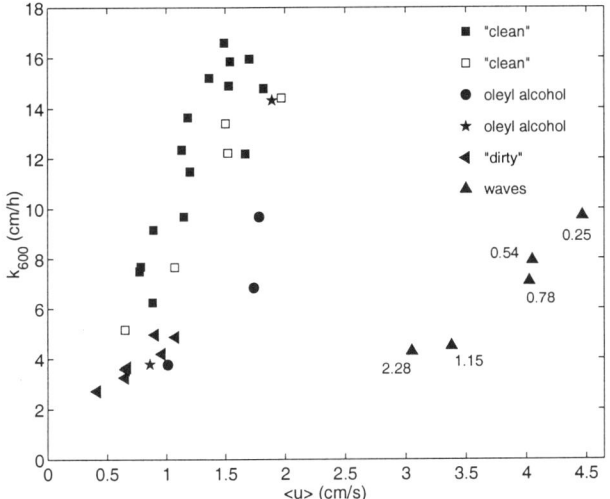

Figure 4. Dependence of the gas-transfer velocity on surface RMS velocity fluctuations for grid-stirred turbulence and mechanically generated surface waves. The details of the figure key are as in Figure 3. Additionally, black markers correspond to ensemble averages of the four PIV spot locations and gray markers correspond to time averages of the temporally intensive PIV point measurements. Surface pressures for wave runs are indicated next to each marker.

fects of mechanically generated waves on gas exchange. This experiment was conducted in the grid tank and the measurement procedure for the transfer velocity was identical to that for the grid-stirred runs. Surface waves were generated using a plunger-type wavemaker positioned along one wall of the tank and spanning the width of the tank. The wavemaker was forced at a frequency of 3.2 Hz. The resulting wave field was dominated by two-dimensional standing waves having amplitudes of less than 2 cm. Such a wave field is far from ideal, but it was simple to generate and provided a completely different flow regime for study. In an effort to keep the grid tank reasonably well-mixed during the wave runs, the grid was oscillated in a low energy state: $S = 10.16$ cm, $f = 0.5$ Hz. Measurement of the transfer velocity under this forcing alone without waves indicated extremely small enhancement ($k_{600} \approx 2$ cm/h). Surface PIV measurements were made at a single location in the center region of the tank in the ensemble imaging mode. Runs were performed with increasing concentrations of spread oleyl alcohol films. For the grid-stirred turbulence data, relating k to $<u>$ reduces the scatter in the data and suggests a more clear monotonic trend. The R^2 regression statistic for this data was 0.68. However, inclusion of the wave data in the regression reduces R^2 to 0.03. Including the wave data seems to indicate that while the surface velocity fluctuations provide an improved relationship for the gas transfer when compared to a bulk turbulence estimate, such a parameterization appears to be dependent upon the particular flow regime. The velocity fluctuations associated with the waves studied are not as effective in creating surface renewal and enhancing gas exchange.

To explore the role of surface divergence, the gas-transfer data shown in Figure 4 are replotted against a measure of the divergence in Figure 5. For the data collected from the spatially intensive measurements at the four PIV locations, $<a>$ is an ensemble-averaged value. The quantity averaged was the magnitude of the divergence, with the divergence being computed from the first-order finite difference of the velocity field. In the case of the time series data, $<a>$ was estimated from the integrated spectral density function for the divergence, as suggested by *McCready et al.* [1986]. The striking feature of this figure, in addition to the favorable monotonic trend, is the fact that both flow regimes, grid turbulence and surface waves, are reconciled by the surface divergence measurement. For grid turbulence alone, $R^2 = 0.82$, an improvement over the k–$<u>$ relationship, and for both turbulence and waves, the result is nearly identical, $R^2 = 0.78$. Also included in this figure are the simple predictions of *Ledwell* [1984] and *Csanady* [1990].

4. CONCLUSION

Based on coincident laboratory measurements of gas transfer and free-surface hydrodynamics under varying

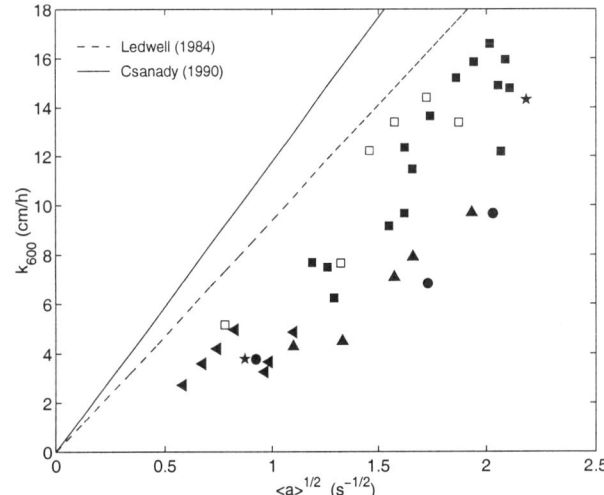

Figure 5. Dependence of the gas-transfer velocity on surface divergence for grid-stirred turbulence and mechanically generated surface waves. Markers are as in Figure 4. Ledwell curve is computed from $k = (2/\pi)\sqrt{Da}$; Csanady curve is computed from $k = \sqrt{2Da/\pi}$.

degrees of interfacial contamination, it appears that the surface divergence is a critical process in air–water gas exchange. Bulk turbulence estimates were unable to provide a unique relationship for the measured gas-transfer velocities. Free-surface turbulence was found to yield an improved relationship, however, there is some evidence that such a relationship is dependent upon the flow regime. It was discovered that surface divergence provided a relationship for the gas transfer that accounted for the presence of surface films and did not depend on flow regime (grid turbulence or waves). The success of the surface divergence is believed to stem from its direct role in surface renewal and its sensitivity to surface film effects. Consequently, this may indicate a meaningful physical relationship between surface divergence and interfacial transport.

Acknowledgments. This work was supported by the Woods Hole Oceanographic Institution Ocean Ventures Fund and the Andrew W. Mellon Foundation for Innovative Research, and by the National Science Foundation under grant no. OCE9724383. (Woods Hole Oceanographic Institution contribution number 10336.)

REFERENCES

Brumley, B. J. and G. H. Jirka, Near-surface turbulence in a grid-stirred tank, *J. Fluid Mech., 183,* 235–263, 1987.

Brumley, B. J. and G. H. Jirka, Air–water transfer of slightly soluble gases: turbulence, interfacial processes and conceptual models, *PhysicoChem. Hydrodyn., 10,* 295–319, 1988.

Chan, W. C. and L. E. Scriven, Absorption into irrotational stagnation flow: a case study in convective diffusion theory, *Ind. Engng. Chem. Fundam., 9,* 114–120, 1970.

Csanady, G. T., The role of breaking wavelets in air-sea gas transfer, *J. Geophys. Res., 95,* 749–759, 1990.

Gulliver, J. S. and A. Tamburrino, Turbulent surface deformation and their relationship to mass transfer in an open-channel flow, in *Air-Water Gas Transfer,* edited by B. Jähne and E. Monahan, pp. 589–600, AEON Verlag, Heidelberg, Germany, 1995.

Hopfinger, E. J. and J.-A. Toly, On mixing across an interface in stably stratified fluid, *J. Fluid Mech., 78,* 155–175, 1976.

Hunt, J. C. R. and J. M. R. Graham, Free-stream turbulence near plane boundaries, *J. Fluid Mech., 84,* 209–235, 1978.

Ledwell, J. R., The variation of the gas transfer coefficient with molecular diffusivity, in *Gas Transfer at Water Surfaces,* edited by W. Brutsaert and G. H. Jirka, pp. 293–302, D. Reidel, Minneapolis, MN, 1984.

McCready, M. J., E. Vassiliadou and T. J. Hanratty, Computer simulation of turbulent mass transfer at a mobile interface, *Am. Inst. Chem. Engrs. J., 32,* 1108–1115, 1986.

McKenna, S. P., Free-surface turbulence and air–water gas exchange, Ph.D. thesis, Massachusetts Institute of Technology, September 2000.

W. R. McGillis, MS 9, Woods Hole Oceanographic Institution, Woods Hole, MA 02543. (email: wmcgillis@whoi.edu)

S. P. McKenna, MS 9, Woods Hole Oceanographic Institution, Woods Hole, MA 02543. (email: smckenna@whoi.edu)

The Critical Importance of Buoyancy Flux for Gas Flux Across the Air-water Interface

Sally MacIntyre

Marine Science Institute, University of California, Santa Barbara, California

Werner Eugster

Institute of Geography, University of Bern, Switzerland

George W. Kling

Department of Biology, University of Michigan, Ann Arbor, Michigan

Physical processes at the air-water interface that affect gas flux include turbulence from wind shear, penetrative convection due to heat loss, micro-wave breaking, and large scale wave breaking. While considerable effort has been expended to parameterize the gas transfer coefficient due to the contributions of wind and surface waves, the contribution of buoyancy fluxes to turbulence at the air-water interface has not received the same attention. In addition, the rate of mixed layer deepening is sensitive to heat loss. Without this deepening, gas concentrations quickly equilibrate with those in the atmosphere leading to lower rates of gas flux. Measured gas flux was up to 5 times higher when heat was being lost from the surface layer of an arctic lake and wind speeds were low than when wind speeds were 5 m s^{-1}. At wind speeds less than 5 m s^{-1}, calculated values of gas transfer velocity based on wind speed alone were 2 to 5 times lower than those calculated using the surface renewal model. These observations indicate the critical importance of including buoyancy fluxes in estimates of gas transfer coefficients.

INTRODUCTION.

Empirical models are frequently used to parameterize the reaeration coefficient k used in computing gas flux across the air-water interface. At present, the three most commonly used equations are based on wind speed (Wanninkhof 1992; Liss and Merlivat 1986; Wanninkhof and McGillis 1999). However, a number of physical processes contribute to gas transfer across the air-water interface (MacIntyre et al. 1995). These include penetrative convection due to heat loss, shear due to wind forcing, microwave breaking at moderate wind speeds (Jessup et al. 1997), and bubbles at high wind speeds (Woolfe and Thorpe 1991). While wind contributes to all of these processes, use of the surface renewal model (Higbie 1935; Danckwerts 1951; Soloviev and Schluessel 1994) to calculate k may prove to be more inclusive. In particular, as heat losses from the air-water interface are due not only to evaporation but also to sensible heat loss and long wave back radiation, the turbulence due to buoyant motions as well as wind may be better modeled by surface renewal. In this paper, we present evidence showing the critical importance of heat loss for gas flux.

Background

Gas flux, when estimated indirectly, is obtained from the equation $F = k(C_w - \alpha C)$ where k is the reaeration coefficient, C and C_w are concentration of the gas in air and water respectively, and α is the Ostwald solubility coefficient. The reaeration coefficient depends upon the Schmidt number Sc, characteristics of the aqueous boundary as given by n, and the physical processes at the interface: $k = Sc^{-n} f(u,l)$. Here, the turbulent velocity and length scales u and l are used to represent the physical processes at the interface.

In the large eddy version of the surface renewal model, $k = a_1 (D\, u/l)^{1/2}$. Here, the gas flux depends explicitly on the turbulent velocity and length scales. D is molecular diffusivity of the gas. In the small eddy version, $k = a_2 D^{1/2} (\varepsilon/\nu)^{1/4}$ where ε is the rate of dissipation of turbulent kinetic energy and ν is kinematic viscosity. ε is related to u and l through the expression $\varepsilon = u^3/l$ (Taylor 1935). When written in terms of the Schmidt number and turbulent Reynolds number Re_t, these expressions become $k\, Sc^{1/2} = c_1\, u\, Re_t^{-1/2}$ and $k\, Sc^{1/2} = c_2\, u\, Re_t^{-1/4}$ for the large and small eddy versions respectively. $Re_t = ul/\nu$. ε, u, l, and Re_t can all be obtained from profiles of turbulent microstructure or from surface energy budgets (MacIntyre et al. 1995). The coefficients a_1, a_2, c_1, and c_2 are determined empirically. The time scales for the upper meter to overturn, u/l, is ~ 5 minutes. Within large eddies are many smaller eddies. Overturning of large eddies puts smaller eddies in contact with the diffusive sublayer allowing them to exchange gases at the interface.

A major difference in our calculations of surface energy exchange is that fluxes are calculated not for the aqueous boundary layer, but for the surface layer, defined as that part of the upper mixed layer in which temperatures are within 0.02 °C of the surface temperature. By this approach, heat gains and losses in the upper water column are identified.

DIURNAL MIXED LAYERS AND GAS FLUX

Our understanding of the dynamics of the upper mixed layer of lakes and oceans has changed dramatically in the last 15 years due to higher resolution temperature profiles and determination of locations of active mixing (Imberger 1985; Shay and Gregg 1986; Brainerd and Gregg 1993; MacIntyre 1993; MacIntyre 1998). The upper mixed layer is not continuously mixing. Instead, during the day, there is an upper, surface layer that may be actively mixing with a diurnal thermocline below, or the water column may be thermally stratified to the surface. When only a thin upper layer is mixing, the concentrations of gas within the layer may quickly equilibrate with the atmosphere. Not until the upper most layer begins losing heat does the surface layer begin to deepen (MacIntyre et al. *In submission*). The larger turbulent eddies may then bring water whose concentration is different from the atmosphere to the surface where gas exchange can resume.

The importance of weakening of the temperature gradient in the upper part of the water column for gas flux was first noted by Crill et al. (1988). Gas flux from L. Calado, Brazil, was five times higher at sunrise and sunset than at noon. The higher fluxes occurred when the temperature difference in the upper 3 m was reduced. Engle and Melack (2000) show the importance of episodic mixed layer deepening for enhanced gas flux. Surface energy budgets obtained from L. Calado during a two-month period when the lake was seasonally stratified show that buoyancy fluxes were 50 to 100% of the total energy flux at night into the lake (MacIntyre, unpublished data). Buoyancy fluxes ranged from 20 to 100% of the total energy flux at sunrise and sunset, and between 0 and 50% during the day. Typically, the upper mixed layer was 0.5 m deep in the day and deepened at night to 6 m depth. Wind speeds were higher in the day. These data indicate the critical importance of heat loss for mixed layer deepening, circulation, and gas flux.

Comparisons of Diurnal Cycles of u_* and w_*

The diurnal periodicity in mixed layer depth and diurnal variations in u_* and w_* are illustrated over two diurnal periods for a tropical lake (Figure 1). The shear velocity scale $u_* = (\tau/\rho)^{1/2}$ where τ is wind stress and ρ is density of water; the convective velocity scale $w_* = (Bh)^{1/3}$ where B is buoyancy flux and h is depth of the actively mixing layer. w_* is non-zero when a water body is losing heat.

The mixed layer is shallow in the day and deepens beginning in late afternoon. Wind speeds increased above 8 m s^{-1} four times during the 60 hour period ($u_* > 0.008$ ms^{-1}). The first three of these events occurred at night causing the thermocline to downwell. Downwelling does not imply mixed layer deepening. Deepening of the mixed layer and entrainment of thermocline waters occurred beginning in late afternoon when w_* increased and exceeded 0.003 ms^{-1}. Afternoon winds ranged from 4 to 6 m s^{-1} (u_* ranged from 0.004 to 0.007 m s^{-1}), but mixed layer deepening did not occur until the lake began losing heat. The rapid shoaling of the mixed layer at 1000 h on 26 April occurred once incident short wave energy exceeded heat losses. The shoaling occurred despite winds of 7 m s^{-1} and $u_* = 0.009$ m s^{-1}. The upper mixed layer is shallow when w_* equals zero, and deepens as w_* increases. Because values of w_* exceed u_* by a factor of 2 to 3 much of the night, it is essential that they be included in computations of the rearration coefficient.

Comparison of Gas Transfer Coefficients Based on Wind Speed Alone and on Surface Renewal

At moderate wind speeds (6-8 ms^{-1}), when evaporation is likely to be included in a wind based formulation of k_{600}, estimates of k based on wind alone and the surface renewal model give comparable results (Figure 2). However, at wind speeds < 5 ms^{-1}, Wanninkhof and McGillis' (1999) regression equation gives lower estimates of k_{600} relative to Cole and Caraco (1998) and the surface renewal model. The equation developed by Cole and Caraco is based on tracer studies from several lakes (Wanninkhof 1992; MacIntyre et al. 1995; Cole and Caraco 1998). Scatter was high and unexplained in these earlier studies. The data from Pilkington Bay, where w_* is often three times higher than u_* at wind speeds < 6 ms^{-1}, suggest that turbulence due to heat loss may explain a large part of the scatter.

Importance of w* for Gas Flux

Gas flux measurements in Toolik Lake, Alaska, further corroborate the importance of buoyancy fluxes. An eddy

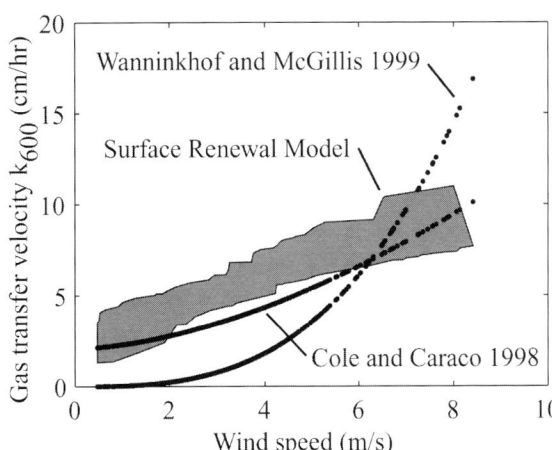

Figure 2. Reaeration coefficient k_{600} computed using the surface renewal model (shaded), a wind based model (••) (Wanninkhof and McGillis 1999), and an empirical model based on tracer studies from several lakes (-) (Cole and Caraco 1998). The small eddy version of the surface renewal model (k Sc$^{1/2}$ = c_2 u Re$_t^{-1/4}$) using $c_2 = 0.56$ (Crill et al. 1988) was used to calculate k_{600}. The total energy flux into the surface layer $F_T = (w_*^3 + 1.33^3 u_*^3)$; $\varepsilon = 0.82$ F_T/h (Imberger 1985); h, u, and Re$_t$ were defined in *Background*. Meteorological data from Pilkington Bay (data from MacIntyre et al. *In submission*. Data are 5 min. averages.)

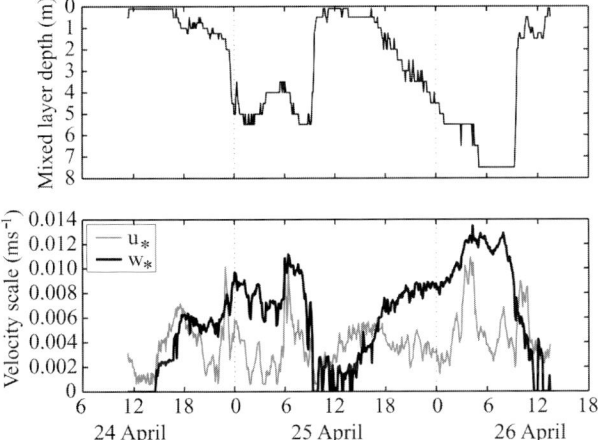

Figure 1. Upper mixed layer depth (upper panel), shear and convective velocity scales (lower panel) over two diurnal cycles in Pilkington Bay, Lake Victoria, East Africa. (from MacIntyre et al. *In submission*). Thermocline downwelling occurs when winds are high, but mixed layer deepening occurs as heat is being lost.

covariance tower with sonic anemometer was deployed at Toolik Lake, Alaska, in July 1995. Methods are described in Eugster and Senn (1995) and Eugster et al. (1997). We computed buoyancy flux following MacIntyre et al. (1995). As no thermistor data were available to determine depth of the mixed layer, we assumed it was 1 m deep at all times. Given that time series data of mixed layer depth in other years show mixed layer depths increasing to 6 m at night, this assumption could cause w_* to be a factor of 2 too low. Eddy covariance data were sampled at 20 Hz and 30 minute averages computed from the raw data; data are presented as 9 point running averages of the 30 minute fluxes.

During this study, wind speeds ranged from 1 to 6 m s^{-1} and u_* ranged from 0 to 0.007 m s^{-1} (Figure 3). w_* was non-zero for only a few hours each day. Flux of CO_2 into or out of the lake occurred at all times of day (Figure 3), but gas fluxes were maximal while heat was being lost from the lake. Fluxes increased over the cooling period suggesting that mixed layer deepening was replenishing surface waters.

SUMMARY

Fluxes of dissolved gases are appreciably higher when the surface layer is cooling. Heat loss contributes to turbu-

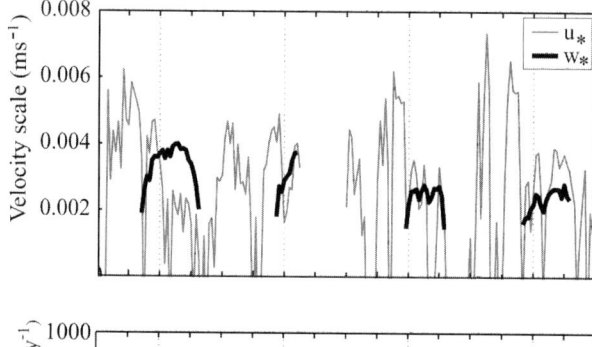
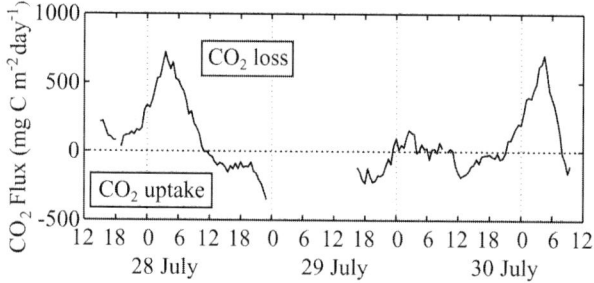

Figure 3. Shear and convective velocity scales (upper panel) and gas fluxes (lower panel) from Toolik Lake, Alaska, during a four day period in July 1995. Maximum gas fluxes are not correlated with wind speed, as represented by u_*, but by whether or not heat was being lost from the surface layer ($w_* > 0$). Gas fluxes are less than 100 mg C m^{-2} day^{-1} at highest wind speeds, but increase above 500 mg C m^{-2} day^{-1} when buoyancy flux led to convective overturning.

lence within the aqueous boundary layer and to mixed layer deepening which entrains dissolved gasses.

The surface area of the world's great lakes totals 997,000 km^2; of the world's large lakes, 686,000 km^2; and of the world's small to moderate sized lakes, 620,000 km^2. However, many of the smaller lakes have higher concentrations of CO_2 and CH_4 and a net flux of CO_2 and CH_4 to the atmosphere (Cole et al. 1994). For instance, in the Canadian Shield, lakes whose surface area is less than 10 km^2 have P_{CO2} values above 500 µatm. Concentrations decrease linearly with lake size, with Lake Superior having P_{CO2} of 100 µatm. If these numbers are extrapolated world wide, the potential for gas flux would be comparable in the three size classes of lakes. While gas flux due to heat loss occurs in all stratified water bodies, it is likely to be especially important in fetch-limited small to moderate sized lakes and wetlands. These data stress the need to incorporate the surface renewal approach into models of gas transfer.

Acknowledgements. Financial support was provided by National Science Foundation Grants DEB93-17986 and DEB97-26932 to SM, DEB93-18085 to GWK and SM, and DEB95-53064 to GWK. WE was supported by NSF grant OPP-9318532 and a Hans-Sigrist Fellowship grant from the University of Bern. We thank J.M. Melack and two anonymous reviewers for critically reading the manuscript and Lorenz Moosmann for graphics.

REFERENCES

Brainerd, K.E., M.C. Gregg, Diurnal restratification and turbulence in the oceanic surface mixed layer .1. Observations, *J. Geophys. Res.* 98: 22645-22656. 1993.

Cole, J.J., N. Caraco, G.W. Kling, and T. Kratz, Carbon dioxide supersaturation in the surface waters of lakes, *Science* 265, 1568-1570, 1994.

Cole, J. J., and N. F. Caraco, Atmospheric exchange of carbon dioxide in a low-wind oligotrophic lake measured by the addition of SF_6, *Limnol. Oceanogr.*, 43, 647-656, 1998.

Crill, P.M., K.B. Bartlett, J.O. Wilson, D.I. Sebacher, R.C. Harriss, J. M. Melack, S. MacIntyre, L. Lesack, and L. Smith-Morrill, Tropospheric methane from an Amazonian floodplain lake, *J. Geophys. Res.* 93, 1564-1570, 1988.

Danckwerts, P.V., Significance of liquid-film coefficients in gas absorption. *Industrial Engineering Chemistry,* 43, 1460-1467, 1951.

Engle D. and J.M. Melack, Methane emissions from an Amazon floodplain lake: Enhanced release during episodic mixing and during falling water, *Biogeochemistry,* 51, 71-90, 2000.

Eugster, W. and W. Senn, A cospectral correction model for measurement of turbulent NO_2 flux., *Boundary-Layer Meteorology,* 74, 321-340, 1995.

Eugster, W., J. P. McFadden and F. S. Chapin III, A comparative approach to regional variation in surface fluxes using mobile eddy correlation towers, *Boundary-Layer Meteorology,* 85, 293-307, 1997.

Higby, R., The rate of absorption of a pure gas into a still liquid during short periods of exposure, *Transactions of the American Institute of Chemical Engineers,* 31, 365-388, 1935.

Jessup, A.T., C.J. Zappa, H. Yeh, Defining and quantifying microscale wave breaking with infrared imagery, *J. Geophys. Res.*, 102, 23145-23153, 1997.

Imberger, J. The diurnal mixed layer. *Limnol. Oceanogr.*, 30, 737-770, 1985.

Liss, P.S., and L. Merlivat. Air-sea gas exchange rates: Introduction and synthesis, in The Role of Air-Sea Exchange in Geochemical Cycling edited by P. Buat-Menard, pp. 113-129. Reidel, Boston, Massachusetts, 1986.

MacIntyre S., Vertical mixing in a shallow, eutrophic lake – possible consequences for the light climate of phytoplankton. *Limnol. Oceanogr.*, 38, 798-817, 1993.

MacIntyre, S. Turbulent mixing and resource supply to phytoplankton, in *Physical processes in lakes and oceans. Coastal and Estuarine Studies 54*, edited by J. Imberger, pp. 539-567, AGU, 1998.

MacIntyre S. and J.M. Melack, Vertical and horizontal transport in lakes – linking littoral, benthic, and pelagic habitats. *J. North Amer. Benth. Soc.,* 14, 599-615, 1995.

MacIntyre, S., R. Wanninkhoff, and J.P. Chanton, Trace gas exchange in freshwater and coastal marine systems, in *Methods in Ecology: Trace Gases*, edited by P. Matson and R. Harriss,. pp. 52-97. Blackwell, 1995.

MacIntyre, S., J.R. Romero, and G.W. Kling, Diel Surface energy fluxes, internal waves, and horizontal transports in a freshwater equatorial bay. *Limnol. Oceanogr.* In submission.

Shay, T. and M. C. Gregg., Convectively driven mixing in the upper ocean. *J. Phys. Oceanogr.*, 16, 1777-1798, 1986

Soloviev, A.V., and P. Schluessel, Parameterization of the temperature difference across the cool skin of the ocean and of the air-ocean gas transfer on the basis of modeling surface renewal. *J. Phys. Oceanogr.*, 24, 1319-1332, 1994.

Taylor, G.I., Statistical theory of turbulence. *Proceedings of the Royal Society of London. A*, 151, 421-478, 1935.

Wanninkhof, R., Relationship between gas exchange and wind speed over the ocean. *J. Geophys. Res.*, 97, 7373-7382, 1992

Wanninkhof, R. and W.R. McGillis, A cubic relationship between air-sea CO_2 exchange and wind speed. *Geophysical Research Letter,*. 26, 1889-1892, 1998.

Woolfe, D.K., and S.A. Thorpe,. Bubbles and the air-sea exchange of gases in near-saturation conditions. *Journal of Marine Research,* 49, 435-466, 1991.

Sally MacIntyre, Marine Science Institute, University of California, Santa Barbara, CA 93106-6150

Werner Eugster, Institute of Geography, University of Bern, Hallerstrasse 12, CH-3012 Bern, Switzerland

George W. Kling, Department of Biology, University of Michigan, Ann Arbor, Michigan 48109-1048

A Model of Air-Sea Gas Exchange Incorporating the Physics of the Turbulent Boundary Layer and the Properties of the Sea Surface

Alexander Soloviev

Oceanographic Center, Nova Southeastern University, Dania Beach, Florida

Peter Schluessel

EUMETSAT, Darmstadt, Germany

The model presented contains interfacial, bubble-mediated, ocean mixed layer, and remote sensing components. The interfacial (direct) gas transfer dominates under conditions of low and—for quite soluble gases like CO_2— moderate wind speeds. Due to the similarity between the gas and heat transfer, the temperature difference, ΔT, across the thermal molecular boundary layer (cool skin of the ocean) and the interfacial gas transfer coefficient, K_{int}, are presumably interrelated. A coupled parameterization for ΔT and K_{int} has been derived in the context of a surface renewal model [Soloviev and Schluessel, 1994]. In addition to the Schmidt, Sc, and Prandtl, Pr, numbers, the important parameters are the surface Richardson number, Rf_0, and the Keulegan number, Ke. The more readily available cool skin data are used to determine the coefficients that enter into both parameterizations. At high wind speeds, the Ke-number dependence is further verified with the formula for transformation of the surface wind stress to form drag and white capping, which follows from the renewal model. A further extension of the renewal model includes effects of solar radiation and rainfall. The bubble-mediated component incorporates the Merlivat et al. [1993] parameterization with the empirical coefficients estimated by Asher and Wanninkhof [1998]. The oceanic mixed layer component accounts for stratification effects on the air-sea gas exchange. Based on the example of *GasEx-98*, we demonstrate how the results of parameterization and modeling of the air-sea gas exchange can be extended to the global scale, using remote sensing techniques.

1. INTRODUCTION

The flux of gases like CO_2 across the air-sea interface is an important part of the global climate and its changes. Quantifying the air-sea gas exchange is a complex, turbulent boundary-layer problem [Soloviev and Schluessel, 1998]. The development of a realistic parameterization of the air-sea gas exchange must therefore include a comprehensive analysis of the main physical processes involved.

The gas transport at the air-sea interface consists of the interfacial (direct) and bubble mediated components. Following Woolf and Thorpe [1991], the net air-sea gas flux, F, can be defined as follows,

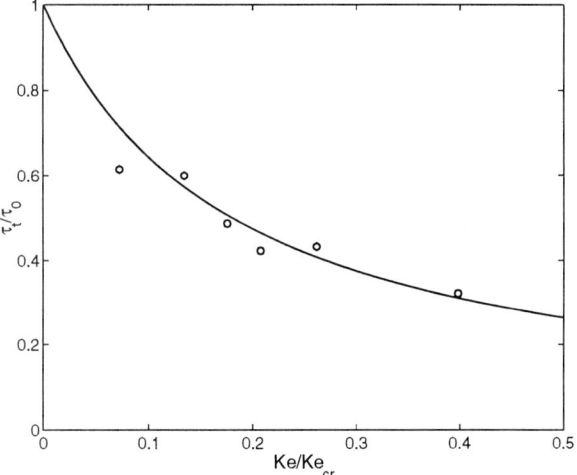

Figure 1. Transformation of the surface wind stress to form drag and white capping at high wind speeds. The line is relationship (6), the circles represent the experiment of Banner and Peirson [1998]. Here: $Ke = u_*^3/(g\nu)$ is the Keulegan number, $Ke_{cr} = 0.18$; τ_0 is the total wind stress, and τ_t is its tangential component.

$$F = K_0[C_w - Sp_a(1+\Delta_e)], \quad (1)$$

where $K_0 = K_{int} + K_b$, K_{int} is the interfacial (direct) transfer velocity, K_b is the bubble-mediated transfer velocity, p_a is the gas partial pressure, S the solubility of the gas, C_w is the gas concentration in the bulk of the water, and Δ_e is the equilibrium supersaturation defined as the fractional increase in partial pressure of the gas in the water caused by bubble overpressure and dissolution.

The equilibrium supersaturation was parameterized by Woolf and Thorpe [1991] in the following way:

$$\Delta_e^i \approx 0.01(U/U_i)^2, \quad (2)$$

where U is the wind speed, U_i is the wind speed for a particular gas at which the equilibrium supersaturation is 1% (nitrogen, 7.3 ms^{-1}; oxygen, 9 ms^{-1}; argon, 9.6 ms^{-1}; carbon dioxide, 49 ms^{-1}). According to (2), the equilibrium supersaturation may reach several percent for poorly soluble gases such as oxygen, but is relatively insignificant for more soluble gases, including carbon dioxide.

For those cases when the equilibrium supersaturation can be neglected, formula (1) reduces to

$$F = K_0(C_w - Sp_a). \quad (3)$$

The aim of this paper is to develop a model of the air-sea gas exchange, based on the physics of the turbulent boundary layers and the properties of the sea surface, which is compatible with the remote sensing techniques. For the parameterization of interfacial gas transfer, we will use a surface renewal model, which is briefly described in Section 2. For the bubble-mediated gas exchange, we will employ the model of Asher and Wanninkhof [1998]. In Section 3, the combined parameterization will be used to model the CO_2 air-sea exchange during *GasEx-98*.

Under low wind speed conditions, values of C_w and S in Eqs. (1) and (3) may depend on the surface mixed layer properties (such as re-stratification of the near-surface layer of the ocean because of diurnal warming and/or precipitation effects). A mixed-layer component has therefore been included in the model. The mixed-layer component is considered in detail in Soloviev et al. [2001].

Section 4 describes the modeling of the air-sea gas transfer velocity in *GasEx-98* Leg 2 and Leg 3. The remote sensing component (Section 5) is included to suggest how the results of modeling can be extended to a global scale. Section 6 contains the conclusions.

2. INTERFACIAL COMPONENT

Interfacial gas transfer at the air-sea interface is curried out via the diffusion sublayers. The thermal and diffusion molecular boundary layers exist near the air-sea interface since the turbulence is inhibited near the interface. Because of similarity between the molecular gas and heat transfer, the temperature difference across the thermal sublayer (cool skin of the ocean), ΔT, and the interfacial (direct) gas transfer velocity (K_{int}) are related to each other [Hasse, 1990]. Soloviev and Schluessel (1994) used a renewal type model to obtain coupled parameterizations for ΔT and K_{int}:

$$\Delta T/T_* = \Lambda_0 \Pr{}^{1/2}\left(1+Rf_0/Rf_{cr}\right)^{-1/4}\left(1+Ke/Ke_{cr}\right)^{1/2}, \quad (4)$$

$$K_{int} = u_* A \Lambda_0^{-1} Sc^{-1/2}\left(1+Rf_0/Rf_{cr}\right)^{1/4}\left(1+Ke/Ke_{cr}\right)^{-1/2}, \quad (5)$$

where $T_* = -q_0/u_*$, $q_0 = -(Q_T + Q_L + Q_H)/(c_p\rho)$ is the normalized heat flux, and u_* is the friction velocity in the near-surface layer of the ocean; $\Lambda_0 = 13.3$, $A = 1.85$, $Rf_{cr} = 1.5\times10^{-4}$, and $Ke_{cr} = 0.18$; $Pr = \nu/\kappa$ is the Prandtl number, and $Sc = \nu/\mu$ is the Schmidt number, μ is the coefficient of molecular gas diffusion, and ν is the kinematic molecular viscosity. $Rf_o = -\alpha g q_o \nu/u_*^4$ is the surface Richardson number introduced by Kudryavtsev and Soloviev [1985] (here α_T is the coefficient of the water thermal expansion, g is the acceleration due to gravity, ρ is the density, and c_p the specific heat capacity of seawater). Rf_0 is responsible for the transition from free to forced convection. $Ke = u_*^3/(g\nu)$ is the Keulegan number introduced by Csanady [1978]; it is an important parameter in the dynamics of the wave-viscous sublayer of the ocean.

An interpretation of the Ke-number dependency in (4)-(5) is that under high wind-speed conditions the tangential wind stress, τ_t, controlling the renewal process at the air-sea interface may be expressed in the following way [Soloviev and Schluessel, 1996]:

$$\tau_t = \tau_0(1 + Ke/Ke_{cr})^{-1}, \quad (6)$$

where $\tau_0 = \rho u_*^2$ is the total momentum flux at the air-sea interface. Equation (6) expresses the transformation of the surface wind stress to form drag and white capping at high wind speeds. According to Figure 1, Eq. (6) is consistent with the momentum flux data available from the air-water interface studies [Banner and Peirson, 1998].

The more readily available cool skin data were used to verify the values of Λ_0, Rf_{cr}, and Ke_{cr}, which appear in both parameterizations (4) and (5); however, the Keulegan-number dependence has been demonstrated with a relatively small number of cool skin observations taken under high wind-speed conditions [see Soloviev and Schluessel, 1994]. The consistency of Eq. (6) with experimental data, as demonstrated in Figure 1, can therefore be considered as an additional verification for the Keulegan-number dependence in (4) and (5).

A further extension of the renewal model included the effects of solar radiation and rainfall on the molecular boundary layers [Soloviev and Schluessel, 1996; Schluessel et al., 1997].

3. BUBBLE-MEDIATED COMPONENT

Woolf [1997] proposed the parameterization of the bubble-mediated gas transfer based on the asymptotes at large and small values of the solubility coefficient α:

$$K_b = V\alpha^{-1}\left[1 + \left(14\alpha Sc^{-0.5}\right)^{-0.83}\right]^{-1.2}, \quad (7)$$

where V is the volume of air entrained as bubbles per unit area per unit time by the breaking waves, defined as $V = bW_b$, W_b is the whitecap coverage defined using the data of Cipriano and Blanchard [1981], and b is an empirical constant. Unfortunately, the empirical constants, entering (7), have been determined only for clean bubbles.

We will therefore use the Merlivat et al. [1993] parameterization with empirical coefficients determined for non-clean bubbles by Asher and Wanninkhof [1998]:

$$K_b = W_c\left(-37\alpha^{-1} + 6120\alpha^{-0.37}Sc^{-0.18}\right), \quad (8)$$

where α the non-dimensional Ostwald solubility coefficient, W_C the fractional area coverage of actively breaking whitecaps defined according to Monahan [1993]. The fractional area coverage of actively breaking whitecaps defined according to Monahan [1993], $W_c = c_1(U_{10} - c_0)^3$, U_{10} is the wind speed at a 10-m elevation, $c_1 = 2.56 \times 10^{-6}$ s^{3}m^{-3}, and $c_0 = 1.77$ ms^{-1}.

4. MODELING THE AIR-SEA GAS EXCHANGE IN GASEX-98

The modeling of the CO_2 exchange in *GasEx-98* Leg 2 and Leg 3 is shown in Plates 1 and 2 respectively. The model includes the interfacial component of Soloviev and Schluessel [1996] and the bubble-mediated component of Asher and Wanninkhof [1998]. The model is forced with the air-sea heat and momentum fluxes measured by Edson et al. [1999]. It is only the blue shaded regions in the top panels of these figures that represent the bubble contributions, and only the red shaded regions in the middle panels that represent the suppression due to solar radiation.

Moderate wind speed conditions prevailed during *GasEx-98* with several storms and a few periods of calm weather in Leg 2 and with several calm weather periods and a few cases of high wind speeds in Leg 3. According to Plates 1 and 2 (bottom panels), the contribution of bubble-mediated transport is relatively small for both legs. Though it achieves the peak values of 20% during Leg 2 and 10% during Leg 3, the averaged values are 8.1 % for Leg 2 and 4.0% for Leg 3. The peak values of the relative suppression due to solar radiation are −65% during Leg 2 and −85% during Leg 3; while, the corresponding averaged values are −2.5% for Leg 2 and −4.9% for Leg 3.

In Plate 3, the contributions of the interfacial and bubble-mediated components to the CO_2 air-sea gas exchange are shown as a function of the wind speed. The scatter of the gas-exchange velocity under low wind speed conditions is explained by its dependence on the buoyancy flux at the air-sea interface (see details in Soloviev and Schluessel [1996]). Note that in the wind speed range 5 to 15 ms^{-1}, the total gas exchange exhibits near linear wind speed dependence. Under very high wind speeds (>15 ms^{-1}), the bubble-mediated component, which is proportional to the cube of wind speed, dominates.

In Figure 2, our model is compared with historical field data on the air-sea gas transfer and Asher's and Wanninkhof [1998] parameterization given by,

$$K_0 = K_b + \left[47U_{10} + W_c(115{,}200 - 47U_{10})\right]Sc^{-1/2}. \quad (9)$$

where K_b is represented by (8). (Note that formula (9) differs from (8) by the presence of the additional term repre-

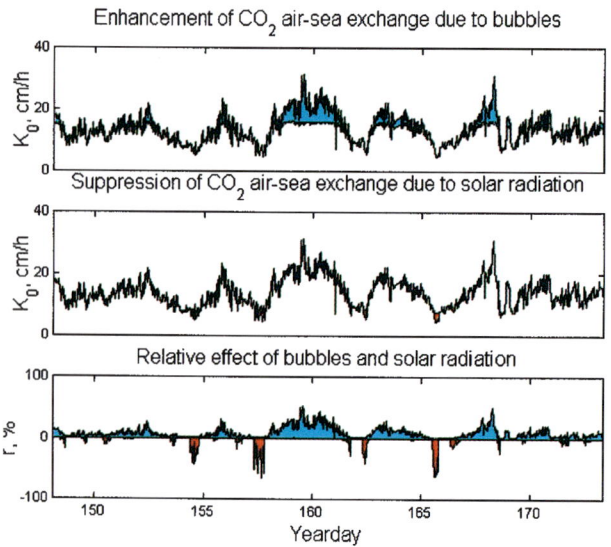

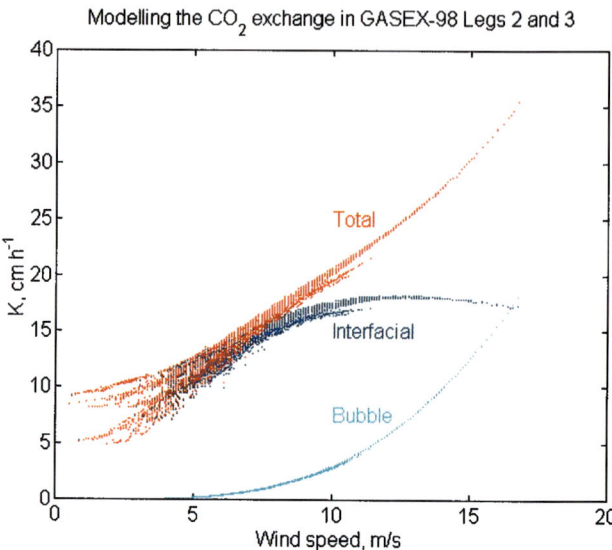

Plate 3. Contribution of interfacial and bubble mediated components into the total CO_2 air-sea transfer velocity at different wind speeds.

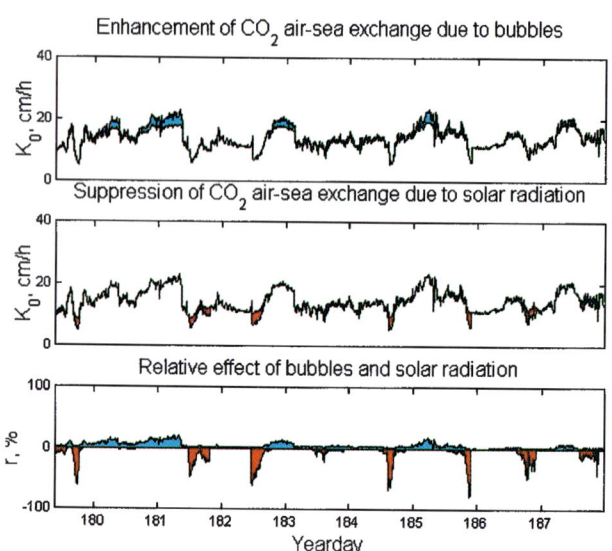

Plate 2. Same as in Plate 1 but for *GasEx-98* Leg 3.

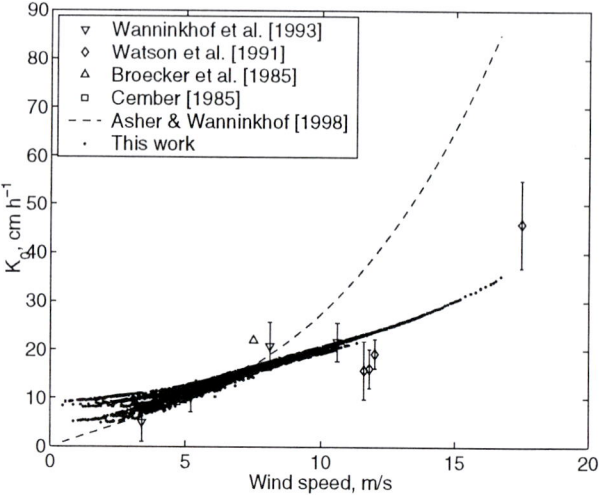

Figure 2. CO_2 transfer velocity according to modeling (which is done in this work for the *GasEx-98* Leg 2 and Leg 3) in comparison with the Asher and Wanninkhof [1998].

senting the interfacial gas transfer in the form suggested by Asher and Wanninkhof [1998].

At wind speeds from 3 to 8 m s^{-1}, both parameterizations predict similar CO_2 air-sea transfer velocities, coinciding reasonably well with the field data (Figure 2). The difference between the two parameterizations is observed under very low and very high wind speed conditions. Compared to Asher and Wanninkhof [1998], our model predicts non-zero gas transfer velocities during calm weather (due to convection in the near-surface layer of the ocean) and smaller values under high wind speed conditions (due to the fact that the part of momentum flux that goes to the wave drag rather than to the tangential stress increases with wind speed). There is, however, not enough field data to fully validate the theoretical frameworks presented.

5. REMOTE SENSING COMPONENT

A broader overview about the exchange processes and its spatial and temporal variability in the vicinity of the ex-

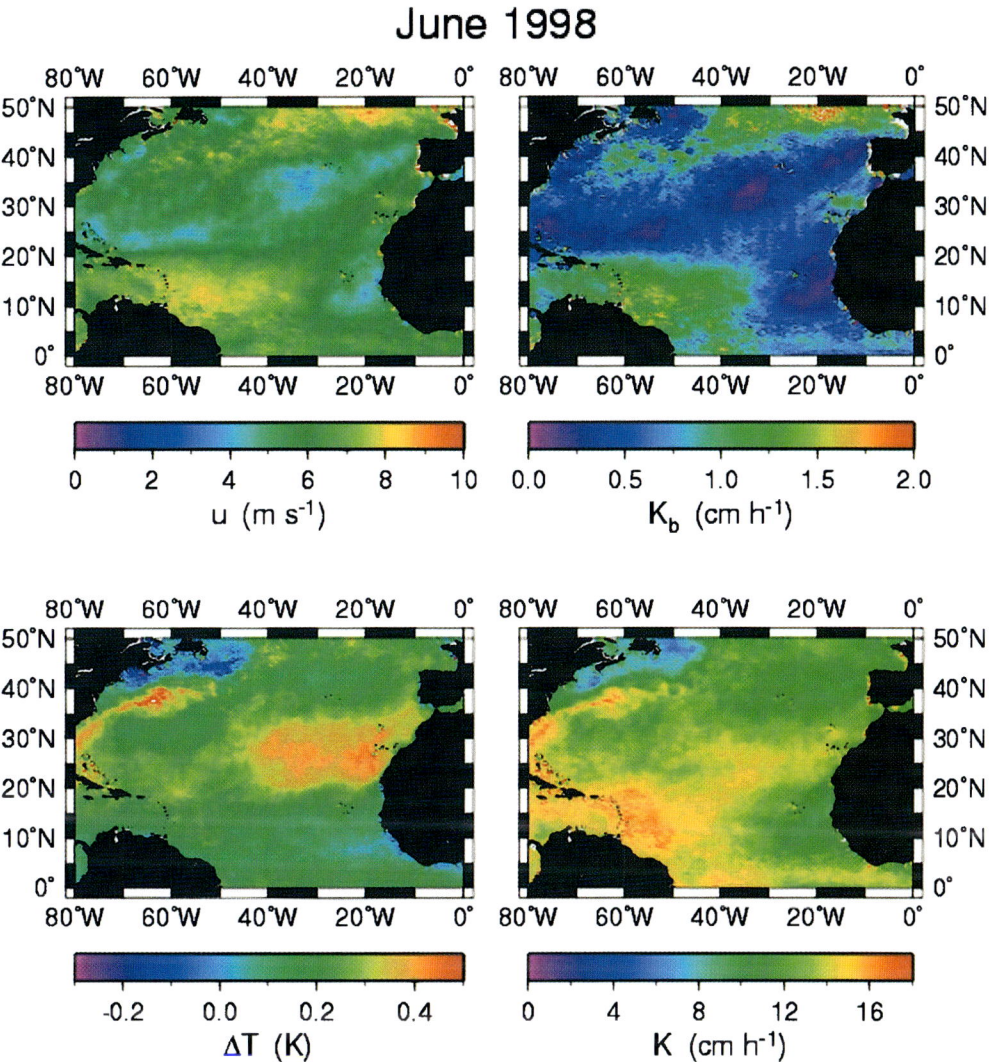

Plate 4. Wind speed (u), cool skin (ΔT), and bubble-mediated (K_b) and total ($K=K_{int}+K_b$) CO_2 air-sea gas transfer velocity for *GasEx-98* period as calculated from the satellite data and the parameterization developed in this work.

perimental area and beyond can be obtained from satellite data. Space-borne infrared and microwave imagery from the Advanced Very High Resolution Radiometer and from the Special Sensor Microwave/Imager has been used to retrieve boundary layer parameters for the time period corresponding to *GasEx-98*. These are the sea surface temperature, surface friction velocity, low-level atmospheric humidity, near-surface stability, and the atmospheric back radiation. These parameters are used to calculate energy and momentum fluxes which in turn are used together with surface renewal modeling to parameterize the temperature difference across the thermal molecular boundary layer of the upper ocean and the air-sea gas exchange transfer velocity (Plate 4).

Surface films can dramatically reduce the air-sea gas exchange through modification of the capillary waves field [Frew et al., 1997]. According to Bock et al. [1999] and Jaehne et al. [1987 the gas transfer velocity shows a reasonable correlation with the mean square slope regardless of the surfactant concentrations. Due to the fact that the remotely sensed wind velocity (like that shown in Plate 4) is determined from the mean square slopes, these wind velocities have in effect been adjusted for the influence of surface films, thus the use of these adjusted wind velocities in estimating the gas transfer velocity substantially eliminates the need to make further adjustments to these velocities for the presence of such films.

A global estimate of the air-sea gas flux with the parameterization suggested in this work will certainly require insight into the distribution of heat and momentum fluxes and the areal extent of very low and very high wind speed regions as well as gas concentration differences. The effect will be biggest where there are big gas concentration differences together with very high wind speed conditions (e.g., North Atlantics or Southern Ocean during Winter time). This additional information is the subject to future research.

6. CONCLUSIONS

The main conclusions of the work are the following:
1) Identifying the similarities between the heat, momentum, and gas exchange in the near-surface layer of the ocean is a useful approach to refining the gas exchange parameterizations and models.
2) The proposed model shows a non-zero gas transfer velocity under calm weather, its almost linear increase with wind at wind speeds of 5 to 15 ms^{-1}, and the cubic wind-speed dependence at very high wind speeds. According to the modeling for *GasEx-98* Leg 2 and Leg 3, the interfacial CO_2 air-sea gas exchange dominates up to the wind speed of about 15 ms^{-1}.
3) Our model is compatible with the remote sensing techniques; the results of *GasEx-98* have been extended to the basin scale using satellite data.

Acknowledgments. We thank Jim Edson, Wade McGillis, Chris Fairall, and Rik Wanninkhof for providing data from *GasEx-98*. This work was done under the auspices of the NOAA Ocean-Atmosphere Carbon Exchange Study (Grant NA 86GP0233) and the National Science Foundation (Grant OCE-9730643).

REFERENCES

Asher, W.E. and R. Wanninkhof, The effect of bubble-mediated gas transfer on purposeful dual-gaseous tracer experiment, *J. Geophys. Res.*, 103, 10,555-10,560, 1998.

Banner, M.L. and W.L. Peirson, Tangential stress beneath wind-driven air-water interfaces, *J. Fluid Mech.*, 364, 107-137, 1998.

Bock, E.J., T. Hara, N. M. Frew, and W.R. McGillis, Relationship between air-sea gas transfer and short wind waves, *J. Geophys. Res.*, 104, 25,821-25,831, 1999.

Broecker, W.S., T.H. Peng, G. Ostlund, and M. Stuiver, The distribution of bomb radiocarbon in the ocean, *J. Geophys. Res.*, 90, 6953-6970, 1985.

Cember, R., Bomb radiocarbon in the Red Sea: A medium scale gas exchange experiment, *J. Geophys. Res.*, 94, 2111-2123, 1989.

Cipriano, R.J., and D.C. Blanchard, Bubble and aerosol produced by a laboratory "breaking wave", *J. Geophys. Res.*, 86, 8085-8092, 1981.

Csanady, G.T., Turbulent interface layers, *J. Geophys. Res.*, 83, 2329-2342, 1978.

Edson, J.B., W.R. McGillis, J. Ware, J.E. Hare, and C.W. Fairall, Direct CO_2 flux measurement at the Air-Sea Interface, supplement to *EOS, Transactions*, April 27, S45, 1999 (abstract only).

Frew, N.M., E.J. Bock, W.R. McGIllis, A.V. Karachintsev, T. Hara, T. Muensterer, and B. Jaehne, Variation of air-water gas transfer with wind stress and surface viscoelasticity, in *Air-Water Gas Transfer*, edited by B. Jaehne and E.C. Monahan, Hanau, 529-541, 1995.

Hasse, L., On the mechanisms of gas exchange at the air-sea interface, *Tellus*, 42B, 250-253, 1990.

Jaehne, B., K.O. Muennich, R. Rosinger, A. Dutzi, W. Huber, and P. Libner, On the parameterization of air-water gas exchange, *J. Geophys. Res.*, 92, 1937-1949, 1987.

Kudryavtsev, V.N. and A.V. Soloviev, On thermal state of the ocean surface, *Izv. Acad. Sci. USSR, Atmos. Ocanic Phys.*, 17, 1065-1071, 1985.

Liss, P.S. and L. Merlivat, Air-sea gas exchange rates: introduction and synthesis, in *The Role of Air-Sea Exchange in Geochemical Cycling*, edited by P. Baut-Menard, Reidel, Dordrech, 113-127, 1986.

Merlivat, L, L.emery, and J. Boutin, 1993: Gas exchange at the air-sea interface. Present status: The case of CO_2, paper presented at the *Fourth International Conference on CO_2 in the Ocean*, Inst. Natl., des Sci. de l'Univers. Cent. Natl. de Rech. Sci., Carqueiranne, France, Sept. 13-17, 1993.

Monahan, E.C., Occurrence and evolution of acoustically relevant sub-surface bubble plumes and their associated, remotely monitorable, surface whitecaps, in *Natural Physical Sources of Unerwater Sound*, edited by B.R. Kerman, pp. 503-517, Kluwer Acad., Norwell, Mass., 1993.

Schluessel , P., W. J. Emery, and A.V. Soloviev, Cool and freshwater skin of the ocean during rainfall, *Boundary-Layer Meteorlogy*, 82, 473-472, 1997.

Soloviev, A., J. Edson, W. McGillis, P. Schluessel, and R. Wanninkhof, Fine thermohaline structure and gas exchange in the near-surface layer of the ocean during *GasEx-98*, in *Geophysical Monograph Series, American Geophysical Union, Washington, D.C.*, this volume, 2001.

Soloviev, A.V. and P. Schluessel, Parameterization of the cool skin of the ocean and of the air-ocean gas transfer on the basis of modeling surface renewal, *J. Phys. Oceanogr.*, 24, 1339-1346 and 1965, 1994.

Soloviev, A.V. and P. Schluessel, Evolution of cool skin and direct air-sea gas transfer coefficient during daytime, *Boundary-Layer Meteorology*, 77, 45-68, 1996.

Soloviev, A. V. and P. Schluessel, Comments on "Air-Sea Gas Transfer: Mechanisms and Parameterization", *J. Phys. Oceangr.*, 28, 1643-1645, 1998.

Wanninkhof, R., W.E. Asher, R. Weppernig, H. Chen, P. Schlosser, C. Langdon, and R. Sambrotto, Gas transfer experiment on Georges Bank using two volatile deliberate tracers, *J. Geophys. Res.*, 98, 20,237-20,248, 1993.

Watson, A.J., R.C. Upstill-Goddard, and P.S. Liss, Air-sea gas exchange in rough and stormy seas measured by a dual-tracer technique, *Nature*, 349, 145-147, 1991.

Woolf, D.K., Bubbles and their role in gas exchange, in *The Sea Surface and Global Change*, edited by R.A. Duce and P.S. Liss, pp. 173-205, Cambridge Univ. Press, New York, 1977.

Woolf, D.K. and S.A. Thorpe, Bubbles and the air-sea exchange of gases in near-saturation conditions, *J. Marine Research*, 49, 435-466, 1991.

Alexander Soloviev, Oceanographic Center, Nova Southeastern University, 8000 N. Ocean Dr., Dania Beach, FL 33004, U.S.A.

Peter Schluessel, EUMETSAT, Am Kavalleriesand 31, 64295 Darmstadt, Germany

Air-Lake Interaction and Surface Layer Processes

G.N. Panin

Institute of Water Problems RAS, Moscow, Russia

A.E. Nasonov

Institute of Water Problems RAS, Moscow, Russia

S.G. Sarkisian

Institute of Water Problems RAS, Moscow, Russia

The near-surface boundary layer, whose thickness may vary from millimeters to centimeters, is formed as a result of energy and mass exchange between air and water at or close to the surface. To define the thickness of this near-surface mixing layer the proportionality to the Kolmogorov dissipation scale is used. Research of the near-surface water layer heat balance reveals temperature distribution anomalies corresponding with cold, warm and isothermal conditions. These temperature distributions are parameterized and form the foundation for the development of a method for calculating near-surface water layer temperature differences based on standard hydrometeorological data. It is concluded that the intensity of gas exchange through the surface may vary by a factor of two, which could influence the hydrobiological regime of the water body. To test this hypothesis, measurements of the biological and chemical processes within the water surface layer, considering hydrological factors, were made in a lake near Moscow, Russia. The results show that during daytime in spring and summer there forms a positive gradient of near-surface water temperature. Oxygen saturation at the surface is lower than that of the bulk water. Amounts of bacterial cells in the surface water layer and concentration of organic substances were higher than those of the deeper layers. From these observations we suggest a model describing the influence of a near-surface mixing layer on the temporal evolution of a spring bloom and the formation of potentially lethal conditions for fish which may develop in summer and autumn.

1. INTRODUCTION

Gas exchange through the water surface has an important and often determining part in forming the oxygen regime of the water bodies. The intensity of this exchange might, to certain degree, be regulated by thermodynamic processes going on in the near-surface layer of the water body. Existence of the near-surface boundary layer with distinct temperature and concentration gradients is only possible under weak and moderate winds. Destruction of this near-surface water boundary layer happens somewhere near surface wind velocities of 7 to 10 m/s [*Panin*, 1985]. Under most natural conditions the near-surface wind above

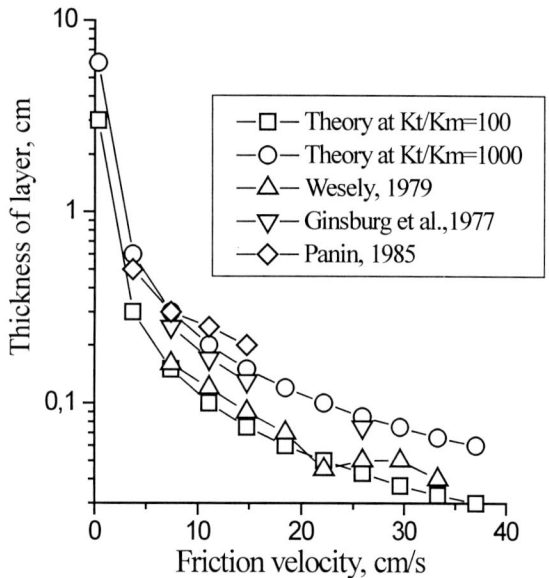

Figure 1. Thickness layer dependence on the friction velocity.

minor lakes, seldom exceeds this critical velocity range. Surface winds above seas and oceans are usually higher than above small lakes. Therefore the processes considered here apply mostly to lakes.

2. THE THICKNESS OF THE NEAR-SURFACE LAYER δ_W AND ITS TEMPERATURE GRADIENT ΔT_W

Natural and laboratory researches have shown that the thickness of the near-surface layer, and magnitude and sign of the temperature drop in it vary strongly depending on the meteorological conditions, time of the day and the season of the year. Most often there is the so-called cold film with the thickness on the order of a millimeter. At the same time the warm layer with the thickness of under 1 cm. and more was registered in experiments on lakes during the day-time in spring and summer [*Ilyin* et al., 1986]. To estimate the thickness of the boundary layer δ_W we used a parameterization in terms of the Kolmogorov's scale (l_0).

As a basis for this hypothesis it was assumed that when approaching the water surface from below the turbulent eddies scale should decrease to the minimum equal to the Kolmogorov's scale. Accordingly, we may write the equality:

$$\delta_W \approx l_0 = \left(K_m^3 / \varepsilon_W \right)^{1/4} \quad (1)$$

In (1) K_m is the kinematic viscosity coefficient, and ε_W is the viscous dissipation rate of turbulent kinetic energy.

To estimate ε_W we can use the equation of the energy budget under conditions of stationarity and horizontal homogeneity:

$$\varepsilon_W \approx U^2_{*w} \frac{\partial U_W}{\partial z} \approx \frac{1}{K_T} \left(\rho_a / \rho_W \right)^2 U^4_{*a} \quad (2)$$

In (2) K_T is the turbulent exchange coefficient; ρ_a, ρ_W are the density of the air and the water respectively, and U_{*a}, U_{*w} are the friction velocity of the air and the water.

Regarding (2) we can rewrite (1) in form

$$\delta_W \approx \frac{K_m \xi^{1/4}}{\left(\rho_a / \rho_W \right)^{1/2} U_{*a}} \approx 10^3 \frac{K_m \xi^{1/4}}{U_{10}} \quad (3)$$

Using characteristic magnitudes for K_T and K_m in Fig 1, the dependence of the near-surface layer thickness on U_{*a} may be presented for two values of relationship:

$$\xi = \frac{K_T}{K_m} = 10^2 \text{ and } 10^3.$$

Figure 1 shows that the theoretical estimates of the near-surface water layer thickness match the data of natural experiments *Wesely* [1979] and *Panin* [1985] as well as the laboratory experimental data of *Ginsburg et al.* [1977] well.

To analyze the thermal regime of the near-surface layer of the water body we consider the equation of the heat balance in it.

$$\rho_W c_p K_m \frac{\partial T_W}{\partial z} = Rn - \lambda E \pm Q_t = Q \quad (4)$$

In (4) c_p - specific heat of air at constant pressure, Rn - net radiation, λE - latent vertical heat flux, Q_t - vertical sensible heat flux, Q - resulting vertical flux of heat in the air.

A linear vertical temperature distribution is assumed:

$$\frac{\partial T_W}{\partial z} \approx \frac{\delta T_W}{\delta z} \approx \frac{T_{W_0} - T_\delta}{\delta_W} \approx \frac{T_{W_0} - T_{W_{ST}}}{\delta_W} \quad (5)$$

When accounting for the fact that the main temperature drop in upper water layer (20-30 cm.) mostly occurs in the near-surface layer $\delta T_W \approx 0.75 \Delta T_W$, we analyze the thermal regime of the water body near-surface layer within the expression $\Delta T_W = F(Q)$. It follows that depending on direction of the resulting heat flux Q in the upper aquatic layer there might be formed either cold (A) or warm near-surface layer (B), as well as the one with the neutral temperature distribution (C):

(A) $T_{W_0} < T_{W_{ST}}$ at $Q > 0$

(B) $T_{W_0} > T_{W_{ST}}$ at $Q < 0$ (6)

(C) $T_{W_0} \approx T_{W_{ST}}$ at $Q \approx 0$

From (6), it follows that the difference between the bulk ($T_{W_{ST}}$) and the surface measurements (T_{W_0}) of the water surface temperature are determined by the water heat balance of the near-surface layer of the water body.

For parameterization of the temperature change ΔT_W we could present the resulting heat flux in form of the sum of two components $Q = Q_1 + Q_2$, where Q_1 is responsible for the turbulent component of the resulting flux, and Q_2 accounts for its solar radiation component.

If we use the stratification parameter of Monin-Obukhov z/L as a characteristic of Q_1:

$$Q_1 \sim z/L \quad (7)$$

and the Berlyand relationship [1956] as a characteristic of Q_2:

$$Q_2 \sim \frac{\sin^2 h}{\sinh + \text{const}} f_1(n) + \left(\gamma_W \sigma T_{W_0}^4 - \gamma_a \sigma T_a^4\right) f_2(n) \quad (8)$$

then the temperature drop in the near-surface water body layer might be defined in the form

$$\Delta T_W = F[z/L; \ \frac{\sin^2 h}{\sinh + 0.07} f_1(n) - k_1 k_2 f_2(n)] \quad (9)$$

In (8-9) h is the sun height in degree, n is the fractional cloud cover, σ is the Stefan-Boltzman constant, γ_W, γ_a are the relative radiation ability of the water surface and the atmosphere, respectively, and k_1 and k_2 are coefficients:

$$(k_1 \approx 1; \quad k_2 = \frac{\sigma \gamma_W T_W^4}{S_0}),$$

where S_0 is the solar constant.

Expression (9) allows calculation of the temperature drop in the near-surface water layer according to the standard hydrometeorological data on the water along with the air temperature, wind velocity, air humidity, cloud cover, and measurement of time and place. The functional form of F was determined on the basis of experimental data and as a whole dependence (9) presents a nomogram. From this nomogram in particular, it follows that under unstable stratification of air (z/L<0), the temperature change ΔT_W in most cases is negative (that is, a cold near-surface layer). For stable stratification (z/L>0) there is mostly a positive temperature change, $\Delta T_W > 0$. In fact, with increased stability ΔT_W becomes greater. It should be noted that under conditions of slight instability (-0.01<z/L<-0.1) the temperature drop changes from 0 to –1°C. These stratification conditions are characteristic for the near-water air above the ocean and our calculation of ΔT_W conform, for example, with generalization of experimental research of cool skin of the ocean.

3. GAS EXCHANGE THROUGH THE NEAR-WATER BOUNDARY LAYER OF THE WATER BODY AND HYDROBIOLOGICAL APPLICATIONS

Panin [1985] described the possible influence of the thermal regime of the near-surface water body layer on the gas exchange between the water and atmosphere and correspondingly on its biological status. The warm near-surface layer in day-time during spring could intensify

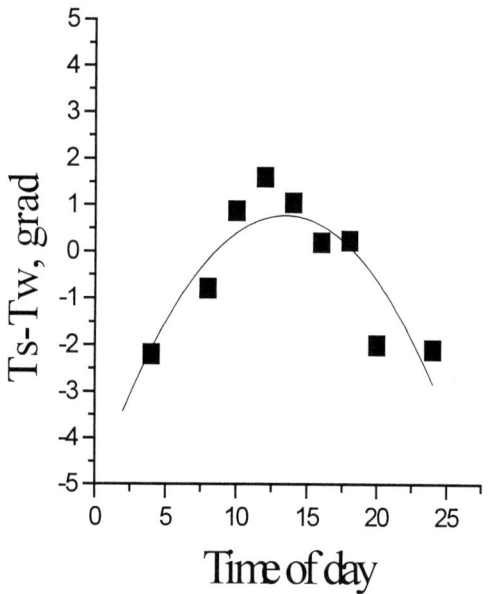

Figure 2a. Change of the temperature ΔT_W during daytime (lake Ivankovo, may 1996).

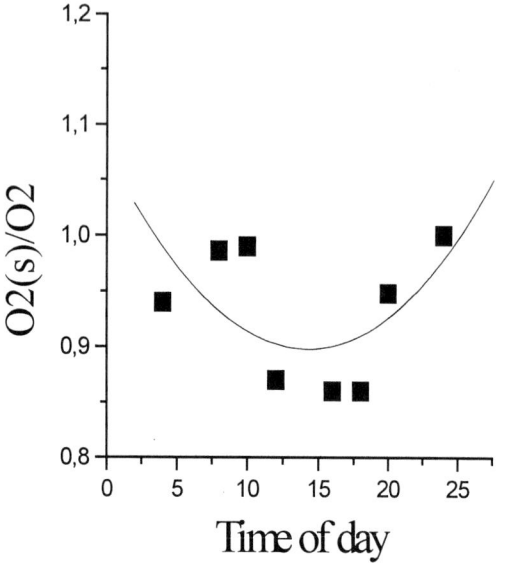

Figure 2b. Relation of the oxygen content in the near-surface layer relatively to the content inside the water (depth 30 cm.) during daytime (lake Ivankovo, may 1996).

reproduction of bacteria and retard gas exchange through it. During night time the near-surface layer cools and a vertical convection develops. This should redistribute the bacteria and transfer hem the inner part of the water body. The warm near-surface layer in summer and beginning of autumn also retards the gas exchange and may cause fish kills. The gas flux (for example, oxygen) can be estimated from: $q_c = K_L \Delta C$, where K_L, is the mass transfer coefficient ($K_L = \mu_q / \delta_W$), μ_q is the coefficient of molecular diffusion, and ΔC is the concentrations difference.

As the thickness of the warm near-surface layer exceeds thickness of the cold near-surface layer (skin) according to experiments this causes a decrease of K_L and the gas flux between the atmosphere and the water body. Estimates of *Brekhovskikh et al.*, [1991] showed that in natural water bodies the gas exchange through the warm near-surface layer may be reduce several fold and may cause fish kills in water bodies.

For an experimental check of the given hypothesis the simultaneous measurement of hydrophysical and hydrochemical characteristics of the near-surface layer of the water body were carried out at Ivankovo lake (located at a distance of 100 km. to the north-west of Moscow). The assays of water for chemical and bacteriological researches were taken from 2 horizons: from a surface ~ 0.3 cm and from a depth of 30 to 50 cm. The sampling from a surface was conducted with a flat glass cone, from depth with a bathometer. The assays were analyzed immediately after collection of the water samples. Dissolved oxygen was determined by the Winkler method. For filtration of bacterial cells nucleopore filters with pores size of 0.17-0.20 microns were used. The cells were counted with the help of a luminescent microscope.

The hydrometeorological measurements were conducted using standard equipment.

These studies revealed that formation of a warm near-surface layer in a water body decreases the oxygen content in the near-surface layer relative to 30 cm depth. (Figure. 2a,b). The bacteria content increases in the warm surface layer (Figure 3a,b). In case of a cold near-surface layer there are no significant differences oxygen and bacteria content compared with the bulk water.

CONCLUSIONS

In two experiments (May 1995 and 1996) the effect of the temperature gradient between the near-surface layer and bulk water is investigated. The experiment shows a smaller dependence of the gas exchange on the temperature regime in the near-surface layer than the theoretical estimates. However, theory and observations were of comparable magnitude. Predicted dependence of the amount of bacteria on temperature of the near-surface water layer also were similar to observations. The temperature stratification and thickness of the near surface water are important both for gas exchange and bacteria .

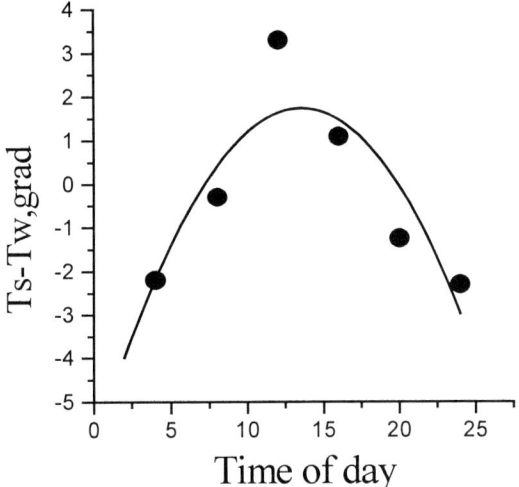

Figure 3a. Change of the temperature change ΔT_w during daytime (lake Ivankovo, may 1995).

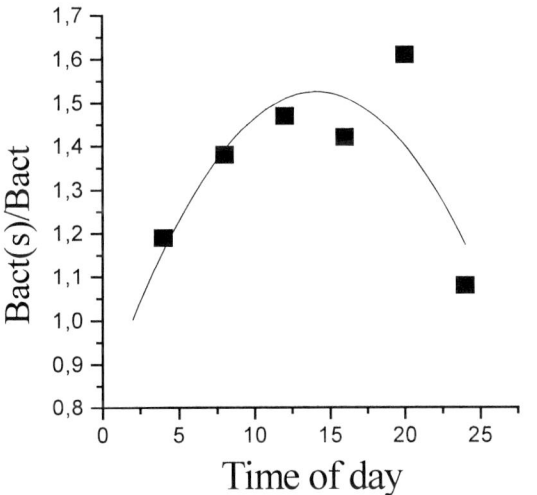

Figure 3b. Relation of the bacteria content in the near-surface layer relative to water at a of depth 30 cm.) during daytime (lake Ivankovo, may 1995).

The qualitative agreement of experimental results with theoretical estimations shows that in natural waters the state of near-surface layer may actually strongly influence the gas concentration in the water body and can cause fish kills.

The majority of the earlier studies of the near-surface layer [e.g. *Carlucci* et al., 1991, *Williams et al.*, 1986] were performed on seas or oceans where conditions typical for large scale vertical convection are predominant. This is the main cause that in the marine and oceanic experiments the phenomena described by us are seldom observed. Our two experiments should be considered a first step requiring further development and thorough confirmation.

REFERENCES

Berlyand, M.E., *Prediction and control of the Earth layer of atmosphere*, 271pp., Gidrometeoizdata Publishers, Leningrad, 1956 (in Russian).

Brechovskikh V.F., Volkova Z.V., Panin G.N., Tarakanov O.Yu., The role of the thermal stratification of the near-the-surface water layer in the gas exchange with the atmosphere, *Water Resources*, N5, 30-36, 1991(in Russian).

Carlucci A.F., Craven D.B., Wolgast D.M., Microbial-populations in surface-films and subsurface waters – amino-acid-metabolism and growth, *Mar Biol* 108, 329-339, 1991.

Ginzburg A.I., Zatsepin A.G., Fedorov K.N., The fine structure of the thermal boundary water layer at the water-air interface. *Izv. AN SSSR, FAO*, 13, 1268-1277, 1977 (in Russian).

Ilyin Yu.A., Panin G.N., Popov N.N., Experimental studies of thermal structure of the water body near-surface layer, *Water Resources*, 97-101, 1986 (in Russian).

Panin G.N., *Heat- and mass exchange between the water and the atmosphere in the nature*, 206 pp Nauka, Moscow, 1985 (in Russian).

Wesely M.L., Heat transfer through the thermal skin of a cooling pond with waves, *J. Geophys. Res.*, V.84, C7, 3696-3700, 1979.

Williams P.M., Carlucci A.F., Henrichs S.M., Van Vleet E.S., Horrigan S.G., Reid F.M.H., Robertson K.J, Chemical and microbiological studies of sea-surface films in the southern gulf of California and off the west-coast of Baja-California, *Mar. Chem.* 19, 17-98, 1986.

G.N. Panin, Institute of Water Problems RAS, Moscow, 117971, Gubkin str. 3, Russia; E-mail:panin@aqua.laser.ru

Spatial Variations in Surface Microlayer Surfactants and Their Role in Modulating Air-Sea Exchange

Nelson M. Frew, Robert K. Nelson, Wade R. McGillis, and James B. Edson

Woods Hole Oceanographic Institution, Woods Hole, Massachusetts

Erik J. Bock

Interdisciplinary Center for Scientific Computing, University of Heidelberg, Heidelberg, Germany

Tetsu Hara

Graduate School of Oceanography, University of Rhode Island, Narragansett, Rhode Island

The potential role of surface microlayer surfactants in modulating air-sea exchange processes in the low wind regime is examined. Variations in surfactant concentration and surface microlayer enrichments are compared with variations in the small-scale wave slope spectrum on large and small spatial scales over large gradients in biological productivity and organic matter level. Surfactant films are shown to strongly reduce the one-sided omnidirectional wave slope $S(k)$ at wave numbers $k = 100 - 400$ rad m^{-1}. The degree of reduction is dependent on the excess microlayer surfactant relative to the underlying water rather than bulk surfactant concentration. The observed wave slope reductions imply corresponding reductions in mass and momentum exchange.

1. INTRODUCTION

The sea surface microlayer plays an important role in air-sea interactions. The boundary conditions set by this interfacial region have a significant impact on many air-sea processes of interest to ocean-atmosphere modelers, including exchange of heat, mass and momentum. In particular, surfactant films have long been thought to be a factor in modulating physical transfer processes. The molecular organization of these films is an important factor governing interfacial processes. Early work [*e.g. Jarvis et al.*, 1962; *Garrett*, 1971] focused on the static or 'barrier' effect of close-packed monolayers in retarding interfacial transfer of heat and water vapor. However, the expanded films formed by natural biogenous surfactants at the air-sea interface modulate surface roughness and near-surface turbulence hydrodynamically, by introducing a viscoelastic modulus [*Bock and Mann*, 1989; *Katsaros et al.*, 1989]. Recent lab studies in wind-wave flumes have yielded new insights on the inhibiting effect of surfactants on gas exchange [*Frew et al.*, 1995]. For a thorough discussion of dynamic monolayer film effects on air-water gas transfer, the reader is referred to a recent review [*Frew*, 1997] and references therein.

A correlation between gas transfer velocity and mean square wave slope has been demonstrated for wave numbers in the short gravity-capillary range [*Bock et al.*, 1999]. During two cruises sponsored by the NSF Coastal Ocean Processes (CoOP) program to study the major forcing parameters that control gas exchange in coastal waters, we measured waves and surfactants in the marine microlayer and near-surface waters. Our goal was to determine whether spatial variations in surfactant levels on various scales were correlated with surface roughness and ultimately with gas exchange rates. Here we present data showing the variability in microlayer surface enrichments under different wind stress conditions and compare small-scale wave properties at sites having different surfactant levels.

2. STUDY AREAS

The study areas included (1) Monterey Bay and the California Bight, during a cruise on the R/V New Horizon in April-May, 1995 and (2) several U. S. east coast transects made on the R/V Oceanus during June-July, 1997 and extending from Woods Hole to the Sargasso Sea. We also include data obtained from the Middle Atlantic Bight (MAB), including multiple transects from Delaware Bay southeast to the Sargasso Sea during several cruises on the R/V Cape Henlopen in 1993-1994. The MAB cruises were limited to collection of large volume near-surface seawater samples for chemical analysis and wind-wave tank gas exchange studies, whereas the east and west coast CoOP cruises involved a complete set of meteorological, surface roughness and surface film measurements.

3. INSTRUMENTATION AND METHODS

3.1 Microlayer Sampling

The sea surface was continuously sampled with a surface microlayer skimmer (SMS) of a design similar to that of *Carlson et al.* [1988]. The SMS consisted of a partially submerged, rotating glass cylinder supported by a small catamaran. The rotating cylinder collected a thin layer of water (40-60 µm thickness) by viscous retention. The theoretical basis for the sampling mechanism has been described by *Levich* [1962] and experimentally verified by *Cinbis* [1992]. Collection efficiencies of the glass cylinder are comparable to those obtained with a glass plate [*Carlson, et al.*, 1988; *Carlson*, 1982; *Harvey and Burzell*, 1972]. The sampler was nested within a larger instrumented catamaran (LADAS) directly aft of a scanning laser slope gauge, which measured the small-scale wave field. The sampler produced a 100 ml min^{-1} flow of microlayer water; a second sampling line supplied subsurface water from a nominal depth of 10 cm. Both flow streams were routed to a fluorometry package mounted on LADAS.

3.2. CDOM Fluorescence Measurements

Colored dissolved organic matter (CDOM) was used as a proxy for surfactants in seawater. CDOM fluorescence is shown here to correlate strongly with surfactants in seawater. CDOM fluorescence was measured using a Turner Designs 10-AU field fluorometer equipped with a 25 mm pathlength, continuous flow quartz cell. The excitation wavelength was 355 nm; the emission wavelength was 450 nm. Fluorescence was calibrated with quinine sulfate standards. The fluorometer alternately measured surface microlayer and subsurface fluorescence over 8 and 2 minute intervals, respectively, to give surface enrichment estimates.

3.3. Ancillary Chemical Measurements

Salinity and sea surface temperature were taken from the ship=s SAIL system. Chlorophyll was measured using a WETLabs WETStar fluorometer mounted on the ship=s continuous flow seawater line. Fluorometer response was calibrated with discrete GF/F filter samples and coproporphyrin methyl ester as a standard. Other chemical parameters were measured on discrete samples taken both from the ship=s flow line and a second SMS deployed by a small boat. Dissolved organic carbon (DOC) was measured by a high temperature catalytic oxidation method [*Peltzer and Brewer*, 1993]. Surface-active organic matter (SAOM) was measured by a polarographic method using Triton-X-100 as a standard [*Hunter and Liss*, 1980]. CDOM fluorescence spectra were measured on an SLM SPF spectrofluorometer using a 1 cm cell. The fluorescence intensity at 450 nm was normalized to the water Raman band to give f450/r.

3.4. Wind and Wave Measurements

Wind speed was measured with a sonic anemometer and converted to wind at 10 m height. Small-scale wave measurements were made using a scanning laser slope gauge (SLSG) [*Bock and Hara*, 1995] mounted on LADAS. During deployment, LADAS was remotely piloted directly into the wind to avoid distortion of the wave field. The SLSG gave full three-dimensional frequency-wave number spectra from 50-800 radian m^{-1} for an undisturbed patch of water directly ahead of the SMS. The data are reported as the one-sided, directionally integrated slope spectrum S(k) for wave numbers k = 100, 200 and 400 radian m^{-1}.

4. RESULTS AND DISCUSSION

4.1. Large-Scale Variability

As a prelude to the observations made during the CoOP field experiments, we first discuss results from the MAB seasonal cruises, in which water samples collected along a strong productivity gradient were analyzed for surfactant concentration, dissolved organic carbon and CDOM fluorescence. The results are shown in Figure 1 as plots of surfactants versus DOC and CDOM fluorescence, respectively. Spatially, up to twenty-fold variations of near-surface seawater surfactant concentration were observed (from 1.5-3 mg l^{-1} in coastal areas to 0.05-0.15 mg l^{-1} in oligotrophic waters). Surfactant concentrations were observed to be correlated with both DOC and CDOM fluorescence. However, the relationships between these parameters appeared to vary significantly with time of year. Samples collected during the spring cruises tended to show much higher SAOM/DOC and SAOM/CDOM ratios than did those collected in late summer and fall. Thus, the fraction of surface-active material (and presumably composition) in the dissolved organic matter pool appears to vary during the year, either due to changes in autocthonous production or differences in quality and quantity of riverine organic matter runoff.

A subset of these MAB samples was used in gas transfer experiments in a small annular wind-wave tank. The tank and methodology have been previously described in *Frew et al.* [1995]. Gas transfer velocities were determined from oxygen evasive fluxes measured in the MAB samples at a nominal wind speed of 5 m s^{-1}. A single point laser slope gauge was used to estimate mean square slope from the wave frequency spectrum of the small-scale waves. Measured transfer velocities for oxygen at 20°C are plotted as a function of bulk surfactant concentration in Figure 2A. A power law dependence on SAOM was observed, with most of the decrease in transfer velocity occurring at SAOM concentrations < 0.5 mg l^{-1}. The approach to an asymptotic limit may be interpreted in a hydrodynamic sense

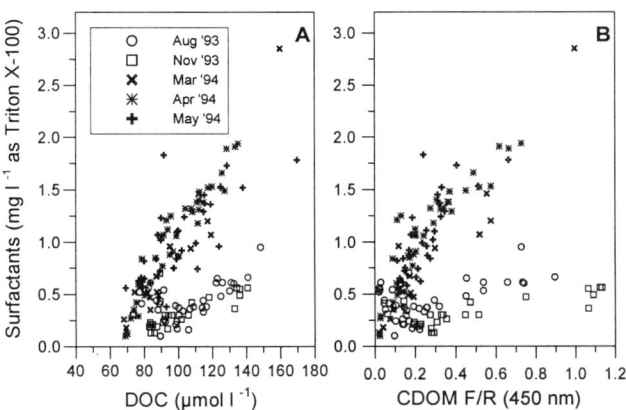

Figure 1. Correlations of surfactant concentration with (A) dissolved organic carbon (DOC) and (B) Raman-normalized CDOM fluorescence (f450/r), for near-surface (1-2 m depth) seawater samples collected during a seasonal series of cruises from Delaware Bay to the Sargasso Sea on the R/V Cape Henlopen during 1993-1994.

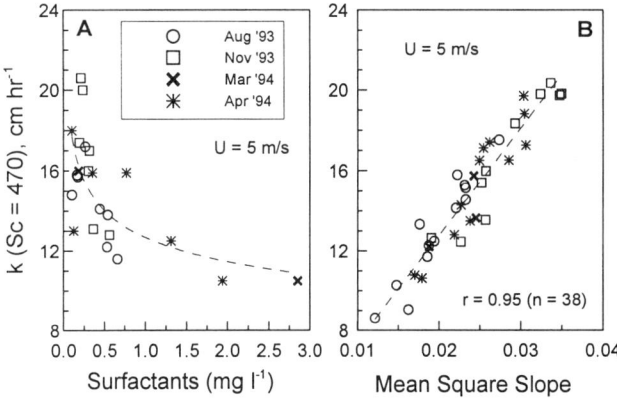

Figure 2. Variation in gas transfer velocity for O_2 at 20°C as a function of (A) surfactant concentration and (B) mean square wave slope, measured in an annular wind-wave tank for a subset of near-surface seawater samples collected during the R/V Cape Henlopen cruises.

as the gradual extinction of turbulent surface renewal events that promote transfer as the films become coherent monolayers. The transfer velocity-mean square slope correlation is shown in Figure 2B. The reduction in gas transfer rate with increasing surfactant concentration and the linear relationship between transfer velocity and the total mean square slope of small-scale waves in the tank experiments suggested to us that *in situ* air-sea exchange rates of mass and momentum at low to moderate winds might also vary with surface-active organic matter distributions. However, since the normal dynamics of surface film accumulation and dispersal could not be reproduced in the wave tank experiments, it was not clear whether *in situ* exchange would depend on bulk SAOM concentration or on the actual surface excess of SAOM, which is dynamically controlled by advective and convective processes, breaking waves, and bubble production, in addition to wind stress.

Next we compare *in situ* chemical and wave slope data at four different stations occupied during the 1997 R/V Oceanus cruise in the western North Atlantic (Figure 3). These four stations, A, B, C and D cover a gradient in biological productivity and organic matter levels as shown in Table 1. The contoured climatological CZCS ocean color data for July are shown in Figure 3 for comparison. The chlorophyll observations made during the cruise (Table 1) are in good agreement with historical CZCS pigment data. The surface waters at the coastal stations, A and D, were highest in chlorophyll, DOC, SAOM, and CDOM fluorescence. Sta. C, in the Sargasso Sea, exhibited the lowest values for these parameters and the waters at Sta. B had intermediate values. Surface enrichment represented by the difference in microlayer and 10 cm CDOM fluorescence (ΔCDOM) was lowest at Stations A and C, and highest for Sta. D near Georges Bank. The ratio of microlayer

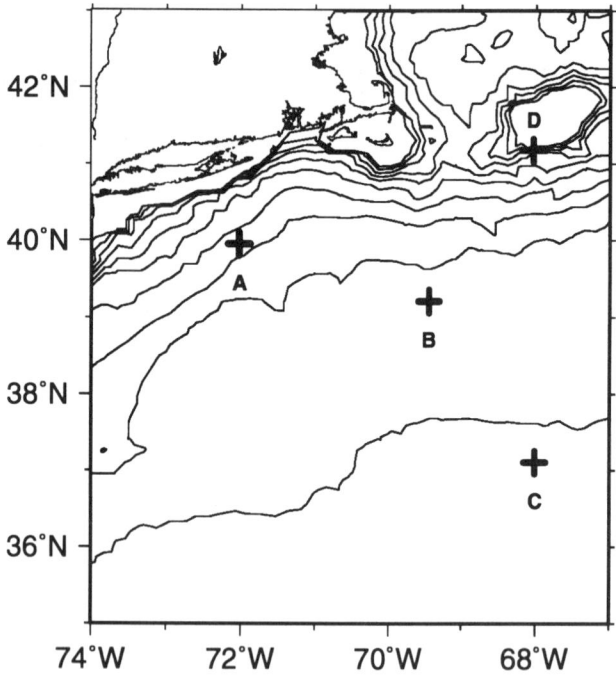

Figure 3. July climatological CZCS chlorophyll distributions (in mg m^{-3}) off the U.S. east coast and four stations occupied during the July, 1997 NSF CoOP coastal gas exchange cruise.

to subsurface fluorescence or CDOM enrichment factor (CDOM E.F.) was lowest at Sta. A and highest at Sta. B.

Wind speed, CDOM, and surface wave slope are first compared for periods of overall low wind stress at these stations (Table 1). We make two comparisons: cases of similar winds and contrasting organic matter levels (Stations B, C and A, D) and contrasting winds and organic matter levels (Stations A, C and D). Mean wind speeds at Stations B and C during these periods were nearly the same (~2.4 m s^{-1}). However, average wave slopes were lower by an order of magnitude at Sta. B. The relative reduction in average wave slope increased strongly with increasing wave number, characteristic of the strong wave slope roll-off caused by surfactants [*Frew*, 1997; *Bock et al.*, 1999]. SAOM levels at Sta. B were three times higher than at Sta. C. Winds at Stations A and D were modestly higher than those at Sta. C (by a factor of 1.3), but average wave slopes at Sta. A for 100, 200, and 400 radian m^{-1} were similar in magnitude to those at Sta. C, suggesting that the higher organic matter levels at Sta. A may have suppressed growth of short gravity waves. Since the actual surface enrichment (i.e. ΔCDOM) at Sta. A was low, the effect is not large. Wave slope at Sta. D, with winds identical to Sta. A, but with high organic matter and the highest ΔCDOM, was reduced by an order of magnitude.

A comparison of the wave slope data (for k = 200 radian m^{-1}) as a function of wind speed for all four of these low wind deployments is given in Figure 4. Wave slope did not depend strongly on wind over this range of wind speeds, possibly because all of these sites were surfactant-influenced, based on CDOM enrichment factors (Table 1); considerable scatter is observed at a given wind speed. Data points for which microlayer CDOM fluorescence was above an arbitrary threshold (0.15 F.U.) are shown as triangles in Figure 4. The effect of surfactants as represented by CDOM fluorescence is highlighted by the low slope, high fluorescence points.

Based on a limited number of stations, these data suggest that the reduction in the amplitude of the wave slope spectrum S(k) for k ≤ 400 rad m^{-1} is more closely related to the surface microlayer excess relative to subsurface water (as reflected by ΔCDOM) than to absolute surfactant concentration (as reflected by either bulk SAOM concentration or CDOM fluorescence). In previous wind-wave tank experiments [*Bock et al.*, 1999], the wave slope roll-off and

Table 1. Chemical and physical parameters for four stations occupied during the CoOP 1997 gas exchange cruise.

STATION	A	B	C	D
Year day (1997)	194	188	190	189
Latitude (N)	39° 57'	39° 13'	37° 06'	41° 10'
Longitude (W)	72° 02'	69° 27'	68° 01'	68° 02'
SST (C)	22.3	23.7	27.7	13.5
Salinity (PSU)	30.28	33.88	35.64	31.88
DOC (μM)	108±7	87±4	76±4	82±3
Chlorophyll (mg m^{-3})	0.30	0.12	0.07	2.07
SAOM (mg l^{-1})	1.24 ±.06	0.68 ±.07	0.13 ±.08	1.0 ±0.1
CDOM at 10 cm (f.u.)	1.38 ±.19	0.19 ±.01	0.06 ±.01	0.72 ±.02
Δ CDOM, μlayer-10 cm	0.05	0.14	0.03	0.24
CDOM E.F.	1.04	1.74	1.50	1.33
f450/r	0.22	0.09	0.04	0.24
Mean wind (m s^{-1})	3.3±0.4	2.3±0.4	2.5±0.7	3.2±0.3
Avg. S(k), k=100 rad m^{-1}	2.1±0.6 E-07	3.3±1.8 E-08	1.8±0.6 E-07	3.1±3.5 E-08
Avg. S(k), k=200 rad m^{-1}	3.2±1.1 E-08	1.4±1.0 E-09	2.4±1.1 E-08	1.8±1.3 E-09
Avg. S(k), k=400 rad m^{-1}	8.9±2.5 E-09	2.0±0.1 E-10	7.3±2.2 E-09	4.6±1.9 E-10

consequent reductions in gas transfer due to surfactants was first detectable for k > 800 rad m^{-1} as bulk surfactant concentration increased. Due to uncertainties in processing the SLSG data for higher wave numbers, we are unable to comment on differential slope reductions for k > 400 rad m^{-1}, where we would most expect to see the impact of the observed order of magnitude spatial changes in bulk surfactant concentration for the different stations. This is likely be resolved with future measurements of gas transfer rates on short time scales in combination with improved wave number-frequency slope measurements.

4.2. Small-Scale Variability

In addition to investigating large-scale variability, deployment of the SMS coupled with the CDOM fluorometry package allowed us to gain a picture of small-scale variability in microlayer surfactant distributions. Figure 5 illustrates two LADAS tows during the 1995 R/V New Horizon cruise off the California coast. The upper panel shows microlayer and subsurface variations in CDOM fluorescence over a 5 hr period when the air-side friction velocity (u_*) ranged from 0.015 to 0.15 m s^{-1} (wind speed, 0.5-4.5 m s^{-1}). The open symbols represent 1 minute averages. The surface microlayer was enriched over the subsurface water during the entire record. The CDOM fluorescence very clearly shows the complex, small-scale features on the ocean surface, ranging from tens to hundreds of meters wide. The largest enrichments, for the period 130.57-130.62, corresponded to a series of intense banded slicks. These may have been associated with current shear at the boundary of different water masses, as indicated by the abrupt drop in subsurface fluorescence. The lower

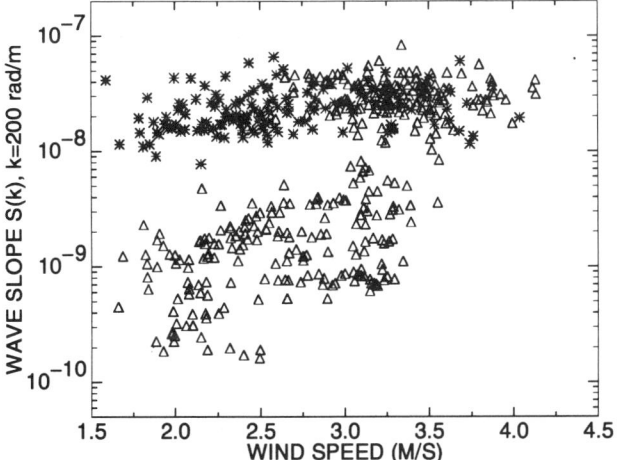

Figure 4. Wave slope at wave number 200 rad m^{-1} versus wind speed (one minute averages) for Stations A-D. Triangles are data where microlayer CDOM fluorescence is above 0.15 F.U.

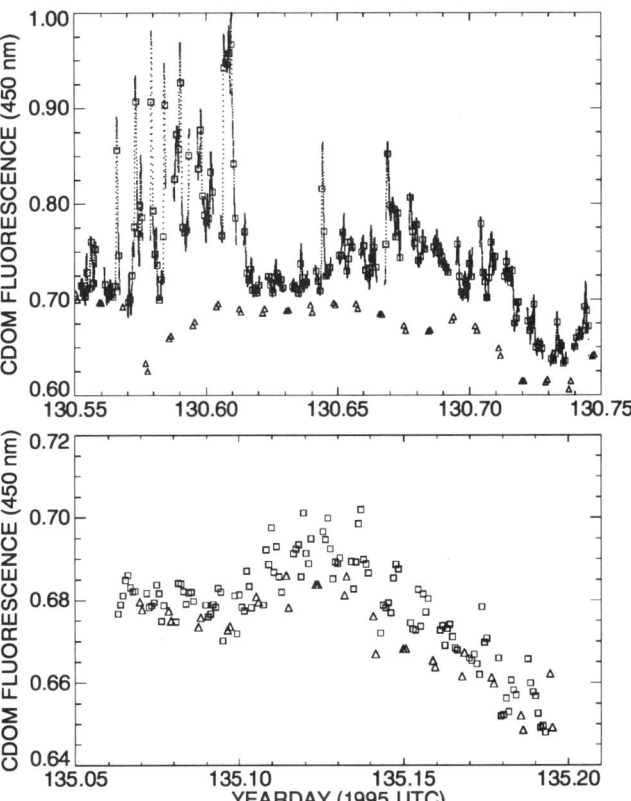

Figure 5. Variation of surface microlayer (9) and subsurface (()) CDOM fluorescence for two stations occupied in the California Bight in 1995. Upper: yearday 130, winds 0.5-4.5 m s^{-1}. Lower: yearday 135, winds 3-7 m s^{-1}. Symbols are one minute averages.

panel represents a 3 hr tow near Santa Catalina Island during a period of moderate wind stress. Wind speeds ranged from 3 to 7 m s^{-1} (0.1 < u_* > 0.3 m s^{-1}). Small CDOM fluorescence enrichments (~3-5%) were observable. These surface enrichments may be underestimated due to occasional washover of the SMS sampling cylinder and collection cup.

Small-scale changes in surface enrichments can be related to both wind and wave slope. In Figure 6, we present a detailed comparison of wind speed, CDOM fluorescence, and wave slope records during a deployment at Sta. B in the western Atlantic. During this period, winds slowly increased from 2 m s^{-1} to 3 m s^{-1}. Although no surface tension measurements were made, numerous slick patches, streaks, and bands with decreased surface roughness were observed visually, indicating the non-homogeneity of the surface. The CDOM fluorescence indicated several regions of surface chemical enrichment (numbered 1-8 in Figure 6). During the period 188.45-188.47, several intense film bands (5-8) were traversed, corresponding to major reductions in slope. (The peak maximum of film band 5 is presumed to have been missed during a subsurface sampling

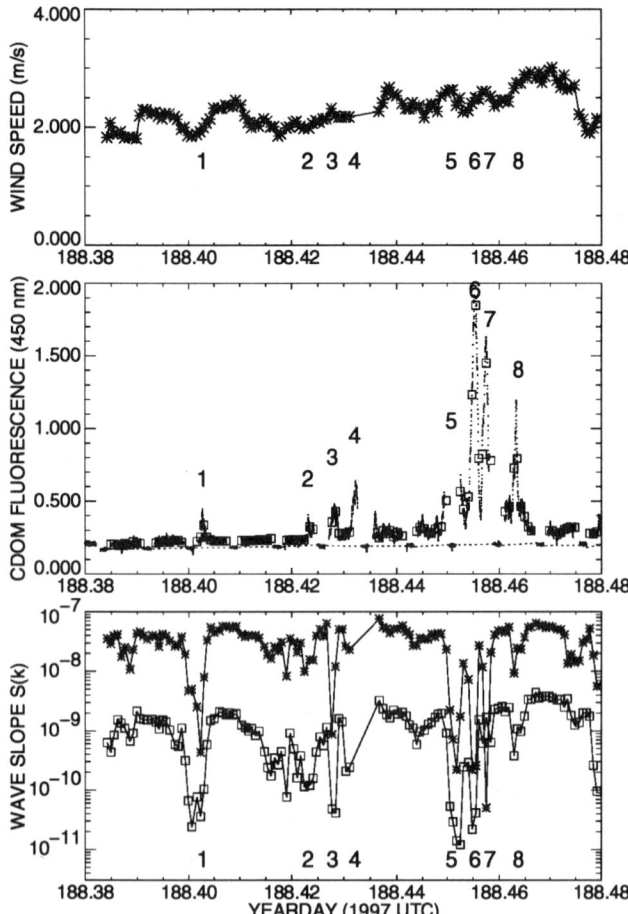

Figure 6. Comparison of wind speed, CDOM fluorescence, and wave slope time series for a LADAS transect at Sta. B during the 1997 CoOP cruise. Top: wind speed; middle: microlayer CDOM (solid; 9) and subsurface CDOM (dashed); bottom: S(k) at k=100 rad m^{-1} ($*$) and 200 rad m^{-1} (9). For clarity, the 200 rad m^{-1} data have been offset by −1.5E−9. Symbols are one minute averages.

interval.) The wave slope record also shows fluctuations that appear to be a response to the combined effects of wind and surfactants. Bands 1-4, for example, appear to grow in during the latter part of periods when the wind speed dropped slightly, reinforcing the effect on wave slope. Maximum wave damping occurred after the accumulation of surface material, suggesting that there is a complex, dynamic interplay between wind, surface films and capillary wave formation in this low wind stress regime.

5. CONCLUSIONS

These unique observations clearly show the effects of differing levels of adsorbed surface-active organic matter on the slope spectrum of the small-scale waves. Since capillary wave slope is strongly correlated with gas exchange rate, these results provide a possible explanation for the high variance commonly observed for gas transfer velocity measurements under low wind stress conditions. These results also have important implications with respect to drag variability (measured versus parameterized drag) as well as remote sensing of low wind states. A study of higher wind stress cases is in progress.

Acknowledgments. The authors gratefully acknowledge the support of this research by NASA (Award NAGW 2431) and by the NSF Coastal Ocean Processes (CoOP) program under Grants OCE-9410534 and OCE-9711285. This is Contribution No. 10337 of the Woods Hole Oceanographic Institution.

REFERENCES

Bock, E. J., T. Hara, N. M. Frew, and W. R. McGillis, Relationship between air-sea gas transfer and short wind waves. *J. Geophys. Res.*, 104, 25821-25831, 1999.

Bock, E. J., and T. Hara, Optical measurements of capillary-gravity wave spectra using a scanning laser slope gauge, *J. Atmos. Ocean. Technol.*, 12, 395-403, 1995.

Bock, E.J., and J.A. Mann, Jr., On ripple dynamics,II. A corrected dispersion relation for surface waves in the presence of surface elasticity, *J. Colloid Interface Sci.*, 147, 422-432, 1989.

Carlson, D. J., A field evaluation of plate and screen microlayer sampling techniques. *Mar. Chem.*, 11, 189-208, 1982.

Carlson, D. J., J. L. Canty, and J. J. Cullen, Description of and results from a new surface microlayer sampling device, *Deep-Sea Res.*, 35, 1205-1212, 1988.

Cinbis, C., Noncontacting techniques for measuring surface tension of liquids, Doctoral Dissertation, Stanford University, E. L. Ginzton Laboratory Report No. 4931, 1992.

Frew, N. M., E. J. Bock, W. R. McGillis, A. V. Karachintsev, T. Hara, T. Münsterer, and B. Jähne, Variation of air-water gas transfer with wind stress and surface viscoelasticity, in *Air-Water Gas Transfer*, edited by B. Jähne and E. Monahan, pp. 529-541, AEON Verlag & Studio, Hanau, Germany, 1995.

Frew, N. M., The role of organic films in air-sea gas exchange, in *The Sea Surface Microlayer and Global Change*, edited by P. S. Liss and R. A. Duce, pp. 121-172, Cambridge University Press, Cambridge, England, 1997.

Garrett, W. D., Retardation of water drop evaporation with monomolecular surface films. *J Atmos. Sci.*, 28, 816-819, 1971.

Harvey, G. W. and Burzell, L. A., A simple microlayer method for small samples. *Limnol. Oceanogr.*, 19, 162-165, 1972.

Hunter, K. A., and P. S. Liss, Polarographic measurement of surface-active material in natural waters, *Water Res.*, 15, 203-215, 1980.

Jarvis, N. L., C. O. Timmons, and W. A. Zisman, The effect of mono-molecular films on the surface temperature of water, in *Retardation of Evaporation by Monolayers*, edited by V. K. LaMer, pp. 41-58, Academic Press, New York, 1962.

Katsaros, K.B., H. Gucinski, S.S. Ataktηrk, and R. Pincus, Effects of reduced surface tension on short waves at low wind speeds in a fresh-water lake, In *Radar Scattering from*

Modulated Wind Waves, edited by G. J. Komen and W. A. Oost, pp. 61-74, Kluwer Academic Publishers, Dordrecht, 1989.

Levich, V. G., *Physico-Chemical Hydrodynamics*, Prentice-Hall International, Englewood Cliffs, N. J., 1962.

Peltzer, E. T., and P. G. Brewer, Some practical aspects of measuring DOC - sampling artifacts and analytical problems with marine samples, *Mar. Chem.*, 41, 243-252, 1993.

Erik J. Bock, Interdisciplinary Center for Scientific Computing, University of Heidelberg, Im Neuenheimer Feld 368, D-69120 Heidelberg, Germany.

James B. Edson and Wade R. McGillis, Department of Applied Ocean Physics and Engineering, Woods Hole Oceanographic Institution, Woods Hole MA, 02543.

Nelson M. Frew and Robert K. Nelson, Department of Marine Chemistry and Geochemistry, Woods Hole Oceanographic Institution, 360 Woods Hole Road, Woods Hole, MA, 02543. (email: nfrew@whoi.edu)

Tetsu Hara, Graduate School of Oceanography, University of Rhode Island, South Ferry Road, Narragansett, RI, 02882.

Thermal Profiling of the Sea Surface Skin Layer Using FTIR Measurements

Jennifer A. Hanafin & Peter J. Minnett

Division of Meteorology & Physical Oceanography, Rosenstiel School of Marine & Atmospheric Science,
University of Miami, Miami, Florida

Sea surface spectral emissivity and the depth of the thermal skin boundary layer were determined using high spectral resolution measurements of the sea surface and the atmosphere taken in the field measurements by the Marine-Atmosphere Emitted Radiance Interferometer. In order to determine the sea surface emissivity, the effective incidence angle was found by minimizing the variance in the brightness temperature spectrum retrieved from the corrected upwelling radiance spectrum. Certain wavelength regions have different absorption characteristics, allowing the temperature at different levels to be retrieved from different spectral regions. In this way, the temperature gradient of the thermal boundary layer was determined. The depth of the skin layer was then calculated by determining the depth at which the thermometrically measured bulk temperature intersects this gradient. At low wind speeds, the skin layer can be up to 0.2mm deep, getting shallower with increased wind speed and becoming very shallow (0.01-0.07mm) above wind speeds of 8ms^{-1}. These results are encouraging for application of this method to determine air-sea heat and gas fluxes in the field.

INTRODUCTION

Fourier Transform Infrared (FTIR) interferometry has proven to be a very powerful tool in many fields of the observational sciences, but only recently has it been used in oceanic field studies (Smith et al. 1996). Oceanographic applications to date have included validation of satellite-derived sea surface temperatures (Kearns et al. 2000). A geographically extensive dataset has been collected by the RSMAS Remote Sensing Laboratory with the Marine-Atmosphere Emitted Radiance Interferometer (M-AERI) over the last three years (Minnett et al., 2001). The following are results of a study to interpret some of the information present in the IR spectra.

Air-sea heat flux measurement has always been problematic. This is partly due to the difficulty of obtaining good quality flux data over the ocean, both for direct measurement and for use with the bulk aerodynamic formulae parameterizations (e.g. Edson et al. 1998). To further complicate this objective, wind and wave fields interact in ways which alter the stability and turbulence of the marine-atmosphere boundary layer, but are currently not well understood (Donelan, 1990). The presence of a thermal 'skin layer' at the sea surface is a result of heat transfer between the ocean and atmosphere. Turbulent transfer is suppressed at the interface due to the density discontinuity between the two media. In calm conditions, a conduction zone exists at the interface in which the temperature, T, varies almost linearly with depth, z, and where heat flux, Q, is given by

$$Q = \kappa \frac{dT}{dz} \qquad (1)$$

where κ is the thermal conductivity of water. This zone is known as the skin layer. As wind speeds increase above 1ms^{-1}, convective surface renewals begin to take place and these replace fluid in the skin layer with bulk fluid. At higher wind speeds (>5-7ms^{-1}) surface renewal events due to wind and wave action begin to dominate the heat transfer regime. Understanding the processes that affect this skin layer will help us to understand air-sea heat transfer. One application of this research may be more direct measurements of the ocean-atmosphere heat flux.

Physical processes affecting heat transfer also affect the air-sea transfer of certain gases. For example, studies such as that of Soloviev and Schluessel (1994) have considered both processes simultaneously using a surface renewal model. The results presented here regarding the thickness of the thermal boundary layer may be applied to determine the thickness of the concentration boundary layer, which is a parameter that can be used as a proxy for ocean-atmosphere gas fluxes (McKeown and Asher, 1997).

While the emissivity of a still water surface is known (Pinkley and Williams, 1976), that of a natural sea surface, roughened by wind and waves and holding surfactants, has not been determined definitively. Theoretical models of its dependence on sea state are available (Masuda et al., 1988; Watts et al., 1996) and are used operationally in field studies involving SST measurement by calibrated infrared radiometers. Validation of these models in the field using radiometry is difficult as the radiometers deployed are generally relatively broadband and of limited accuracy. With an FTIR spectroradiometer, such as the M-AERI, high spectral resolution and complementary sea and sky measurements provide a unique opportunity to study sea surface emissivity across the IR spectrum. Such studies are timely, as sea-surface temperature retrievals by infrared radiometry are very sensitive to emissivity assumptions.

DATA

The data used in this study were collected off Baja California, on the R.V. Melville during the Marine Optical Characterization Experiment 5 (MOCE5) cruise, the track of which is shown in figure 1. This cruise was specifically designed for the study of bio-optical properties, so some data were from areas of high biological productivity and very high spatial variability. As the effects of biology (e.g. biogenic slicks) on surface properties such as the emissivity are not well known, these data were removed from the subsequent analysis.

The M-AERI is a well calibrated FTIR spectroradiometer (Minnett et. al. 2001). It is a passive instrument, which measures the radiance emitted by both the sea surface and the atmosphere at complementary angles, and by the atmosphere at zenith. Every sequence of scene views was sandwiched between two calibration sequences, each of which consisted of two blackbody views, one at ambient air temperature and the other at 60°C. To increase the signal-to-noise ratio, radiance measurements were averaged over 1-3 minute intervals, the longest interval being for the zenith sky view. The measured emitted radiance spectra (with units of W m^{-2} sr^{-1} cm^{-1}) are in the wavenumber range 500-3000cm^{-1} (approximately 3-20μm wavelength) and have a spectral resolution of 1cm^{-1}. The accuracy of the derived brightness temperatures in this wavelength region has been determined to be <0.02K at 20°C and <0.04K at 30°C (Minnett et al., 2001). Sample spectra of the measured upwelling (oceanic) and downwelling (atmospheric) radiance are shown in figures 2a and 2b, with the spectral intervals used in this work marked between dashed lines.

Segelstein (1981) published values of the complex refractive index of water compiled from literature. These data were derived for pure water and a small correction for density changes due to temperature and salinity was applied to them (Pinkley and Williams, 1976) before being applied to the problem at hand.

The atmospheric transmission data used to determine clear channels were computed using the HITRAN database (Rothman et al., 1987) at 296K and 1atm. Heat fluxes were calculated using the algorithm described in Fairall et al. (1996).

THEORETICAL BASIS

The 2000-3000cm^{-1} wavenumber range was identified in laboratory studies by McKeown et al. (1995) as being useful for determining thermal profiles in the surface skin layer of water. Requirements for an appropriate window include minimal atmospheric absorption and a variable, wavenumber-dependent penetration depth in water. The former is desirable so that most of the radiance being

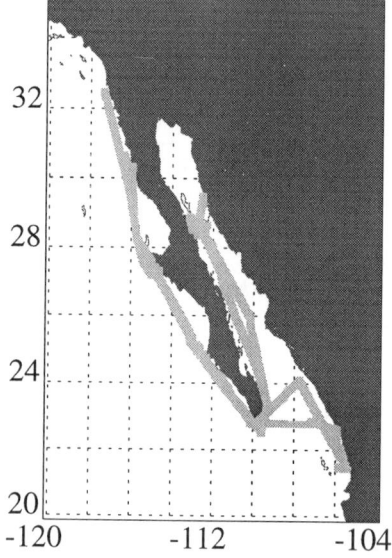

Figure 1. Track of the R.V. Melville during the Marine Optical Characterization Experiment.

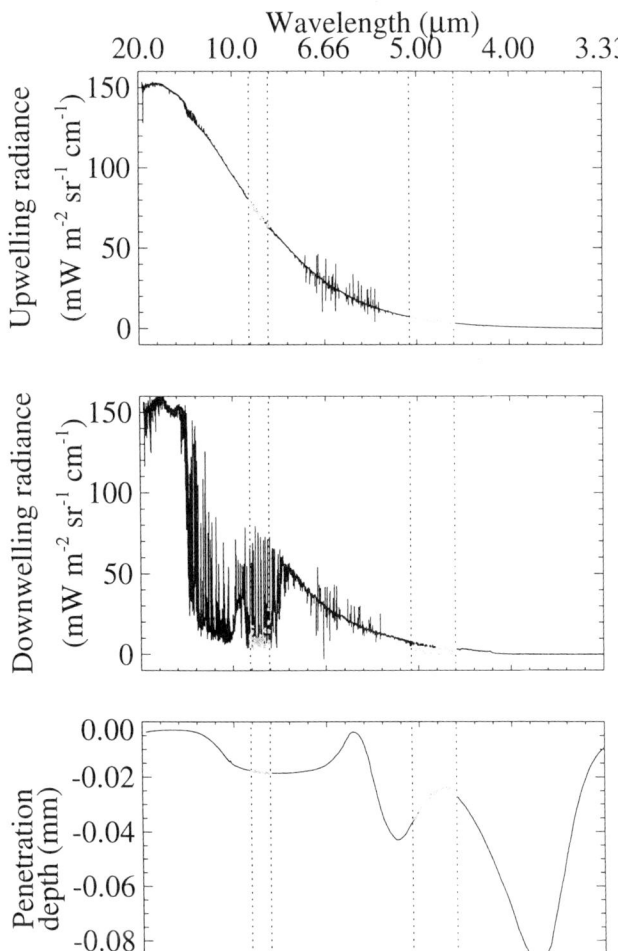

Figure 2. Samples of (a) upwelling and (b) downwelling radiance measured by the Marine-Atmosphere Emitted Radiance Interferometer and (c) penetration depth calculated from the absorption coefficient of water.

measured is emanating from the sea surface. The latter is a result of the fact that the index of refraction of water is complex and the imaginary part of the refractive index determines the absorption/emission characteristics. As these characteristics depend on wavelength, water is more transparent at certain wavelengths than at others, so radiation emitted at these wavelengths carries information from slightly deeper levels than that emitted at those where water is less transparent. Figure 2c shows how the penetration depth changes with wavelength. The channels used in this study are marked in the figure. The signal to noise ratio of the M-AERI above 2200 cm^{-1} is not high enough in field data to use this region, but the wavelengths used here also have depth information, albeit at shallower depths.

The Lambert absorption coefficient, α, is defined in the equation:

$$I(z) = I_0 e^{-\alpha z} \quad (2)$$

where I(z) is radiant intensity at depth z and I_0 is radiant intensity at the surface. The absorption coefficient can be determined from the attenuation coefficient, k, through the relation:

$$\alpha = 4\pi k \nu \quad (3)$$

where ν is wavenumber. The penetration depth is the reciprocal of the absorption coefficient and this supplies an estimate of the depth from which the measured radiation is being emitted.

To retrieve the brightness temperature at these wavelengths, the true sea surface radiance, R↑, must is found from:

$$R(\nu) = \varepsilon(\nu) R\uparrow(\nu) + \rho(\nu) R\downarrow(\nu) \quad (4)$$

where R is the measured upwelling radiance, R↓ is downwelling sky radiance and ε and ρ are the sea surface emissivity and reflectivity, respectively. The second term on the right hand side represents the atmospheric radiation which is reflected at the sea surface. All of the quantities are wavenumber dependent. The atmospheric absorption and emission between the instrument and the sea surface is assumed to be negligible. This is justified by noting that the path length between the instrument and the sea surface was very short (~10m) and the wavelengths used are coincident with high atmospheric transmittance.

The wavelength dependence of each of the surface characteristics mentioned above had to be determined before accurate temperatures can be retrieved. As these corrections are applied before the radiance is converted to temperature, using Planck's equation, small variations in the corrections have a large effect on the temperatures retrieved. With this in mind the reflectivity of the sea surface, ρ, was computed from the parallel and perpendicularly polarized components, $\rho_\parallel$ and $\rho_\perp$, using Fresnel's equations in the form:

$$\rho = (\rho_\perp + \rho_\parallel)/2$$
$$\rho_\parallel = r_\parallel^2 = \left(\frac{n_t \cos\theta_i - n_i \cos\theta_t}{n_i \cos\theta_i + n_t \cos\theta_t} \right)^2 \quad (5)$$
$$\rho_\perp = r_\perp^2 = \left(\frac{n_i \cos\theta_i - n_t \cos\theta_t}{n_i \cos\theta_i + n_t \cos\theta_t} \right)^2$$

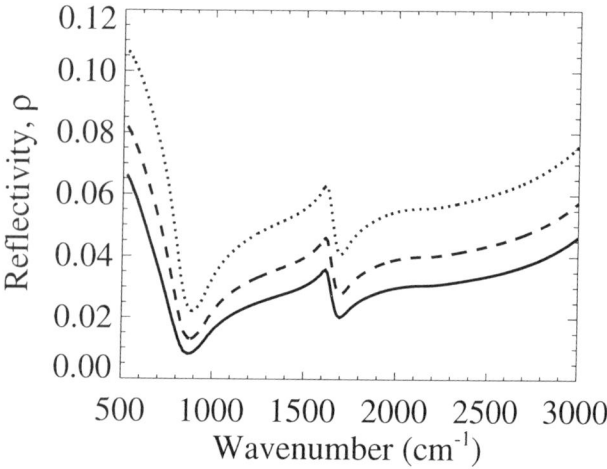

Figure 3. Dependence of reflectivity calculated from Fresnel's equations on wavelength, for incidence angles of 50° (solid line), 55° (dashed line) and 60° (dotted line). Reflectivities are corrected for typical sea water density.

where θ_i and θ_t are the incidence and transmission angles and n_i and n_t are the refractive indices of air and water. The wavelength and incidence angle dependence for the reflectivity of water with salinity of 35psu is shown in figure 3.

As the data used here were measured from a ship, a number of other factors must be taken into account. The ship's motion and the presence of waves on the surface will both change the effective incidence angle (EIA) of the measurements. To account for this, the EIA was first calculated from the radiance data. This is made possible by the spectral resolution of the M-AERI measurements, which is high enough to resolve individual atmospheric absorption lines. These absorption features are also seen in the upwelling radiance spectra as a result of reflection from the sea surface. As both the signal and its reflected amplitude are measured, the magnitude of the reflectivity can be determined from equation 2, given $\varepsilon = 1 - \rho$ (Kirchhoff's Law).

In this analysis, the incidence angle was treated as a variable and a range of plausible values were selected. The reflectivity for each value was then computed from Fresnel's equations and the resulting radiance spectrum converted into a brightness temperature spectrum. A distinct minimum occurred in the variance of the spectral brightness temperature for a particular value of incidence angle. So, by minimizing the variance of brightness temperature along parts of the spectrum, a value for ρ can be determined directly from the data, which also gives the spectral emissivity, ε.

In this study, the depth of the aqueous boundary layer was determined by computing the temperature at wavelengths that have different depth assignments, fitting a line to these temperature data and determining where the measured bulk temperature intersects that line. Three different channels were used: 1092-1195cm^{-1}, 1960-2100cm^{-1} and 2100-2200cm^{-1} (see figure 2c). In the first waveband, 'microwindows' were used to deduce the temperature at the corresponding depth. These temperatures were subsequently averaged and the average temperature and depth were utilized. In the 1960-2100 and 2100-2200cm^{-1} intervals, the brightness temperature (BT) at each wavenumber was measured and then an average over the depth interval was taken. The final data point was the temperature whose spectral Planck function radiance best fitted the observed upwelling sea surface radiance. By assuming that the best-fit Planck function temperature is the shallowest measurement and that the gradient in the conduction layer is linear we now have four temperatures, which may be used to calculate the gradient.

RESULTS

Sea Surface Emissivity

Values of the effective incidence angle computed using this technique are shown in figure 4a. The mean value is 55.72°, and the discrepancy between this and the nominal pointing angle of 55° can be explained by a small ship tilt or skewness in the distribution of the sea surface slope. Leveling of the instrument during installation on the deck took place before the ship was refueled, which may have resulted in a change in the level. While the scatter (+/- 5°) is acceptable considering that the data were collected from a vessel at sea, a number of environmental parameters were explored to determine its cause.

There is no significant dependence on wind speed (figure 4b), which is contrary to previous models, which include a specific wind-speed dependence. This may be due to the fact that the instrument has a 1minute averaging time, so the effects of individual waves, and ship motion, are averaged out. There is a dependence on cloud cover, however. The atmospheric transmittance in the 10-10.2µm interval is high, so measurements here are responsive to the presence of cloud; low radiance implying clear skies. Using this proxy, figure 4c shows that the higher values of EIA occur in clear skies and there is much more scatter when clouds are present. As EIA is proportional to the sea surface reflectivity (and inversely proportional to the emissivity), large EIA values mean that the atmospheric component of the correction is relatively large. The presence or absence of clouds has two separate effects on the correction. The first is the magnitude of the correction, which is larger in clear skies. The second is heterogeneity, as the upwelling and downwelling measurements are not

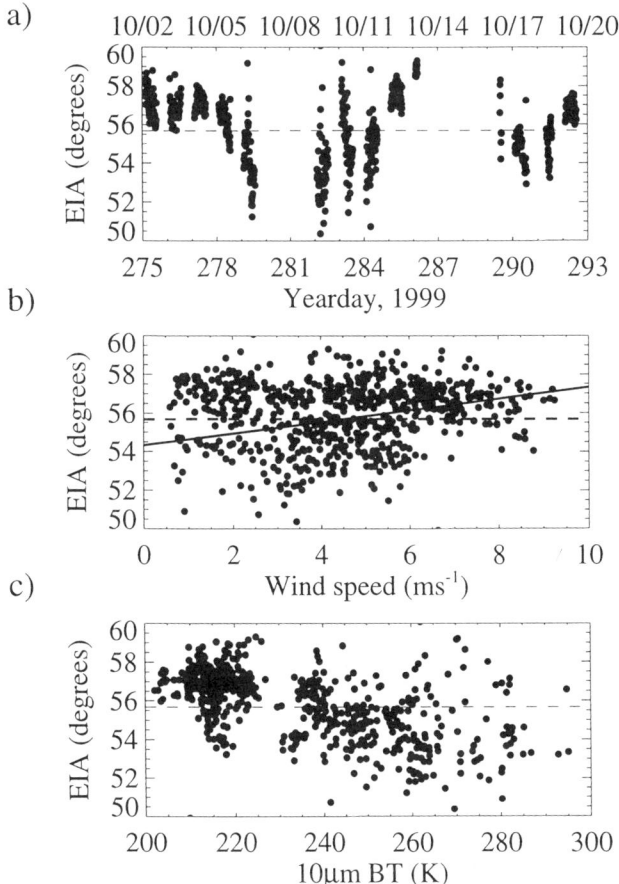

Figure 4. Effective incidence angle as a function of (a) time, (b) wind speed and (c) 10-10.2μm sky brightness temperature.

taken simultaneously, a temporal change in downwelling radiation, such as caused by the movement of clouds, means that the correction is less accurate. As the field of view of the instrument is so small (45mrad full-angle beam width), the problem of spatial heterogeneity is reduced. However, the presence of clouds does introduce variance in the reflectivity results (for a discussion of this effect, see Donlon and Nightingale, 2000).

Skin Layer Depth

The magnitude of the computed skin layer (figure 5a) is an order of magnitude less than the 0.5mm expected from inserting average values in equation 1. A previous study (Katsaros, 1980) noted that the skin depth decreases rapidly from 0.5mm to about 0.1mm with onset of winds between 1-5ms^{-1} and postulated that the skin disappears completely at wind speeds greater than 8ms^{-1}. It is likely that the skin layer does not disappear at these moderately high wind speeds, on average, however, as some heat will be transferred by conduction even if there is a large increase in surface renewal frequency. Unfortunately, no completely calm periods occurred during this cruise, so no skin layer depths greater than 0.2mm were observed. The low and moderate wind regimes do behave as expected, however, and a line fitted to these data (figure 5b) has a zero crossing at a wind speed of 8.3ms^{-1}. The values of skin depth above 7ms^{-1} are finite but very small (0.01-0.07mm). It should be noted that as the wavelengths used in this study have very shallow penetration depths (~0.02-0.03mm), a small error in the depth assignation of temperature would result in a bias in the magnitude of the calculated skin depth. Further studies using higher wavenumber regions with larger penetration depths will help to reduce this problem.

When compared to heat fluxes in figure 5c, the skin layer depth goes to zero as the magnitude of the net heat flux decreases towards zero. This indicates that, as the net heat transfer across the interface decreases, the depth of the thermal boundary layer required to support that flux also

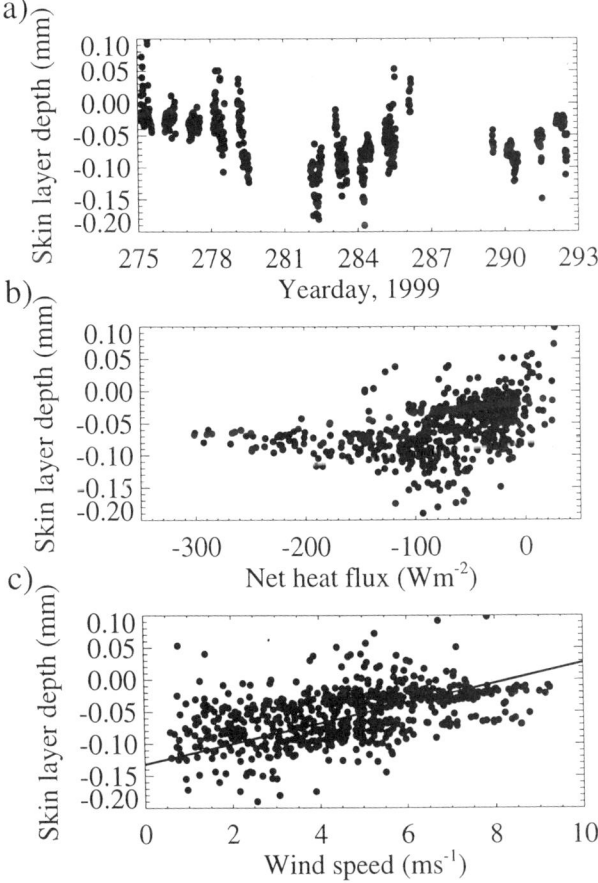

Figure 5. Skin layer depth as a function of (a) time, (b) net heat flux and (c) wind speed. The solid line in (c) is a least-squares fit to the data. Negative fluxes indicate heat flow is from ocean to atmosphere.

decreases. Positive skin depth values imply that the bulk temperature is warmer than the skin temperature, and these occur primarily when the heat flux is very small or positive, *i.e.* net heat flow is from atmosphere to ocean. The skin depth appears to decrease with increasing net flux at higher heat fluxes. This may be due to a lack of data or due to increased atmospheric instability in a more convective regime.

CONCLUSION

This multiple wavelength approach to determine both the spectral sea surface emissivity and the depth of the thermal boundary layer has the potential of being a very valuable tool for air-sea heat and gas transfer research. One of the problems with bulk formulae heat fluxes is the flow distortion around the ship, which perturbs standard measurements. The M-AERI takes measurements away from the ship's influence and is robust enough to make good measurements in moderately high wind speeds, except when rain or large amounts of spray are present. Atmospheric instability may affect the measurements by increasing spatial and temporal inhomogeneity, however. The problem of explicitly determining the effective incidence angle of measurements of the sea surface, which is skewed due to the effects of ship motion and sea state, is not trivial. Over the 0.5-1 minute averaging interval of the M-AERI it would require an entire suite of complementary measurements for its solution. This variable can significantly affect the temperature retrieval through the magnitude of the reflectivity correction and emissivity value. Calculating these quantities by utilizing the information present in the high spectral resolution of M-AERI measurements simplifies this problem greatly by implicitly taking these convolved variables into account. Additionally, using a single, well-calibrated instrument is preferable to using a suite of independently calibrated sensors. The ability to calculate a variable such as the marine thermal boundary layer depth with a degree of confidence is very encouraging for the extension of this work into air-sea flux studies.

Acknowledgements. This work was supported by NASA contracts NAS5-31361 to Dr O. B. Brown and NAG-56577 to Dr P. J. Minnett. Many thanks to the Captain and crew of the R. V. Melville, to Drs Ed Kearns and Bob Evans for data collection and Dr Walt McKeown for helpful discussion.

REFERENCES

Donelan, M.A., Air-Sea Interaction, in The Sea: Ocean Engineering Science, edited by B. LeMehaute, and D.M. Hanes, John Wiley & Sons, New York, 1990

Donlon, C. J. and T. J. Nightingale, The effect of atmospheric radiance errors in radiometric sea surface skin temperature measurements, *Appl. Optics.*, 39, 15, 2387-2392, 2000

Edson, J.B., A.A. Hinton, K.E. Prada, J.E. Hare, and C.W. Fairall, Direct covariance flux estimates from mobile platforms at sea. *J. Atm. Ocean. Tech*, 15, 547-562, 1998.

Fairall, C. W., E. F. Bradley, D. P. Rogers, J. B. Edson, G. S. Young, Bulk parameterization of air-sea fluxes for TOGA-COARE, *J. Geophys. Res.*, 101, C2, 3747-3764

Katsaros, K.B., The Aqueous Thermal Boundary Layer, *Bound.-Layer Meteor.*, 18, 107-127, 1980

Kearns, E. J., J. A. Hanafin, R. E. Evans, P. J. Minnett, O. B. Brown, An independent assessment of Pathfinder AVHRR sea surface temperature algorithm performance, *Bull. Am. Met. Soc.* 81, 7, 1525-1536, 2000

Masuda, K., T. Takashima, Y. Takayama, Emissivity of pure and sea waters for the model sea surface in the infrared window regime, *Remote Sens. of the Environ.*, 24, 313-329, 1988

McKeown, W. & W. Asher, A radiometric method to measure concentration boundary layer thickness at an air-water interface, *J. Atmos Ocean Tech.*, 14, 6, 1494-1501, 1997

McKeown, W., F. Bretherton, H. L. Huang, W. L. Smith, & H. L. Revercomb, Sounding the skin of water: sensing air-water interface temperature gradients with interferometry, *J. Atmos. Ocean. Tech.*, 12, 1313-1327, 1995

Minnett, P. J., R. O. Knuteson, F. A. Best, B. J. Osborne, J. A. Hanafin, O. B. Brown, Marine-Atmosphere Emitted Radiance Interferometer (M-AERI): a high-accuracy, sea-going, infrared spectroradiometer. *J. Atmos. Ocean. Tech.* In press, 2001

Moller, K. D., *Optics*, University Science Books, California, 1988.

Pinkley, L. W. and D. Williams, Optical properties of sea water in the infrared, *J. Opt. Soc. Am.*, 66, 6, 554-558, 1976

Rothman, L. S., R. R. Gamache, A. Goldman, L. R. Brown, R. A. Toth, H. M. Pickett, R. L. Poynter, J.-M. Flaud, C. Camy-Peyret, A. Barbe, N. Husson, C. P. Rinsland, and M. A. H. Smith, The HITRAN database: 1986 Edition, *Appl.Opt.*, 26, 4058-4097, 1987

Segelstein, D. J., M. Sc. Thesis, *The complex refractive index of water*, University of Missouri, Kansas City, 1981

Smith, W. L, R. O. Knuteson, H. E. Revercomb, W. Feltz, H. B. Howell, W. P. Menzel, N. Nalli, O. B. Brown, J. Brown, P. J. Minnett, W. McKeown, Observations of the IR properties of the ocean – implications for the measurement of SST via satellite remote sensing, *Bull. Am. Met. Soc.*, 77, 41-51, 1996

Soloviev, A.V and P. Schluessel, Parameterization of the cool skin of the ocean and of the air-ocean gas transfer on the basis of modeling surface renewal, *J. Phys. Ocean.*, 24, 1339-1346, 1994

Watts, P. D., M. R. Allen, T. J. Nightingale, Wind speed effects on sea surface emission and reflection for the along-track scanning radiometer, *J. Atmos. Ocean. Tech.*, 13, 1, 126-141, 1996

Jennifer A. Hanafin & Peter J. Minnett, Division of Meteorology and Physical Oceanography, Rosenstiel School of Marine & Atmospheric Science, University of Miami, 4600 Rickenbacker Causeway, Miami, FL 33139-1098

An Autonomous Profiler for Near Surface Temperature Measurements

Brian Ward

Geophysical Institute, Bergen, Norway.[1]

Peter J. Minnett

Meteorology and Physical Oceanography, Rosenstiel School of Marine and Atmospheric Science, Miami, Florida

This paper describes the profiling instrument SkinDeEP (Skin Depth Experimental Profiler), which measures the temperature of the water column from a depth of about 6 meters to the surface with high resolution thermometers. The instrument operates in an autonomous mode as it has the capability to change buoyancy by inflating a neoprene bladder attached to the body of the profiler. Measurements are recorded only during the ascending phase of the profile so as to minimize disturbances at the surface. Results from deployment of the profiler show strong temperature gradients within the bulk waters under conditions of high insolation. These data were compared to the skin temperatures as measured by the M-AERI (Marine – Atmospheric Emitted Radiance Interferometer), a high accuracy infrared spectroradiometer. The corresponding bulk - skin temperature differences, ΔT, were shown to have strong dependence on the depth of the bulk measurement during the daytime with low wind speeds, but at higher wind speeds, the depth dependence vanishes. One set of profiles under nighttime conditions is also presented, showing the presence of overturning and thus a heterogeneous temperature structure within the bulk.

INTRODUCTION

As long as there is a net heat flux out of the ocean and no breaking waves present, a cool skin exists on the ocean surface. The residual inaccuracies in the correction of the effects of the atmosphere in the measurement of sea surface temperature (SST) by infrared radiometers on satellites are comparable to the skin effect [*Kearns et al.*, 2000; *Minnett*, 1991]. The skin temperature is also important for air-sea interaction studies, as the atmosphere cannot be directly affected by waters below this layer. Current heat flux algorithms use skin temperature measurements to improve the accuracy of the computed fluxes [*Fairall et al.*, 1996]. Gas exchange across the air-sea interface is affected by the presence of the cool skin because of the temperature dependence of the solubility of gases. The results of Van Scoy et al. [1995] show that failure to account for the skin effect leads to an underestimation of the global air-sea CO_2 flux of between 0.17 and 0.4 $GtCyr^{-1}$, which is a significant part of the total exchange, estimated to be in the range of ~0.6 to ~2.2 $GtCyr^{-1}$ [*Tans et al.*, 1990].

The existence of the temperature gradient is explained by the way heat is transferred: within the bulk waters, heat transfer occurs due to turbulence, but as the surface is approached, viscous forces dominate and molecular processes prevail. Because heat transfer by molecular conduction is less efficient than by turbulence, a strong tempera-

---
[1] Now at Cooperative Institute for Marine and Atmospheric Studies, Miami, Florida.

Gas Transfer at Water Surfaces
Geophysical Monograph 127
Copyright 2002 by the American Geophysical Union

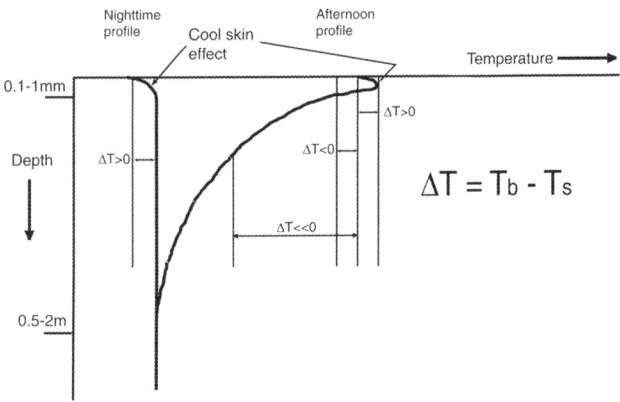

Figure 1. Idealized daytime and nighttime profiles within the near surface waters and molecular boundary layer.

ture gradient is established across the boundary layer. The temperature difference between that of the skin layer and the base of the molecular boundary layer is called the sea surface bulk - skin temperature difference, or ΔT, and is defined as:

$$\Delta T = T_b - T_s$$

where T_b and T_s are the bulk and skin temperatures, respectively.

Figure 1 shows schematically a daytime and nighttime profile incorporating the molecular boundary layer and the surface layer. The molecular boundary layer is shown to have a thickness of 0.1-1 mm, and is most likely to depend on wind speed [Robinson et al., 1984]. The nighttime profile shows a well-mixed bulk temperature with a cool skin and a resulting depth independent ΔT. In contrast, the daytime profile shows a warming of the surface layer due to insolation, which can lead to complications when deriving ΔT. The absorption of shortwave radiation in the molecular boundary layer is not enough to overcome the heat loss due to the sensible and latent heat fluxes, and the temperature gradient across the molecular boundary layer retains a positive value. However, if the bulk temperature is taken at a depth below this, the magnitude of ΔT is similar but the sign is reversed, indicating an apparent warm skin. Taking a bulk measurement deeper within the surface layer results in a much larger negative ΔT. Thus, the magnitude and sign of the apparent ΔT is strongly dependent on the structure of the near surface vertical temperature gradient, and the depth of the in situ measurement.

This short paper briefly describes the SkinDeEP autonomous profiling instrument and then goes on present a selection of data that were acquired in the Gulf of California in October 1999 during the MOCE-5 (Marine Optical Characterization Experiment-5) SeaWiFS validation cruise. The data presented include fixed-depth bulk temperature measurements, near-surface profiles, high quality skin temperature measurements, as well as some meteorological data.

INSTRUMENT

SkinDeEP is an autonomous profiler that carries high resolution temperature sensors to provide a record of the bulk temperature. A schematic of the instrument is shown in Figure 2. The transport vehicle for the sensors and associated instrumentation is an anodized aluminum cylinder with two detachable, hemispheric end-caps. Each endcap has a groove to accommodate an o-ring where it meets the cylinder, thus providing a watertight seal. When assembled, the total length of the instrument shell is 1.1 m. Debler and Vest [1977] have shown a distortion in the stratification within one diameter of a body, and thus the sensors protrude 30 cm from the endcap, which is twice the diameter of the profiler. A protection guard of maximum

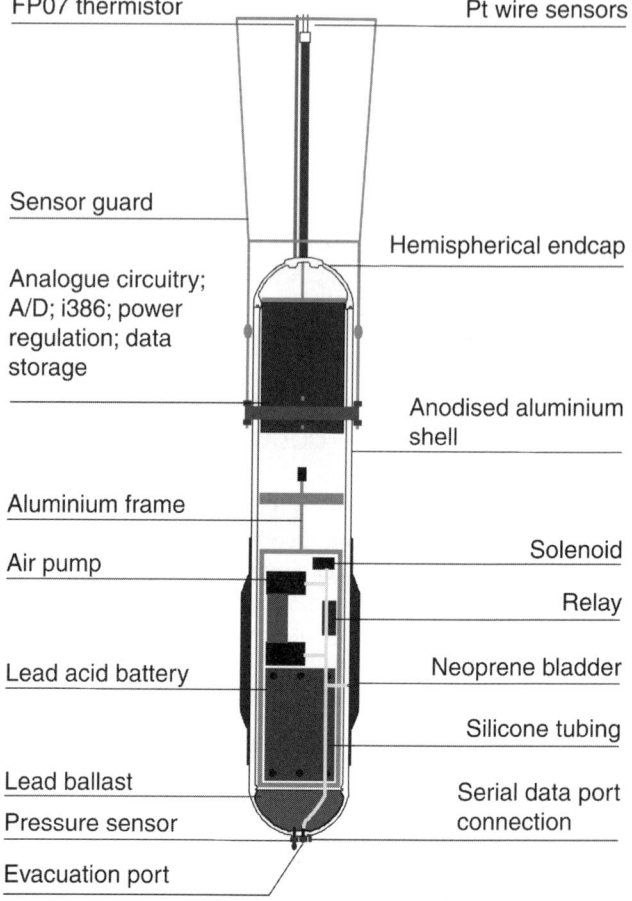

diameter 35 cm is situated just below the sensors to prevent damage to the probes.

There are four ports available in each endcap for attachment of sensor supports or where access to the interior is required. The lower endcap contains the pressure sensor, a communications connection for access to the on-board computer, and a valve. The upper endcap accommodates the FP07 thermistor and Pt wire sensor. Data from the latter were not available when the instrument was deployed during the MOCE-5 cruise. The primary objective of the Pt sensor is to determine the temperature structure within the molecular boundary layer, as it will have a much higher temporal and spatial resolution than the thermistor. All SkinDeEP data presented in the next section are from the FP07 thermistor.

The profiler has the capability to change its density so that it can rise and sink autonomously. Once sealed, the instrument is negatively buoyant by only few grams when the neoprene bladder, attached to the outside of the profiling cylinder, is deflated. Positive buoyancy is achieved by inflation of the bladder, accomplished by pumping air from within the vehicle through a port, thereby expanding the sleeve. Deflation is accomplished by allowing this air to return into the interior. Before deployment, the profiler must be partially evacuated to allow a pressure gradient to exist between the bladder and the interior of the cylinder.

When deployed, the instrument is attached with 50 m of synthetic line to a spar buoy that is equipped with a flashing beacon and a VHF transmitter. Should the buoyancy system fail, the buoy will hold the profiler at the end of its tether line.

BULK TEMPERATURE PROFILES

A map of the Gulf of California showing the cruise track is presented in figure 3. The data described in this section are from stations 9 (Bahia de Altata) and 10 (T/S Irwin) at the southern end of the Gulf, and station 15 (Isla San Esteban) near the Gulf's Mid-Rift.

Skin temperature was provided by M-AERI, a passive infrared radiometric interferometer which makes radiance measurements in the 500 - 3000 cm^{-1} wavelength range with a resolution of 0.5 cm^{-1}. A rotating gold-plated mirror allows for both sea and sky views at complementary angles to nadir and zenith. Real-time calibration is accomplished by viewing two internal blackbody cavities, one at 60°C and one at ambient temperature. Measurements are integrated over a few tens of seconds to obtain a satisfactory signal-to-noise ratio, and a typical measurement cycle including two view angles of the atmosphere, one to the ocean, and calibration, takes about five minutes. The accuracy of the derived SSTs from M-AERI is better than 0.05°C [*Minnett and Ward*, 2000]. A more detailed description of M-AERI can be found in Minnett et al. [2001].

Fixed-depth bulk temperatures were provided by the HardHat and the ship's water intake. The former is an inverted hard plastic helmet filled with foam with a thermistor mounted just below the waterline providing a measurement of the temperature within the top 10 cm while on station. It has an accuracy of better than 0.05°C, and a time response of approximately 10 seconds. Data from the HardHat were logged at a frequency of 1 minute, which was an average of 60 measurements. The temperature measurement from the ship intake was taken at approximately 3 m below the surface. No attempt was made to correct for any warming effects from the ship's structure. Data were logged at a 1 minute interval.

SkinDeEP made profiles continuously once deployed. A complete profile from surface to re-surface took approximately 1 minute, but measurements were made for only 16 seconds during this time. The profile data are displayed as a time series in Plate 1, even though the data are not continuous, but rather discrete profiles acquired during the deployment time. The blank vertical bars reflect this discontinuity and represent a time of 1 minute.

Plate 1 presents data taken on three days, with station 10 shown in panel a (upper), station 9 in panel b (middle), and station 15 in panel c (lower). The data within each panel consist of subplots, showing the following: the skin temperature (T_s) and the two bulk temperatures from the HardHat (T_{bh}) and ship intake (T_{bi}); SkinDeEP profiles

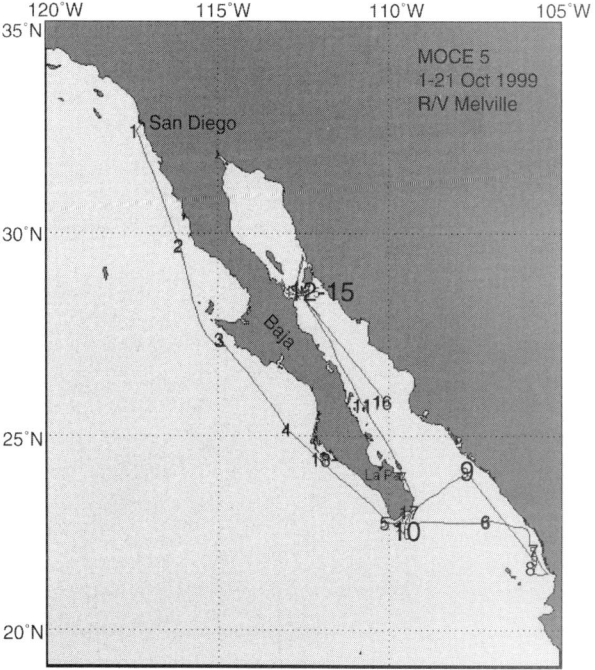

Figure 3. Map of the Gulf of California showing the location of the MOCE-5 stations where the data was acquired (courtesy Stephanie Flora).

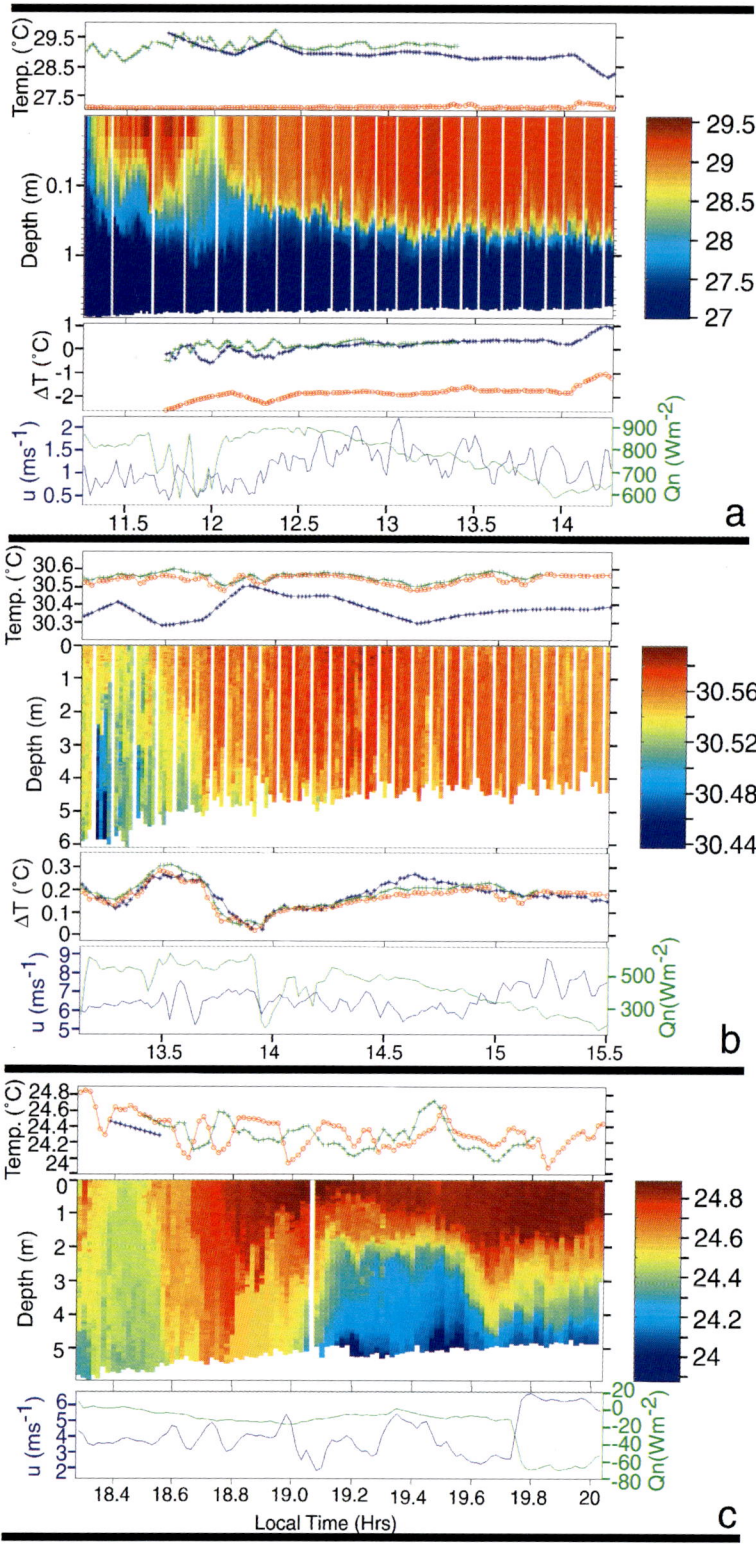

Plate 1. Time series of three deployments of SkinDeEP during the MOCE-5 cruise. For each station, there are subplots of (i) M-AERI, HardHat and ship intake temperatures; (ii) SkinDeEP profiles; (iii) ΔT values; (iv) wind speed and net heat flux. Panel a.: Station 10. Panel b.: Station 9. Panel c.: Station 15 (Note: No ΔT data shown for station 15 due to lack of skin temperatures.

from a depth of 6 - 7 m to the surface; ΔT computed from the M-AERI skin temperature and the bulk temperatures taken from: (i) surface SkinDeEP measurement (ii) Hard-Hat measurement (iii) ship intake measurement; wind speed (u) at 10 m and computed net heat flux (Q_n).

Panel a in Plate 1 reflects the most benign conditions encountered during the cruise, with the wind speed reaching a maximum of only 2 ms^{-1}. The SkinDeEP profiles are expanded with depth using a log scale as there is very little structure below 0.6 m, as seen from T_{bi} which is almost constant during this period. Much of the heat was trapped in the upper 0.6 m resulting in a temperature difference of 2°C. The drop in net heat flux just before noon, probably due to a decrease in net longwave radiation from a cloud, is reflected in the profile data with a cooling at the surface by 0.7°C, but the warm surface is soon re-established. The implications for the derived ΔT, as well as the potential errors of in-situ validation of satellite-derived SST, when using the bulk measurements at 1 m and below, are significant.

Panel b shows a well-mixed surface layer, resulting from the higher wind speeds of minimum value 5.2 ms^{-1}. Both fixed-depth bulk temperatures are within 0.02°C of each other. The profile data show some structure up to 1330 hours and may be the result of a weak front, but the temperature range within this structure is less than 0.1°C. The homogeneous temperature field allowed for a depth independent bulk measurement when deriving the ΔT values. The average ΔT was 0.18 ± 0.06°C, which agrees with Donlon et al. [1999] who used a critical wind speed of 6 ms^{-1} above which they found a ΔT bias of 0.14 ± 0.1°C.

Due to problems with M-AERI at station 15, there is practically no skin temperature data available, and thus the ΔT subplot is omitted for this period. This is the only available nighttime data i.e. nighttime in as much as there is a zero downwelling shortwave component, but the effects of diurnal heating within the bulk waters are still very apparent. Low wind speeds are once again prevalent and the direction of the net heat flux is from ocean to atmosphere. The T_{bh} and T_{bi} temperatures are de-coupled from each other as compared to the other two stations. The profile data show some large gradients - up to 1°C, but there is not the same constant layering process that is found on station 10. Rather, there is a strong over-turning process that is responsible for the heterogeneous temperature structure.

CONCLUDING REMARKS

The SkinDeEP profiler has proven itself to be a useful instrument for studies of near sea surface processes, in this case bulk temperature and resulting ΔT values. The data acquired by the instrument have highlighted a temperature structure within the near surface layers of the ocean that is inaccessible to fixed-depth temperature measurements, and that exhibit large temporal and spatial as well as vertical and horizontal variation. The results have shown the depth dependence of ΔT under diurnal conditions of low winds, and that this dependence decreases as the mixing in the upper waters increases with wind. The results have also emphasized the necessity for taking the bulk measurement close to the surface if the correct sign and magnitude of ΔT is to be determined.

The low wind speed and high insolation conditions presented here are not wholly representative of the global oceanic environment, but these are exactly the kind of cloud-free conditions under which successful satellite SST measurements can be made. The profile data highlight the potential errors that can be introduced into in-situ validations of satellite-derived SSTs if anything other than skin temperature is used, thus emphasizing the need for high quality radiometric instruments such as the M-AERI. The free-floating nature of SkinDeEP meant that it was not in the same local vicinity as the other temperature measurements, thus preventing any decoupling of the horizontal temperature structure from the vertical. However, this would have little effect for satellite validation where the footprint is often greater than 1 km.

With current plans within the GODAE (Global Ocean Data Assimilation Experiment) framework to deploy some thousands of ARGO (Array for Real-Time Geo-strophic Oceanography) surface-piercing floats, the data from the SkinDeEP instrument are both timely and appropriate considering that SkinDeEP shell and buoyancy engine is the same as those to be used on the ARGO floats. Continuous hourly profiles over several days will be undertaken during the GasEx-2001 experiment in the Pacific Ocean, where repeated high-resolution measurements of temperature, conductivity, and oxygen will be made in order to study small scale variability within the top 10 m of the water column.

Acknowledgements. Much gratitude to Denis Clark for providing the opportunity to deploy SkinDeEP during the MOCE-5 cruise. Also to the crew of the R/V Melville who's assistance in deployment and recovery was invaluable. Drs. R. Evans and E. Kearns provided valuable at-sea support for the M-AERI operations. Funding for the development of SkinDeEP was supplied by the Norwegian Research Council (Prosjektnr. 127872/720). Brian Ward was supported by the European Commission under the Marie Curie Fellowship contract ERBFMBICT983162. The M-AERI development and deployment were funded by NASA (NAS5-3161 and NAG56577).

REFERENCES

Debler, W.R., and C.M. Vest, Observations of a stratified flow by means of holographic interferometry, *Proc. Royal Soc. London*, A358, 1-16, 1977.

Donlon, C.J., T.J. Nightingale, T. Sheasby, J. Turner, I.S. Robinson, and W.J. Emery, Implications of the oceanic thermal skin temperature deviations at high wind speed, *Geo. Res. Lett.*, 26, 2502-2508, 1999.

Fairall, C.W., E.F. Bradley, D.P. Rogers, J.B. Edson, and G.S. Young, Bulk parameterization of air-sea fluxes of the Tropical Ocean-Global Atmosphere Coupled Ocean-Atmosphere Response Experiment, *J. Geophys. Res.*, 101, 3747-3764, 1996.

Kearns, E.J., J.A. Hanafin, R.H. Evans, P.J. Minnett and O.B. Brown, An independent assessment of Pathfinder AVHRR sea surface temperature accuracy using the Marine-Atmosphere Emitted Radiance Interferometer (M-AERI), *Bull. Am. Met. Soc.*, 81, 1525-1536, 2000.

Minnett, P.J., Consequences of sea surface temperature variability on the validation and applications of satellite measurements, *J. Geophys. Res.*, 96, 18475-18489, 1991.

Minnett, P.J., R.O. Knuteson, F.A. Best, B.J. Osborne, J.A. Hanafin, and O.B. Brown, The Marine-Atmosphere Emitted Radiance Interferometer (M-AERI): a high-accuracy, seagoing infrared spectroradiometer, *J. Atmos. Oceanic Tech.*, 18, 994-1013, 2001

Minnett, P.J. and B. Ward, Measurements of near-surface ocean temperature variability - consequences on the validation of AATSR on ENVISAT, Proc. ERS-ENVISAT Symposium, ESA SP-461, 2000.

Robinson, I.S., Wells, N.C., and Charnock, H., The sea surface thermal boundary layer and its relevance to the measurement of sea surface temperature by airborne and spaceborne radiometers, *Int. J. Rem. Sens.*, 5, 19-45, 1984.

Tans, P.P., I.Y. Fung, and T. Takahashi, Observational constraints on the global atmospheric CO_2 budget, *Science*, 247, 1431-1438, 1990.

Van Scoy, K. A., K. P. Morris, J. E. Robertson and A. J. Watson, Thermal skin effect and the air-sea flux of carbon dioxide: A seasonal high-resolution estimate, *Global Biogeochem. Cycles*, 9, 253-262, 1995.

B. Ward, NOAA/AOML/OCD, 4301 Rickenbacker Causeway, Miami, Florida 33149.

P. J. Minnett, Rosenstiel School of Marine and Atmospheric Sciences, 4600 Rickenbacker Causeway, Miami, Florida 33149.

Water Column CO$_2$ Measurements During the Gas Ex-98 Expedition

R. A. Feely,[1] R. Wanninkhof,[2] D. A. Hansell,[3] M. F. Lamb,[1] D. Greeley,[1] and K. Lee[2]

During the recent GasEx-98 cruise in the North Atlantic aboard the NOAA ship *Ronald H. Brown*, carbon measurements were performed in the area of 46°N, 20.5°W. This process study followed a warm core ring tagged with the deliberately introduced tracer, SF$_6$. Continuous surface water measurements were combined with vertical profiles sampled daily to depths up to 1000 m for carbon mass balance studies. Dissolved inorganic carbon (DIC) and fCO$_2$ measurements were conducted onboard in both underway and discrete analysis modes. During the 25-day experiment in the tagged patch surface water fCO$_2$ values averaged 275 ± 9 µatm, providing a constant condition of undersaturation and flux of CO$_2$ into the ocean. Using the *Wanninkhof* [1992] exchange coefficient, the estimated CO$_2$ flux ranged from approximately 1–27 mol m^{-2} yr^{-1}. The largest CO$_2$ flux occurred during a large wind event beginning on June 6. After the event, DIC and fCO$_2$ values decreased for a few days, as a result of increased productivity associated with the strong mixing event. The DIC results were combined with the TOC, TON, and nutrient data to provide a mass balance for carbon within the patch. The results for the 25-day period indicate DIC increases in the mixed layer ranging from 0.2–1.8 µmol kg^{-1} d^{-1} due to gas exchange.

1. INTRODUCTION

Large uncertainties in the air-sea CO$_2$ flux prevent us from verifying the partitioning of fossil fuel CO$_2$ between the atmosphere, ocean, and the terrestrial biosphere with a high degree of certainty. Although the magnitude of the CO$_2$ uptake by the ocean presently is constrained in the range between 1.2 to 2.8 GtC per year [*Quay et al.,* 1992; *Sarmiento and Orr,* 1991; *Siegenthaler and Sarmiento,* 1993], the interannual variability of the CO$_2$ flux is still unknown. The mechanisms for transfer of heat, momentum, and matter between the atmosphere and oceans are fundamental to our understanding of air-sea exchange processes. Our present understanding of the flux of CO$_2$ and other trace gas across the air-sea interface are primarily based upon measurements of their degree of supersaturation or undersaturation in the surface ocean mixed layer, combined with estimates of the gas transfer velocity, k. The flux of a CO$_2$ across the air-sea boundary is given by:

$$\text{Flux} = ks(\text{pCO}_{2\text{sw}} - \text{pCO}_{2\text{a}}) \quad (1)$$

where s is the Henry's law of solubility of the gas in mol m^{-3} atm^{-1}, pCO$_{2\text{sw}}$ is the partial pressure of CO$_2$ in surface seawater, and pCO$_{2\text{a}}$ is the partial pressure of CO$_2$ in the atmosphere above the interface. The Henry's law solubility for CO$_2$ is well known from the careful work of *Weiss* [1974]. In addition, k and s are temperature dependent but vary in the opposite direc-

---

[1]NOAA, Pacific Marine Environmental Laboratory, Seattle, Washington.
[2]NOAA, Atlantic Oceanographic and Meteorological Laboratory, Miami, Florida.
[3]Bermuda Biological Station for Research, Inc., St. Georges, Bermuda.

Gas Transfer at Water Surfaces
Geophysical Monograph 127
This paper not subject to U.S. copyright
Published in 2002 by the American Geophysical Union

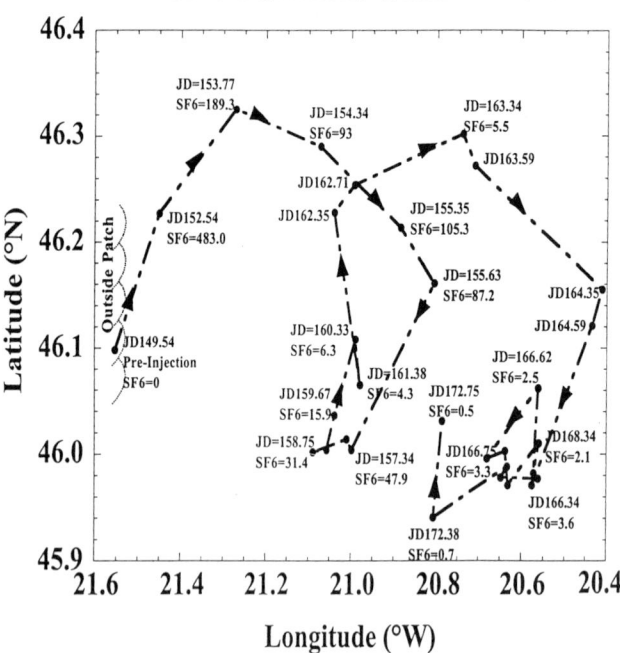

Figure 1. Discrete sample station locations with the tracer-tagged patch during the experiment. The surface SF_6 concentrations are given in pmol kg^{-1}.

ous wind tunnel and open-ocean experiments that the gas transfer velocity is roughly dependent upon wind speed, there are large differences in relationships between gas exchange and wind speed [*Liss and Merlivat*, 1986; *Tans et al.*, 1990; *Wanninkhof*, 1992; *Wanninkhof and McGillis*, 1999].

During the Gas Ex-98 Cruise (May–June, 1998) aboard the NOAA ship *Ronald H. Brown* seawater samples were collected both from 85 discrete CTD/rosette/ 10L PVC bottle hydrocasts collected at least once per day, and from the ship's cleanwater sampling system for continuous sampling of near-surface seawater. More than half of the samples were collected within an anticyclonic warm core eddy of the eastern North Atlantic centered at approximately 46°05'N, 20°46'W.

For the underway fCO_2 measurements samples were collected using a shipboard pumping system which continuously pumped seawater from the ship's sea chest to the oceanographic laboratory at a flow rate of approximately 50 L min^{-1}. The underway measurements of CO_2 mole fractions in dry air were made with a Licor (Model 6251) non-dispersive infrared analyzer linked to the shipboard equilibrator following the methods described in *Wanninkhof and Thoning* [1993] and *Feely et al.* [1998]. Total dissolved inorganic carbon was determined by the coulometric method discussed in *Johnson et al.* [1985, 1987] employing a Single Operator Multoparameter Metabolic Analyzer (SOMMA) automated sample injection system. Water samples taken for organic carbon and nitrogen determinations were not filtered, hence total organic carbon (TOC) and nitrogen (TON) were measured. Samples for TOC and TON analysis were collected using the PVC bottles on the CTD rosette into 125 ml acid-cleaned polyethylene bottles. Bottles and caps were rinsed 3 times, filled full, and frozen (−20°C) for later onshore analysis.

tion such that the temperature variance of their product does not vary by more than ±10% from the value at 20°C [*Etcheto and Merlivat*, 1988]. Both pCO_{2sw} and pCO_{2a} are generally measured with a high degree of precision (±0.5 μatm) and accuracy (±1 μatm). Consequently, the major uncertainty in the local flux estimated along ship tracks is associated with the estimate of the gas transfer velocity. While it is clear from previ-

Table 1. Average Conditions Inside and Outside the Tracer-Labeled Patch Within the Mixed-Layer of the Anticyclonic Warm Core Eddy in the North Atlantic During Gas Ex-98 (May–June, 1998)

Parameter	Inside patch mean ± 1 std. dev.	Outside patch mean ± 1 std. dev.
Temperature (°C)	15.508 ± 0.34	15.596 ± 0.37
Salinity	35.641 ± 0.022	35.645 ± 0.022
fCO_{2sw} (μatm)	276.5 ± 9.8	276.4 ± 13.5
ΔfCO_2 (μatm)	−84.4 ± 9.4	−84.2 ± 13.5
U^{10} (m s^{-1})	7.01 ± 3.0	NA
k (cm hr^{-1})	16.1 ± 13.7	NA
CO_2 Flux (moles m^{-2} yr^{-1})	−4.5 ± 3.9	NA

NA = Not Available

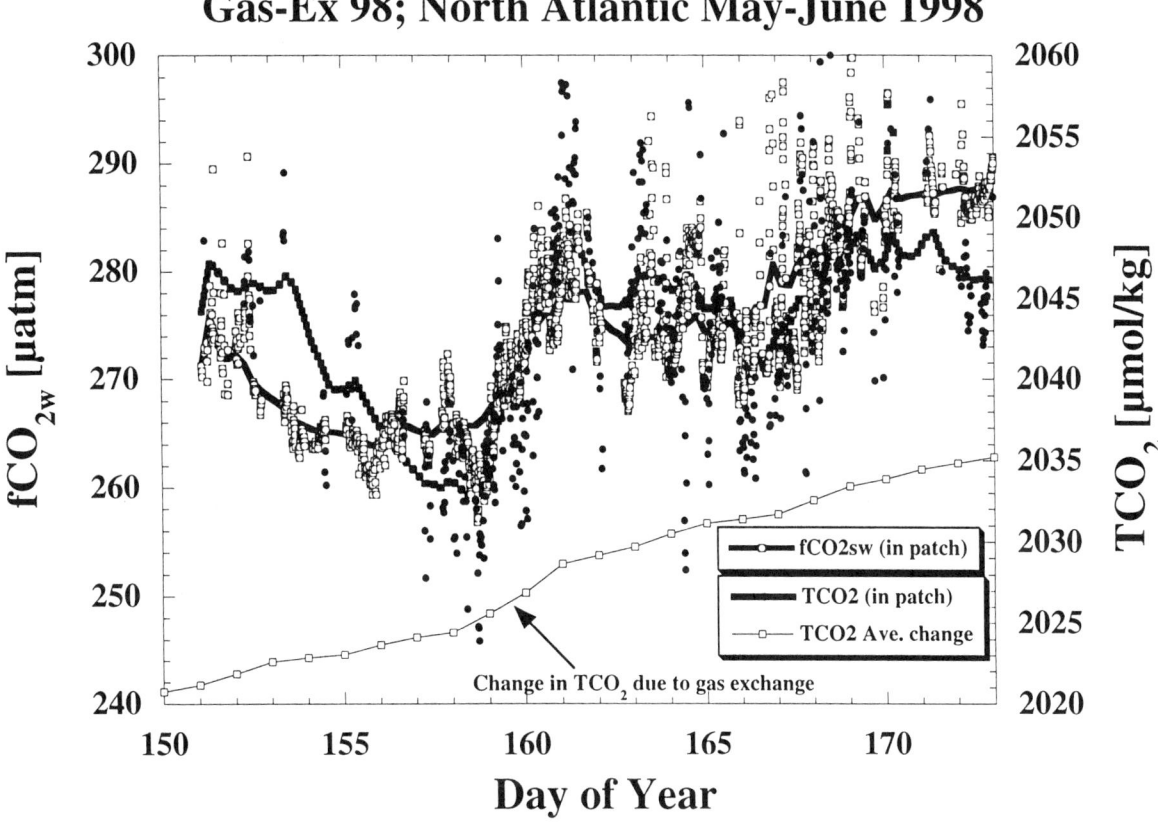

Figure 2. Time series of surface water fCO$_2$ (μatm) and TCO$_2$ (μmol kg^{-1}) within the patch and estimated average daily TCO$_2$ concentration changed in the mixed layer due to gas exchange using the Wanninkhof (1992) relationship.

TOC was analyzed by a high-temperature combustion method slightly modified from that previously described by *Hansell et al.* [1997] and *Carlson and Ducklow* [1995]. tion. Concentrations of TON were determined by UV photooxidation according to the method described by *Walsh* [1989]. In the surface layer, the RSD for TON was approximately 4%.

2. RESULTS AND INTERPRETATIONS

On May 28, 1998 the scientists and crew of the NOAA ship *Ronald H. Brown* arrived in the region of the anticyclonic warm-core eddy centered at approximately 46°05′N, 20°46′W. They injected a seawater patch inside the eddy with SF$_6$ tracer, three WOCE/ARGOS tracer drifters, and two CARIOCA pCO$_2$ buoys on 29 May, and followed the patch until 22 June 1998. The CARIOCA buoys were retrieved twice during the experiment and repositioned in the patch. The drifters remained within the patch, following an anticyclonic circular path. The ship traversed the patch several times per day making air and seawater CO$_2$ measurements and collecting hydrocast samples at least once per day near the center of the lagrangian patch (Figure 1). The warm-core eddy was characterized by a warmwater subsurface core with a cold mixed layer. Outside the eddy, surface waters were warmer and fCO$_2$ values were close to equilibrium with respect to the atmosphere; whereas, inside the eddy SST values averaged 15.508°C and fCO$_2$ values averaged about 276.5 ± 9μatm (Table 1). Within the eddy the fCO$_2$ values were nearly the same inside and outside of the patch (Table 1).

The shipboard winds, which were strongly correlated with the decreases in atmospheric pressure, show six separate wind events having periods of about 4 days. The largest of these was a 4-day event occurring between 6–10 June (Day of Year: 158-162), in which wind speeds ranged from 2-18 m s^{-1}. Significant increases in fCO$_2$ and DIC occurred on 7-8 June and 15-16 June, respectively, shortly after the beginnings of two wind events (Figure 2). These enrichments were caused by several factors, including enhanced gas exchange and

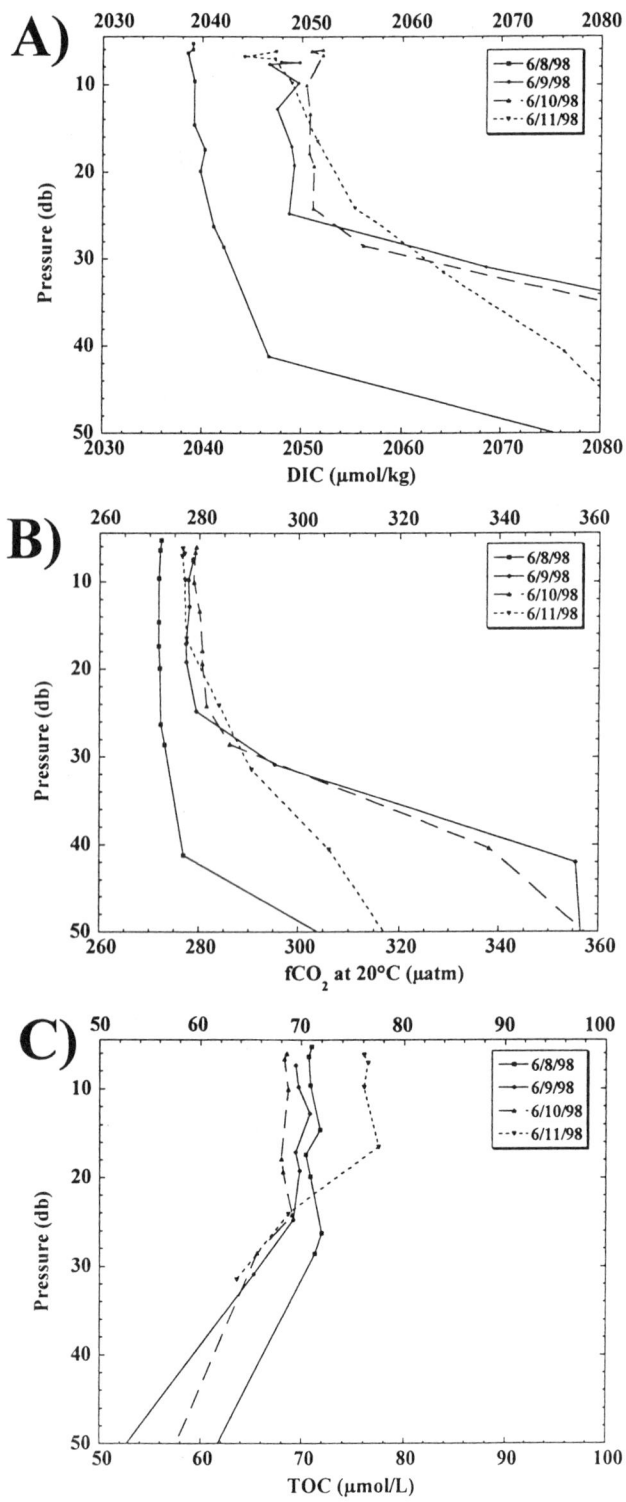

Figure 3. Vertical profiles of (a) DIC in μmol kg^{-1}, (b) fCO$_2$ in μatm, and (c) TOC in μmol/l during the main wind event (8–11 June, 1998).

turbulent mixing caused by the wind events. The figure shows an estimate of the DIC change that is due to gas exchange using the *Wanninkhof* [1992] relationship for high-frequency winds. During the wind events the estimated increase of DIC due to gas exchange averaged 0.61 (range = 0.2–1.8) μmol kg^{-1} d^{-1}, with the highest single day increases occurring on June 9 and June 15. The CO$_2$ flux into the ocean ranged from 0.05 to 27.3 mol m^{-2} yr^{-1}, with a mean of 4.5 mol m^{-2} yr^{-1}. The wind events contributed to most of the CO$_2$ flux into the ocean.

Vertical profiles of the distributions of fCO$_2$, DIC, and TOC through the top 50 m of the water column are given in Figures 3 and 4. The data are presented in two groups by the day of year: during the major wind event, and post major wind event. For reference, the major wind event began on day 158 and ended on day 162. The pre-wind event data (not shown) indicate significant decreases in DIC, fCO$_2$, and TOC after the injection of the tracer on day 150. During the 8 days following the addition of the tracer, the ship sampled a region spanning over 120 km distance with widely varying SF$_6$ concentrations (range = 31–483 pmol kg^{-1}). In contrast, during and after the major wind event the ship remained in two separate localized regions in which the ship remained within a 30 km radius and the SF$_6$ concentrations were less variable (range = 0.5–6.5 pmol kg^{-1}). Both the during-wind event and post-wind event data sets show increases in DIC over time, which is due, in large part, to vertical mixing and gas exchange. During the wind event, the DIC increase is about 5–8 μmol kg^{-1} throughout the mixed layer. Similarly, fCO$_2$ also increased throughout the mixed layer at the same time. The TOC data remained steady at about 70 ± 2 μmol l^{-1} until day 162, when it increased to 76 μmol l^{-1}.

The post-wind event data set from day 167–170 also show increases of DIC of 7–10 μmol kg^{-1} and fCO$_2$ (8–18 μatm) in the mixed layer, followed by a sharp positive gradient below the mixed layer. In addition, TON increased slightly over the same period. These results suggest that gas exchange, biogeochemical reactions, and mixing processes are all important during this period.

3. DISCUSSION

The changes in the concentration of DIC in the mixed layer are a function of air-sea gas exchange, biological uptake/respiration, and horizontal and vertical mixing processes. *Chipman et al.* [1993] studied DIC distributions in a similar eddy from the same region of the North Atlantic in 1990 and found that, under normal conditions, biological utilization of carbon and air-sea gas exchange were the predominant processes controlling CO_2 distributions in the mixed layer. Horizontal mixing and vertical entrainment of DIC from depth into the mixed layer were negligibly small. One possible exception to this conclusion would be during strong wind events when the mixed layer deepens and entrainment of deeper sub-layers into the mixed layer occurs. Under these conditions, small but significant changes in DIC concentrations due to vertical mixing processes can be estimated from the DIC concentration gradients in the sub-layers. We used the mixing coefficients derived from the SF_6 data of *Lee et al.* [in preparation] in a one-dimensional advection-diffusion model [*Craig*, 1969; *Craig and Weiss*, 1970] to estimate the DIC changes in the mixed layer due to vertical mixing shortly after a wind event. Using a vertical eddy diffusion rate of 1.0 ± 0.4 cm^2 s^{-1} [*Lee et al.*, in preparation], we estimate an average DIC increase of 3.5 μmol kg^{-1} d^{-1} during the part of the wind event when the mixed-layer depth increases the most. On average, the DIC changes due to vertical mixing during the Wind event are about 43% of the total. The remainder must, therefore, be due to biological uptake/respiration processes and air-sea gas exchange.

Within the mixed layer, the daily changes in total DIC can be expressed as the sum of change in total organic carbon (ΔTOC), the particulate organic carbon export (POCexport), the change in DIC due to air-sea exchange (ΔDICair-sea exchange), and the change in DIC due to vertical mixing (ΔDICmixing) according to equation 2:

$$\Delta\text{DICtotal} = \Delta\text{TOC} + \text{POCexport} \quad (2)$$
$$+ \Delta\text{DICair-sea exchange} + \Delta\text{DICmixing}$$

In the present case, we can assume that ΔDICmixing is due only to entrainment of sublayers during wind events and can be estimated using the SF_6 mixing coefficients (see above). The total changes in the concentration of the nitrogen species is equivalent to the amount of nitrogen that is exported from the mixed layer according to equation 3:

$$\Delta\text{Total Nitrogen(TN)} = \Delta\text{NO}_3^- + \Delta\text{NO}_2^- \quad (3)$$
$$+ \Delta\text{TON} = \text{PONexport}$$

Since the POCexport and the PONexport are related by the Redfield ratios [*Anderson and Sarmiento*, 1994], the POCexport can be estimated from ΔTN using the equation:

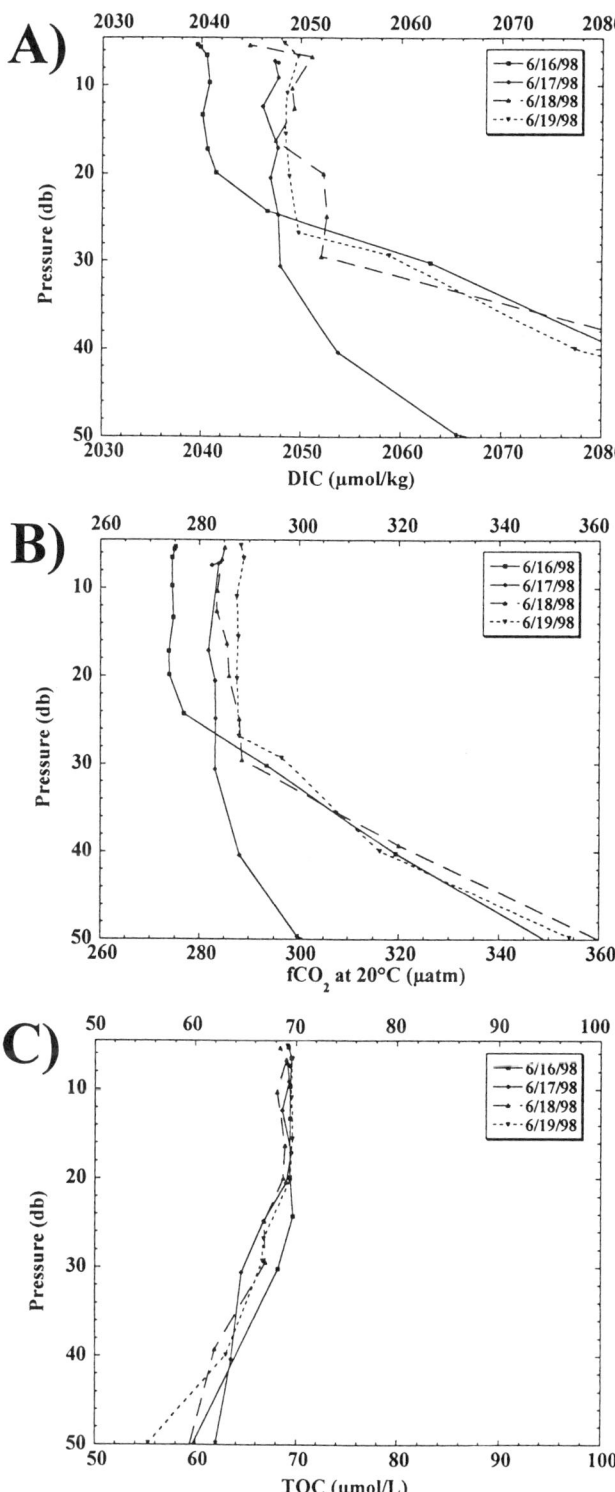

Figure 4. Vertical profiles of (a) DIC in μmol kg^{-1}, (b) fCO$_2$ in μatm, and (c) TOC in μmol/l after the main wind event (15–18 June, 1998).

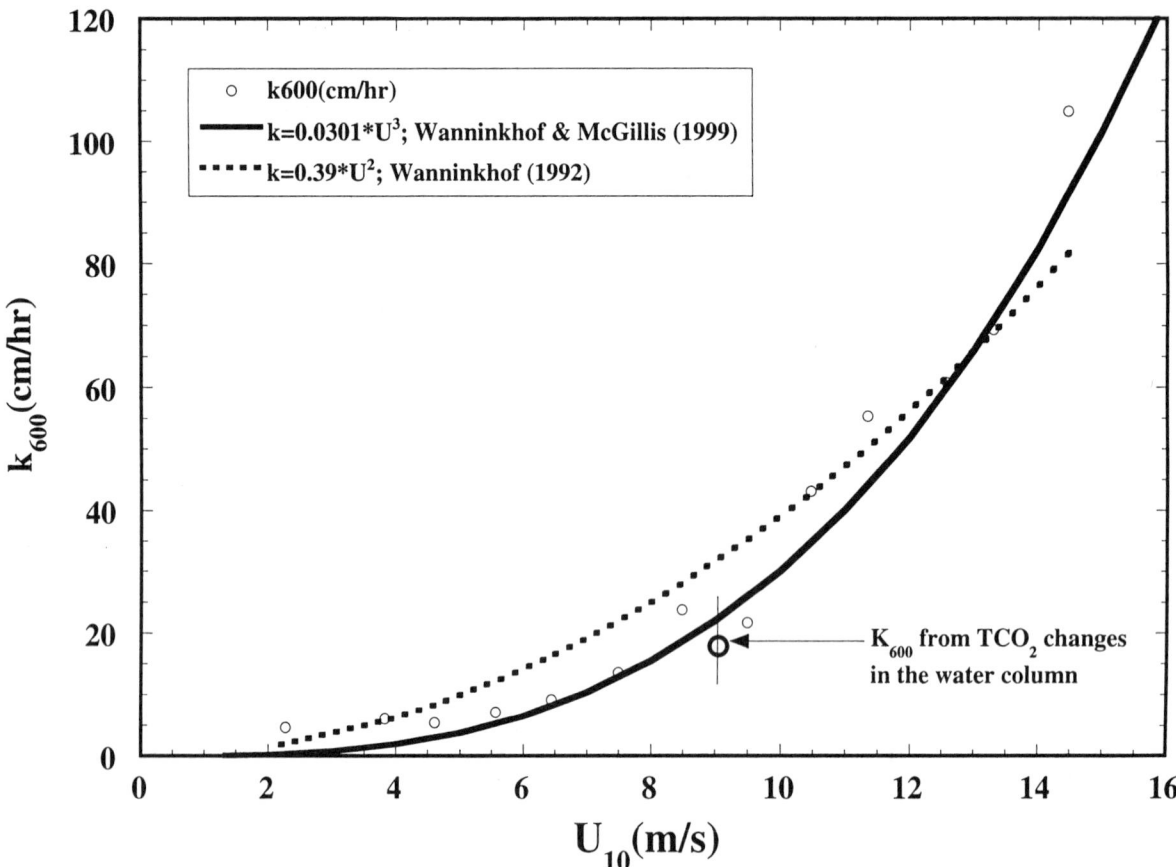

Figure 5. Gas transfer velocity (large open circle) based on carbon mass balance in the mixed layer. The eddy correlation results of McGillis and Edson as reported in Wanninkhof and McGillis [1999] are also shown.

$$POCexport = 6.6\Delta TN \quad (4)$$

Thus, by substitution of equation 4 into equation 2 and rearranging we arrive at:

$$\Delta DIC\text{air–sea exchange} + \Delta DIC\text{mixing} = \Delta DIC\text{total} - (\Delta TOC + 6.6\Delta TN) \quad (5)$$

Therefore, if ΔDICmixing is independently determined, ΔDICair- sea exchange can be estimated from measured parameters in the water column. An example of this calculation is given in Table 2 for the period from 16–19 June, 1998. The results indicate that approximately 43% of the total change in DIC is due to mixing, 30% is due to biological uptake and carbon export flux, and the remaining 27% is due to air-sea exchange of CO_2 (i.e., 0.84 μmol kg^{-1} d^{-1}). This estimate is slightly higher than estimates obtained by *Chipman et al.* [1993] for a warm-core eddy from the same region studied in the spring of 1990. Since both the wind speeds were higher and ΔpCO_2 values were more negative during the Gas Ex-98 study, this increase is to be expected. The uncertainties in this calculation are primarily controlled by the uncertainty in ΔTN (± 0.1 μmol kg^{-1}) and the uncertainty in ΔDICmixing (± 0.2 μmol kg^{-1}), giving a total uncertainty of about $\pm 60\%$ for air-sea exchange. Using these values for the 3-day period from June 16–19, 1998, we then estimate a gas exchange velocity of 19 ± 11 cm/hr based on the carbon mass balance in the water column. This estimate is consistent with the earlier estimates based upon the eddy correlation studies of *McGillis et al.* [submitted] as shown in Figure 5. However, it should be noted that since the total uncertainty of these measurements ($\pm 60\%$) is still quite large, other wind speed-gas exchange relationships are also possi-

Table 2. Example Estimate of $\Delta DIC_{sea-air\ exchange}$ During Gas Ex-98

Example: June 16–19, 1998

$\Delta DIC_{sea-air\ exchange} + \Delta DIC_{mixing} = \Delta DIC_{total} - (\Delta TOC + 6.6*\Delta TN)$
$\Delta DIC_{sea-air\ exchange} + 3.49 = 8.20 - (-0.26 + 6.6*0.37) = 2.56\ \mu mol\ kg^{-1}$
$\frac{2.56\ \mu mol\ kg^{-1}}{3\ days} = 0.84\ \mu mol\ kg^{-1}\ day^{-1}$ due to gas exchange
This results in $k = 19$ cm/hr for wind speeds of 9 m s^{-1}

ble. The usefulness of this approach is limited to periods when the ship was consistently in the center of the patch and the SF$_6$ concentrations were relatively constant. Nevertheless, the combination of strong negative ΔpCO_2 values and moderate-to-strong wind conditions within the warm-core eddy made it possible to obtain these results. As similar kinds of results from the other experiments for estimating CO$_2$ exchange from the Gas Ex-98 expedition become available, we'll be able to provide a clearer picture of the relationship between wind speed and air-sea exchange of CO$_2$ at the ocean surface.

4. CONCLUSIONS

Under limited conditions the water column carbon mass balance approach can be utilized to provide an estimate of gas exchange in warm core eddies. Our results for a 25-day experiment in a warm core eddy in the North Atlantic indicate DIC increases in the mixed layer ranging from 0.2–1.8 μmol kg^{-1} d^{-1} due to gas exchange. This range is consistent with the estimated gas transfer velocity from the eddy correlation results for the same cruise. Further studies are needed to delineate the wind speed-gas exchange relationship in strongly outgassing regions, such as the Equatorial Pacific.

Acknowledgments. This work was sponsored by the NOAA/OGP Ocean-Atmosphere Carbon Exchange Study under the leadership of Dr. Lisa Dilling. Nutrient values were provided by Jia-Zong Zhang of NOAA/AOML and wind data were provided by Jim Edson of Woods Hole Oceanographic Institution. We thank Wade McGillis and Jim Edson of Woods Hole Oceanographic Institution for the eddy correlation results. We also thank the officers and crew of the NOAA ship *Ronald H. Brown* for logistics support. PMEL contribution 2112.

REFERENCES

Anderson, L., and J. Sarmiento, Redfield ratios of remineralization determined by nutrient data analysis, *Global Biogeochem. Cycles, 8,* 65–80, 1994.

Carlson, C. A., and H. W. Ducklow, Dissolved organic carbon in the upper ocean of the central equatorial Pacific Ocean, 1992: Daily and finescale vertical variations, *Deep Sea Res. II, 42*(2–3), 639–656, 1995.

Chipman, D. W., J. Marra, and T. Takahashi, Primary production at the 47°N and 20°W in the North Atlantic Ocean: A comparison between the 14°C incubation method and the mixed layer carbon budget, *Deep-Sea Res. II, 40*(1/2), 151–169, 1993.

Craig, H., Abyssal carbon and radiocarbon in the Pacific, *J. Geophys. Res., 74,* 5491–5506, 1969.

Craig, H., and R. F. Weiss, The GEOSECS 1969 intercalibration station: Introduction, hydrographic features, and total CO$_2$–O$_2$ relationships, *J. Geophys. Res., 75,* 7641–7647, 1970.

Etcheto, J., and L. Merlivat, Satellite determination of the carbon dioxide exchange coefficient at the ocean-atmosphere interface: A first step, *J. Geophys. Res., 93*(C12), 15,669–15,678, 1988.

Feely, R. A., R. Wanninkhof, H. B. Milburn, C. E. Cosca, M. Stapp, and P. P. Murphy, A new automated underway system for making high precision pCO$_2$ measurements onboard research ships, *Anal. Chim. Acta, 377,* 185–191, 1998.

Hansell, D. A., C. A. Carlson, N. Bates, and A. Poisson, Horizontal and vertical removal of organic carbon in the equatorial Pacific Ocean: A mass balance assessment, *Deep-Sea Res. II, 44,* 2115–2130, 1997.

Ho, D. T., R. Wanninkhof, J. Masters, R. A. Feely, and C. E. Cosca, Measurement of underway fCO$_2$ in the eastern equatorial Pacific on NOAA ships *Baldridge* and *Discoverer*, *NOAA Data Report ERL AOML-30*, 52 pp., NTIS, Springfield, Ill., 1997.

Hood, E. M., R. Wanninkhof, and L. Merlivat, The effects of wind-induced mixing on short timescale surface variability of fCO$_2$ and fluorescence: Results from the GASEX98 CARIOCA buoy data, *J. Geophys. Res.,* in press.

Johnson, K. M., A. E. King, and J. McN. Sieburth, Coulometric DIC analyses for marine studies: An introduction, *Mar. Chem., 16,* 61–82, 1985.

Johnson, K. M., P. J. Williams, L. Brandstrom, and J. McN. Sieburth, Coulometric total carbon analysis for marine studies: Automation and calibration, *Mar. Chem., 21,* 117–133, 1987.

Lee, K., R. Wanninkhof, and J.-Z. Zhang, Vertical diffusion rates of the upper thermocline determined from a Lagrangian SF6 tracer study of GASEX-98, in preparation.

Liss, P. S., and L. Merlivat, Air-sea gas exchange rates: Introduction and synthesis, in *The Role of Air-sea Exchange in Geochemical Cycling,* edited by P. Buat-Menard, pp. 113–129, Reidel, Boston, 1986.

McGillis, W. R., J. B. Edson, J. E. Hare, and C. W. Fairall, Direct covariance air-sea CO2 fluxes, *J. Geophys. Res.*, submitted.

Quay, P. D., B. Tilbrook, and C. S. Wong, Oceanic uptake of fossil fuel CO_2: Carbon-13 evidence, *Science*, *256*(5053), 74–79, 1992.

Sarmiento, J. L., and J. C. Orr, Three-dimensional simulations of the impact of the Southern Ocean nutrients depletion on atmospheric CO_2 and ocean chemistry, *Limnol. Oceanogr.*, *36*(8), 1928–1950, 1991.

Siegenthaler, U., and J. L. Sarmiento, Atmospheric carbon dioxide and the ocean, *Nature*, *365*(6442), 119–125, 1993.

Tans, P. P., I. Y. Fung, and T. Takahashi, Observational constraints on the global atmospheric CO_2 budget, *Science*, *247*(4949), 1431–1438, 1990.

Tsunogai, S., H. Yamahata, and O. Saito, Calcium in the Pacific Ocean, *Deep-Sea Res.*, *20*, 717–726, 1973.

Walsh, J. J., Arctic carbon sinks: present and future, *Global Biogeochem. Cycles*, *3*(4), 393–411, 1989.

Weiss, R. F., Carbon dioxide in water and seawater: The solubility of a non-ideal gas, *Mar. Chem.*, *2*, 203–215, 1974.

Wanninkhof, R., Relationship between gas exchange and wind speed over the ocean, *J. Geophys. Res.*, *97*, 7373–7381, 1992.

Wanninkhof, R., and W. R. McGillis, A cubic relationship between air-sea CO_2 exchange and wind speed, *Geophys. Res. Lett.*, *26*(13), 1889--1892, 1999.

Wanninkhof, R., and K. Thoning, Measurement of fugacity of CO_2 in surface water using continuous and discrete sampling methods, *Mar. Chem.*, *44*, 189–205, 1993.

R. A. Feely, M. F. Lamb, and D. Greeley, NOAA, Pacific Environmental Laboratory, 7600 Sand Point Way NE, Seattle, WA, 98115-6349. (e-mail: feely@pmel.noaa.gov, lamb@pmel.noaa.gov, greeley@pmel.noaa.gov)

R. Wanninkhof and K. Lee, NOAA, Atlantic Oceanographic and Meteorological Laboratory, 4301 Rickenbacker Causeway, Miami, FL, 33149. (e-mail: wanninkhof@aoml.noaa.gov, lee@aoml.noaa.gov)

D. A. Hansell, Bermuda Biological Station for Research, Inc., 17 Biological Lane, St. Georges, GE-01, Bermuda (e-mail: dennis@bbsr.edu)

Fine Thermohaline Structure and Gas-Exchange in the Near-Surface Layer of the Ocean During *GasEx-98*

Alexander Soloviev[1], Jim Edson[2], Wade McGillis[2], Peter Schluessel[3], and Rik Wanninkof[4]

During the *GasEx-98* field campaign, observations of the upper ocean structure were performed to identify relationships between the fine thermohaline structure, turbulence, and gas exchange in the near-surface layer of the ocean. The upper ocean dynamics were then simulated using a 1-D mixed layer model with the mixing parameterization developed during TOGA Coupled Ocean-Atmosphere Response Experiment (COARE). The model was initialized with the temperature, salinity, and velocity profiles in the upper 50 m thick layer of the ocean obtained from the Conductivity-Temperature-Depth (CTD) and Acoustic Doppler Current Profiler (ADCP) measurements and was forced with the air-sea heat and momentum fluxes measured by Edson et al. [1999]. The model produced a set of parameters including the time and depth dependent mixing coefficient and the depth of the mixed layer. The simulated mixed layer depth is consistent with the depth of the actively mixed layer determined from the turbulence profiles taken occasionally during *GasEx-98 leg 2* with a free-rising profiler. Moderate wind speed conditions prevailed during *GasEx-98 leg 2* with several storms and a few periods of calm weather. Both the modeling and experimental results demonstrate that under conditions of low wind speed, the surface-generated turbulence is constrained within a relatively thin surface layer of the ocean. In the near-surface layer, appreciable temperature, salinity, and gas concentration differences are formed because of diurnal warming or precipitation effects. These results are applied to the estimation of the effect of mixed layer processes on the bulk-flux formulation for the air-sea exchange of gases.

---

[1]Oceanographic Center, Nova Southeastern University, Dania Beach, Florida

[2]AOP&E, Woods Hole Oceanographic Institution, Woods Hole, Massachusetts

[3]EUMETSAT, Darmstadt, Germany

[4]NOAA/Atlantic Oceanographic and Meteorological Laboratory, Miami, Florida

1. INTRODUCTION

In the commonly used bulk-flux formulation for the net air-sea gas flux,

$$F = k(C_w - Sp_a), \quad (1)$$

the gas concentration, C_w, in the bulk of water is usually taken from the ship's thermosalinograph intake or CTD measurements at a 3-5 m depth. The thermosalinograph or CTD data from the same depth are also usually used to calculate the gas solubility, S, which is a function of temperature and salinity. (The other parameters in formulation (1)

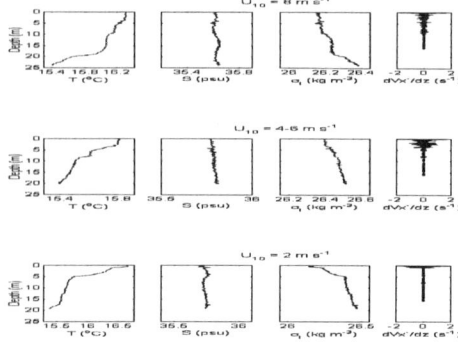

Figure 1. Examples of the near-surface microstructure and turbulence under different wind speed conditions obtained with free-rising profiler during *GasEx-98* expedition.

are the total bulk air-sea gas transfer velocity k and gas partial pressure in air p_a).

The surface mixed layer is not always well mixed. Bruce and Firing [1975], Soloviev and Vershinsky [1982], Atkinson et al. [1987], Soloviev and Lukas [1997], Soloviev and Schluessel [1999] reported appreciable temperature, salinity and gas concentration differences in a few upper meters of the ocean caused by the diurnal cycle and precipitation effects under low wind speed conditions.

The aim of this work is to find relations between the fine thermohaline structure in the near-surface layer of the ocean and air-sea gas exchange, using the *GasEx-98 leg 2* data and a numerical model simulation. These results will then be applied to the estimation of how the mixed layer processes may affect the bulk-flux formulation for the air-sea gas exchange.

2. OBSERVATIONS

Only the actively mixed layer is involved in the local air-sea heat and mass exchange. Figure 1 demonstrates three examples of the measurements with a free-rising profiler taken during *GasEx-98 Leg 2* under different wind speed conditions. According to Fig. 1, when the wind speed drops below 4-6 m s^{-1}, the depth of the diurnal mixed layer reduces and an appreciable temperature gradient is established across the diurnal thermocline.

Under calm weather conditions and strong solar insolation, the diurnal mixed layer is localized just near the ocean surface; the temperature difference across the diurnal thermocline can reach as much as several °K. An example from *GasEx-98* is shown in Figure 2. This measurement was done in *GasEx98*, using the sensors mounted in front of the bow of the vessel. The bow sensors were "scanning" the upper 3 m layer of the ocean because of the ship pitching in the surface wave field. The vertical profile of temperature shows a strong diurnal thermocline in the upper ½ m layer of the ocean. There is also a salinity increase and dissolved oxygen (DO) depression in the upper ½ m of the ocean. The excess salinity in the near-surface layer is accumulated within the diurnal mixed layer because of surface evaporation; while the turbulent mixing is suppressed by the stable stratification [Soloviev and Lukas, 1997].

The DO depression observed in the upper ½ m layer of the ocean (Figure 2) can be explained by the O_2 evasion form ocean to atmosphere. (During this observation, the DO in the bulk of the mixed layer water was at a 117-% saturation level). This near-surface depression of O_2 could not be mixed with the underlying water mass; because, the turbulence is inhibited by stratification. The near-surface anomalies in temperature and salinity affect the gas solubility, S, as well.

The example shown in Fig. 2 suggests that under low wind speed conditions the thermohaline structure and gas exchange in the surface mixed layer of the ocean are strongly coupled with each other. The effect of mixed layer

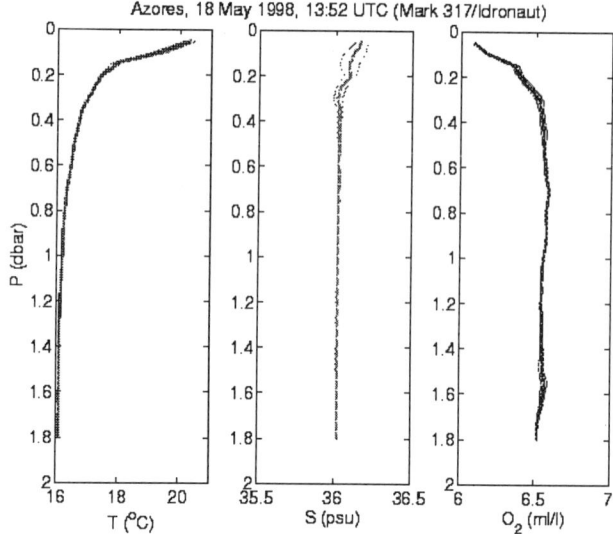

Figure 2. Vertical profiles of temperature T, salinity S, and dissolved Oxygen concentration O_2 during a strong diurnal warming event observed near Azores in *GasEx-98 leg 1* (low wind conditions). The vertical profiles are calculated from the bow sensors (undulating because of the surface waves and vessel's pitching) by averaging the corresponding records within 5-cm bins over 1 min. time period. Note the O_2 concentration depression in the upper half a meter of the ocean, which is supposedly related to the gas evasion through air-sea interface. The bulk of the mixed layer is over-saturated.

dynamics on the air-sea gas exchange will be evaluated in the next paragraph using a 1-D mixed layer model and the data sets obtained during *GasEx-98 leg 2*. The microstructure and turbulence profiles taken occasionally with a free-rising profiler in *GasEx-98 leg 2* are then used to validate the results of the numerical model simulation.

3. MODELING

This study employs a 1-D model of the same class as that of Price et al. [1986]. With the new mixing parameterization scheme developed during TOGA COARE [Soloviev et al., 2001], the model is able to resolve the relatively small but dynamically important gradients within the surface mixed layer. The equation for gas diffusion (no biology effects) is added to estimate the influence of the mixed layer dynamics on the bulk-flux formulation for the air-sea gas exchange.

3.1 Mixing Parameterization

To model the mixed layer effects on the air-sea gas exchange, we use the TOGA COARE parameterization for the exchange coefficients for momentum and heat transfer [Soloviev et al., 2001]:

$$K_m = \kappa u_* z (a_m - c_m Ri)^{1/3} \theta(Ri_m - Ri) + \\ + \kappa u_* z (1 - \alpha Ri)^{1/4} \theta(-Ri) \theta(Ri - Ri_m) + \\ + k u_* z (1 - Ri/Ri_{cr}) \theta(Ri) + K_{mt} \quad (2)$$

$$K_t = \kappa u_* z (a_s - c_s Ri)^{1/3} \theta(Ri_s - Ri) + \\ + \kappa u_* z (1 - \alpha Ri)^{1/4} \theta(-Ri) \theta(Ri - Ri_s) + \\ + k u_* z (1 - Ri/Ri_{cr}) \theta(Ri) + K_{mt} \quad (3)$$

where $\kappa = 0.4$ is the Von Karman constant, u_* is the friction velocity in water, z the depth, Ri the gradient Richardson number, $Ri_m = -0.20$, $Ri_s = -1.0$, $Ri_{Cr} = 0.25$, $\alpha = 16$, $a_m = 1.26$, $a_s = -28.86$, $c_m = 8.38$, and $c_s = 8.38$. $\theta(x)$ is the step function so that $\theta(x) = 0$ at $x \leq 0$, $\theta(x) = 1$ at $x > 0$, K_{mt} is the thermocline value of the mixing coefficient determined according to Peters et al. [1988].

3.2 Equations

The equations of for the heat, salinity, gas concentration (no biology), and momentum balance are as follows:

$$c_p \rho \partial_t T = -\partial_z Q - \partial_z Q_R \quad (4)$$

$$\rho \partial_t S = -\partial_z F \quad (5)$$

$$\partial_t C = -\partial_z G \quad (6)$$

$$\rho \partial_t u = -\partial_z \tau_x + fv \quad (7)$$

$$\rho \partial_t v = -\partial_z \tau_y - fu \quad (8)$$

where $Q = -c_p \rho K_t(Ri) \partial_z T$, $F = -\rho K_S(Ri) \partial_z S$, and $G = -K_C(Ri) \partial_z C$ are the heat, salinity, and gas flux, respectively; $\tau_x = -\rho K_m(Ri) \partial_z u$ and $\tau_y = -\rho K_m(Ri) \partial_z v$ are the components of the shear stress, all in the upper ocean; ρ is the density of sea water, c_p is the specific heat of sea water, $Ri = g\rho^{-1} \partial_z \rho / [(\partial_z u)^2 + (\partial_z v)^2]$, f is the Coriolis parameter.

K_m and K_t are determined from parameterizations (2)-(3) respectively, and $K_C = K_S = K_t$. The absorption of solar radiation with depth (Q_R) is parameterized by an exponential sum,

$$Q_R = (1-A)Q_{R0} \sum_{i=1}^{9} a_i \exp(-\beta_i z),$$

according to Soloviev and Schluessel [1996], where Q_{R0} is the short wave solar radiation flux, α_i and β_i the empirical coefficients. To calculate the albedo, A, for the shortwave radiation forcing, we used the Fortran program written by Peter A. Coppin (CSIRO Centre for Environmental Mechanics, Australia), employing the Payne [1972] model.

The surface boundary conditions are as follows:

$$Q = Q_T + Q_E + Q_L, \quad (9)$$

where Q_T, Q_E, and Q_L are the sensible, latent, and effective long wave radiative flux density (note that the solar radiation doesn't enter the surface boundary condition because it is treated as the volume source of heat),

$$\tau_x = \tau_{x0}, \tau_y = \tau_{y0}, \quad (10)$$

where τ_{x0} and τ_{y0} are the surface east- and northward components of the momentum flux, and

$$G = k(C_{w0} - S_0 p_a), \quad (11)$$

where $k = k_{int} + k_b$, k_{int} and k_b are the interfacial and bubble mediated air-sea gas transfer coefficients correspondingly, C_{w0} and S_0 are the surface gas concentration and solubility (in the first bin). Note that the difference between Eqs. (11) and (1) is that (1) only holds if there is no near-surface gradient. The interfacial gas transfer coefficient is parameterized according to Soloviev and Schlues-

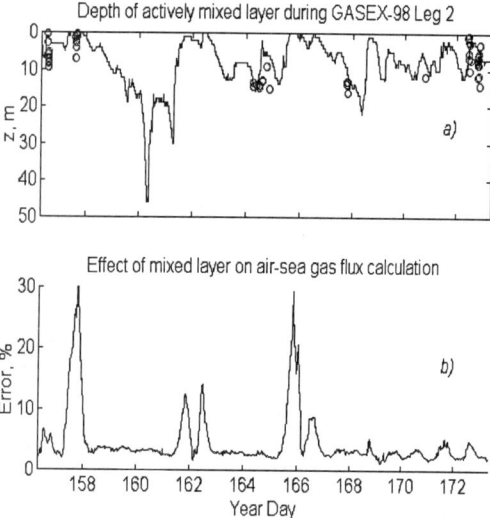

Figure 3. Effect of the mixed layer dynamics on the bulk-flux formulation for the air-sea exchange of O_2: (a) Depth of the actively mixed layer according to modeling (contiguous line) and turbulence measurements (circles) in *GasEx-98 leg 2*; (b) Relative error in the gas flux calculation when the bulk gas concentration and solubility are taken from a 4-m depth (see explanations in the text).

sel [1996]; while, the bubble-mediated gas transfer coefficient is taken in the form suggested by Merlivat et al. [1993] with the empirical coefficients determined by Asher and Wanninkhof [1998].

A grid with 50 evenly spaced points within the top 50 m of the ocean is used for the calculation. The initial temperature and salinity profiles are taken from the CTD profiles that were obtained during *leg 2* of *GasEx-98*. For simplicity, the O_2 concentration is assumed invariant with depth and at 120-% saturation as the initial condition. The model is forced with the air-sea heat and momentum fluxes measured during *GasEx-98* Leg 2 by Edson et al. [1999]. The results of the actively mixed layer calculation are shown in Figure 3a. The actively mixed layer depth is defined using the gradient Richardson number criteria, $Ri = Ri_{cr} = 0.25$.

4. DISCUSSION

According to Figure 3a, the simulated mixed layer depths is consistent with the depth of the actively mixed layer as determined from the turbulence profiles taken during *GasEx-98 leg 2* with a free-rising profiler [Soloviev et al., 1999]. Moderate wind speed conditions prevailed during *GasEx-98 leg 2* with several storms and a few periods of calm weather. Both the modeling and experimental results demonstrate that under conditions of low wind speed, the surface-generated turbulence is constrained within a relatively thin near-surface layer of the ocean. As a result, appreciable temperature, salinity, and gas concentration differences are formed in the near-surface layer because of diurnal warming and/or precipitation effects.

In *GasEx-98*, the measurements of T, S, and C_W were routinely taken from a 4-m depth (the thermosalinograph intake on the bow of the ship), assuming that the near-surface layer of the ocean is well mixed. To evaluate at what extent the mixed layer processes may influence the bulk-flux formulation (1) for the air-sea gas exchange, the lower plate in Figure 3b shows the relative error that occurs if the mixed layer dynamics is ignored. This error is calculated as follows:

$$Error = (F_0 - F_4)/F_0, \quad (12)$$

where $F_0 = k(C_{W0} - S_0 p_a)$, $F_4 = k(C_{W4} - S_4 p_a)$; C_{W0} and C_{W4} are the gas concentration at the surface (upper bin) and 4 m depth; S_0 and S_4 are the gas solubility at the surface (upper bin) and 4 m depth respectively.

Note that a small gas concentration difference between the surface and 4 m depth exists even under high wind speed conditions that results in a few-percent error in the bulk flux formulation. This difference substantially increases under calm weather conditions because of stratification effects. During a few periods of low wind speed conditions observed during *GasEx-98 leg 2*, the relative error for O_2 reaches 30 % (Figure 3). The averaged error for *leg 2* is, however, relatively small (~4%).

Note that for supersaturated gases (like O_2 in *GasEx-98 leg 2*), the temperature dependence of solubility increases the gas saturation level, thus enhancing the effect of diurnal cycle on the air-sea gas exchange. For the undersaturated gases (like CO_2 in *GasEx-98 leg 2*), the temperature dependence of solubility reduces the gas saturation level, thus suppressing the effect of diurnal cycle on the air-sea gas exchange. The modeling for CO_2 is, however, complicated due to the fact that in seawater the CO_2 is a component of the carbonate system.

5. CONCLUSIONS

The results of this study suggest that the fine thermohaline structure in the near-surface layer of the ocean, which develops under low wind speed conditions, cause near-surface gas concentration anomalies. These gas concentration anomalies affect both the air-sea gas exchange and the bulk formula for gas-flux calculation.

Acknowledgments. This work is done under the auspices of the NOAA Ocean-Atmosphere Carbon Exchange Study (OACES).

REFERENCES

Asher, W.E. and R. Wanninkhof, 1998: The effect of bubble-mediated gas transfer on purposeful dual-gaseous tracer experiment, *J. Geophys. Res.*, 103, 10,555-10,560.

Atkinson, M.J., T. Berman, B.R. Allanson, and J. Imberger, Fine-scale oxygen variability in a stratified estuary: patchiness in aquatic environments, *Mar. Ecol. Prog. Ser.*, 36, 1-10, 1987

Bruce, J.C. and E. Firing, Temperature measurements in the upper 10 m with modified expendable bathythermograph probes. *J. Geophys. Res.*, 79, 4110-4111, 1974.

Edson, J.B., W.R. McGillis, J. Ware, J.E. Hare, and C.W. Fairall, Direct CO_2 flux measurement at the air-sea interface, supplement to *EOS, Transactions*, April 27, S45, 1999 (abstract only).

Mervilat, L., L. Memery, and J. Boutin, 1993: Gas exchange at the air-sea interface. Present status: The case of CO_2, paper presented at the *Fourth International Conference on CO2 in the Ocean*, Inst. Natl. des Sci. de l'Univers., Cent. Natl. de la Rech. Sci., Carqueiranne, France, Sept. 13-17, 1993.

Payne, R.E., Albedo of the sea surface. *J. Atmos. Sci.*, 29, 959-970, 1972.

Price J., R. Weller and R. Pinkel, Diurnal cycling: Observations and models of the upper ocean response to diurnal heating, cooling and wind mixing. *J. Geophys. Res.*, 91, 8411-8427, 1986.

Peters, H., M.C. Gregg, and J.M. Toole, On the parameterization of equatorial turbulence. *J. Geophys. Res.*, 93, 1199-1218, 1988.

Soloviev, A. and R. Lukas, 1997: Observation of large diurnal warming events in the near-surface layer of the western equatorial Pacific warm pool. *Deep-Sea Research*, 44, 1055-1076.

Soloviev, A., R. Lukas, and P. Hacker, An approach to parameterization of the oceanic turbulent boundary layer in the western Pacific warm pool, *J. Geophys. Res.*, 106, 4421-4435, 2001.

Soloviev, A., R. Lukas, P. Hacker, H. Schoeberlein, M. Baker, and A. Arjannikov, A near-surface microstructure sensor system used during TOGA COARE. Part II: Turbulence measurements, *J. Atmos. Oceanic Technology*, 16, 1598-1618, 1999.

Soloviev, A.V. and P. Schluessel, Evolution of cool skin and direct air-sea gas transfer coefficient during daytime, *Boundary-Layer Meteorology*, 77, 45-68, 1996.

Soloviev, A. V. and P. Schluessel, An observation of the fine thermohaline structure and dissolved oxygen profiles in the near-surface layer of the ocean, supplement to *EOS, Transactions*, April 27, S50, 1999 (abstract only).

Soloviev A.V. and N.V. Vershinsky, The vertical structure of the thin surface layer of the ocean under conditions of low wind speed. *Deep-Sea Research*, 29, 1437-1449, 1982.

Jim Edson and Wade McGillis, MS 12, Woods Hole Oceanographic Institution, Woods Hole, MA 02543, U.S.A.

Peter Schluessel, EUMETSAT, Am Kavalleriesand 31, 64295 Darmstadt, Gemany

Alexander Soloviev, Oceanographic Center, Nova Southeastern University, 8000 N. Ocean Dr., Dania Beach, FL 33004, U.S.A. (e-mail: soloviev@ocean.nova.edu)

Rik Wanninkhof, NOAA/Atlantic Oceanographic and Meteorological Laboratory, 4301 Rickenbacker Causeway, Miami, FL 33149, U.S.A.

A Field Study of Whitecap Coverage and its Modulations by Energy Containing Surface Waves

V.A. Dulov, V.N. Kudryavtsev, A.N. Bol'shakov

Marine Hydrophysical Institute (MHI), Sebastopol, Ukraine

A field study of whitecap coverage generated by breaking wind waves has been performed from MHI's Black Sea Research Platform. It is revealed that the main contribution to the whitecap coverage of the sea surface results from breaking of short wind waves, which are more than 3 times shorter than the wavelength of the spectral peak. The energy containing waves strongly modulate the whitecap coverage. Zones of enhanced wave breaking are located on the modulating waves' crests. The effect is described in terms of a modulation transfer function for whitecap coverage. Its magnitude equals about 24, and decreases with the increase of inverse wave age of the energy containing waves.

INTRODUCTION

The breaking of wind waves is one of the most important physical phenomena on the sea surface that results in energy dissipation in wind generated waves, is responsible for turbulent mixing in the sub-surface layer, affects surface drag (supporting air flow separation) and influences air-sea interaction and gas transfer processes. An important application of this study of wave breaking is in the interpretation of radar backscatter from the sea surface, as wave breaking significantly contributes to the radar normalized cross-section.

DESCRIPTION OF THE EXPERIMENT

A field study of wave breaking was performed in October, 1999 from MHI's Research Platform, which is located at a distance of 0.5 km from the Black Sea coast in 30 m of water (Figure 1). The general goals of the experiment were: (i) to understand what spectral range of breaking waves contributes to the total whitecap coverage, and (ii) to analyze the influence of dominant surface waves (waves of the spectral peak and swell) on wave breaking. The latter was treated as a modulation problem.

A view of the platform and plan of the experiment area are shown in Figure 1. Measurements of whitecap coverage were performed by an optical system recording brightness along a 4 meter long linear 'footprint' on the ocean surface. Sampling for brightness of 600 footprint elements was carried out at 10 Hz. The appearance of a whitecap in the footprint was identified if the sea surface brightness exceeded a threshold level. The threshold level was chosen as 3/4 of the averaged (over 3 min.) spikes of brightness. This level automatically follows natural changes in surface illumination (see [*Dulov et al.*, 1998] for details). The optical system footprint was always directed in the main propagation direction of the wave breakers.

The sea surface elevations associated with the wind wave field were registered by four resistance gauges, which formed a spatial grid to assess frequency-directional wave spectra. The footprint of the optical system was located in the vicinity of the wave gauge grid (Figure 1). Wave measurements were accompanied with measurements of wind velocity and air temperature at 19 m height, and seawater temperature.

The experimental conditions are summarized in the Table, where average wind velocity, air and water temperature, significant wave height, frequency of dominant waves and their direction, for each run, are shown.

188 FIELD STUDY OF WHITECAP COVERAGE AND ITS MODULATIONS

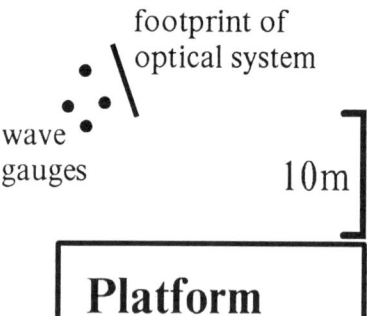

Figure 1. View of the research platform and plan of the layout of the field system.

BACKGROUND CHARACTERISTICS

To analyze the data we have selected two different wind-wave situations. In the first case, the wind-wave field was purely wind generated, and the dominant waves were running along the coast (runs from 4 to 8 in the Table). In the second case, the measurements were done in off-shore wind conditions, when young developing wind waves are running opposite to the swell propagating from the open sea (runs from 1 to 3 in the Table). Examples of the wave frequency spectra for each of these cases are shown in Fig. 2.

A fragment of the spatial-temporal record of the sea surface brightness obtained by the optical system is shown in Figure 3. The vertical axis is distance along the footprint, and the horizontal time-axis is formed by consecutive optical samples. We have used the following procedure of data processing. In those footprint elements where the sea surface brightness exceeds the threshold level, the signal is replaced by 1; in other elements it is replaced by zero. Such a procedure gives explicit mapping for appearance and evolution of individual whitecaps in spatial-temporal images recorded by the optical system (see Figure 3). The number of footprint elements containing ones divided by 600 is treated as an instant fraction of whitecap coverage $q(t)$. This quantity averaged over a certain time gives the mean fraction of the sea surface covered by whitecaps, Q.

In Figure 4 mean values of whitecap coverage Q for all of the runs are shown as a function of wind speed U. The data are consistent with the [Monahan and Woolf, 1989] empirical equation for the water-air temperature difference of 7°C, but exhibit a stronger wind exponent. This fact is not surprising, as we have used the data related to both mixed seas and pure wind seas. The other reason is that some of the points were obtained at low wind conditions, when one might expect a steeper wind dependence than is predicted by a cubic relationship.

As is apparent from Figure 3, the spatial-temporal image of the sea surface can be used to assess a speed of whitecap advance along the footprint direction. This speed is equal to the average slope of the whitecap signature in the tx-plane. It can be defined as a slope of a line fitted by least squares to an individual whitecap area. If we contend that the speed of whitecaps can be related to the scale of breaking waves, then a set of whitecaps sorted in accordance with their speed gives us a mechanism by which to determine what spectral range of wind waves contributes to the total whitecap coverage.

We made calculations of the speed C for each individual whitecap and its area in the tx-plane over all the runs. Note

Table. Conditions of the experiments.

Run	Duration, hours	Wind at 19m height:		Significant height, m	Dominant waves:		Temperature	
		Speed, m/s	Direction (from)		Spectral peak frequency, Hz	Direction (from)	Water, °C	Air, °C
1	1.0	10.8	340°	0.45	0.23	249°	20.5	19
2	0.8	11.1	0°	0.49	0.21	247°	20.5	19
3	2.3	7.9	328°	0.36	0.18	204°	19	12.5
4	1.8	5.9	68°	0.61	0.29	118°	18	11
5	1.4	6.1	58°	0.64	0.25	113°	18	11
6	1.6	7.9	65°	1.69	0.16	115°	18	11
7	2.5	7.3	68°	1.36	0.14	115°	18	11
8	1.5	4.5	60°	1.08	0.16	108°	18	11

Direction is clockwise relative to the North.

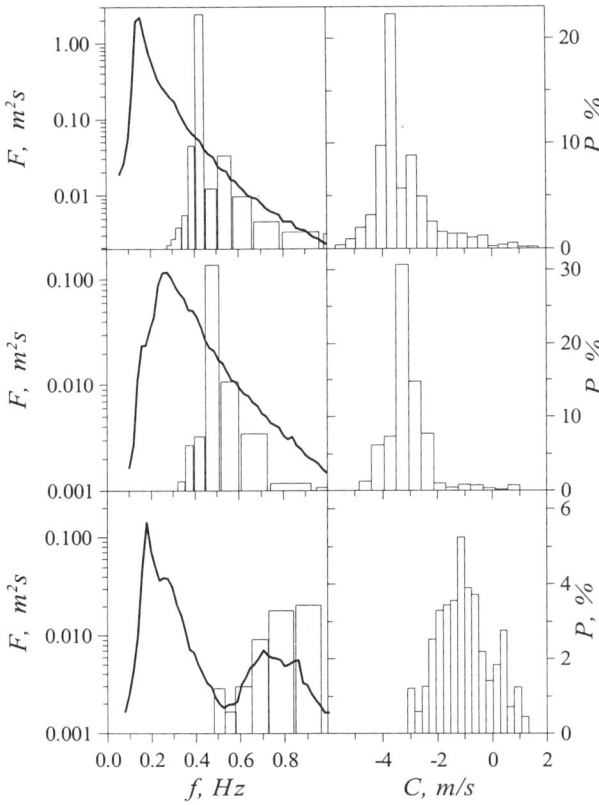

Figure 2. Frequency spectra F (bold line) and contributions in whitecap coverage P (bars) as a function of frequency f and whitecap speed C. The data of two adjacent runs were used for each of plots, the runs shown from the top to the bottom are: 6 and 7; 4 and 5; 1 and 2.

that only those whitecaps were accounted for that consisted of not less than three discrete elements. These data were used to obtain a discrete set Q_j that describes the contribution of whitecaps, which have a velocity of advance in the range from C_j to $C_j+\Delta C_j$, so that the sum of Q_j over all C_j equals the total whitecap coverage Q. Examples of distribution functions

$$P(C_j) = Q_j / Q$$

are shown in the right column of Figure 2.

To relate whitecap speed to a certain scale of breaking waves we assumed following *Phillips* [1985], that C is related to the wave frequency f in accordance with the gravity wave dispersion relation:

$$C = g / (2\pi f)$$

Distribution functions $P(f_j)$ transformed from C to the frequency domain are shown in the left column of Figure 2, along with the wave spectra. The most remarkable conclusion that can be drawn from these plots is that the main contribution of breaking waves to the total whitecap coverage comes not from the dominant waves, but from significantly shorter breaking waves. This fact was reflected in all the data we collected. In cases of pure wind seas (runs from 4 to 8), waves in the wavenumber range $k > 3k_m$ (where k_m is the wavenumber of the spectral peak) contribute more than 90% of the total whitecap coverage. In the case of mixed seas (runs from 1 to 3) wind waves, being the very young waves, support almost the total whitecap coverage.

MODULATIONS OF WAVE BREAKING BY LONG SURFACE WAVES

The fact that breaking of relatively short wind waves is mainly responsible for the total whitecap coverage, results in the question, what is the influence of the energy-containing waves on wave breaking? As a first step we have investigated how dominant surface waves affect spatial distribution of wave breaking.

Whitecaps occur at the sea surface relatively rarely. This means that the temporal variability of the instantaneous values of $q(t)$ measured by our optical system exhibits a noisy signal rather than smooth variations, which could be related to passing dominant waves. That is why to reveal modulation in whitecaps caused by dominant waves, we have used a phase averaging procedure. This procedure has been performed in the following way. As a first step, records of surface elevation have been smoothed to filter out all the surface waves with periods shorter than 2s. Then the smoothed elevations $z(t)$ were split into a number of individual waves. Each individual wave was defined as a tem-

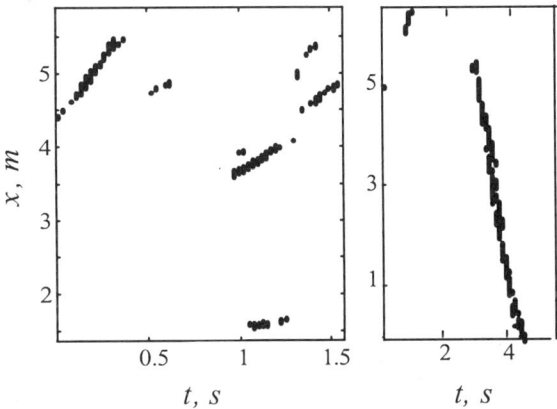

Figure 3. Fragments of the optical record in the time-distance plane. Images of typical whitecaps (left plot) and a large fast whitecap (right plot).

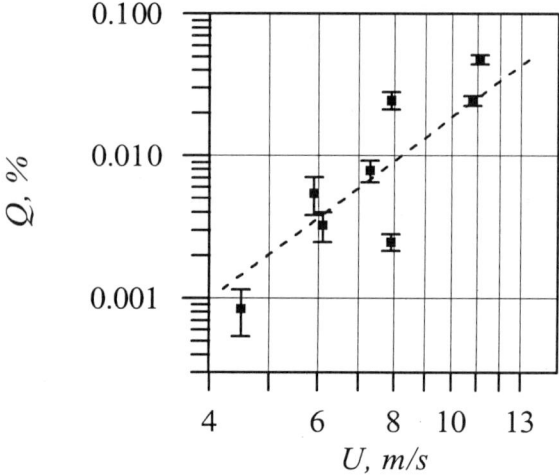

Figure 4. Whitecap coverage percentage as a function of wind speed. Dashed line shows the empirical dependence [by *Monahan and Woolf, 1989*] for water/air temperature difference of 7°C.

poral segment between successive upward zero-crossings of $z(t)$. The length of such temporal segments was treated as a period T_i of individual waves thus defining a wavenumber:

$$K_i = (2\pi/T_i)^2 / g.$$

The surface elevation confined in this interval was considered to be an individual wave profile $z_i(t)$. The series of instantaneous whitecap coverage $q(t)$ has also been split into temporal intervals in accordance with selected individual surface waves, so that each individual wave has been provided with a whitecap coverage fragment $q_i(t)$. Each run consists of 500-1500 individual waves.

As the second step, for each run the set of profiles of individual wave slope $K_i z_i(t)$ and the related set of fragments $q_i(t)$ have been transformed from the temporal domain $0 < t < T_i$ to the phase domain $0 < \tau < 1$, $\tau = t/T_i$ and then have been averaged. Note that only individual waves with frequency less than 0.3-0.4 Hz were taken into account in averaging (as can be seen from Figure 2, this range of dominant waves does not contribute to the whitecap coverage). In the end, we obtained mean profiles of the wave slope $KZ(\tau)$ and corresponding mean distributions of the whitecap coverage $Q(\tau)$ for each run listed in the Table. In Figure 5 some examples of mean slope profiles and distributions of whitecap coverage along them are shown. Vertical bars in Figure 5 show confidence intervals that hereinafter correspond to a single standard error. The remarkable feature is that whitecap coverage is significantly increased on the wave crest and damped in the wave trough vicinity. In cases of pure wind seas (runs 4 to 8), this effect is most pronounced. In the cases when the dominant waves relate to swell (runs from 1 to 3), the effect is also present, but not as evident as in the pure wind sea case.

These data have been analyzed in terms of the modulation transfer function (MTF). The MTF describes the linear response of a sea surface parameter to long surface wave steepness (such quantities are widely used in radar observations, see, e.g. [*Plant, 1987*]). In our case the complex amplitude of the whitecap coverage MTF is defined as

$$M = Q_1 / (Q_0 KA) \qquad (1)$$

where Q_0 is a mean of $Q(\tau)$, Q_1 is a Fourier amplitude of the whitecap coverage variations $Q(\tau)$ along a long wave,

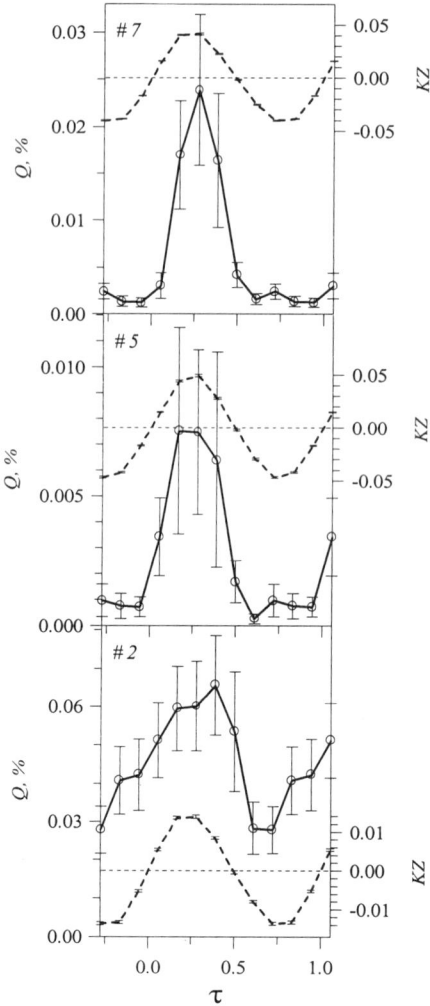

Figure 5. Whitecap coverage distribution Q (solid line with error bars) on a long wave for runs 7, 5 and 2. Dashed line is an elevation profile multiplied by wavenumber, KZ. τ is wave phase divided by 2π.

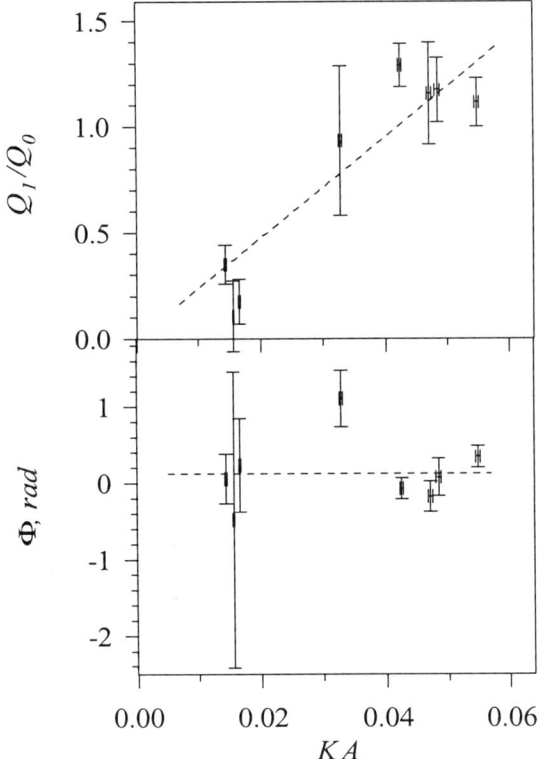

Figure 6. Normalized magnitude Q_1/Q_0 and phase Φ of whitecap modulations as a function of long wave steepness KA. Dashed lines show the estimate (2).

KA is a Fourier amplitude of the profile $KZ(\tau)$ of the long wave steepness. To apply MTF analysis to our data, we have used only the first Fourier harmonics. Dependencies of the magnitude of Q_1 and its phase Φ on KA shown in Figure 6 demonstrate that our data are consistent with equation (1). A least squares estimate of the MTF obtained from these data is

$$|M|=23.8\pm1.3 \qquad \Phi=0.12\pm0.08 \qquad (2)$$

where Φ is in radians. Positive values of the MTF's mean phase show that the zone of enhanced wavebreaking is shifted toward the rear slope of the dominant wave.

The experimental estimates of whitecap coverage MTF were obtained in a relatively wide range of wind-wave conditions (see Table). This makes it possible for us to assess the dependence of $|M|$ on wind speed and frequency of the dominant waves. The latter was used in normalized form, as an inverse wave age, U/C_m, where C_m is phase velocity of the spectral peak. These dependencies are shown in Figure 7. In spite of the scatter, the data exhibit a discernible trend for both wind and age dependencies. As

these parameters are correlated for our data, both the trends can be caused by the real MTF dependence on a single parameter. The fact, that the age trend is more clearly expressed in the Figure, advocates the conclusion, that the whitecap MTF decreases with the increase of inverse wave age.

CONCLUSIONS

In this paper we have presented results of an observational study of whitecaps generated by breaking waves and their modulations caused by dominant surface waves. It was revealed that breaking of short wind waves with wavenumbers upwards of 3 times that of the spectral peak are responsible for more than 90% of the whitecap coverage. Contribution of energy-containing waves to whitecap coverage is negligible. The influence of the dominant waves on wave breaking results in very strong modulations in white-capping. Breaking of waves is significantly enhanced at the long wave crests, and is suppressed in the

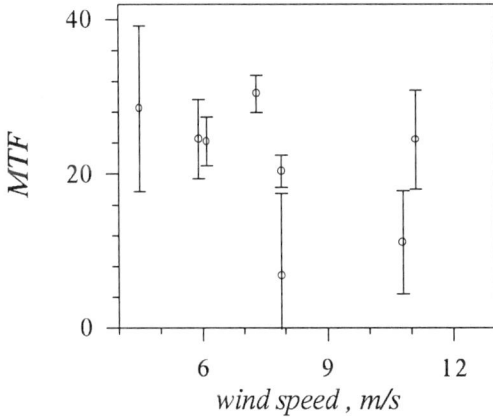

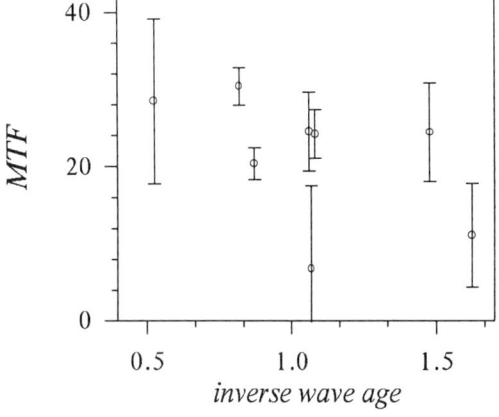

Figure 7. Whitecap MTF as a function of wind speed and inverse wave age.

vicinity of their troughs. The amplitude of whitecap coverage modulations (in terms of the MTF) is approximately equal to 24. This means that the magnitude of the relative variations of whitecap coverage is 24 times larger than the steepness of modulating waves. The whitecap coverage MTF exhibits a dependence on the inverse wave age of the modulating waves: the "older" the dominant waves, the higher the modulations in whitecap coverage.

Acknowledgments. This research was sponsored by the Office of Naval Research (ONR grant N00014-98-10653), and INTAS International Association (grant INTAS/CNES 97-0222).

REFERENCES

Dulov, V.A., Kudryavtsev, V.N., Sherbak, O.G. and Grodsky, S.A. Observations of Wind Wave Breaking in the Gulf Stream Frontal Zone. *The Global Atmosphere and Ocean System*, 1998, Vol. 6, pp.209-242.

Monahan, E.C. and Woolf, D.K. Comments on "Variations of Whitecap Coverage with Wind Stress and Water Temperature". *J. Phys. Oceanogr.*, 1989, Vol. 19, pp. 706-709.

Phillips, O.M. Spectral and statistical properties of the equilibrium range in wind-generated gravity waves. *J. Fluid Mech.*, 1985, Vol. 156, pp. 505-531.

Plant, W.J. The Modulation Transfer Function: Concept and Applications. In *Radar Scattering from Modulated Wind Waves*, Eds. G.J. Komen and W.A. Oost, Kluwer Academic Publishers, 1989, pp. 155-172.

Dulov V.A., Kudryavtsev V.N. and Bol'shakov A.N. Marine Hydrophysical Institute, 2 Kapitanskaya str., Sevastopol, Ukraine 99011.
E-mail: odmi@alpha.mhi.iuf.net

The Physical and Practical Implications of a CO_2 Gas Transfer Coefficient that Varies as the Cube of the Wind Speed

Edward C. Monahan

Department of Marine Sciences, University of Connecticut at Avery Point, Groton, Connecticut

Wanninkhof and McGillis [1999] recently presented further observational evidence supporting the contention that there is a cubic relationship between wind speed and the air-sea gas transfer ("piston") velocity for CO_2. They acknowledged that *Monahan and Spillane* [1984] were the first to propose that "the gas transfer coefficient is proportional to whitecap coverage and that whitecap coverage scales approximately as u^3." While Wanninkhof and McGillis' findings give further support to the contention that at moderate and high wind speeds it is the fraction of the sea surface covered by Stage-A whitecaps (spilling wave crests) that determines the magnitude of the air-sea gas transfer coefficient, it should be noted that a cubic expression is only an approximation of the true dependence of whitecap coverage on wind speed [*Monahan and O'Muircheartaigh*, 1980], and that other factors such as wind duration need to be taken into account [*Monahan and O'Muircheartaigh*, 1986]. Since global maps of fractional oceanic whitecap coverage can now be generated using satellite-borne microwave radiometers, it follows that the potential exists for routinely producing similar global maps of CO_2 gas transfer coefficient ("piston velocity").

INTRODUCTION

Two motives lie behind the efforts of various investigators to determine the dependence of the gas transfer coefficient on wind speed. There is the practical desire to be able, by parameterizing the gas transfer coefficient in terms of the wind speed, to estimate the effective average magnitude of this coefficient over certain time periods, and certain portions of the world ocean, from the appropriate wind statistics. The second rationale for these studies of the influence of wind speed on the magnitude of the gas transfer coefficient, or "piston velocity," is to gain insights into the actual physical/chemical mechanisms that control the exchange of particular gases across the air-sea interface. One such discussion of the wind dependence that would follow from various gas transfer models can be found in the paper of *Yu et al.* [1984] that appeared in the proceedings of the First International Symposium on Air-Water Gas Transfer. All of the power-law descriptions of k_L that appear on Figure 1 as straight lines in log-log space, except the two steepest, are taken from Figure 3 of *Yu et al.* [1984].

PLACING RECENT RESULTS IN CONTEXT

In the recent article by *Wanninkhof and McGillis* [1999] in which these authors conclude from recent field results that "gas transfer velocities can be well quantified with a cubic relationship," they acknowledge that *Monahan and Spillane* [1984] were the first to propose that "the gas transfer coefficient is proportional to whitecap coverage and that whitecap coverage scales approximately as u^3," where u is the 10 m-elevation wind speed. There are two points that might appropriately be made in clarification of this statement. First, it should be noted that the same proceedings volume, that stemming from the First International Symposium on Air-Water Gas Transfer that contained the Monahan and Spillane paper, also contained an insightful paper by *Kerman* [1984] that likewise focused on the role of breaking waves, whitecaps, and bubbles, in enhancing the air-sea exchange of certain groups of gases. The second point that should be made is that in the Monahan and Spillane model for the gas transfer coefficient k, a simple mechanistic model that considered each whitecap to be a "low impedance vent," the resulting expression (reproduced here as equation 1) for the effective gas

$$k_e = k_m (1 - W) + k_t W \qquad (1)$$

transfer coefficient k_e, in terms of fractional whitecap coverage, W, took the form of a weighted average of two gas transfer coefficients, a k_m of low magnitude associated with the wind-ruffled but whitecap-less portion of the sea surface, and a k_t of much greater magnitude associated with that small fraction of the sea surface actually covered by whitecaps. It is true that for computational convenience they did use a cubic expression to describe the wind dependence of the fraction of the sea surface covered by whitecaps, but only after acknowledging that this was only an approximation of the true wind dependence in the case of a fully developed sea, as described, e.g., in *Monahan and O'Muircheartaigh* [1980]. Now it should be apparent that the k_e described in this early expression becomes proportional to the cube of the wind speed only when the wind speed and, hence, the whitecap coverage, becomes large enough so that the second term overwhelms the first term in Equation 1.

In order to evaluate k_m and k_t in Equation 1, *Monahan and Spillane* [1984] had recourse to some early field data based on radon profiles obtained during the GEOSECS and TTO cruises [*Peng et al.,* 1979; *Smethie,* personal communication]. When these values for k_m and k_t are introduced into Equation 1, one obtains the curve labeled "Monahan 1984" that appears on Figure 1 of *Wanninkhof and McGillis* [1999]. This same curve, now labeled MS84, has been included, along with the other six curves from Wanninkhof and McGillis' Figure 1, on Figure 2 of the present paper.

In reviewing the various $k[u]$ parameterizations in the literature, *Wanninkhof and McGillis* [1999] apparently overlooked the follow-on paper of *Monahan and Torgersen* [1990] that appeared in the volume arising from the Second International Symposium on Air-Water Gas Transfer, as that paper contains expressions for k_e parameterized in terms of the fraction of sea covered by spilling wave crests, W_A (Stage-A whitecaps), and in terms of the portion of the sea surface covered by decaying foam patches, W_B (Stage-B whitecaps). Since it is the spilling wave crests and the associated intense bubble plumes that are most relevant to the mechanics of air-water gas transfer, we will reproduce here as Equation 2 the parameterization of *Monahan and Torgersen* [1990] which is explicitly in terms of Stage-A whitecaps.

$$k_e = 7.88 + 1.52 \times 10^5 \, W_A \text{ (micrometers per sec)} \qquad (2)$$

Given that at winds speeds as high as 20 m s^{-1} W_B is only about 0.10, i.e., 10% of the sea surface, and W_A is typically less than 0.01, i.e., 1% of the ocean surface, the first term of Equation 2 can be simplified by taking the $(1-W)$ of Equation 1 to be unity. It should be noted that the k values associated with the whitecap- and non-whitecap-areas of the sea surface were obtained in the case of *Monahan and Torgersen* [1990] from radon evasion experiments conducted in a "tipping bucket whitecap simulation tank." (While these were the first published findings obtained from the use of such controlled whitecap tank experiments, the insights to be gained from the use of such facilities are manifest from, e.g., the results of *Asher et al.* [1995] and *Leifer et al.* [1995] described in the proceedings of the Third International Symposium on Air-Water Gas Transfer.) If the wind dependence of Stage-A whitecap coverage given in *Monahan* [1993] that appears as Equation 4 of *Monahan and Torgersen* [1990] is used in evaluating Equation 2, the curve labeled MT90 on Figure 2 is obtained.

CONFLUENCE

It should be noted that if the W_A [u] expression that appeared recently in *Andreas and Monahan* [2000] (based on line A1 on Figure 2 of *Monahan* [1989]) is introduced in Equation 2, and if the resulting expression is converted from S.I. units to the cm per hour used by *Wanninkhof and McGillis* [1999], Equation 3 is obtained.

$$k_e = 2.84 + 1.73 \times 10^{-2} \, u^{3.2} \text{ (cm per hour)} \qquad (3)$$

It is instructive to compare the values of k_e predicted by Equation 3 with those calculated using Equation 4,

$$k = 2.83 \times 10^{-2} \, u^3 \text{ (cm per hour)} \qquad (4)$$

the equivalent expression given by *Wanninkhof and McGillis* [1999] (where the small but finite Schmidt number dependence has been omitted). As can be seen from Table 1, except at low wind speeds where the contributions of the "wind-ruffled but

Table 1. Comparison of two k_e parameterizations.

u (m/s)	k_e (W&McG) (cm/hr)	k_e (M&T+) (cm/hr)	Diff. (%)
7	9.70	11.59	19.5
8	14.49	16.27	12.3
10	28.30	30.26	6.9
12	48.90	51.98	6.3
14	77.66	83.31	7.3
16	115.92	126.22	8.9
18	165.05	182.69	10.7
20	226.40	254.81	12.5

whitecap-less" regions included in Equation 3 cause significantly higher predicted values than those calculated by Equation 4, the values of k_e obtained using these two expressions do not differ by more than 12.5% over the wind range from 8 to 20 m s^{-1}. The curve described by Equation 3 appears on Figure 1, labeled with its high wind speed slope of 3.2, and on Figure 2, labeled MT90+AM00. Likewise, the curve defined by Equation 4, essentially the Wanninkhof and McGillis formulation, appears as the line of slope 3 on Figure 1, and as the curve labeled WMcG99 on Figure 2.

CONCLUSIONS

The marked agreement of Equations 3 and 4 over the pertinent wind range is remarkable, given that the Wanninkhof and McGillis expression was based in large measure on the field results obtained during the recent Gas Ex-98 program (and on a consideration of the global effective k required to accommodate the oceanic bomb C-14 inventory), while the Monahan and Torgersen coefficients were obtained a decade ago from a simple whitecap simulation tank experiment. Likewise, in the above comparison, no adjustments were made to take account of the fact that the field experiments involved covariance measurements of the invasion of CO_2 into the ocean while the simulation tank study involved the evasion of radon. While remarkable, it can be argued that this agreement is by no means fortuitous, but rather that the findings of *Wanninkhof and McGillis* [1999] represent further confirmation of the basic validity of the simple "low impedance vent" model first set out in *Monahan and Spillane* [1984].

While *Wanninkhof and McGillis* [1999] tested quadratic and cubic $k[u]$ models against their binned CO_2 covariance estimates and demonstrated that a better fit was obtained with the cubic model, it would have perhaps been more insightful if they had determined the best power-law fit to these CO_2 data points. By the simple expedient of replotting the data points representing the binned CO_2 covariance flux results that appear on Wanninkhof and McGillis' Figure 2 on full logarithmic paper, and fitting a straight line to those points corresponding to the findings at the higher wind speeds (where the power-law term in Equation 3 can be expected to predominate), one finds from the geometric slope of this line that the best power-law exponent to describe their data is a number less than 3, indeed it appears to fall in the range of 2.4 to 2.9. Now in an intercomparison of various W_B data sets *Monahan and O'Muircheartaigh* [1986] found that the best power-law fits to those sets collected under conditions where the high wind observations also corresponded to instances

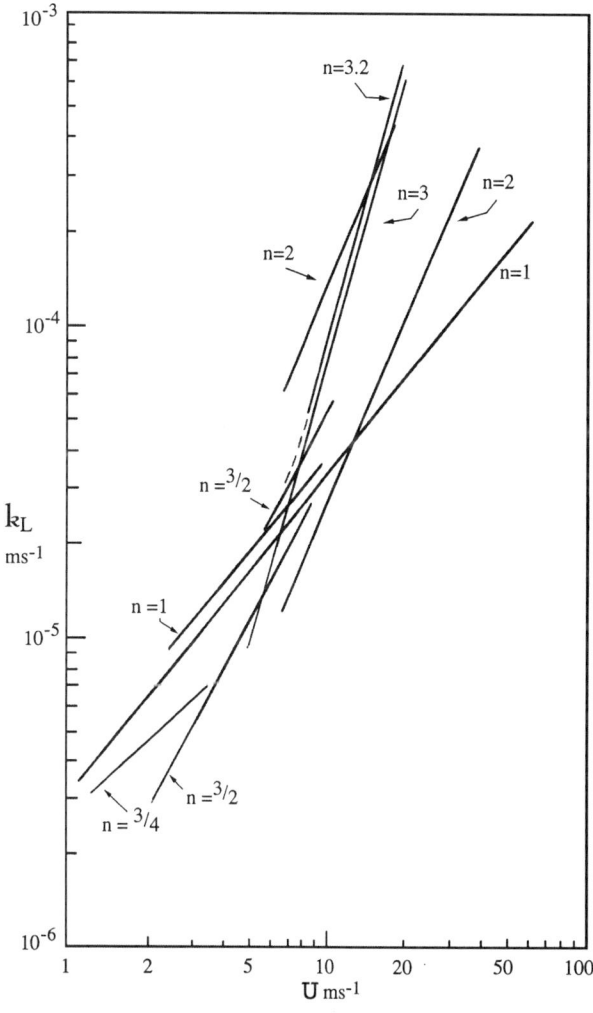

Figure 1. Gas transfer coefficient, k_L, versus 10 m elevation wind speed, U. All but the two steepest lines are from Figure 3 of *Yu et al.* [1984], based on air-water oxygen transfer models and observations. Line labeled 3 from Equation 4, see *Wanninkhof and McGillis* [1999]. Line labeled 3.2 from Equation 3, see text or derivation. Note that all labels correspond to slopes of respective lines on this log-log plot, i.e., to power-law exponents.

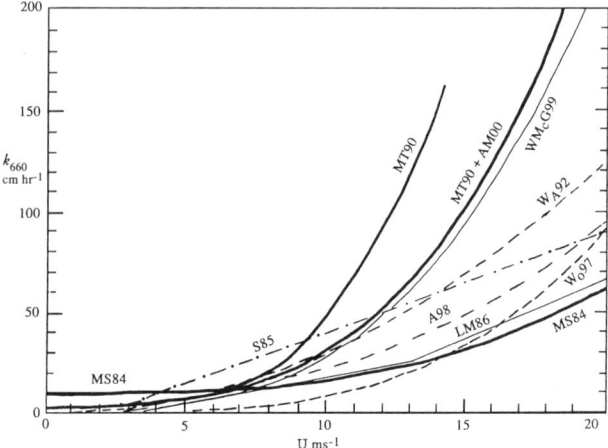

Figure 2. Gas transfer coefficient k_{660}, versus 10 m elevation wind speed, U. First seven of the nine curves identified below transcribed from Figure 1 of *Wanninkhof and McGillis* [1999]. A98–*Asher and Wanninkhof* [1998], Wo97–*Woolf* [1997], MS84–*Monahan and Spillane* [1984], S85–*Smethie et al.* [1985], LM86–*Liss and Merlivat* [1986], Wa92–*Wanninkhof* [1992], WMcG99–*Wanninkhof and McGillis* [1999], MT90–*Monahan and Torgersen* [1990], and MT90+AM00–coefficients from *Monahan and Torgersen* [1990] combined with $W_A(U)$ from *Andreas and Monahan* [2000]. See text for discussion of last two curves.

where the wind duration was adequate for the attainment of the elusive "fully developed sea," could best be described with exponents slightly higher than 3 (but less than the 3.75 proposed by *Wu* [1979] on theoretical grounds), but they also found that the best fit to W_B data sets collected at latitudes where the high wind events were often of relatively short duration called for exponents in the 2.1 to 2.8 range. These lower values for the power-law exponent are a consequence of the fact that the sea state and extent of whitecapping encountered during the typically brief high wind events associated with these data sets were less than would be expected in the case of a fully developed sea, while the observations from lower wind speeds that also contributed to these data sets typically had associated with them fully developed seas and the appropriate complement of whitecaps. Given the season and latitude associated with the Gas Ex-98 observations, and the reference in the *Wanninkhof and McGillis* [1999] letter to "episodic high wind events," and noting that the best $k[u]$ power law descriptor of the data has associated with it an exponent in the 2.4 to 2.9 range while the characteristic power-law exponent in the W_B expression describing the JASIN data collected in the North Atlantic in September in the face of episodic high winds was 2.75 [*Monahan and O'Muircheartaigh*, 1986], lends further credence to the contention that it is the extent of whitecapping present on the sea surface, and not some factor related solely to the wind speed, that controls the air-sea gas exchange coefficient. (Having stressed the significance of W_A, the fraction of the sea surface covered by spilling wave crests in determining $k[u]$, it is nonetheless appropriate to draw the above inference from the similarity of the $k[u]$ and JASIN $W_B[u]$ power law exponents, as the fraction of the sea surface covered by decaying foam patches, W_B is found to be at any instant a fixed multiple of the fraction of the sea covered by spilling wave crests W_A, regardless of the wind speed [see, e.g., *Monahan and Lu*, 1990]).

While concurring with *Wanninkhof and McGillis'* [1999] reservations about the possibility of parameterizing $k[u]$ in terms of wind speed alone, the present author does not agree with their contention that nonetheless "wind is currently the most robust parameter available to estimate global exchange" of CO_2 across the air-water interface. Rather, we would posit that global "maps" of oceanic whitecap coverage can now be generated from the same satellite-borne microwave sensors used to estimate surface wind speeds from the enhancement in microwave brightness temperatures caused by the very presence of whitecaps [e.g., *Schluessel and Luthardt*, 1991], and that whitecap coverage, rather than wind speed, is the most appropriate variable to use in obtaining global estimates of $k[u]$. In conclusion, it is just such global "maps" of the CO_2 gas transfer coefficient that will be used by the climate research community to derive the needed improved estimates of the annual uptake of CO_2 by the world ocean.

Acknowledgments. The support provided for the author's research over the past three decades by the Office of Naval Research is acknowledged with gratitude, as is the support of the Department of Energy. The aid provided by P. Van Patten in the preparation of the figures for this paper is also acknowledged with thanks.

REFERENCES

Andreas, E.L., and E.C. Monahan, The role of whitecap bubbles in air-sea heat and moisture exchange, *J. Phys. Oceanogr.*, 30, 433-442, 2000.

Asher, W.E., L.K. Karle, B.J. Higgins, P.J. Farley, I.S. Leifer, and E.C. Monahan, The effect of bubble plume size on the parameterization of air-seawater gas transfer velocities, in *Air-Water Gas Transfer*, edited by B. Jaehne and E.C. Monahan, pp. 227-238, AEON Verlag, Hanau, 1995.

Asher, W.E., and R. Wanninkhof, The effect of bubble-mediated gas transfer on purposeful dual gaseous-tracer experiments, *J. Geophys. Res.*, 103, 10,555-10,560, 1998.

Kerman, B.R., A model of interfacial gas transfer for a well-roughened sea, in *Gas Transfer at Water Surfaces*, edited by W.

Brutsaert and G.H. Jirka, pp. 311-320, Reidel Publishing Co., Boston, MA, 1984.

Leifer, I.S., W.E. Asher, and P.J. Farley, A validation study of bubble mediated air-sea gas transfer modeling for trace gases, in *Air-Water Gas Transfer,* edited by B. Jaehne and E.C. Monahan, pp. 269-283, AEON Verlag, Hanau, 1995.

Liss, P.S., and L. Merlivat, Air-sea gas exchange rates: Introduction and synthesis, in *The Role of Air-Sea Exchange in Geochemical Cycling,* edited by P. Buat-Menard, pp. 113-129, Reidel Publishing Co., Boston, MA, 1986.

Monahan, E.C., From the laboratory tank to the global ocean, in *Climate and Health Implications of Bubble-Mediated Sea-Air Exchange,* edited by E.C. Monahan and M.A. Van Patten, pp. 43-63, Connecticut Sea Grant, Groton, CT, 1989.

Monahan, E.C., Occurrence and evolution of acoustically relevant subsurface bubble plumes and their associated, remotely monitorable, surface whitecaps, in *Natural Physical Sources of Underwater Sound,* edited by B.R. Kerman, pp. 503-517, Kluwer Academic Publishers, Dordrecht, 1993.

Monahan, E.C., and M. Lu, Acoustically relevant bubble assemblages and their dependence on meteorological parameters, *IEEE J. Oceanic Eng.,* 15, 340-349, 1990.

Monahan, E.C., and I.G. O'Muircheartaigh, Optimal power-law description of oceanic whitecap coverage dependence on wind speed, *J. Phys. Oceanogr.,* 10, 2094-2099, 1980.

Monahan, E.C., and I.G. O'Muircheartaigh, Whitecaps and the passive remote sensing of the ocean surface, *Int. J. Remote Sens.,* 7, 627-642, 1986.

Monahan, E.C., and M.C. Spillane, The role of oceanic whitecaps in air-sea gas exchange, in *Gas Transfer at Water Surfaces,* edited by W. Brutsaert and G.H. Jirka, pp.495-503, Reidel Publishing Co., Dordrecht, 1984.

Monahan, E.C., and T. Torgersen, The enhancement of air-sea gas exchange by oceanic whitecapping, in *Air-Water Mass Transfer,* edited by S.C. Wilhelms and J.S. Gulliver, pp. 608-617, American Society of Civil Engineers, New York, 1990.

Peng, T.-H., W.S. Broecker, G.G. Mathieu, Y.-H. Li, and A.E. Bainbridge, Radon evasion rates in the Atlantic and Pacific Oceans as determined during the GEOSECS program, *J. Geophys. Res.,* 84, 2471-2486, 1979.

Schluessel, P., and H. Luthardt, Surface wind speeds over the North Sea from Special Sensor Microwave/Imager observations, *J. Geophys. Res.,* 96, 4845-4853, 1991.

Smethie, W.M., T. Takahashi, D.W. Chipman, and J.R. Ledwell, Gas exchange and CO_2 flux in the tropical Atlantic Ocean determined from [exp222]Rn and pCO_2 measurements, *J. Geophys. Res.,* 90, 7005-7022, 1985.

Wanninkhof, R., Relationship between gas exchange and wind speed over the ocean, *J. Geophys. Res.,* 97, 7373-7381, 1992.

Wanninkhof, R., and W.R. McGillis, A cubic relationship between air-sea CO_2 exchange and wind speed, *Geophys. Res. Lett.,* 26, 1889-1892, 1999.

Woolf, D.K., Bubbles and their role in gas exchange, in *The Sea Surface and Global Change,* edited by P.S. Liss and R.A. Duce, pp. 173-206, Cambridge University Press, 1997.

Wu, J., Oceanic whitecaps and sea state, *J. Phys. Oceanogr.,* 9, 1064-1068, 1979.

Yu, S.L., J.M. Hamrick, and D.S. Lee, Wind effects on air-water oxygen transfer in a lake, in *Gas Transfer at Water Surfaces,* edited by W. Brutsaert and G.H. Jirka, pp. 357-367, Reidel Publishing Co., Dordrecht, 1984.

E.C. Monahan, Sea Grant Office, University of Connecticut at Avery Point, 1084 Shennecossett Road, Groton, CT 06340-6097.

Fractional Area Whitecap Coverage and Air-Sea Gas Transfer Velocities Measured During GasEx-98

William Asher[1], James Edson[2], Wade McGillis[2], Rik Wanninkhof[3], David T. Ho[4], and Trina Litchendorf[1]

GasEx-98 was an air-sea exchange process cruise conducted aboard the NOAA ship *Ronald H. Brown* in the North Atlantic during May and June of 1998. During the cruise, air-sea gas transfer velocities for carbon dioxide were measured using the direct-covariance method. Because the sampling times for the covariance method are on the same order as the timescales of changes in meteorological forcing, the GasEx-98 results provide a unique data set for investigating whether changes in different forcing mechanisms correlate with changes in gas transfer. In particular, fractional area whitecap coverage, W_C, was measured during daylight hours using a dual-camera video system mounted on a bow tower. Several high wind speed events occurred during the cruise, and the resulting correlation between wind speed and W_C is consistent with previous oceanic measurements. The whitecap coverage data were combined with the wind speed records and these data were used in a parameterization of whitecap-mediated gas transfer to predict transfer velocities. These predicted transfer velocities are in good agreement with the transfer velocities derived from the direct-covariance data.

1. INTRODUCTION

The flux, F, of CO_2 is determined by the product of the transfer velocity, k_L, and the air-sea concentration difference of CO_2, ΔpCO_2, as

$$F = k_L s \Delta pCO_2 \tag{1}$$

[1]Applied Physics Laboratory, University of Washington, Seattle, Washington
[2]Woods Hole Oceanographic Institution, Woods Hole, Massachusetts
[3]NOAA/AOML, Miami, Florida
[4]Lamont-Doherty Earth Observatory, Columbia University, Palisades, New York

where s is the solubility of CO_2, k_L parameterizes the dependence of the flux on the physical forcing mechanisms and ΔpCO_2 represents the thermodynamic driving potential. Increasing wind speeds, which are associated with the presence of waves and whitecaps, generate turbulence and bubbles. Both bubbles and turbulence have been shown to increase the gas flux. Therefore, at high wind speeds it is possible that increases in k_L are correlated with increases in fractional area whitecap coverage, W_C. However, previous attempts to relate k_L to W_C or to develop a model for k_L that incorporates the effects of whitecaps have been hampered by the relatively long averaging times inherent in oceanic measurements of gas transfer made using water-side tracers [*Asher and Wanninkhof*, 1998].

In May and June of 1998, the research cruise GasEx-98 aboard the NOAA Ship *Ronald H. Brown* conducted gas transfer process studies in the North Atlantic. During GasEx-98, k_L for CO_2 was measured using the direct covariance method [*McGillis et al.*, 2000], which provided hourly

Figure 1. Sea surface video image taken during GasEx-98 at 09:54:49 GMT on YD 160. The wind speed was 13.4 m s^{-1}.

averaged measurements of the transfer velocity. *Wanninkhof and McGillis* [1999] found that when measured over this relatively short timescale, k_L was correlated with the cube of the wind speed, U. They suggested that one explanation for this cubic dependence was breaking waves increasing gas transfer.

In addition to the gas transfer measurements, the fractional area coverage of stage-A whitecaps [*Monahan*, 1993], W_C, was measured using video photography. Because both measurements are derived over the same measurement timescales, the combination of the whitecap coverage and gas transfer data sets has allowed study of the relationship of k_L with W_C. In this paper, results from the analysis of the whitecap video photography are presented. Whitecap coverages derived from this analysis are used in the whitecap gas transfer parameterization proposed by *Asher and Wanninkhof* [1998] to predict k_L.

2. METHODS

Whitecap coverage during daylight hours was estimated from images of the sea surface acquired using two video cameras (Model 2122, Cohu Electronics, San Diego, California) mounted on the port and starboard sides of the topmost railing of the bow tower on the *Brown*. The height of the cameras with respect to the mean water level was approximately 17 m, each mounted at an incidence angle of 50°. The vertical field of view of each camera was 30° so that they did not image the bow wake of the ship except during large rolls. Water surface images were digitized every 30 s at a resolution of 640×480 pixels, with alternate images taken from the port and starboard cameras.

Figure 1 shows an image taken during GasEx-98 on Year Day (YD) 160 at 09:54:49 GMT. Areas of active wave breaking are readily identified as the white areas in the lower half of the image. A gray-scale threshold procedure similar to that described by *Monahan* [1993] and *Asher and Wanninkhof* [1998] was used to analyze the images. Because the average brightness of an image increased with incidence angle, in general the top half of the image could not be analyzed for whitecap coverage. Therefore, a region-of-interest (ROI) was defined as a rectangle centered horizontally within the image beginning at row 275. The size of the ROI was 350 pixels wide by 115 pixels high. The ROI is shown as the white rectangle in Figure 1.

Five days with relatively high wind speeds, YD 155, YD 158, YD 159, YD 160 and YD 163, were chosen for analysis. Images from each day were evaluated to reject large time periods with contamination from sun glint, specular reflection of sky and clouds, or water droplets on the camera housing window. Whitecap grayscale thresholds were set by manual inspection. Using these manually determined thresholds, the images were automatically segmented and the fractional area containing breaking waves was measured. Figure 2 shows a time series from YD 159 of hourly averaged W_C values. Data points where the W_C value was greater than three times the average value at that time were noted and the individual images examined to ensure ship wake, sun glint, specular reflection, or other bright objects were not counted as whitecaps. Once the time series for W_C had been validated, hourly averaged values for W_C were re-computed.

Also shown in Figure 2 are the grayscale thresholds determined for the port-side camera and the hourly averaged

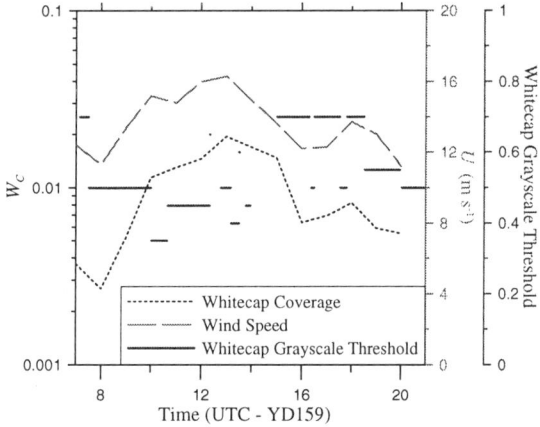

Figure 2. Plot of hourly averaged whitecap coverage, W_C, wind speed, U, and whitecap grayscale threshold from the GasEx-98 data set.

values for U. The hourly averaged values for W_C are seen to correlate quite well with changes in U, despite the variations in the grayscale threshold. This is significant because the thresholds were chosen blind from wind speed. This shows that even when whitecap thresholds are chosen by subjective analysis, it is possible to obtain data for W_C that objectively tracks changes in whitecap coverage.

3. RESULTS

Figure 3 shows $W_C^{1/3}$ plotted versus U. As found by *Monahan* [1993] and *Asher and Wanninkhof* [1998], $W_C^{1/3}$ increases linearly with increasing U. This linear relationship implies that W_C can be estimated from U using a relationship of the form

$$W_C = a(U - b)^3 \qquad (2)$$

where a and b are constants that can be determined from a linear regression of the data in Figure 3. This regression is shown on the figure as the solid line, and from the regression it was determined that $a = 3.7 \times 10^{-6}$ and $b = 1.2$. The results from similar regressions of the data from *Monahan* [1993] and *Asher and Wanninkhof* [1998] show that the W_C values from GasEx-98 are consistent with previous whitecap data sets.

The uncertainty in the measurement of W_C was estimated by comparing the hourly averaged values of whitecap coverage from the port and starboard cameras. The average ratio of concurrent port to starboard W_C values was found to be 1.09 with a standard deviation of 0.71. This showed that the port and starboard cameras gave consistent measurements and that the relative uncertainty in an average value of W_C was 0.71.

Asher and Wanninkhof [1998] developed an empirical parameterization for predicting k_L for any liquid-phase rate-controlled gas forced predominately by wind that includes the effects of turbulence and bubbles generated by breaking waves. Defined in terms of U and W_C, this relationship has the form

$$k_L = \left(47U + \left(1.5 \times 10^5 W_C - 47U\right)\right) Sc^{-1/2} + W_C \left(-\frac{37}{\alpha} + 10{,}440\, \alpha^{-0.41} Sc^{-0.24}\right) \qquad (3)$$

where Sc and α are the Schmidt number and Ostwald solubility, respectively, of the gas. The first term on the

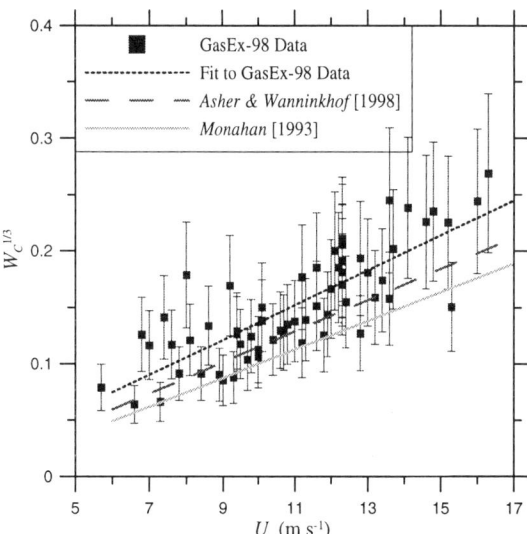

Figure 3. Plot of the cube root of the hourly averaged fractional area whitecap coverage, W_C, plotted versus wind speed, U. The error bars represent ±1 standard deviation. Also shown on the graph are the least-squares regressions of the GasEx-98 data (from which it is found that $W_C = 3.7 \times 10^{-6}(U-1.2)^3$ ($r^2=0.65$)), the *Asher and Wanninkhof* [1998] ($W_C = 2.6 \times 10^{-6}(U-1.7)^3$) data, and the data sets in *Monahan* [1993] ($W_C = 2.0 \times 10^{-6}(U-2.2)^3$).

right-hand side of (3) parameterizes the effect of the wind-generated and whitecap-generated turbulence on k_L and the second term parameterizes the effects of solubility-dependent bubble-mediated exchange. *Asher et al.* [1996] have shown that a relation of this form can be used to describe k_L for invasion or evasion, provided the air-water concentration difference is not close to equilibrium. During GasEx-98, sea surface equilibrators and air-phase measurements showed $\Delta pCO_2 \approx -90$ µatm, which implies a large net ocean sink. *Asher et al.* [1996] measured k_L for invasion of CO_2 in the presence of simulated breaking waves with $\Delta pCO_2 \approx -150$ µatm. Because of the similarity of ΔpCO_2 between the data from *Asher et al.* [1996] and GasEx-98, the values for the coefficients used in (3) are those from *Asher et al.* [1996] for invasion in seawater.

Using (3), k_L for invasion of a gas can be estimated provided U and W_C are known. The GasEx-98 data set allows two methods for predicting k_L. Figure 4 shows $k_L(600)$ plotted as a function of U where $k_L(600)$ (i.e., k_L for a gas with $Sc=600$ and $\alpha=0.87$) has been estimated using (3), U as measured during GasEx-98, and W_C calculated using (2) with the values for a and b given above. For comparison, the hourly averaged $k_L(600)$ values measured at the same time as W_C as reported by *Wanninkhof and McGillis* [1999] are

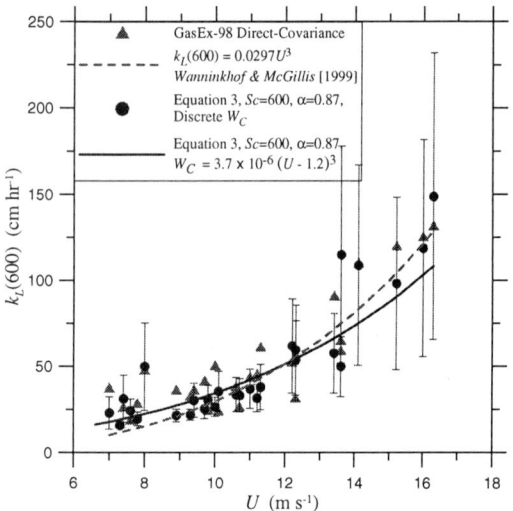

Figure 4. Plot of the gas transfer velocity at a Schmidt number of 600, $k_L(600)$, versus wind speed, U. The figure shows $k_L(600)$ measured using the direct-covariance method [*McGillis et al.*, 2000], $k_L(600)$ predicted from (3) with whitecap coverage, W_C, estimated using (2) and the coefficients for the GasEx-98 whitecap data, and $k_L(600)$ predicted using (3) and the measured values for U and W_C. The dashed line is $k_L(600)$ calculated using the cubic relation of *Wanninkhof and McGillis* [1999] at $Sc=600$.

plotted in Figure 4 as a function of U. Equation 3 closely follows the observed cubic relationship of $k_L(600)$ with respect to U.

An alternative method of applying (3) to the GasEx-98 data set is to predict $k_L(600)$ at discrete times from the directly measured values of W_C and U. Then, the predicted values of $k_L(600)$ can be compared to the concurrent direct-covariance measurement of $k_L(600)$. These discrete values of $k_L(600)$ are shown plotted as a function of U in Figure 4. Comparison of the predicted and measured transfer velocities shows there is good agreement.

Figure 5 shows the discrete values for $k_L(600)$ predicted using (3) as described above, $k_L(600\text{-predicted})$, plotted versus the directly measured $k_L(600)$, $k_L(600\text{-measured})$. Although there is some scatter, the direct comparison demonstrates that (3) works well in estimating the measured transfer velocities. This suggests that whitecap related processes may be the cause of the observed cubic dependence of $k_L(600)$ on U.

Finally, the influence of whitecaps on k_L for CO_2 should not be taken to imply that bubble-mediated transfer processes are the major gas transfer pathway. One of the benefits of using (3) is that the relative importance of solubility-dependent bubble processes in determining the overall transfer velocity can be estimated as

$$\Phi_B = \frac{W_C\left(\frac{-37}{\alpha} + 10{,}440\alpha^{-0.41}Sc^{-0.24}\right)}{k_L} \quad (4)$$

where Φ_B is the fraction of k_L due to bubble processes, where k_L is estimated using (3). Figure 6 shows Φ_B calculated from the discrete $k_L(600)$ values plotted versus U. The model implies that the fractional contribution of bubbles to the total transfer velocity is limited to a maximum value of 25%. This is consistent with accepted theories of bubble-mediated gas transfer, which state that bubble processes are less important for relatively soluble gases such as CO_2 [*Memery and Merlivat*, 1985; *Woolf*, 1993].

4. CONCLUSIONS

The whitecap coverage analysis has shown that objective values of W_C can be obtained from video images of the sea surface. Furthermore, the GasEx-98 whitecap results provide further support that W_C scales with U^3 as proposed by *Monahan* [1993]. The GasEx-98 whitecap data are in good agreement with previous oceanic estimates of W_C.

There is excellent agreement between the transfer velocities measured by the direct-covariance technique and k_L values calculated using the *Asher and Wanninkhof* [1998] whitecap parameterization with directly measured values for

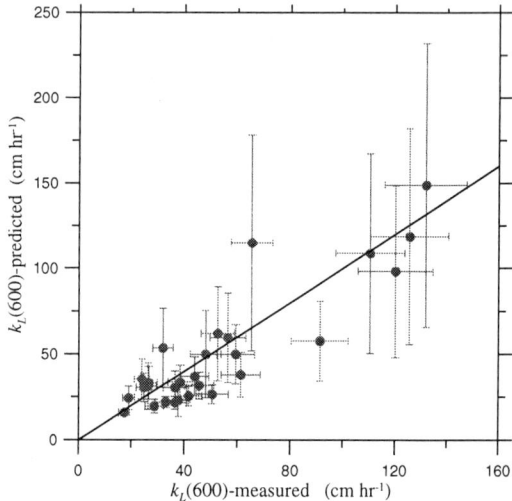

Figure 5. Plot of the gas transfer velocity at a Schmidt number of 600, $k_L(600)$, predicted from (3) using the measured values for wind speed and whitecap coverage, $k_L(600\text{-predicted})$, plotted versus $k_L(600)$ determined by the direct-covariance method, $k_L(600\text{-measured})$.

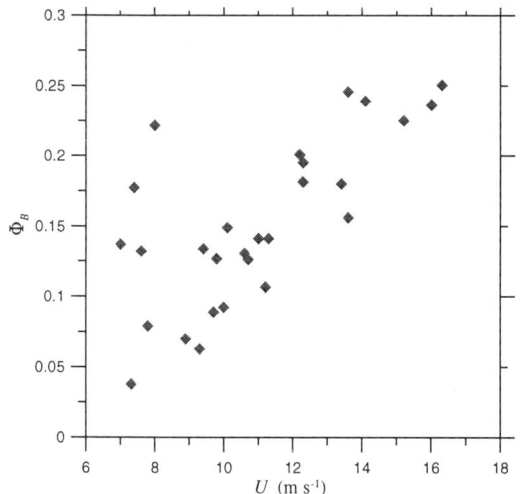

Figure 6. The fraction of the total gas transfer velocity that is due to bubble mediated processes, Φ_B, as estimated using (4).

W_C and U. This agreement provides support for the hypothesis of *Wanninkhof and McGillis* [1999] that an observed cubic dependence of k_L on U could be due to the effects of breaking waves. However, the model used here implies that the primary influence of breaking waves on the air-sea exchange of CO_2 is via turbulence, not bubbles.

Acknowledgments: This work was supported by the National Science Foundation under Grant OCE-9633423 and Grant OCE-9711218 and the NOAA/OGP Ocean-Atmosphere Carbon Exchange Study. (Woods Hole Oceanographic Institution contribution number 10339.)

REFERENCES

Asher, W.E., L.M. Karle, B.J. Higgins, P.J. Farley, E.C. Monahan, and I.S. Leifer, The influence of bubble plumes on air-seawater gas transfer velocities, *J. Geophys. Res.*, *101*, 12,027-12,041, 1996.

Asher, W.E., and R. Wanninkhof, The effect of bubble-mediated gas transfer on purposeful dual gaseous-tracer experiments, *J. Geophys. Res.*, *103*, 10,555-10,560, 1998.

Memery, L., and L. Merlivat, Modeling of the gas flux through bubbles at the air-water interface, *Tellus, Ser. B*, *37*, 272-285, 1985.

McGillis, W.R., J.B. Edson, J.E. Hare, and C.W. Fairall, Direct covariance air-sea CO2 fluxes, Submitted to *J. Geophys. Res.*, 2000.

Monahan, E.C., Occurrence and evolution of acoustically relevant sub-surface bubble plumes and their associated, remotely monitorable, surface whitecaps, in *Natural Physical Sources of Underwater Sound*, edited by B.R. Kerman, pp. 503-517, Kluwer Academic Publishers, Norwell, 1993.

Wanninkhof, R., and W.R. McGillis, A cubic relationship between air-sea CO2 exchange and wind speed, *Geophys. Res. Lett.*, *26*, 1889-1892, 1999.

Woolf, D.K., Bubbles and the air-sea transfer velocity of gases, *Atmos.-Ocean*, *31*, 517-540, 1993.

W. Asher and T. Litchendorf, Applied Physics Laboratory, University of Washington, 1013 NE 40th Street, Seattle, WA 98105; asher@apl.washington.edu

J. Edson and W. McGillis, Dept. of Applied Ocean Physics and Engineering, Woods Hole Oceanographic Institution, Woods Hole, MA 02543; jedson@whoi.edu and wmcgillis@whoi.edu

D.T. Ho, Lamont-Doherty Earth Observatory, Columbia University, Route 9W, Palisades, NY 10964; daveho@ldeo.columbia.edu

R. Wanninkhof, NOAA/AOML, 4301 Rickenbacker Causeway, Miami, FL, 33149; wanninkh@aoml.noaa.gov

Gas Transfer in Energetic Conditions

David Kevin Woolf

James Rennell Division, Southampton Oceanography Centre, Southampton, United Kingdom

Surface kinetics and surface deformation and break-up (waves, bubbles and droplets) are identified as factors in determining air-water exchange coefficients. Both factors are likely to be highly significant in the most energetic conditions (e.g., storms at sea). Mixing across boundary layers usually limits exchange, but interfacial processes will be limiting at very high levels of stirring. For a planar interface, surface kinetics will set an asymptotic limit on exchange coefficients. An enhancement in interfacial area due to surface deformation and break-up will enhance exchange irrespective of surface kinetics. Bubbles are likely to be the dominant form of enhanced interfacial area for natural wind-driven air-water exchange. A combination of surface kinetics and surface extension may explain deviations of exchange rate from the behaviour expected from "stirring only" models for highly energetic conditions.

1. INTRODUCTION

1.1. Mixing, Surface Kinetics and Deformation

One description of air-water exchange of an unreactive gas is simply as a mixing process involving two media. For example, transfer of a gas molecule from the "bulk" of the atmosphere to the bulk of the upper ocean can be achieved by a combination of stirring and molecular diffusion, first across the marine atmospheric boundary layer and then across the marine microlayer. Given that the solubility of the gas and molecular diffusion in the two media are understood, in this view our interests narrow to understanding the stirring process (i.e., turbulent transfer) in the two boundary layers. Further, many of the models of these stirring processes treat the air-water interface as a simple plane whose sole influence is to impose boundary conditions on the turbulent motions in the boundary layers.

This paper looks beyond the stirring processes (I shall introduce only the simplest generic model of mixing) in order to consider additional influences of the surface on exchange coefficients. Firstly, I discuss the kinetics of molecules in the surface. It is proposed that while these kinetics do not have a discernible influence in most air-water gas exchange experiments, it is likely that in energetic conditions (e.g., marine storms), the influence of surface kinetics will be manifested at least for some gases. Secondly, I consider the effect of deformation and break-up of the interface on exchange simply through the extension of interfacial area. In air-sea exchange, most of the additional interface will take the form of bubbles and drops. The influence of bubbles in air-water gas exchange has been identified before. It is reviewed again here, in the context of its role in modifying and masking the effect of surface kinetics. Laboratory measurements of gas transfer velocity at high wind stress are interpreted in terms of the combined effects of stirring, surface kinetics and deformation. Finally we summarise and discuss the role of surface kinetics in air-water gas exchange in energetic conditions.

1.2. Resistance Models

The air-water transfer of gases is commonly described by the equation:

$$F = -K_T \Delta C$$

The net air-water flux, F (mol m^{-2} s^{-1}), is related to the concentration difference ΔC by a constant describing the exchange rate: the transfer velocity, K_T. The "force", $\Delta C = C_w - C_a/H$, is analogous to the electromotive force in an electrical circuit, with F corresponding to the current [*Danckwerts*, 1951; *Liss*, 1983]. (C_w is the concentration in the bulk of the water (mol m^{-3}), C_a is the concentration in the air and H is the Henry's law constant). Just as in an electrical circuit the potential may be dropped over a number of resistances, the concentration difference will be dropped between the water and the air according to the local exchange characteristics, with the greatest drop where exchange is slowest (or the resistance to exchange is greatest). Where a number of processes must occur in series, they may be represented by a series of resistances, R_j. A transfer velocity, K_j, may be defined for each process, where at steady state (a constant gas flux), this transfer velocity is related to the concentration drop across this process by $F = -K_j \Delta C_j$. For a series of processes:

$$R_j = 1/K_j$$

$$1/K_T = \sum_j 1/K_j = \sum_j R_j$$

If the exchange of gas may occur by more than one independent pathway, the exchange characteristics of each pathway may be represented by one or more resistances in series, and the separate pathways may be treated like resistances in parallel. This simple framework is quite powerful, and is only inadequate where processes are not independent.

The most common representation of air-sea exchange is one in which exchange is limited by mixing across boundary layers on each side of the interface. In this case, it is common to represent the exchange characteristics in each boundary layer by a resistance, R_a and R_w. The resistance on the air-side, R_a, depends on mixing in the air-side boundary layer, but also depends in our formalism on the Henry's law constant of the gas. Generally, the air-side resistance is more substantial compared to the water-side resistance for more soluble gases. Where one of a series of resistances is much greater than the sum of the other resistances in that series, this single resistance will dominate and the associated process is described as limiting. In air-water exchange, particularly if mixing processes on each side of the interface are coupled (e.g., wind-driven mixing in air-sea exchange), many gases can be readily identified as limited by mixing on only one side of the interface. A large number of important and poorly soluble gases are generally assumed to be controlled by water-side mixing [*Liss*, 1983].

2. SURFACE KINETICS

2.1. Kinetics of Molecules in the Surface

Surface processes include all processes, additional to ordinary molecular or turbulent diffusion, that must or can occur during migration between a free gas molecule above the interface and a dissolved molecule, remote from surface forces. These processes also affect the solubility of a gas in a liquid [*Pollack*, 1991], but here we are concerned primarily with the kinetic rates rather than the resulting equilibrium.

Consider first the rate of transfer of an unreactive gas if there are no special barriers to exchange. From the kinetic theory of gases, the flux of molecules striking the water from above, dn/dt, is related to the partial pressure of the gas, p, by

$$dn/dt = p\,(2\pi m k T)^{-1/2}$$

Here, k is the Boltzmann constant, T is absolute temperature and m is the mass of one molecule [*Davies and Rideal*, 1963]. At a maximum, we may expect this flux to enter the surface, balanced at saturation by desorption from the surface into the gas phase. If diffusion from the absolute surface, to the adjacent layers of molecules is identical to diffusion within the interior of the water, then no additional processes are involved in exchange.

Inhibited exchange of an unreactive gas may involve the rejection of a large fraction of the gas molecules striking the surface, or reduced mobility between the surface and adjacent layers [*Danckwerts*, 1951]. At the core of most theories of inhibited exchange (especially for chemically inert gases) is the fact that the opening of cavities in the liquid, as is necessary for migration, requires a lot of energy. Pollack notes that the energy necessary to open a cavity of molecular dimensions in water is typically a factor of 30 greater than the kinetic energy of a monomolecular gas molecule at room temperature. If the "reversible work" required for a gas molecule to enter the liquid is W_a, then the probability of each gas molecule entering the liquid on impact is $\alpha_a = \exp(-W_a/kT)$, which will be very small if $W_a \gg kT$, and will reduce the flux to:

$$dn/dt = \alpha_a\, p\, (2\pi m kT)^{-1/2}$$

This model of exchange is used in studies of atmosphere-particle gas dynamics [Knipping et al., 2000], where α_a is termed the "mass accomodation coefficient". Reduced mobility between the surface and adjacent layers in the water is less well defined. A number of factors can be identified. There may be differences in diffusion very close to the surface simply because of the altered geometry. Note, for instance, a molecule in the interior may "diffuse" without ever moving relative to its immediate neighbours, simply as a result of the mobility of more distant molecules; but in order to reach or leave the surface, the molecule must displace an immediate neighbour. There are also alterations in intermolecular forces and the arrangement of water molecules as the surface is approached [Horne, 1969]. The influence of these changes is unclear but some reduction in mobility is likely. For surface-active molecules, an inhibition of migration between the surface and the adjacent layer is associated with the energy of desorption. Thus, the average time for migration, t, over an intermolecular distance, λ, is usually related to the molecular diffusion coefficient, D, by $t = \lambda^2/D$, but if W_w is the energy of desorption, then $t = \lambda^2/(D\, \alpha_w)$ where $\alpha_w = \exp(-W_w/KT)$. This relationship can be used as a simple model of reduced mobility applicable to unreactive gases, where α_w is a general inhibition factor for mobility between the surface and adjacent layers.

The models of inhibited exchange described above can be manipulated to calculate resistances (R_{ia} and R_{iw}):

$$R_{ia} = (1/\alpha_a H)\,(2\pi m/kT)^{1/2}$$

$$R_{iw} = \lambda/(\alpha_w\, D)$$

A significant effect on air-sea exchange may be expected where the resistance exceeds 1s/mm (see next section), which requires major inhibition. In the case of R_{ia}, the resistance will generally be greater for more soluble gases. (Note, however, that α_a is related to the energy required to create a surface cavity, while this energy partly determines the solubility of the gas [Pollack, 1991]). For $H = 1$, an inhibition $\alpha_a \leq 10^{-5}$ is required. Alternatively, inhibited mobility within the water with $\alpha_w \leq 3 \times 10^{-4}$ will have a comparable effect.

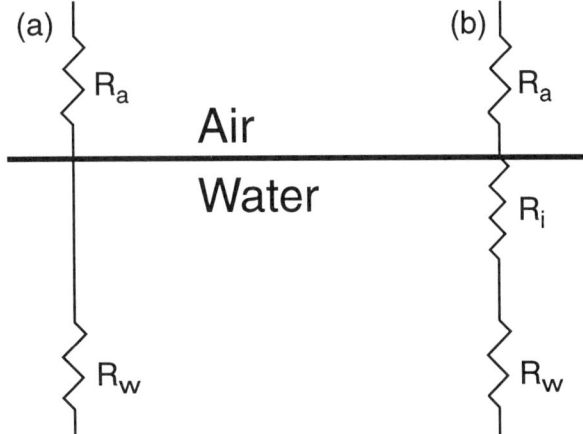

Figure 1. Standard (a) and modified (b) resistance models of air-water gas exchange

2.2. Air-Water Exchange Including Interfacial Resistance

Let us consider the effect on air-water exchange coefficients of small but finite values of interfacial resistance. The standard model of air-water exchange is modified by adding an additional interfacial resistance (Figure 1). This model has been used before, usually to model diffusion across a surface film [Davies and Rideal, 1963], but here we apply it to the inhibition described in the previous section. A total interfacial resistance $R_i = R_{ia} + R_{iw}$ is supposed to be independent of stirring. For simplicity, let us consider only gases that are usually considered to be controlled by liquid-side processes, i.e., $R_a \ll R_w$ [Liss, 1983]. Also, let us assume a simple case of wind-driven stirring where $K_w = a u_{*w}$ where u_{*w} is the friction velocity in water (and is proportional to the turbulent velocity fluctuation in the liquid boundary layer) and "a" is a constant. The value of "a" will depend on the Schmidt number of the gas and probably some additional characteristics of the stirring, but $a = 0.01$ is an approximate value. We can calculate the relationship $K_T(u_{*w})$ for a number of fixed values of R_i (Figure 2). At moderate values of wind forcing ($u_{*w} = 10$mm/s corresponds to a wind speed of ≈ 8m/s over the sea), low or moderate values of interfacial resistance (≤ 1s/mm) slightly reduce transfer coefficients. However, the departure from a simple proportional relationship to friction velocity is so slight that it may easily be overlooked in experiments. Clear experimental evidence of the influence of interfacial resistance may be limited to experiments at high wind stresses (or other energetic stirring) on gases with relatively high interfacial resistance (> 1 s/mm).

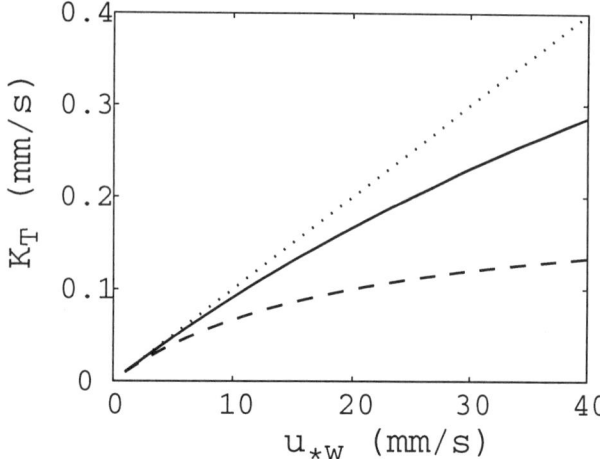

Figure 2. The relationship of transfer velocity to friction velocity in water for a single boundary layer model but different values of interfacial resistance. (dotted line, $R_i = 0$; unbroken curve, $R_i = 1s/mm$; dashed, $R_i = 5s/mm$).

3. DEFORMED AND BROKEN INTERFACES

3.1. Wavy Surfaces

It is common to make an association between surface waves and air-sea gas exchange. Aside from the practical advantages of using surface waves to parameterise gas exchange, we need to be clear about the direct relationship between surface waves and surface exchange. As already discussed air-sea exchange is a result of mixing and surface kinetics. The relationship of waves (especially breaking waves) to turbulence and mixing below the interface is beyond the scope of this paper. Instead, the direct effect of extension of the air-water interface by surface waves on exchange is examined.

It is fundamental to mixing that the process is completed at the molecular level. The role of stirring (or turbulence) can be thought of solely as continually generating contact between fluid of different properties, thus assisting molecular diffusion in eroding gradients in those properties. This process will occur in the interior of the fluids by turbulence, but surface waves themselves slightly increase contact area between the air and water. In the case of surface kinetics, exchange will be proportional to the interfacial area. The effect on mixing rates is less obvious. An "effective boundary layer thickness" must be identified, i.e., the layer that crucially determines exchange rates. If, for example, the entire marine atmospheric boundary layer determines mixing rates, surface waves will not noticeably enhance mixing. On the other hand, if the layer thickness is very thin (much thinner than the radius of curvature of the interface), then the mixing rate will be closely proportional to the interfacial area. The latter approximation should be adequate for the exchange of poorly soluble gases (controlled by transport in the marine microlayer). We can differentiate between interfacial transfer velocity, J, which describes the local exchange per unit interfacial area, and K_T, which describes exchange per unit planar area. J depends on mixing through the boundary layers and surface kinetics. K_T depends additionally on the fractional increase of interfacial area. For poorly soluble gases, the approximation $K_T = A\ J_m$ is reasonable, where J_m is the mean value of J, and A the area of interface per unit planar area. For a wavy surface of small slope, A will be related to the mean square slope, sq, by $A \approx 1 + sq/2$. The relationship of mean square slope of the sea surface to the wind stress has been investigated in several field and laboratory studies. An approximate description is given by (u_{*w} in m/s):

$$A \approx 1 + 1.5\ u_{*w}$$

The enhancement in interfacial sea surface area by surface waves is very small at reasonable wind speeds, and can be neglected without large error. The enhancement of interfacial area by the generation of bubbles and drops at the sea surface is more significant, but the resulting air-sea exchange is relatively complicated. The basic properties are considered in the next section.

3.2. Bubbles and Droplets

Air-sea exchange is enhanced by transport across the surface of the bubbles and drops. However, there are a number of complications for this indirect ("mediated") exchange between the atmosphere and ocean, which we describe below.

The concentration difference across the bubble or drop surface will not be the same as that across the sea surface. The gas concentration in bubbles is modified by their compression and by exchange of gas through their lifetime. In the case of a drop, it may not begin with the composition of the bulk water, and it will evolve in the atmosphere. Each gas in a bubble or drop will tend towards equilibrium with the surrounding fluid, and generally this will suppress the transfer. For the exchange of poorly soluble gases by bubbles, the equilibration places a maximum limit on the contribution to the transfer velocity of a trace gas, $K_b < V_b H$ [*Woolf*, 1997], where V_b is the "volume flux" ($m^{-2}s^{-1}$) of bubbles flushed through the sea surface. Similarly, for drops, $K_d < V_d$. The interfacial area associated with small (< 10μm radius) brine droplets is quite large, but the volume flux is far too small for them to make a significant contribution to the air-sea exchange of gases. The volume or mass flux of large spray drops can be more substantial, but they will only be airborne for a short period. In summary, only bubbles will significantly enhance the exchange of poorly soluble gases, and this enhancement will be smaller for more soluble gases.

Bubble generation will be very sensitive to wind forcing. A simple model of bubble-mediated gas exchange [*Woolf*, 1997] describes a contribution to the transfer velocity of the gas proportional to the cube of the friction velocity. The wind dependence arises from the sensitivity of the bubble statistics to the wind. The constant of proportionality will depend on the solubility and diffusion coefficient of the gas, and may also depend on interfacial resistance.

4. EXCHANGE IN ENERGETIC CONDITIONS

4.1. A Composite Model

I present a simple model of the air-sea exchange of poorly soluble gases that incorporates the principal effects of wind stirring, interfacial resistance and bubble-mediated transfer. This model is not expected to accurately describe any gas, but is useful to examine the net influence of the three main elements, and how this influence depends on values of coefficients. The simple model is given by:

$$K_T = K_o + K_b$$

$$1/K_o = 1/K_w + R_i$$

$$K_w = au_{*w}$$

$$K_b = bu_{*w}^3$$

The coefficients a, b and R_i determine the influence of wind stirring, surface extension (here assumed to be due solely to bubbles) and interfacial resistance respectively. The influence of wind stirring is common to all gases with relatively little variation. The influences of bubbles and of interfacial resistance are both likely to vary greatly between gases. We can explore the resulting variation with wind stress by comparing predictions resulting from different sets of coefficients (Figure 3). Where the interfacial resistance is quite small, and bubble-mediated exchange is significant, the inhibition by interfacial resistance is "hidden" by the enhancement associated with bubbles. If the interfacial resistance is large, its inhibiting effect will be apparent, but surface extension will enhance exchange at sufficiently high wind stress.

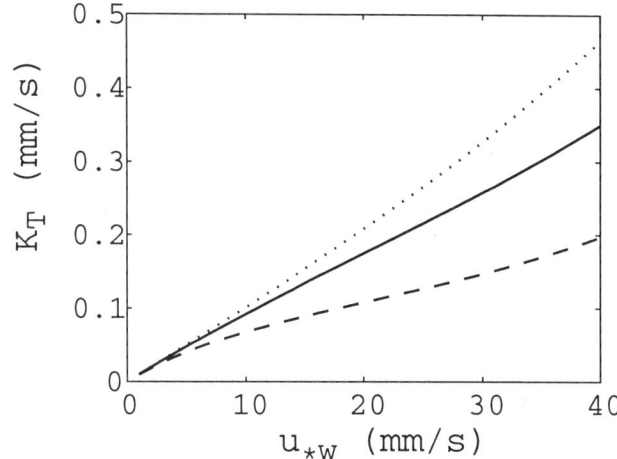

Figure 3. Transfer velocity plotted against friction velocity. (a = 0.01; b = 10^{-6}; dotted curve, R_i = 0; unbroken curve, R_i = 1s/mm; dashed curve: R_i = 5s/mm).

4.2. Laboratory Experiments

The primary role of wind-driven stirring is established. There is also strong evidence for bubble-mediated transport in some laboratory experiments. Here, we note that some laboratory results also suggest a role for interfacial resistance. Indeed, the incentive for this study arose from laboratory results on the exchange of several gases at high wind stress; part of the "Luminy" study [*de Leeuw et al.*, 2001]. Two features of the relationship between transfer velocity and wind stress may imply a significant interfacial resistance for at least some of the gases. Firstly, the transfer velocity of gases with a high interfacial resistance may exhibit low sensitivity to friction velocity at high wind stress. Secondly, if the ratios of the transfer velocity of different gases at high wind stress can not be understood in terms of stirring and bubble-mediated exchange, then another process such as surface kinetics is implied. Unless the interfacial resistance is very large, experimental evidence will be difficult to gather. Very accurate estimates of transfer velocities and friction velocity are necessary.

Measurements of the transfer of sulfur hexafluoride and several chlorinated volatiles [*Alaee et al.*, 1995] are interesting since these are relatively large and immobile compounds. Also, though this study does not supply wind stress values, I have calculated friction velocity using the relationship to wind speed reported in a previous study in this facility [*Ocampo-Torres et al.*, 1994]. The results can be described (Figure 4) by a simple "interfacial resistance" model of the form:

$$(600/Sc)^{0.5}/K_T = 1/K_{600} = 1/0.01u_{*w} + R_i$$

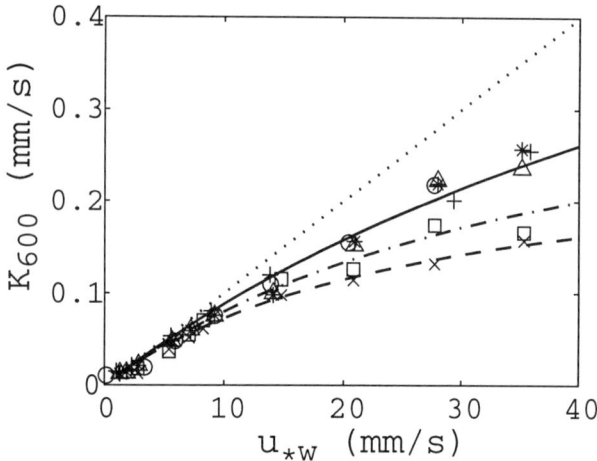

Figure 4. Transfer velocity plotted against friction velocity. Results from a laboratory study, together with a model fit. (SF_6, +; CH_3Cl, o; C_2Cl_4, x; CCl_4, squares; $CHCl_3$, asterices; CH_2Cl_2, triangles; $R_i = 0$, dotted line; $K_i = 0.75$mm/s, unbroken curve; $K_i = 0.4$mm/s, dash-dotted; $K_i = 0.27$mm/s, dashed).

5. CONCLUSIONS

5.1. Summary

Treatment of air-water exchange as a mixing process ignores the kinetics of molecular transfer at the interface. A finite kinetic rate at the surface can be modelled by including an "interfacial resistance" in series with, and intermediate to, the resistances associated with mixing across the two boundary layers. A constant interfacial resistance is more significant where the resistances of the boundary layers are lower. Interfacial resistance may ultimately control exchange rates at extreme rates of stirring.

For poorly soluble gases, deformations of the interface (surface waves) enhance mixing proportionally to the extension of interfacial area. At sea and on lakes, the increase in interfacial area associated with waves is slight. An enhancement in interfacial area is also associated with bubble and aerosol generation. The enhancement of the air-sea exchange of poorly soluble gases by surface extension is dominated by bubble-mediated transport.

Models of the air-sea gas exchange of poorly soluble gases in storm conditions should include wind-driven stirring, surface kinetics and bubble-mediated transfer. The simplest models of wind-driven stirring on both sides of the sea surface predict a simple proportionality between gas transfer velocity and friction velocity. In the absence of any extension of the sea surface, exchange would approach a maximal rate set by the interfacial resistance. An increase in interfacial area (primarily as bubbles) in storm conditions will counteract the suppression by interfacial resistance, and must eventually cause a sharp rise in exchange in extreme conditions irrespective of the interfacial resistance.

5.2. Discussion

A set of processes occurring at air-water interfaces has been identified which can be represented in models of gas transfer by including a term for "interfacial resistance". The mobility of gas molecules is critical to the interfacial resistance. In the absence of reactions or polar tendencies, mobility should be related inversely and nonlinearly with the size of cavity required by a dissolved gas molecule. A high degree of variability in interfacial mobility between gases may be expected. Interfacial resistance may be inversely correlated to the molecular volume or weight. Unfortunately, beyond these vague expectations, I cannot predict values of interfacial resistance. Interfacial resistance is closely related to molecular diffusion in the bulk liquid and (especially) the solubility of the gas. If the various contributions to the solubility of a gas [*Pollack*, 1991] were understood, we should be fairly close to also understanding the surface kinetics of the gas. Direct computation of the interfacial resistance is a future goal.

There is empirical evidence for a significant interfacial resistance only for a few gases. The situation is clouded by our limited understanding of stirring processes at high wind stress and bubble-mediated processes. There is a case for seeking evidence of interfacial resistance in small-scale experiments, where bubbles and droplets may be eliminated and the stirring characteristics carefully measured.

Evidence of interfacial resistance is currently limited to only a few gases. There is a broad inference that interfacial resistance needs to be considered at least for gases of large molecular volume, but for most gases there is no estimate of the interfacial resistance. Also, estimates of interfacial resistance may be biased by inaccurate estimates of stirring or bubble-mediated exchange. All the experiments were carried out in large tanks with contaminated fresh water. Interfacial resistance does not require contamination of the surface, but it is inevitable that surface-active material will influence the surface kinetics. Dissolved salts also have an effect on the properties of water, including its structure near an interface [*Horne*, 1969]. Interfacial resistance is likely to vary significantly with both salinity and temperature.

Interest in the role of surface kinetics in air-water gas exchange should be revived. In particular, understanding of the air-sea exchange of gases at high wind speeds, and the exchange of large-molecular-weight gases in all conditions may be critically dependent on establishing the interfacial resistance.

Acknowledgments. This research was largely carried out in the framework of the LUMINY project. I am grateful to the support and encouragement of my fellow participants in this project, and to the European Commission for financial support (ENV4-CT95-0080).

REFERENCES

Alaee, M., M.A. Donelan and W.M.J. Strachan, Wind and wave effects on mass transfer velocities of halomethanes and SF_6 measured in a gas transfer flume, in *Air-Water Gas Transfer*, edited by B. Jähne and E. C. Monahan, 617-626, AEON Hanau, Germany, 1995.

Danckwerts, P.V., Significance of liquid-film coefficients in gas absorption. *Ind. Engng. Chem.*, 43, 1460-1467, 1951.

Davies, J.T. and E. K. Rideal, *Interfacial Phenomena*, 2nd ed., Academic, New York, 1963.

De Leeuw, G., G. J. Kunz, G. Caulliez, D.K. Woolf, P. Bowyer, I. Leifer. P. Nightingale, M. Liddicoat, T.S. Rhee, M.O. Andreae, S.E. Larsen, F.Aa. Hansen and S. Lund. LUMINY: An overview. In *Gas Transfer at Water Surfaces*, edited by M.A. Donelan, W.M. Drennan, E.S. Saltzman and R. Wanninkhof, pp. xxx-xxx, AGU, this volume, 2001.

Horne, R. A., *Marine Chemistry*, Wiley, New York, 1969.

Knipping, E.M., M.J. Lakin, K.L. Foster, P. Jungwirth, D.J. Tobias, R.B. Gerber, D. Dabdub and B.J. Finlayson-Pitts, Experiments and simulations of ion-enhanced interfacial chemistry on aqueous NaCl aerosols, *Science, 288*, 301-306, 2000.

Liss, P.S., Gas transfer: experiments and geochemical implications, in *Air-Sea Exchange of Gases and Particles*, edited by P.S. Liss and W.G.N. Slinn, D. Reidel, Dordrecht, 241-298, 1983.

Ocampo-Torres, F.J., M.A. Donelan, N. Merzi and F. Jia, Laboratory measurements of mass transfer of carbon dioxide and water vapour for smooth and rough flow conditions, *Tellus, 46B*, 16-32, 1994.

Pollack, G.L., Why gases dissolve in liquids, *Science, 251*, 1323-1330, 1991.

Woolf, D.K., Bubbles and their role in gas exchange, in *The Sea Surface and Global Change* edited by P.S. Liss and R.A. Duce, Cambridge University Press, 173-205, 1997.

D. Woolf, James Rennell Division, Southampton Oceanography Centre, European Way, Southampton, SO14 3ZH, United Kingdom. (e-mail: dkw@soc.soton.ac.uk)

Estimation of Whitecap Coverage Percentage Using Shallow Grazing-Angle Video and FMICW Radar

Craig L. Stevens, M.J. Smith and J.A. McGregor

National Institute of Water and Atmospheric Research, Wellington, New Zealand.

Independent S-band radar and video estimates of active whitecap percentage are derived. The video estimate uses thresholding based on statistical normalization whilst the radar uses a combined velocity-power criterion. Both estimates are influenced by subjective thresholds. The techniques are applied to shore-based and open-ocean situations. Video estimates of whitecap coverage are generally larger than the radar estimates. The broad agreement of the S-band radar data with other published results suggests that radar at this frequency may be more useful than higher-frequency radars for the task.

1. INTRODUCTION

The near-surface turbulence, bubble and aerosol production, resulting from deep-water wave breaking contributes significantly to air-sea exchange [*Melville*, 1996; *Monahan*, 1989]. Additionally, the instantaneous proportion of whitecap coverage of the sea surface is important for modelling wave energy dissipation [*Donelan and Yuen*, 1994]. An important parameter in the evaluation of these processes is the percentage of the water surface covered by breaking events and their residue at any time. This is termed the whitecap coverage percentage (W).

Since the occurrence of breaking at a single point is rare, most techniques utilize some spatial as well as temporal sampling (e.g. video or photographic techniques). The relatively low cost and ease of deployment of visual techniques is attractive and has been employed in a number of studies [*Kraan et al.*, 1996; *Monahan*, 1989]. It is well recognized that a fundamental difficulty in this approach is separating active whitecapping from residual sea-foam. *Monahan* [1989] describes the separation of W into active whitecapping W_A, and residual sea-foam, W_B components and show that they are at least an order of magnitude different. W_A is the focus of this study since it is the parameter most directly related to wave dynamics and gas transfer.

A number of difficulties present themselves to visual shallow grazing angle thresholding techniques. (1) Specular reflection can provide an erroneous signal. (2) The active whitecaps may simply have the same intensity as the sea-foam. (3) Resolution effects may reduce the intensity of the whitecap regions at long ranges. (4) Low grazing-angles mean that breaking waves generate an overestimate since shadowed water behind a breaker will be marked as broken to the observer. (5) Any significant platform motion will change the field of view and modulate the effects of (1), (3) and (4). (6) Finally, numerically, W_A is generally quite small. For example *Bortkovskii and Novak* [1993] found values no greater than 2.5% for wind speeds up to 20 m s^{-1}.

An alternate approach to estimating W is through the use of microwave radar observations whereby horizontally polarized backscatter exhibits sea-spikes that are related to steep or breaking waves [*Jessup et al.*, 1991]. The relationship between radar sea-spikes and breaking is complicated by the fact that the radar signature

Gas Transfer at Water Surfaces
Geophysical Monograph 127
Copyright 2002 by the American Geophysical Union

can sometimes precede the visible breaking [*Loewen and Melville* 1991], and at X-band (3cm) wavelengths small-scale steep surface features often give rise to sea-spikes [*Liu et al.*, 1998]. At intermediate incidence angles *Jessup et al.* [1991] also observed specular type echoes of equal backscatter on vertical and horizontal polarization.

The goal of the work is to develop and compare independent radar and video estimates of W_A for a range of environmental conditions. The radar technique is a more consistent approach because it is less affected by external conditions and so has the potential to improve understanding of the parameterization of W_A. The techniques are applied to a variety of field situations including both stationary and moving platforms in harbours and the open ocean.

2. METHODS

Table 1 describes environmental parameters from nine harbour and open ocean datasets which cover wind speeds from 6 to 18 m s^{-1}, with the stronger wind data experienced at the ocean sites. The local wind speed had been corrected to a height of 10m assuming a logarithmic wind profile and the friction velocity u_* has been derived assuming a neutral density drag coefficient [*Smith*, 1980]. In the present results the radar and video were aligned into the wind-wave direction and mounted at heights of 4-25m giving grazing angles that can be as low as 0.9°.

2.1. Video

The video configuration recorded a monochrome charge-coupled device (CCD) camera output onto SVHS-format media. Video sequences were captured at three different frame-rates, in sequences of 300 frames that matched simultaneous radar data. The sequences, sampled at frame-intervals of 0.16s, 0.32s and 0.64s, had image dimensions of 256 by 240 pixels, with each pixel digitized at 8 bits.

The image sequence is denoted $I(x, y, t)$ where x and y are coordinates in the image plane and t is time. Timestamps on the image, the horizon, and in some cases registration floats, were masked from the images to form a new sequence of reduced images I_R containing only passed pixels. Fig. 1(a) shows a single image with the masked region reduced in intensity. Image sequence equalization was applied to remove variations in average intensity which arise from a variety of causes, including: solar variations, changes in viewing angle and video intensity variations. The 1st two moments for the passed data for each image, $m_0 = \mathrm{E}[I_R]$, and $m_1 = (\mathrm{VAR}[I_R])^{1/2}$ were then calculated. Each individual image in I_R was then normalized by its own m_1. The (x, y) pixel was assumed to contain a breaker if $(I_R - m_0)/m_1 > \alpha$ where α is a subjective operator-defined threshold that is constant for each experimental video sequence. The video estimate of W_A is denoted W_V with α selected to be above the lower intensity foam. The estimated sensitivity to this threshold is 0.008 variation in W_V per 1% variation in α.

The operator-selected value of α appears to be repeatable within a range of around 5%, giving a typical uncertainty of 0.04 in W_V. Fig. 1(b) shows an example of masked and thresholded data. Attempts were made to determine the threshold based on an objective criterion related to the slope of the cumulative intensity distribution. However, it was clear that an operator could determine the threshold to eliminate residual foam more accurately, especially when considering the temporal evolution of the wave field. Repeating the thresholding for all t generates a timeseries of instantaneous W_V (Fig. 1c), the mean of which is used for comparison with other studies.

The geometry has some bearing on W_V. *Kraan et al.* [1996] argue that "the diminishing contribution to the picture of areas further away from the camera then does not affect the whitecap fraction, because the whitecap area in the picture is distorted in the same way as the remaining wave field". It is possible, though, that at some point in the imaged range, the whitecap size becomes comparable to the pixel size. At this point broken and unbroken water is averaged and active breaking will potentially appear as sea-foam within the thresholded image.

Table 1. Experimental parameters: including date and location, platform type (S - harbour shore based, T - harbour tower, M - ocean mooring and V - ocean vessel), water depth (m), U_{10} (m s^{-1}), the video normalization parameter α and the radar maximum whitecap area fraction parameter β.

Name	type	d (m)	U_{10}	α	β
VRD	S	15	11.1	1.4	0.04
APA	T	2	6	1.25	0.013
DBB	S	7	10	1.5	0.09
FLIP1	M	>1000	10.5	1.0	0.25
FLIP2	M	>1000	11.5	1.5	0.13
OV0	V	>1000	11	2.2	0.2
OV1	V	>1000	8	1.65	0.02
OV2	V	>1000	18	3.2	0.38
OV3	V	>1000	14	2.5	0.08

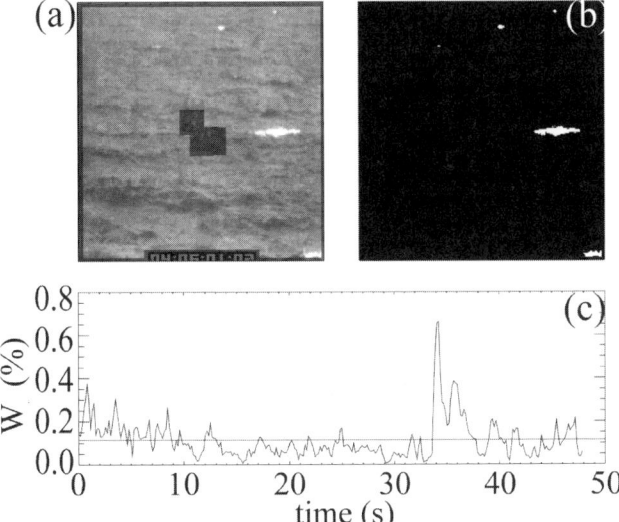

Figure 1. Single images of (a) raw video with shaded masking, (b) thresholded whitecaps and (c) a timeseries of W_V.

This is illustrated by artificial 'dilation' of images of foam and active breaking. Fig. 2a shows an expanded subsection of an image containing a foam patch. Averaging over a varying domain where all pixels in a given box are replaced by the box average (Fig. 2b is a single example) is akin to moving the foam patch away from the CCD camera. This is repeated with a breaking wave (Figs. 2c-d). The timeseries (Fig. 2e) of the resulting intensity at one location for both breaker and foam, centered over the original point of most interest (approximately the centre, 400, 370) shows that, once the resolution is reduced by a factor of 4 or so, the intensity starts to drop. It does so in a noisy fashion since, at different resolutions, the averaging box encounters other bright foam regions. This illustrates how a breaker at one range can have the same intensity as foam at another range.

2.2. Radar

The radar observations were recorded using a 3 GHz frequency modulated, interrupted continuous wave (FMICW) microwave radar [Poulter et al., 1995]. This simultaneously records data polarized in both vertical (VV) and horizontal (HH) axes. The data typically contain Doppler-shifted velocity spectra from range cells at distances extending between 50 to 300m away from the radar. Range cell sizes are typically 1.5-5m in length with an angular width of $\approx 5°$, while the velocity spectra are acquired at intervals of between 0.5 and 1 seconds. For frequencies near the 3 GHz center-frequency (wavelength 10cm) it is assumed that VV backscatter is dominated by Bragg scatter from short (5 cm) gravity-capillary waves, with a velocity dominated by the orbital motion of the advecting wave field. In contrast the HH backscatter is dominated by intermittent breaking or near-breaking waves [Smith et al., 1996]. The velocity of these features approaches the phase speed of the breaking wave, which is substantially larger than the orbital speed. Both the sudden increase in backscattered power and the high velocities are used as criteria for identifying whitecap features, described as 'fast-scatterers' by Lee et al. [1996].

Fig. 3a shows a range-time-intensity (RTI) image of the temporal evolution of radar-derived line-of-sight velocities measured with VV polarization from the RP FLIP. The power levels are arbitrary. Fig. 3b shows the thresholded HH polarization RTI image indicating range-cells associated with wave breaking [Smith et al., 1996]. W_R is the radar-derived estimate of W_A, and is calculated using the following steps. First, in order to remove the effects of the antenna pattern, the intermittent HH data is normalized in range based on the average of the VV data. Then platform motion (if any)

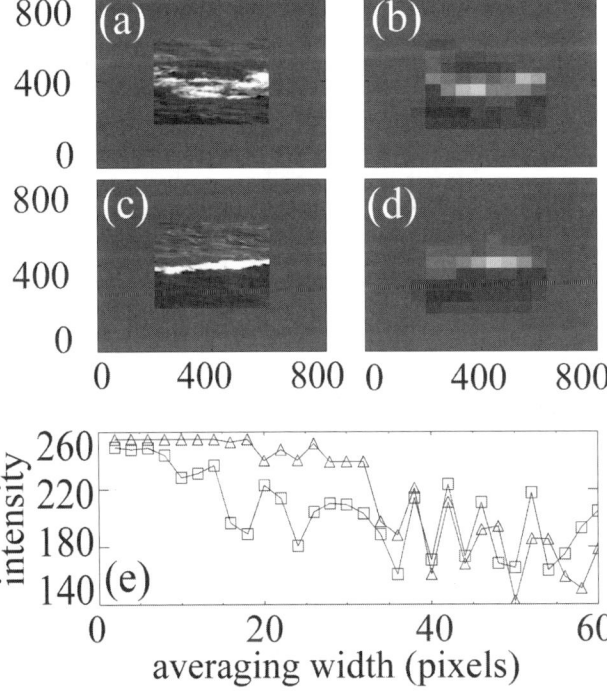

Figure 2. Images of a foam patch (a) before and (b) after image-dilation and repeated for a breaking wave, (c) before and (d) after dilation. The intensity variation from a location (400,370) for both images at different dilation scales is shown in panel (e). The squares are from the foam image and the triangles are from the breaker image.

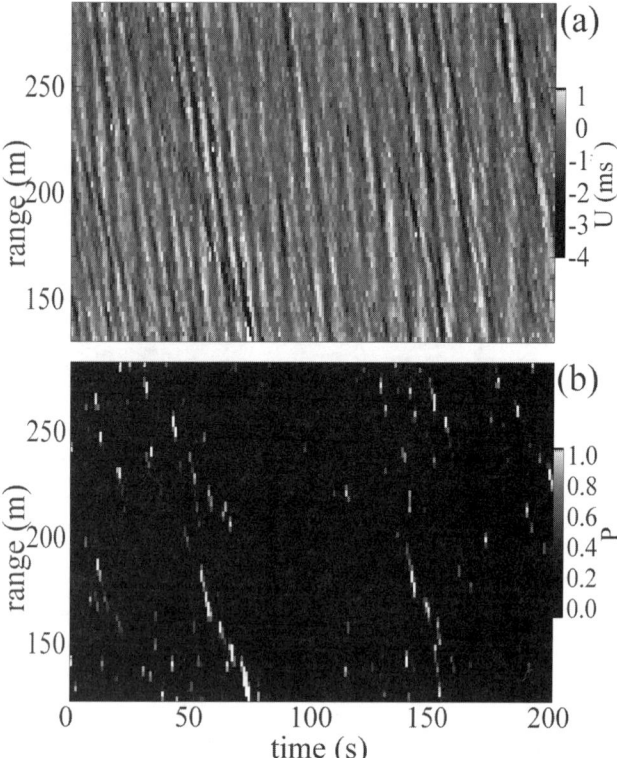

Figure 3. Range-time intensity (RTI) images of radar Doppler velocity recorded aboard RP FLIP. The images are of (a) vertical polarization velocity and (b) horizontal polarization backscattered power.

is removed by subtracting the range-averaged VV velocity from each range in both the VV and HH velocity data. Only HH bins moving towards the observer at a speed greater than the local average of VV (i.e. orbital) velocity are selected. The local average VV velocity is used as the exact VV bin will be contaminated by plume velocities. The lower panel of Fig. 4 plots the velocity ratio as a function of HH power.

From the HH bins passing these velocity criteria, only those that are above a threshold level of HH power (HHP_{min}) are accepted as containing a breaking wave. Since breaking need not fully occupy a range cell, the backscatter power levels from the breaking-wave range-time bins are linearly scaled between 0 and β where 0 is given to those bins with power at the HHP_{min} and β is given to the range-cell with the maximum power. The factor β represents the areal proportion of the range-cell occupied by the largest breaker and is constant for each radar sequence. A range cell is typically 15 m wide and 2.5 m long in range, much bigger than most breakers. The β proportion is estimated by viewing close range raw video and estimating the width and length of the largest breakers seen over a reasonable period. The present FMICW radar configuration generates two dimensional data (range and time) so that, assuming the antenna is pointing into the waves, the width of any breaking event is ambiguous. Use of a focused array [*Frasier et al.*, 1998] removes this ambiguity although it is still possible that a breaker will be smaller than the radar pixel size. W_R is then calculated as the sum of area-adjusted range-time cell sizes as a percentage of total range-time. The factor β, which should be reliably determined to within a factor of 2, is directly proportional to the radar-derived W_R. The sensitivity to the HHP_{min} for a sea-spike is much less but the value is not known. Here we have set HHP_{min} to the power at one standard deviation above the mean power for all experiments.

The upper panel of Fig. 4 shows the normalized cumulative contribution to W_R from each range cell as a function of power. The dashed vertical line indicates the threshold HHP_{min}. The area-normalized proportions in each power band are shown as a series of symbols and the cumulative sum is plotted as the rising line that asymptotes to the final radar W_R value.

3. RESULTS AND DISCUSSION

The present W_V and W_R estimates of W_A are plotted as a function of u_* in log-log space in Fig. 5. This plot

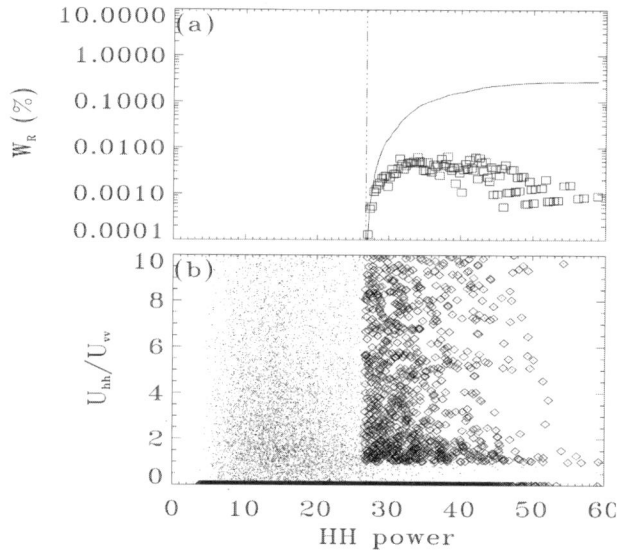

Figure 4. (a) A distribution of the cumulative contribution to W_R (rising line) from individual HH range cells (squares). The vertical line is the HHP_{min} power cutoff. (b) The distribution of HH velocity relative to VV velocity as a function of HH power (dots) in arbitrary units. Accepted breakers are shown as superimposed diamonds.

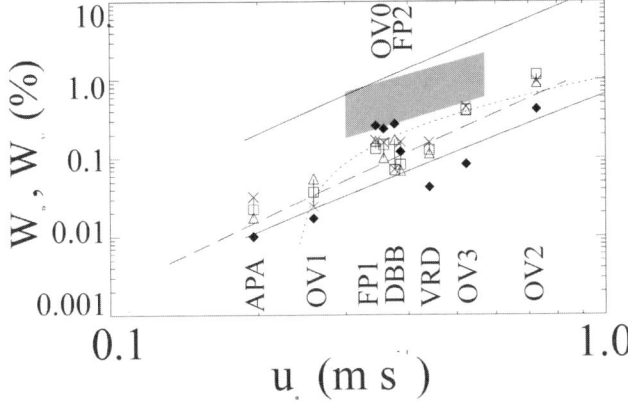

Figure 5. W_V and W_R as a function of friction velocity. The triangles, squares and crosses are video realizations at the different sampling rates whilst the solid diamonds are radar realizations, all from the present study. The solid lines are the upper (W_B) and lower (W_A) bounds due to Monahan [1989], the shaded area represents the middle and lower bounds of X-band data due to *Frasier et al.* [1998], the dotted line is due to *Bortkovskii and Novak* [1993] and the dashed line is due to *Hanson and Phillips* [1999].

summarizes 8100 separate video realizations and 4500 radar space-series. The use of different video sample frequencies, shown as symbols in Fig. 5 (cross, square and triangle), illustrates how the sampling can affect the W_V by up to a factor of 2. There is no evidence of any systematic bias and this gives an indication in the inherent variability in the technique. The empirical power law expressions for W_A and W_B used in *Monahan* [1989] are shown, together with the results of *Bortkovskii and Novak* [1993] and *Hanson and Phillips* [1999]. Finally, the range of *Frasier et al.*'s [1998] X-band radar data between their middle and lower-threshold (from their Fig. 14b) is included. Our present data fit the established W_A relationship reasonably well. The three high radar points are from quite different situations and there is no suggestion of some systematic effect. In all cases the W_V were higher, possibly consistent with a failure to discriminate against foam.

The suggestion from Fig. 5 is that the radar approach provides values indicative of the lower bound (W_A) of Monahan's [1989] work - that which is associated with the active phase of breaking events. The video results are somewhere between W_A and W_B, but closer to W_A. Replotting these data as a function of estimated wave age (data not shown) indicated no clear relationship - this is at odds with the theoretical development of *Kraan et al.* [1996].

The majority of the radar values, W_R, are closer to the W_A relationship than the video data. This suggests that S-band radar has the potential to provide a reasonable estimate of W_A without the lighting restrictions of video techniques. In contrast the X-band results lie significantly higher than W_A, indicating the sensitivity to short surface features at that higher frequency [*Frasier et al.*, 1998].

A cross-comparison, Fig. 6, shows that the differences between the radar and video techniques are generally less than a factor of 5. The present radar configuration generates a significantly lower estimate than the video in 5 of 9 cases, with 3 of the remaining 4 somewhat higher than the video. Only in one case does the W_R fall between the W_V's generated at different sampling rates by the video (FLIP2).

Here we note that while selection of threshold parameters was iterative, this was only in consideration of image and power distributions. The processing in this paper requires the user to set one parameter in the video analysis and two parameters in the radar analysis. The video parameter α separates water considered to be breaking from the background. Variations in lighting and sea-state mean that this parameter must be set by an operator for each image sequence analyzed but for a good timeseries is easily and reliably resolvable. An important difference between analysis of still photographs and video is that video enables the operator to consider the time-history of a patch evolution. It is possible that foam is swept up by subsequent waves to appear indistinguishable from a breaking crest. Consideration of the previous imagery helps identify this.

The radar free parameters HHP_{min} and β are more difficult to determine. The maximum breaker fractional area β is reasonably easily determined after close examination of visual observations. HHP_{min} is less readily set. At present this is simply set in a way that the RTI image (e.g. Fig 3b) is similar to visual descriptions - especially the duration of group breaking sequences. This value has been left constant here but is the focus of our future work in this area.

4. CONCLUSIONS

Video thresholding based on statistical normalization and dual velocity-power thresholding for S-band radar have been used to generate independent estimates of active whitecap coverage of the sea surface (W_A). Both estimates are influenced by non-arbitrary thresholds. Temporal averaging is important because the timeseries of W_A were intermittent. The main limitation to the video technique is the effect of lighting conditions on image contrast. On a minority of occasions foam could not be discriminated against on the basis of an inten-

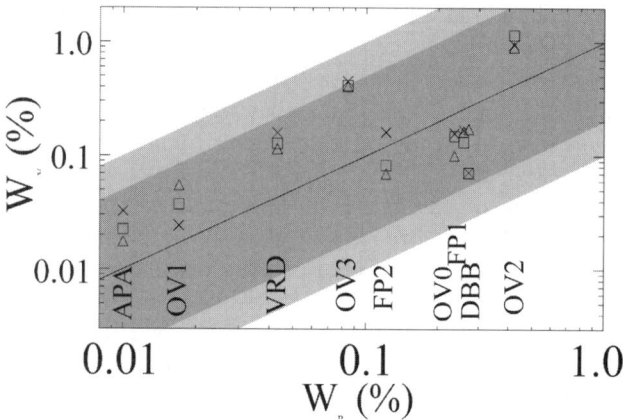

Figure 6. W_V as a function of W_R, the symbols are as for the previous plot. The bands indicate deviations from a one-to-one relationship by factors of 5 and 10 respectively.

sity threshold. This resulted in slightly higher values of W_V than predicted by the relationship of *Monahan* [1989]. The radar results were limited in the ability to determine the dimensions of the breaking; this was parameterised in terms of backscattered intensity. The video estimates of W were generally larger than the radar estimates by up to a factor of 5. On the majority of occasions the radar estimate lay close to *Monahan's* [1989] active whitecap relationship. This is in contrast to X-band data which lay well above, possibly indicating greater sensitivity to small, steep surface features [*Frasier et al.*, 1998]. This adds weight to the suggestion that S-band radars may be able to give a closer estimate of active whitecap coverage than higher frequency radars.

Acknowledgments. This work was funded by the New Zealand Foundation for Research, Science and Technology. Murray Poulter and Bill Ireland are thanked for their support.

REFERENCES

Bortkovskii, R.S. and V.A. Novak, Statistical dependence of sea state characteristics on water temperature and wind-wave age, *Jour. Mar. Syst.*, 4, 161-169, 1993.

Donelan, M.A. and Y. Yuen, Wave dissipation by surface processes, in *Komen et al.* (eds.) *Dynamics and modelling of ocean waves*, Cambridge Univ. Press, 1994.

Frasier, S.J., Y. Liu and R.E. McIntosh, Space-time properties of radar sea-spikes and their relation to wind and wave conditions, *J. Geophys. Res.*, 103, 18745-18757. 1998

Hanson, J.L. and O.M. Phillips Wind sea growth and dissipation in the open ocean, *J. Phys. Oceanogr.*, 29, 1633-1648, 1999.

Jessup, A.T, W.K. Melville and W.C. Keller, Breaking waves affecting microwave backscatter 1. detection and verification, *J. Geophys. Res.*, 96, 20547-20559, 1991

Kraan, C., W.A. Oost and P.A.E.M. Janssen, Wave energy dissipation by whitecaps, *J. Atmos. Oceanic Technol.* 13, 262-267, 1996.

Lee, P.H.Y., J.D. Barter, E. Caponi, M. Caponi, C.L. Hindman, B.M. Lake and H. Rungaldier, Wind-speed dependence of small-grazing-angles microwave backscatter from sea surfaces, *IEEE Trans. Antennas Propagat.* 44, 333-340, 1996.

Liu. Y., S.J. Frasier and R.E. McIntosh, Measurement and classification of low-grazing-angle radar sea spikes, *IEEE Trans. Antennas Propagat.*, 46, 27-40, 1998.

Loewen, M.R. and W.K. Melville, Microwave backscatter and acoustic radiation from breaking waves, *Jour. Fluid. Mech.* 224, 601-623, 1991.

Melville, W.K., Wave breaking in deep water, *Annu. Rev. Fluid Mech.* 28, 279-321, 1996.

Monahan, E.C. From the laboratory tank to the global ocean, in *The climate and health implications of bubble-mediated sea-air exchange*, E.C. Monahan and M. A. Van Patten (eds.), Conn. Sea Grant, 43-63, 1989.

Poulter, E.M., M.J. Smith, and J.A. McGregor, S-band FMCW radar measurements of ocean surface dynamics, *Jour. Atmos. Oc. Tech.* 12, 1271-1286, 1995.

Smith, M.J., E.M. Poulter and J.A. McGregor, Doppler radar measurements of wave groups and breaking waves, *Jour. Geophys. Res. 101*, 14269-14282, 1996.

Smith, S.D., Wind stress and heat flux over the ocean in gale force winds, *J. Phys. Oceanogr.*, 10, 709-726, 1980.

C.L. Stevens, M.J. Smith and J.A. McGregor, National Institute of Water and Atmospheric Research, Greta Point, P.O. Box 14-901 Kilbirnie, Wellington, New Zealand. (e-mail: c.stevens@niwa.cri.nz)

ASGAMAGE, the Air-Sea Gas Exchange/MAGE Experiment

Wiebe Oost

Royal Netherlands Meteorological Institute, the Netherlands

ASGAMAGE participants[1]

The ASGAMAGE project addressed the problem of the large discrepancy between the chemistry based and micrometeorological methods and aimed to determine any geophysical parameters apart from the wind speed that affect air-sea gas exchange in an effort to reduce the uncertainty in the global carbon balance. Experiments were performed in the spring and fall of 1996 at and near a research platform off the Dutch coast and two surface layer models were developed for the gas exchange process. The results gave a reduction of the difference between the two types of methods from an order of magnitude to a factor of two as well as indications for the causes of the remaining difference.

BACKGROUND AND OBJECTIVES

The ASGAMAGE project formally started on March 1, 1996 and lasted until March 1, 1999. The name is a contraction of ASGASEX (for Air Sea Gas Exchange, an earlier project with partially the same participants) and MAGE (for Marine Aerosol and Gas Exchange), activity 1.2. of IGAC, the International Global Atmospheric Chemistry project, which in turn is part of the International Geosphere-Biosphere Programme, IGBP. Three problems lay at the root of the ASGAMAGE project:

- Given the importance of the carbon balance with respect to the global climate, the sizable share of the oceans in this balance and the size of the "missing sink", the difference between the sum of the known sources and sinks cannot be attributed to a known process or depository.
- The cause of the order-of-magnitude discrepancy between the air-sea transfer velocity k_w for CO_2 found with chemistry oriented methods (based on e.g. ^{222}Rn, ^{14}C, deliberate tracers) and the newly developed micrometeorological techniques, especially the eddy correlation method, is poorly understood and
- There are large uncertainties in the values for the transfer velocity (even without the micrometeorological results) in combination with the expectation that there are more geophysical parameters affecting air-sea gas exchange beside wind speed. Micrometeorological methods, with their measurement times in the order of half an hour, could be of prime importance for the study of other geophysical parameters, (e.g. wave slope, wave breaking, bubbles, atmospheric stability etc.) because their effects are largely averaged out during the measurement periods of the chemistry oriented methods, which are of the order of a day or more.

The project focused on four objectives:
1. To find relationships between the transport coefficients for the gas fluxes and any relevant geophysical parameters.
2. To intercompare different methods and systems to measure the transfer velocity of trace gases over the sea.
3. To find out whether and, if at all, under what conditions, there can be significant carbon dioxide stratification in the upper meters of the water column.

[1] See Table 1.

Table 1. Participating institutes, departments and principal scientists   Nationality

1	Royal Netherlands Meteorological Institute (KNMI), Department of Oceanography (Project leader) Wiebe Oost, Cor Jacobs, Wim Kohsiek	The Netherlands
2	Department of Harbours and Public Works (RWS-DNZ), North Sea Directorate Guus Goossens, Jaap van der Horn	The Netherlands
3	Max Planck Institut für Chemie (MPIC), Abteilung Biochemie Detlev Sprung, Spyros Rapsomanikis, Thomas Kenntner, Thomas Reiner	Germany
4	University College Galway (UCG), Department of Oceanography Peter Bowyer	Ireland
5	Risø National Laboratory (Risø), Department of Meteorology and Wind Energy Søren Larsen	Denmark
6	TNO Physics and Electronics Laboratory (TNO-FEL) Gerrit de Leeuw, Gerard Kunz	The Netherlands
7	Southampton Oceanographic Centre (SOC), George Deacon Division Alan Hall	United Kingdom
8	University of East Anglia (UEA), School of Environmental Sciences Peter Liss, Gill Malin	United Kingdom
9	University of Newcastle upon Tyne (NUT), Department of Marine Sciences and Coastal Management Rob Upstill-Goddard	United Kingdom
10	Southampton University, Department of Oceanography (SUDO) David Woolf, Angus Graham	United Kingdom
11	Plymouth Marine Laboratory (PML) Phil Nightingale	United Kingdom
12	NOAA Environmental Technology Laboratory, Boulder, CO (NOAA-ETL) Chris Fairall, Jeff Hare	USA.
13	NOAA Climate Monitoring and Diagnostics Laboratory, Boulder, CO (NOAA-CMDL) Richard Dissly, Pieter Tans	USA
14	Bedford Institute of Oceanography, Dartmouth N.S. (BIO) Bob Anderson, Stu Smith	Canada

4. To test new methods and new equipment for the measurement of air-sea fluxes of CO_2, N_2O, CH_4 and DMS.

Objectives 1, 2 and 4 can be directly understood from the motives for the project. Objective 3 resulted from the earlier ASGASEX experiment in which strong fluctuations of the CO_2 fugacity with the tide had been found, that could be seen as an indication of vertical concentration gradients. This was of special importance for the second objective because in the analysis of the differential tracer method the assumption is generally made that concentration gradients within the water column can be neglected and that for air-sea gas transfer processes only the surface is important. A model study of this subject is presented during the present conference by Jacobs ("Comparison of deliberate tracer techniques and eddy covariance measurements to determine the air-sea transfer velocity of carbon dioxide").

THE PROJECT

The ASGAMAGE participants and their nationalities are given in Table 1. There were two experimental periods, the first one from May 6 - June 7, 1996, the second from October 7 - November 8, 1996. During the spring period measurements were only made at and around Meetpost Noordwijk (MPN), a research platform 9km off the Dutch coast, owned and operated by RWS-DNZ (Figure 1). The fall campaign was again at MPN, but this time the UK RRS "Challenger" also participated, operating in the wider neighbourhood of MPN. These latter activities, made simultaneously with micrometeorological experiments at MPN, were primarily aimed at determining air-sea gas transfer coefficients with the differential tracer method. A full description of the instrumentation used during both experiments is well outside the scope of this paper. Figure

2 gives an impression of the sensors operated at MPN. The activities on the "Challenger" were devoted to the differential tracer experiment (tracer preparation, tracer release, sampling, analysis) and meteorological observations.

After the experimental phases the analysis and interpretation of the data were started. In order to interpret the experimental results two models were developed, a two-dimensional one, primarily aimed at estimating the effects of horizontal inhomogeneities on the results of the experiments and a one-dimensional one, specifically intended to estimate the effects of the various processes affecting air-sea gas transfer.

The one-dimensional model was developed at KNMI [*Jacobs et al.*, in prep.] and specifically aimed at the study of the possible causes of the remaining discrepancy between the direct ec results and the tracer results on k_w. A crucial assumption in the differential tracer method is a well-mixed water column. Simulations with the model suggest, however, that near-surface tracer gradients below the skin layer may cause the tracer method to underestimate k_w by some 15% under conditions of strong turbulent mixing. A study with the two-dimensional model, independently developed at Risø [*Kjeld*, 1999], confirms the results of the one-dimensional model. If turbulence is suppressed - for example, by penetration of light into the water column - tracer gradients and therefore the difference with the true value of k_w could become even greater.

RESULTS

The following conclusions can be drawn at the end of the project.

Concerning OBJECTIVE 1:
- Of the two main contenders for the relationship between the transfer velocity and the wind, the Liss-Merlivat and the Wanninkhof parameterizations, the one due to Wanninkhof agrees best with the experimental results, especially at higher wind speeds. This conclusion was independently supported by results from GASEX-98, a later, US supported, air-sea gas exchange experiment in the Atlantic Ocean.
- Despite a strongly improved accuracy of the micrometeorological results, reflected in an order of magnitude improvement in the correspondence between the results of various measurement methods, the uncertainty in the data is still too high to detect dependencies that are more subtle than the one with the wind.
- The accuracy of the data is nevertheless sufficiently high to allow the statement that there seems to be no significant dependence of k_w on atmospheric stability.

Concerning OBJECTIVE 2:
- The order-of-magnitude difference between the transfer velocities found with the micrometeorological and with the differential tracer experiments is reduced to about a factor two (Figure 3).
- The remaining difference can largely be attributed to near surface gas concentration gradients (compare the conclusions concerning Objective 3) and reflects the idiosyncrasies of the measurement methods.
- Another important cause of differences between the eddy correlation and differential tracer results is a daily trend in the data, which is averaged out in the differential tracer outcomes, due to its 24 hour (and longer) measurement times, but visible in the eddy correlation results.

Concerning OBJECTIVE 3:
- During ASGAMAGE no vertical carbon dioxide concentration gradients of the size seen in the 1993 experiment have been detected, neither at the platform, nor from the ship. *Kunz et al.* [1998] detected concentration fluctuations of only 15ppm, related to the tide.
- As noted below modelling results indicate a gas concentration gradient just below the water surface. This gradient is different for different gases. If indeed present it would have a small effect on the eddy correlation results. The differential tracer results would be affected to a much higher degree, causing an estimated 15% change in the values for the transfer velocity compared to the uniform situation.
- The one dimensional model indicates the existence of a permanent vertical concentration gradient close to the water surface. The main part of this gradient is too

Figure 1. The Meetpost Noordwijk platform with the outrigger extended.

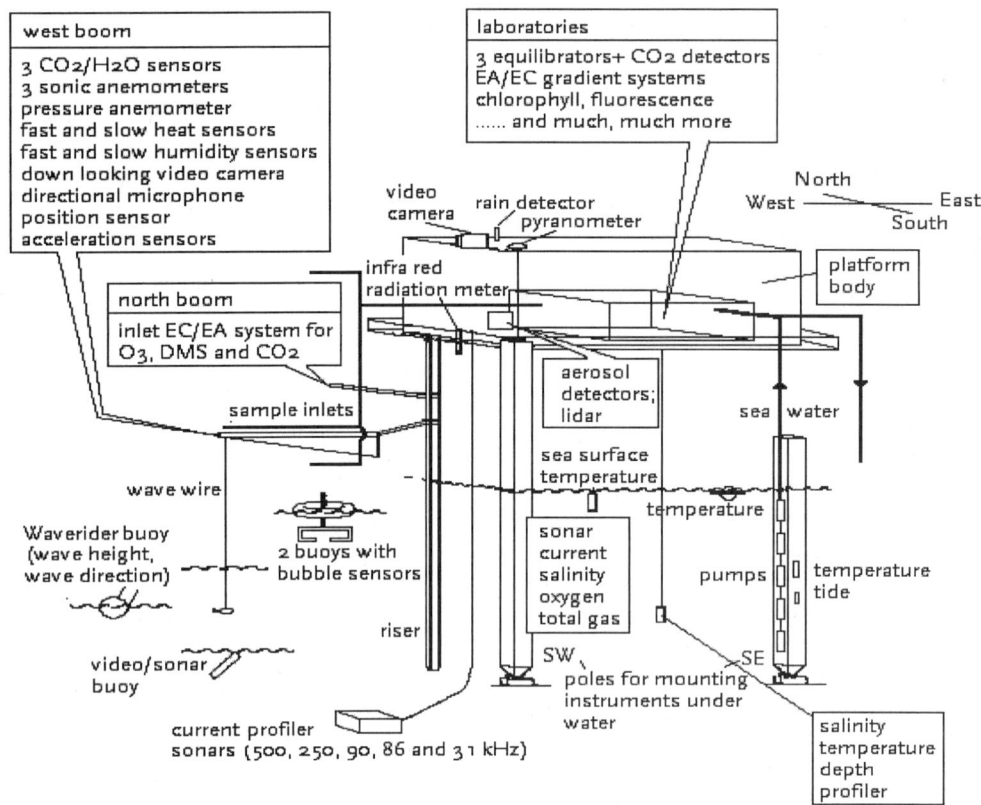

Figure 2. A schematic of Meetpost Noordwijk during ASGAMAGE, giving an impression of the instruments used during the experiments. "EA" (for eddy accumulation) and "EC" (for eddy correlation) are methods used to measure fluxes.

close to the surface, however, to have been detected with the instrument configurations used during ASGAMAGE.

Concerning OBJECTIVE 4:

New methods that have been applied during ASGAMAGE are
- The differential tracer method using non-volatile tracers and a large overdetermination of the system through the use of five different tracers. This method has led to a higher accuracy of the final result. (NUT, PML, UEA) [*Nightingale et al.*, 2000]
- The eddy correlation technique for DMS and its application at sea (MPIC).
- The relaxed eddy accumulation technique for DMS (MPIC)
- The inertial dissipation method (TNO-FEL and Risø) [*Kunz et al.*, 1998]
- The simultaneous use of two CO_2 fluctuation sensors to reduce experimental noise (TNO-FEL and Risø) [*Kunz et al.*, 1998].

New equipment used during ASGAMAGE:
- A closed ultrasensitive CO_2 detection system, used for eddy correlation, relaxed eddy accumulation and gradient measurements of the CO_2 flux designed, owned and operated by NOAA/CMDL.
- The latest version of the Infrared Fluctuation Meter for CO_2/H_2O eddy correlation measurements (KNMI, *Kohsiek*, 1998).
- Equipment for the new techniques indicated above, e.g. the APIMS (Atmospheric Pressure Ionization Mass Spectrometer) of MPIC

New models developed and used during and after ASGAMAGE:
- A one-dimensional model specifically aimed at the discrepancy between the direct ec results and the tracer results on k_w (*Jacobs et al.*, in prep.).
- A two-dimensional model, primarily aimed at the effects of horizontal inhomogeneities

Lessons learned:
- The eddy correlation and relaxed eddy accumulation measurements with the NOAA/CMDL instrument have only yielded upper limits for the transport coefficients, due to the long inlet tube, an insufficient flow

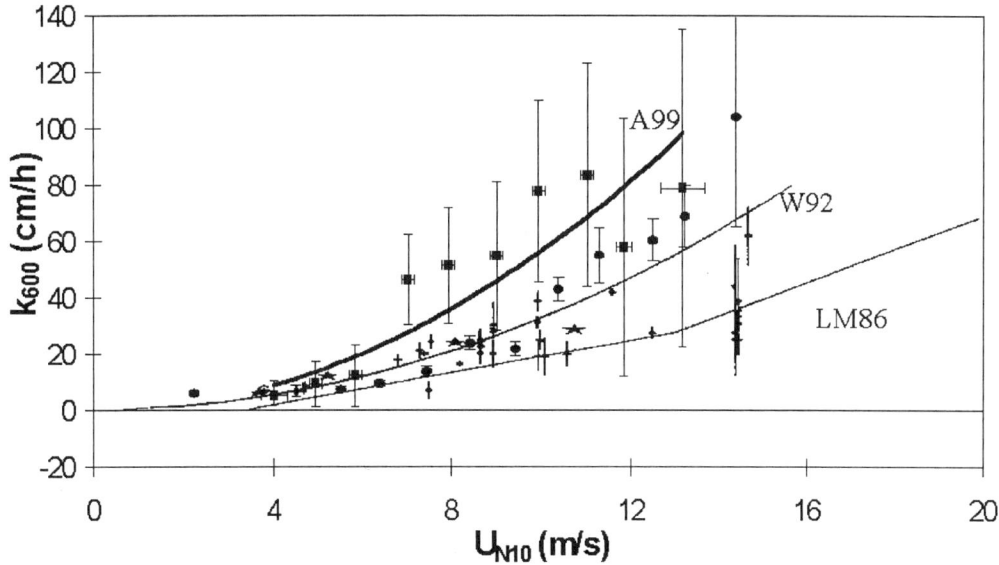

Figure 3. Air-sea transfer velocity of CO_2 normalised to a Schmidt number of 600 (k600) versus wind speed at a height of 10m. Open squares: results from ASGAMAGE direct observations: averages over wind speed bins of 1 m/s (error bars denote 95% confidence limits); bold line, labeled A99: fit determined by *Jacobs et al.* (1999); dots and stars: results from ASGAMAGE multiple tracer and other field experiments. Solid lines labeled "W92" and "LM86" are fits proposed by *Wanninkhof* (1992) and *Liss and Merlivat* (1986), respectively.

speed in that tube and an unexpected difference in the effective flow speeds of water vapor and CO_2 in the tube. The relaxed eddy accumulation technique (applied by MPIC and NOAA) had not been used over sea before and is notoriously difficult to use there because of its high sensitivity to small misalignments. Much experience has been gathered, although the accuracy of the data was not yet sufficient to make them useful.

More details about the project and its results, including references to the various parts can be found in the final report of the project: "ASGAMAGE, the ASGASEX MAGE experiment", KNMI Scientific report 99-04, available from the institute of the first author and on the Internet at *http://www.knmi.nl/asgamage*.

REFERENCES

Jacobs, C. M. J., W. Kohsiek and W. A. Oost, Air-sea fluxes and transfer velocity of CO2 over the North Sea: results from AS-GAMAGE, *Tellus* 51B, 629-641, 1999.

Kohsiek, W., Measurement of CO_2 fluxes with the IFM during ASGAMAGE. In: *Report of the ASGAMAGE workshop, September 22-25, 1997* (ed. W.A. Oost), *KNMI Scientific Report* 98-02, Royal Netherlands Meteorological Institute, De Bilt, 33-37, 1998.

Kjeld, J. F., A model study of the air-sea exchange of trace gases and the wind flow in complex terrain. Ph.D thesis, University of Odense and Risø National Laboratory, 1999.

Kunz, G. J., S. W. Lund, S. E. Larsen, F. A. Hansen, G. de Leeuw, Air-sea CO_2 gas transfer velocity during ASGAMAGE-B. TNO report FEL-98-C190, 45 pp. 1998.

Liss, P. S. and L. Merlivat, Air-sea gas exchange rates: Introduction and synthesis. *In: P. Buat Ménard (Ed.) The role of air-sea exchange in geochemical cycling*, 113-128, D. Reidel Publishing Company, Dordrecht, 1986.

Nightingale, P. D., G. Malin, C. S. Law, A. J. Watson, P. S. Liss, M. I. Liddicoat, J. Boutin and R. C. Upstill-Goddard, In-situ evaluation of air-sea gas exchange parameterisations using novel conservative and volatile tracers, *Global Biogeochemical Cycles* 14, 373-387, 2000.

Wanninkhof, R., 1992: Relationship between wind speed and gas exchange over the ocean, *J. Geophys. Res.*, 97, 7373-7382, 1992.

___

Dr. Wiebe A. Oost, Department of Oceanography, Royal Netherlands Meteorological Institute, P.O.Box 201, 3730 AE de Bilt, The Netherlands. Email: oost@knmi.nl.

For the addresses of the other participants the reader is referred to http://www.knmi.nl/asgamage on the Internet.

Comparison of the Deliberate Tracer Method and Eddy Covariance Measurements to Determine the air/sea Transfer Velocity of CO_2

Cor Jacobs[1], Phil Nightingale[2], Rob Upstill-Goddard[3], Jørgen Friis Kjeld[4], Søren Larsen[4], and Wiebe Oost[1]

During the international Air Sea Gas Exchange / MAGE (ASGAMAGE) field experiment in the Southern North Sea in 1996, deliberate tracer (DT) releases and simultaneous eddy covariance (EC) measurements of the CO_2 flux were carried out to determine air/sea gas transfer rates. When expressed as a function of wind speed, EC-derived values of gas transfer velocity (k) are a factor of 2-2.5 higher than k-values from the DT experiment. Uncertainties in k, especially for the EC-derived values, are large. Possible reasons for the systematic differences and data scatter were investigated using 1-D and 2-D models of the upper-ocean boundary layer and atmospheric surface layer (ASL). Model outputs suggest that the DT method may underestimate k by 10-30%, due to near-surface vertical concentration gradients. The required correction depends on environmental conditions and sampling position relative to the centre of the tracer patch. Chemical buffering of CO_2 renders the results from EC measurements much less sensitive to vertical gradients. However, horizontal heterogeneities in the aqueous concentration field can cause flux divergence in the ASL. At patch scales of less than a kilometre, this flux divergence might account for scatter of 20-50% in the true flux. Additional scatter of a similar magnitude might arise from instationarity of the aqueous CO_2 concentration, if the sources and sinks that cause the instationarity are located predominantly near the air-sea interface.

---

[1]Department of Oceanography, Royal Netherlands Meteorological Institute (KNMI), De Bilt, The Netherlands
[2]Plymouth Marine Laboratory, Plymouth, United Kingdom
[3]Department of Marine Sciences and Coastal Management, University of Newcastle upon Tyne, Newcastle upon Tyne, United Kingdom
[4]Wind Energy and Atmospheric Physics Department, Risø National Laboratory, Denmark

Gas Transfer at Water Surfaces
Geophysical Monograph 127
Copyright 2002 by the American Geophysical Union

INTRODUCTION

One of the goals of the international Air-Sea Gas Exchange/MAGE project (ASGAMAGE) [*Oost et al*, 2001] was to compare the eddy covariance (EC) technique [*Geernaert*, 1999] and the deliberate tracer (DT) method [e.g. *Watson et al*, 1991] to determine the air-sea transfer velocity of CO_2. During the ASGAMAGE-B experiment in the fall of 1996, EC measurements and a DT release were performed in the southern North Sea. CO_2 fluxes and air/water concentrations were measured at *Meetpost Noordwijk* (MPN), a research platform 9 km off the Dutch coast. The DT release, from RRS *Challenger*, was in the vicinity of

Table 1. Brief characterization and comparison of the models used in this study.

Model 1: one-dimensional	Model 2: two-dimensional
• designed to model OBL structure and fluxes; adjusted and extended from *Large et al* [1994]	• designed to study effects of inhomogeneities on gas flux [*Kjeld*, 1999]
• diurnal evolution: dynamic simulation of water speed, temperature, salinity and passive tracer (inert or CO_2)	• spatial distribution: dynamic simulation of passive tracer (inert or CO_2)
• CO_2 chemistry: slow hydration and chemical equilibrium of bicarbonate and carbonate [*Emerson*, 1995]	• CO_2 chemistry: chemical equilibrium
• non-local, 1st order turbulence closure	• local, 1st order turbulence closure
• diffusive sublayer and k: surface renewal theory	• $k = 12.4\ Sc^{-1/2}\ u_*^2$ [*Coantic*, 1986]
• wave breaking enhances turbulence; it increases diffusivity and decreases surface renewal time scale; parameterization uses significant wave height and wave age [*Terray et al*, 1996]	• roughness length in water adjusted to wave breaking [*Zillitinkevitch and Kreiman*, 1994]
• buoyancy effects and light penetration included	• strictly neutral stratification
• BBL instead of deep-sea formulation of *Large et al* [1994]	• conditions at 5 m (water) fixed
• tidal current + wind induced current	• fixed water current + wind induced current
• flux-profile relationships for atmospheric surface layer (reference level at 10 m)	• ASL concentrations and fluxes resolved with high vertical resolution (logarithmic grid with 102 levels)
• staggered grid with spacing 0.005-0.2 m in the water, 110 levels	• logarithmic grid between the roughness length and reference depth in water, 102 levels

MPN, but located about 15 km to the NW. Oceanic boundary layer (OBL) models have been developed to study air-sea gas exchange theoretically. The models were used to simulate EC and DT measurements at sea. We studied the impact of possible violations of the assumption, pertinent to both methods, that the water is well-mixed. We also attempted to assess the possible consequences of instationarity and horizontal inhomogeneity of the concentration fields.

This paper describes the principal results of our ASGAMAGE EC and DT comparisons. It will be shown that the fact that the DT method utilizes inert tracers while EC results are based on CO_2 might be essential to a proper interpretation of field data.

Only the ASGAMAGE EC measurements are described here. For a description of DT releases and associated tracer measurements, the reader is referred to *Upstill-Goddard et al* [1991] and *Nightingale et al* [2000]. Furthermore, only a phenomenological description of the models is provided here. More detailed model descriptions are given in *Kjeld* [1999] and *Jacobs et al* [in preparation].

METHODS

The EC measurements at MPN were some 6m above Mean Sea Level (MSL). Instruments were mounted at the end of a 21m boom, oriented in a westerly direction, in order to preclude the possibility of flow distortion by the platform. Data were corrected for residual distortion as described in *Oost et al* [1994].

Fluctuations in CO_2 concentration were measured using the Infrared Fluctuation Meter (IFM) described by *Kohsiek* [2000]. The IFM is an aspirated open-path sensor. The aspirator protects the sensor from rain and salt, and has negligible influence on the fluxes. The IFM optics were slightly heated to prevent the formation of liquid water in the optic path. Concentration data were correlated with vertical wind speed measured using a three-dimensional sonic anemometer (*Solent, Gill Instruments Ltd., Lymington, Hampshire, UK*). Heat and water vapor flux data from various participants were combined to get high-quality consensus values for these fluxes [*Oost et al*, 2000]. These data were used to compute the Webb correction [*Webb et al*, 1980] as well as a cross-talk correction [*Kohsiek*, 2000]. Further processing included corrections for instrument tilt and flow distortion by other instruments [*Oost et al*, 1994]. We assumed the effect of sensor separation and line averaging to be negligible [*Jacobs et al*, 1999].

Fluxes were computed from 20 Hz samples, averaged over 18-minute intervals. Only data from runs with a wind direction greater than 220 degrees were accepted. This angle of acceptance excludes runs significantly influenced by flow distortion by the platform, and precludes possible effects of terrestrial CO_2 sources. Data that were possibly affected by dirty IFM optics or by the presence of liquid water in the optical path were rejected. Finally, we checked for consistency between the signals from the IFM and from a NOAA CO_2-sensor [*Hare and Fairall*, 1999]. The raw fluxes from the two sensors were regressed and data outside the 90% prediction interval were excluded from further analysis. The NOAA data were not used quantitatively because the NOAA sensor optics were unheated, raising the possibility of severe cross-talk problems that cannot be corrected for [*Kohsiek*, 2000]. Furthermore, the separation distance from our EC system was rather large, which introduces the problem of flux loss.

Partial pressure of CO_2 in air (height 12-20 m), pC_a, was determined using Infrared Gas Analyzers (*LI-COR, Inc., Lincoln, Nebraska, USA*). CO_2 concentration in water was determined by analyzing water samples, obtained by submerged pumps at 2, 3.5, 5, 7, 11 or 15 m below MSL, in the laboratory at 12m above MSL. Samples were routinely collected at a specific level between 3.5 and 11m over many hours. On some occasions concentration profiles were determined by switching the inlet height along all levels available. The sampled water was sprayed continuously into an equilibrator, placed in a vessel that was flushed with seawater from the sampling level. This isolation and the high pumping rate assured that the difference between the *in situ* temperature and the temperature of equilibrated water was less than 0.1K. The CO_2 partial pressure in the equilibrator air, pC_{eq}, was determined using the same equipment as for pC_a. The values of pC_a and pC_{eq} determined by various institutes were combined into consensus values [*Jacobs et al*, 1998]. CO_2 fugacity of ambient and equilibrator air, $f_{CO2,a}$ and $f_{CO2,eq}$, respectively, were computed from the observed atmospheric pressure, air temperature and seawater temperature, T_w [*DOE*, 1994]. The transfer velocity, normalized to $Sc=660$ (with Sc the Schmidt number), k_{660}, was then computed as:

$$k_{660} = \frac{\Phi_{CO2}}{K_0(T_w, S=31)(f_{CO2,eq} - f_{CO2,a})}\sqrt{\frac{Sc(T_w)}{660}}$$

where Φ_{CO2} is the observed CO_2 flux. K_0 is solubility, which is calculated following *Weiss* [1974], using salinity $S=31‰$, representative of the seawater near MPN during ASGAMAGE. We computed Sc for CO_2 from data in *Wanninkhof* [1992] for seawater.

Gas exchange simulations for CO_2, helium (He) and sulfur hexafluoride (SF_6) were performed using complementary one and two-dimensional models (1-D model, 2-D model, respectively) of the OBL. The 1-D model is suitable for studying diurnal variations, allowing the penetration of light into the water, and it parameterizes the bottom boundary layer (BBL) at the seabed, including mixing by tidal currents, and enhancement of turbulence by wave breaking. The 2-D model is suited to studying the effect of horizontal inhomogeneities on fluxes and concentrations, and it allows detailed resolution of the atmospheric surface layer (ASL). However, the latter model runs for strictly neutral stability

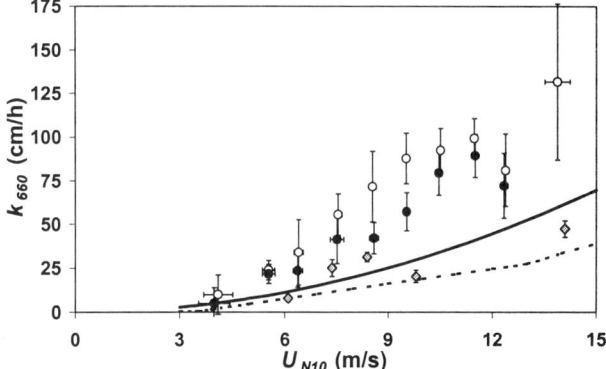

Figure 1. 1m/s-bin averaged transfer velocity from ASGAMAGE-B EC data: all data (open circles) and nighttime data (closed circles) and comparison with DT data (diamonds). Error bars indicate 95% confidence limits. Dotted line: *Liss and Merlivat* [1986]; Dashed line: *Wanninkhof* [1992].

conditions in water and air, and ignores the influence of solar radiation. The main model characteristics are compared in Table 1.

In both models the momentum flux across the air-sea interface and the reference level wind, U_{N10}, are related by a logarithmic wind profile, using roughness length $z_0 = 0.015 u_*^2/g$, where u_* is friction velocity, and g is acceleration due to gravity (9.81 m/s^2) [*Garratt*, 1992]. Initial conditions were chosen to resemble the environmental conditions during ASGAMAGE-B: $T_w=12°C$, $S=31‰$.

In the 1-D model, the amplitude of tidal current was set to 0.5 m/s (two cycles per day). Significant wave height and wave age were set to conform with the ASGAMAGE-B data (range 0.32-4.94m and 2-72, respectively). The net heat flux was driven by conditions at a reference level in air (10 m), taken as constant during a run (temperature 12°C, relative humidity 85%). Bulk formulations were used to compute the sensible and latent heat flux [*DeCosmo et al*, 1996]. Solar radiation absorbed by the water was computed assuming fair weather conditions in The Netherlands [*Holtslag and Van Ulden*, 1983], and assuming an albedo of 0.1.

The 2-D model was driven by concentration changes at the reference levels in air and water (10 and 5m respectively). Such conditions can represent horizontal homogeneity as well as advection. In the examples presented here, u_* was fixed at 0.4 m/s. The water velocity was computed assuming a logarithmic profile, with water roughness length $5.5 u_*^2/g$ [*Zilitinkevitch and Kreimann*, 1994].

RESULTS AND DISCUSSION

Figure 1 shows k_{660} from ASGAMAGE-B as a function of U_{N10}. Also plotted are the parameterisations of *Liss and*

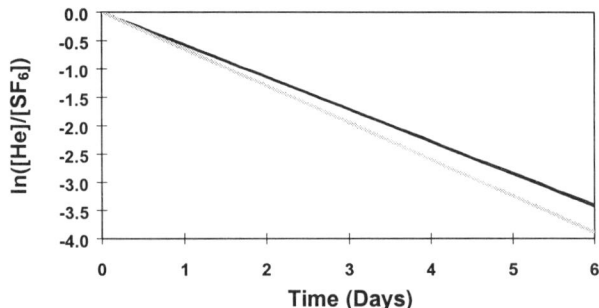

Figure 2. 1-D model simulation of $\ln([He]/[SF_6])$ (bold line) and comparison with theoretical decrease in the case of perfect mixing (grey line). $U_{N,10}$=10m/s, depth = 18m.

Merlivat [1986] and *Wanninkhof* [1992]. The EC-derived k_{660} values are clearly higher than the DT-derived values and the parameterizations, especially at intermediate wind speeds. Coefficients a of a quadratic fit $k_{660}=aU_{N10}^2$ are 0.69 and 0.29, for the EC and DT data, respectively. The DT data are in reasonable agreement with other DT measurements from the same region of the southern North Sea [*Nightingale et al*, 2000].

The scatter in the ASGAMAGE-B EC-data is large (see also Figure 6). However, many of the very high k_{660} values were obtained in daylight conditions. Using only nighttime values (Figure 1) reduces a to 0.59 and reduces the variance in spite of the lower number of data (not shown). The difference between nighttime and daytime data is statistically significant (confidence level, 0.05) around a wind speed of about 9 m/s. We will return to this issue later.

In conclusion, k_{660} values from the ASGAMAGE-B EC data are on average a factor of 2-2.5 higher than from the DT measurements. Furthermore, the data scatter, when scaled with U_{N10}, is large. In our subsequent modeling studies we tried to find possible explanations for these features.

In a DT experiment in the North Sea simulated with the 1-D model, He and SF_6 were "released" into the water in initially equal concentrations with instantaneous and complete mixing. Tracer concentrations in the air were maintained at zero. Tracer concentrations in the water were allowed to gradually decrease by gas exchange and mixing. A "measured" transfer velocity was determined from the change in the concentration ratio of He and SF_6 [e.g. *Watson et al*, 1991], based on their simulated concentrations at 5m depth. These results were then compared to the "true" values of k_{660} determined by the model equations.

EC measurements were also simulated, using CO_2 as the passive tracer. Initial fugacity of the water, $f_{CO2,w}$=46 Pa (well-mixed); $f_{CO2,a}$=36 Pa throughout, which resembles the conditions during ASGAMAGE-B. The model was forced to quasi-stationarity, that is, diurnal averages do not vary significantly anymore. This was achieved using a vertically homogeneous source term for dissolved inorganic carbon chosen so as to balance exactly the rate of CO_2 efflux.

Similar simulations were performed with the 2-D model, for a horizontally homogeneous concentration field and for a Gaussian distribution of the concentrations at the reference level. By definition, the 2-D model assumes stationary conditions. It contains no sources or sinks for inorganic carbon.

Figure 2 shows that the 1-D model simulates a decrease of $\ln(R)=\ln([He]/[SF_6])$ that is about 15% slower than would be expected in the case of perfect mixing. Both models suggest this to be caused by vertical gradients in the He/SF_6 concentration ratio. The steepest parts of the simulated gradients are within 1-2m of the surface. The lack of comparable gradients in simulated $f_{CO2,w}$ reflects chemical buffering. Further details of the simulated profiles may be found in *Jacobs et al* [in preparation].

The relatively slow decrease in simulated $\ln(R)$ implies that DT releases of He/SF_6 could underestimate k_{660} by about 15%. According to the 1-D model the error is relatively insensitive to wind speed at ambient light levels typically experienced in the southern North Sea during October. Using light intensities near MPN typical of early to mid summer increased the error to 18-30% for the wind speed range 4-16 m/s (data not shown). Also, 2-D model runs with a Gaussian tracer concentration distribution at the reference depth showed gradients to be a function of sampling position. An analysis of corresponding errors in k_{660} [*Kjeld*, 1999] revealed that the modeled error at about a quarter of the distance between the center and edge of the patch was ~30% upstream of the center, but only ~10% at the corresponding downstream position. However, we expect that at sea tracer movement backwards and forwards with the tide might equalize these errors somewhat.

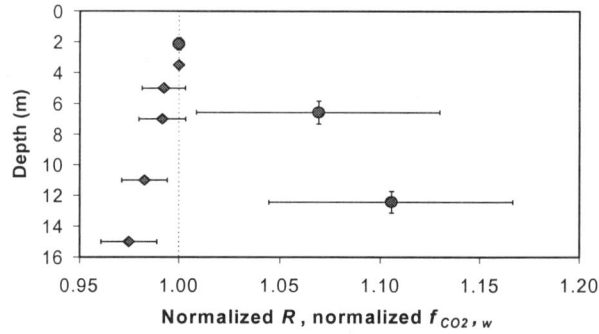

Figure 3. Observed (ASGAMAGE-B) level-averaged concentration ratio $R=[He]/[SF_6]$ (circles) and $f_{CO2,w}$ (diamonds), normalized by their values at the highest level. Error bars indicate the 95% confidence interval.

The errors related to subsurface gradients are determined by the deviation of the concentration just below the surface diffusive sublayer, from the concentration at the sampling level. ASGAMAGE concentrations were not observed as close to the surface as desired to test the model results. However, the data nevertheless indicate the presence of gradients below 2-3m (Figure 3). The level-averaged concentration ratio [He]/[SF_6] normalized by its value at the highest level (2.2m) is significantly greater than 1 at depths of 6.5 and 12.5 m, respectively, whereas normalized $f_{CO2,w}$ is significantly less than 1 at the lower levels. The latter feature can only be explained if additional carbon sources and sinks are assumed (see below).

Unlike the DT simulation results, as a result of chemical buffering (*Kjeld*, 1999) the Gaussian distribution of $f_{CO2,w}$ at the reference depth (length scale 1 km) did not result in *vertical* gradients in the water large enough to affect EC data significantly through the air-sea f_{CO2} difference. However, the model simulated related inhomogeneities in the air, resulting in CO_2-flux divergence in the ASL (Figure 4), a serious violation of a central assumption of the EC technique. The model indicates that EC instruments, typically mounted at a height of 5-10 m, will not be able to capture the true surface flux in such conditions. Depending on the flux footprint, the error may be less than, equal to, or greater than zero. Thus, small scale (~1km or less) inhomogeneities might cause considerable scatter in EC data (flux or k). For the case considered in Figure 4, i.e "measurements" at 9m above MSL, the associated scatter was 20-50% of the "true" flux.

The effect of instationarity of $f_{CO2,w}$ was investigated using the 1-D model. A sinusoidal CO_2 sink/source term was applied to mimic a change in $f_{CO2,w}$ from 44 to ~48 Pa over 3-4 days, with modulations by tidal currents, as observed during ASGAMAGE-B (Figure 5). Several forms of the source/sink function were tested. With the sources and

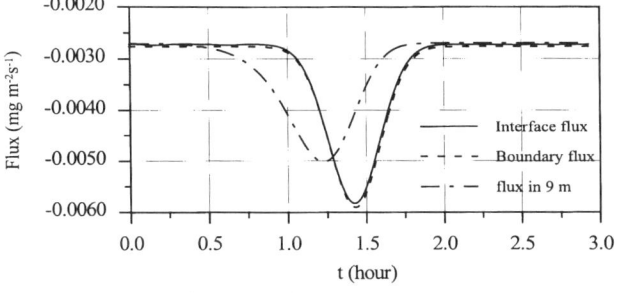

Figure 4. Simulated CO_2 exchange across the air/sea interface, using the 2-D model with a Gaussian concentration distribution at 6 m depth (length scale 1km): "true" interface flux (solid line); atmospheric flux at 9 m high (dashed-dotted line); and flux based on concentration difference between 6 m depth and 9 m height.

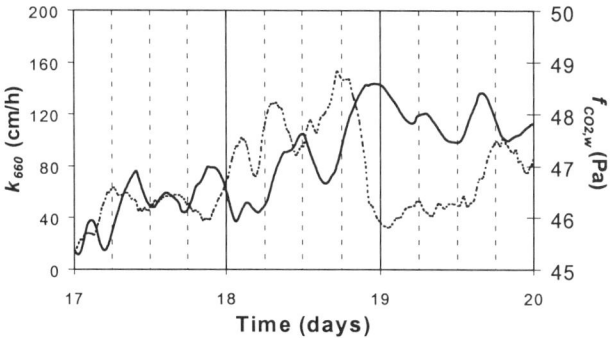

Figure 5. Observed k_{660} (dashed line, left axis) and $f_{CO2,w}$ (solid line, right axis) during day 17-20 of ASGAMAGE-B.

sinks located predominantly in and near the diffusive sublayer, the simulated normalized $f_{CO2,w}$ profiles resembled the observed level-averaged profile (Figure 3). Figure 6 compares the scatter in the simulated "observed k_{660}" to the scatter of the individual EC observations from ASGAMAGE-B. Scatter introduced by instationarity of the $f_{CO2,w}$ field, 30-50%, increases with wind speed and is caused by variations in the $f_{CO2,w}$ profile that lead to a bias in $f_{CO2,w} - f_{CO2,a}$. There is also a slight systematic effect: average k_{660} is over-estimated by 13% at $U_{N10} = 6$ m/s and 2% at $UN_{10}=14$ m/s.

Now, it is informative to speculate on the effect of light on the EC derived k_{660} values described above. Explaining this apparent impact of light is not straightforward. As light penetration acts to stabilize the water column, it is expected to reduce turbulence and therefore k. Because the water was supersaturated in CO_2, net gas fluxes were evasive. Thus, any photosynthesis by organisms in the microlayer that could bypass the physical transfer of CO_2 (cf. Matthews [1999]), would also act to reduce the flux and hence k. However, in our simulations, decreasing $f_{CO2,w}$ was accompanied by increasing k (not shown). The observations showed a similar tendency (Figure 5). Because photosynthesis is one mechanism that can transform an initial surface CO_2 source into a sink, we propose that high light intensity and measured – not true – k_{660} might indeed be correlated through biases in the air-sea f_{CO2} difference. Unfortunately, our data preclude direct testing of this possibility.

CONCLUDING REMARKS

During ASGAMAGE we compared k values derived from direct CO_2 flux measurements, using the EC technique and from a simultaneous DT experiment in adjacent waters. EC derived values of k were a factor of 2-2.5 higher than DT derived values of k. Thus, ASGAMAGE narrowed the gap between the two methods considerably, but the differ-

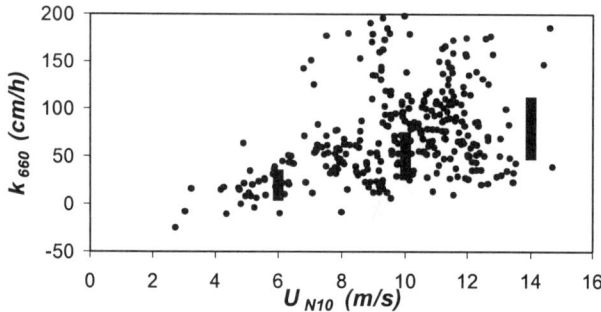

Figure 6. k_{660} versus U_{N10}: individual ASGAMAGE-B observations from EC measurements (dots); modelled values with instationary $f_{CO2,w}$ (bold vertical lines).

ence is still too large. Although data scatter, particularly for the EC measurements, was large, there is no reason to believe either set of data to be technically unreliable.

Our subsequent modeling studies suggest a considerable impact of near-surface gradients on the results from both methods. In the case of the inert tracers, such gradients, maintained by diffusion, could lead to underestimation of k by the DT method ~ 10-30%, dependent on light intensity and sample positions relative to the tracer patch center. Chemical buffering prevents the development of similar vertical gradients in CO_2. However, in contrast to He and SF_6, CO_2 is biologically cycled and is actively discharged to the coastal ocean from rivers [e.g. *Frankignoulle et al*, 1996]. Such inputs could lead to inhomogeneity and instationarity of $f_{CO2,w}$. Our simulations indicate that vertical flux divergence in the ASL might accompany $f_{CO2,w}$ inhomogeneities on scales ~ 1km. Furthermore, instationarity due to sources and sinks located predominantly near the surface can lend a complex fine structure to vertical $f_{CO2,w}$ profiles. For the simulations presented here, each of these mechanisms might account for ~20-50% of the EC data scatter and ~10% of the difference with the DT data.

We speculate that CO_2 sources in the surface diffusive sublayer that bypass the physical path of air-sea CO_2 transfer and contribute to the gas flux directly [cf. *Matthews*, 1999] might also be responsible for part of the unexplained 70-90% bias between the EC and DT results.

Acknowledgments. The research described here was an integral component of the international Marine Aerosol and Gas Exchange (MAGE) initiative, a sub-activity of the International Global Atmospheric Chemistry (IGAC) core project of the International Geosphere-Biosphere Program (IGBP). We acknowledge support from the European Commission's Marine Science and Technology Program (MAST III), contract MAS3-CT95-0044, and the UK Natural Environment Research Council, who provided ship time for the DT measurements as part of its ACSOE thematic program.

REFERENCES

Coantic, M., A model for gas transfer across air-water interfaces with capillary waves, *J. Geophys. Res.*, 91, 3925-3943, 1986.

Decosmo, J., K.B. Katsaros, S.D. Smith, R.J. Anderson, W.A. Oost, K. Bumke and H. Chadwick, Air-sea exchange of water vapor and sensible heat: the Humidity Exchange of the Sea (HEXOS) results, *J. Geophys. Res.*, 101, 12001-12016, 1996.

DOE, *Handbook of methods for the analysis of the various parameters of the carbon dioxide system in sea water, version 2, ORNL/CDIAC-74*, edited by A. G. Dickson and C. Goyet, Carbon Dioxide Information and Analysis Center, Oak Ridge, Tennessee, 1994.

Emerson, S., Enhanced transport of Carbon Dioxide during gas exchange, in *Air-water gas transfer. Selected papers from the Third International Symposium on air-water gas transfer*, edited by B. Jähne and E.C. Monahan, AEON Verlag & Studio, Hanau, pp. 23-35, 1995.

Frankignoulle, F., I. Bourge, C. Canon and P. Dauby, Distribution of surface seawater partial CO_2 pressure in the English Channel and in the Southern Bight of the North Sea, *Continental Shelf Research*, 16, 381-395, 1996.

Garratt, J.R., *The Atmospheric Boundary Layer*, Cambridge University Press, Cambridge, 1992.

Geernaert, G.L., Theory of air-sea momentum, heat and gas fluxes, in *Air-sea exchange. Physics, chemistry and dynamics*, edited by G.L. Geernaert, pp. 25-48, Kluwer Academic Publishers, Dordrecht, 1999.

Hare, J.E. and C.W. Fairall, Contributions of NOAA scientists in ASGAMAGE-B, in *ASGAMAGE: the ASGASEX MAGE experiment. Final report. KNMI Scientific Report 99-04*, Edited by W.A. Oost, pp.175-176, Royal Netherlands Meteorological Institute, De Bilt, 1999. (also: www.knmi.nl/asgamage).

Holtslag, A.A.M. and A.P. van Ulden, 1983: A simple scheme for daytime estimates of the surface fluxes from routine weather data. *J. Climate Appl. Meteor.*, 22, 517-529.

Jacobs, C.M.J., W. Kohsiek and W.A. Oost, Air-sea fluxes and transfer velocity of CO_2 over the North Sea: results from ASGAMAGE, *Tellus* 51B, 629-641, 1999.

Jacobs, C.M.J., G. Kunz, D. Sprung and M.H.C. Stoll, CO_2 in water and air during ASGAMAGE: concentration measurements and consensus data. *KNMI Technical Report TR-209*, Royal Netherlands Meteorological Institute, De Bilt, 1998.

Kjeld, J.F., *A model study of air-sea gas exchange of trace gases and wind flow in complex terrain*, Ph.D. Thesis, Risø, 1999.

Kohsiek, W., Water vapour cross-sensitivity of open path H_2O/CO_2 sensors, *J. Atm. Oceanic Tech.*, 17, 299-311, 2000.

Large, W.G., J.C. McWilliams and S.C. Doney, 1994: Oceanic vertical mixing: a review and a model with a nonlocal boundary layer parameterization. *Rev. Geophys.*, 32, 364-403.

Liss, P.S. and L. Merlivat, Air-sea exchange rates: introduction and synthesis, in *The role of air-sea exchange in geochemical cycling*, edited by P. Buat-Menard, pp. 113-128, D.Reidel Publishing Company, Dordrecht, 1986.

Matthews, B.J.H., *The rate of air-sea CO_2 exchange: chemical enhancement and catalysis by marine microalgae*, Ph.D. Thesis, School of Environmental Sciences, University of East Anglia, Norwich, 1999.

Nightingale, P.D., G. Malin, C.S. Law, A.J. Watson, P.S. Liss, M.I. Liddicoat, J. Boutin, and R.C. Upstill-Goddard, *In situ* evaluation of air-sea gas exchange parameterizations using novel conservative and volatile tracers, *Global Biogeochem. Cycles*, 14, 373-387, 2000.

Oost, W. *et al*, ASGAMAGE, the Air-Sea Gas Exchange/ MAGE experiment. In Gas Transfer at Water Surfaces, edited by M.A. Donelan, W.M. Drennan, E.S. Saltzman and R. Wanninkhof, pp xxx-xxx, AGU, this volume, 2001.

Oost, W.A., C.W. Fairall, J.B. Edson, S.D. Smith, R.J. Anderson, J.A.B. Wills, K.B. Katsaros and J.Decosmo, Flow distortion calculations and their application in HEXMAX, *J. Atmos. Oceanic Technol.*, 11, 366-386, 1994.

Oost, W.A., C.M.J. Jacobs and C. van Oort, Stability effects on heat and moisture fluxes at sea. *Boundary-Layer Meteorol.*, 95, 271-302, 2000.

Terray, E.A., M.A. Donelan, Y.C. Agrawal, W.M. Drennan, K.K. Kahma, A.J. Williams III, P.A. Hwang and S.A. Kitaigorodskii, Estimates of kinetic energy dissipation under breaking waves, *J. Phys. Oceanogr.*, 26, 792-807,1996.

Upstill-Goddard, R.C., A.J. Watson, J. Wood and M.I. Liddicoat, Sulfur hexafluoride and He-3 as seawater tracers – Deployment techniques and continuous underway analysis for sulfur hexafluoride, *Anal. Chim. ActaI,* 249, 555-562, 1991.

Wanninkhof, R., Relationship between wind speed and gas exchange over the ocean, *J. Geophys. Res.*, 97, 7373-7382, 1992.

Watson,A.J., R.C. Upstill-Goddard, and P.S. Liss, Air sea gas exchange in rough and stormy seas measured by a dual tracer technique, *Nature*, 349, 145-147, 1991.

Webb, E.K., G.I. Pearman and R. Leuning, Correction of flux measurements for density effects due to heat and water vapour transfer, *Quart. J. R. Met. Soc.*, 106, 85-100, 1980.

Weiss, R.F., Carbon dioxide in water and seawater: the solubility of a non-ideal gas, *Marine Chemistry*, 2, 203-215, 1974.

Zilitinkevitch, S.S. and K.D. Kreimann, Wind-induced drift of surface films, *in Modeling air-lake interaction*, edited by S.S. Zilitinkevitch, pp. 63-73, Springer Verlag, Heidelberg, 1991.

Dr. Cor M.J. Jacobs, Department of Oceanography, Royal Netherlands Meteorological Institute (KNMI), P.O. Box 201, 3730 AE de Bilt, The Netherlands. Email: cor.jacobs@knmi.nl.

Dr. Jørgen Friis Kjeld, Department of Meteorology and Wind Energy, Risø National Laboratory, P.O. Box 49, DK-4000 Roskilde, Denmark.

Dr. Søren E. Larsen, Wind Energy and Atmospheric Physics Department, Risø National Laboratory, P.O. Box 49, DK-4000 Roskilde, Denmark. Email: soeren.larsen@risoe.dk.

Dr. Phil D. Nightingale, Plymouth Marine Laboratory, Prospect Place, West Hoe, Plymouth PL1 3DH, United Kingdom. Email: PDN@ccms.ac.uk.

Dr. Wiebe A. Oost, Department of Oceanography, Royal Netherlands Meteorological Institute (KNMI), P.O. Box 201, 3730 AE de Bilt, The Netherlands. Email: wiebe.oost@knmi.nl.

Dr. Rob C. Upstill-Goddard, Department of Marine Sciences and Coastal Management, University of Newcastle upon Tyne, Newcastle upon Tyne, NE1 7RU, United Kingdom. Email: rob.goddard@newcastle.ac.uk.

Gas Transfer Velocities for ^{3}He in a Lake at High Wind Speeds

Philippe Jean-Baptiste and Elise Fourré

Laboratoire des Sciences du Climat et de l'Environnement, CEA-CNRS, Gif/Yvette, France

Alain Poisson

Laboratoire de Physique et Chimie Marines, Université Pierre et Marie Curie, Paris, France

A gas exchange experiment was performed in a lake of the Kerguelen Islands by using helium-3 as a tracer. The objective was to study the rate of gas exchange at air-water interface for high wind speeds, up to 20m/s. After injection and initial mixing of the tracer, the escape rate was monitored by collecting water samples for mass spectrometry analysis. After 20 days, the lake had almost returned to the atmospheric equilibrium and a new tracer injection was done, allowing the acquisition of a second ^{3}He time-series. Assuming a well-mixed flat bottom reservoir, the ^{3}He transfer coefficient – wind speed relationship, determined using a power law fit, leads to $k_{3He} = 1.1 W^{1.5}$, where k_{3He} is the ^{3}He transfer velocity in cm/h at the temperature of the lake (T=1°C, Sc=310) and W, the wind speed in m/s at 10 m height. Taking into account the possible vertical heterogeneity of the lake (assuming a vertical diffusion coefficient K_v=50cm/s^2), and a more realistic bathymetry, leads to a 32% increase in the transfer velocity, thus giving $k_{3He} = 1.45 W^{1.5}$ cm/h. This result is consistent within the uncertainties with our recent measurements in a larger lake using ^{3}He and SF$_6$ [*Jean-Baptiste and Poisson*, 2000]. It is significantly above the *Liss and Merlivat* [1986] parametrization.

INTRODUCTION

An extensive work has been done over the last two decades to measure gas transfer at air-sea interface and to parametrize transfer velocities versus wind speed. Most field results are bracketed by the empirical relation proposed by *Liss and Merlivat* [1986] on the lower side and by *Smethie et al.* [1985] on the higher side, implying large uncertainties. Moreover, most data are restricted to low and intermediate wind regimes (<13m/s), a situation which does not help to clarify the shape of the transfer velocity-wind speed relationship.

Thus, the primary goal of our experiment was to document the transfer velocity–wind speed relationship in a lake at high wind speeds, up to 20 m/s.

EXPERIMENTAL DESIGN

Lake Studer, where the experiment was carried out during austral winter, is a fresh water lake of the Kerguelen Islands (southern Indian Ocean) approximately 360 m wide and 2.4 km long, with a mean depth of 23 m (maximum depth=43m) corresponding to a volume of 0.02 km^3. During wintertime, the run-off water supply to the lake as well as water outflow are totally negligible. Owing to the

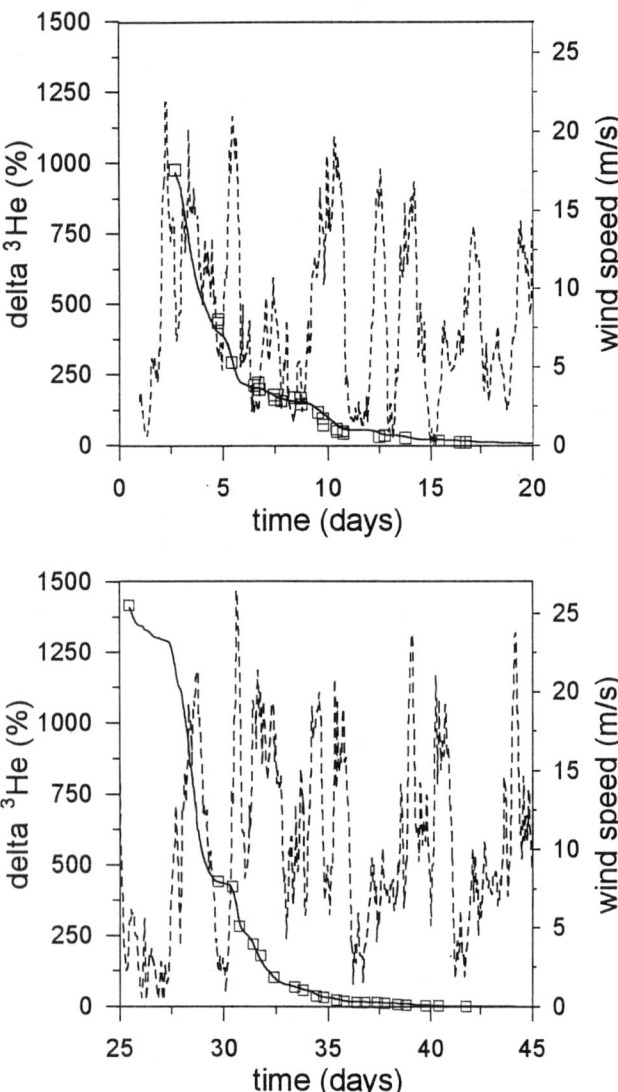

Figure 1. Decline of excess ^{3}He (δ^3He) with time (open squares), and wind speed record corrected to a height of 10 m, for the two successive tracer injections. The solid line corresponds to δ^3He values computed from equations 1 and 2.

local topography, the direction of the wind is mainly parallel to the long axis of the lake. Water temperature during the experiment was constant (1°C) throughout the lake, showing no vertical stratification. ^{3}He was released from a tank containing 0.3 liter STP of ^{3}He mixed with nitrogen to reach a pressure of 150 bars. The tank was towed across the lake at various depths and the tracer was released through a set of diffusing stones following the procedure described in *Jean-Baptiste and Poisson* [2000]. SF$_6$ was also released with the objective of conducting a dual-tracer experiment. Unfortunately, severe problems with the gas chromatography detector prevented any quantitative exploitation of the data. However, the SF$_6$ measurements indicated that the lake had returned to the atmospheric equilibrium after 20 days, so that we were able to make a new tracer injection and obtain a second ^{3}He time-series. Samples for ^{3}He analysis were taken in copper tubes at the centre of the lake using hydrographic bottles. The sampling depth was around z = 1.5 m. Helium extraction and mass spectrometry measurements were performed at Saclay using our standard procedure [*Jean-Baptiste et al.*, 1992] for oceanic samples. Note that since ^{4}He concentration is at its natural background and is basically constant, the ^{3}He concentrations can be expressed as an isotopic ratio R=^{3}He/^{4}He using the delta notation:

$$\delta^3\text{He}(\%) = (R/R_a - 1) \times 100$$

where R_a is the atmospheric helium isotopic ratio. The overall relative uncertainty in the ^{3}He/^{4}He ratio is better than 0.5%.

Wind speed was recorded at a height of 4 m at two different positions on the shore. The time-series of the two anemometers are in good agreement, with differences not exceeding a few percent. To relate gas transfer to wind speed, measured wind speeds were converted to standard wind speeds at 10m height under conditions of neutral stability following *Large and Pond* [1981].

^{3}He TRANSFER VELOCITY VERSUS WIND SPEED

The two ^{3}He time-series are displayed in figure 1 along with the record of the wind speed, which shows that the meteorology of the island is characterized by a succession of depression with wind speeds up to 18-20 m/s. From each time-series, the ^{3}He transfer velocity-wind speed relationship was determined by assuming a power law, $k = \beta W^\alpha$, where k is the transfer velocity and W, the wind speed at 10 m height. Parameters α and β were adjusted to the experimental data by solving the tracer mass balance equation for the lake (1) between time t and $t+dt$, so as to minimize the mismatch between the calculated δ^3He(t) curve and the data.

$$\delta^3\text{He}(t+dt) = \delta^3\text{He}(t) - \Phi_{3He}(t) \times dt/h \quad (1)$$

where h is the mean depth of the lake and $\Phi_{3He}(t)$, the surface ^{3}He evasion flux forced by the measured wind speed using equation (2):

$$\Phi_{3He}(t) = \beta W^\alpha [\delta^3\text{He}(t) - \delta_{eq}] \quad (2)$$

with $\delta_{eq} \approx -1.6\%$ [*Weiss*, 1970].

At the lake temperature (T=1°C), corresponding to a Schmidt number for ^{3}He Sc_{3He}=310 [*Wanninkhof*, 1992], the results are $k_{3He} = (1.15\pm0.05)W^{(1.5\pm0.02)}$ for the first time-series and $k_{3He} = (1.10\pm0.05)W^{(1.5\pm0.02)}$ for the second experiment (figure 2a), with k_{3He} in cm/h and W in m/s The corresponding δ^3He fits appear as solid lines along with the data in figure 1.

^{3}He Vertical Gradient Correction

Like most of the studies published so far, the above results assume a well-mixed reservoir. However, we have shown in a previous experiment in Lake Suisse (another lake, about twice deeper, of the Kerguelen Islands) [*Jean-Baptiste and Poisson*, 2000] that even in non-stratified lakes, tracer vertical homogeneity hypothesis is not fully met. In the Lake Suisse experiment, a vertical tracer gradient was observed, that increased with the wind speed. Especially during strong wind episodes, the tracer appeared to escape preferentially from the surface layers, which could not be resupplied instantaneously with the tracer from deeper layers. We concluded that, even at high wind speeds, wind stirring was not efficient enough as to instantaneously rehomogenize the lake from the surface to the bottom (mean depth = 47m). Thus, the lake behaved as a multi-layered system. In the present experiment, we do not have evidence for these gradients at high wind speed due to the lack of measurements on the vertical. Nevertheless, we applied the 20-layer model developed by *Jean-Baptiste and Poisson* [2000], with the same vertical diffusivity of the tracer in water (K_v=50 cm^2/s), to compute the corrected transfer velocity–wind speed relationship. Taking into account the vertical transport of the tracer leads to a 18% increase in the transfer velocity for the second time-series, with $k_{3He} = 1.30W^{1.5}$ (figure 2a). The corresponding δ^3He fit is shown in figure 2b. (Results for the first time-series are not reported here since they are very similar). Although the occurrence of breaking waves and injected bubbles may affect our conclusions, the above result suggests that the tracer homogeneity issue may have a significant influence on calculated gas exchange coefficients.

Bathymetric Correction

Lake experiments usually assume a flat bottom reservoir with a constant depth equal to the mean depth. To quantify this approximation, we chose to describe the reservoir bathymetric section by a triangle, a shape which represents, in the case of Lake Studer, a significant improvement in the description of the real bathymetry. For a lake mean depth h, the maximum depth will then be 2×h. We computed the

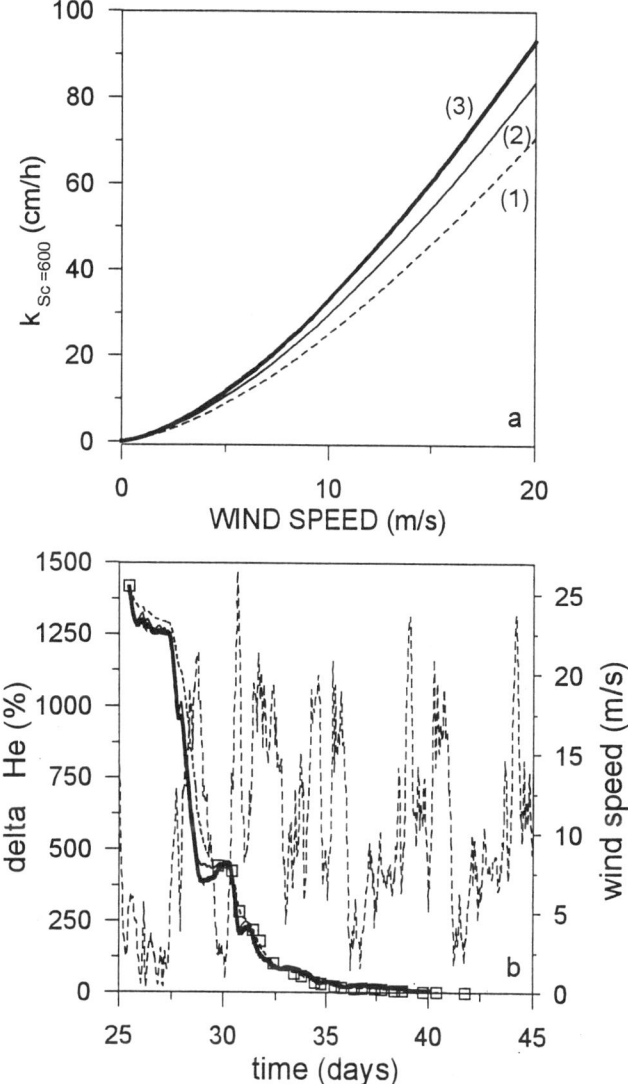

Figure 2. (a) Gas transfer velocity-wind speed relationships $k = \beta W^\alpha$ (normalized to a Schmidt number Sc=600) deduced from the Lake Studer second experiment (1) for a well-mixed reservoir with a flat bottom (dashed curve), (2) for a flat bottom reservoir with vertical diffusion (thin solid curve) and (3) for a triangular bathymetric section with vertical diffusion (thick solid curve). (b) Corresponding δ^3He(t) curves. The measurements are indicated by squares.

transfer velocity-wind speed relationship and the corresponding δ^3He fit using the above 20-layer model (figure 2a and 2b). It shows a further 12% increase in the transfer velocity, with $k_{3He} = 1.45W^{1.5}$.

DISCUSSION

Figure 3 shows the Lake Studer results scaled to CO_2 at 20°C (Schmidt number Sc=600) assuming that transfer

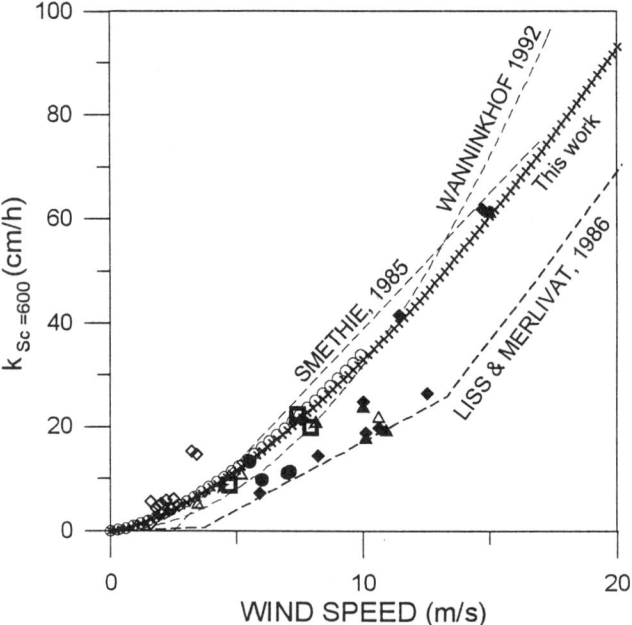

Figure 3. Gas transfer velocities normalized to a Schmidt number Sc=600 assuming n= -0.5. The power law resulting from the Lake Studer experiment (thick solid line in figure 2) is represented by crosses. The dashed curves are the relations of *Smethie et al.*, [1985], *Liss and Merlivat* [1986] and *Wanninkhof* [1992]. ^{3}He-SF$_6$ open-sea experiments are from *Watson et al.*, [1991] (solid triangles), as revised by *Nightingale et al.*, [2000a], *Wanninkhof et al.*, [1993] (open triangles) as reworked by *Asher and Wanninkhof* [1998a], *Nightingale et al.*, [2000a] (solid diamonds), *Nightingale et al.*, [2000b] (solid circles) and *Wanninkhof et al.*, [1997] (star). Lake data are from *Clark et al.*, [1995] (open diamonds) and *Jean-Baptiste and Poisson* [2000] (open circles). Squares correspond to air-sea gas transfer velocities obtained by *Broecker et al.*, [1980, 1985] and *Cember* [1989] from ^{14}C.

velocity is proportional to Sc$^{-0.5}$. Our data are in agreement with oceanic ^{14}CO$_2$ inventories [*Broecker et al.*, 1980, 1985; *Cember*, 1989]. They also support the *Wanninkhof* [1992] relationship at least for wind speeds up to 10 m/s. For stronger winds, owing to the low solubility of helium, scaling to CO$_2$ becomes increasingly problematic due to the greater contribution of bubbles to gas transfer.

Literature Data Dispersion

In addition to the recent dual-tracer data mentioned above, numerous determinations of the gas transfer-wind speed relationship in various environments (including wind tunnels, lakes of various sizes and the open-sea) have been published (see *Liss* [1983]; *Liss and Merlivat* [1986] and *Wanninkhof* [1992] for review), but dispersion among the results remains substantial. Our analysis of the Lake Studer data suggests that significant bias can arise from the simplifying assumptions that are usually made to calculate gas transfer velocities from experimental measurements. In our view, four main reasons can be invoked to explain the observed scatter in the literature data:

a) No direct relationship between wind speed and gas transfer velocity at air-sea interface: wind speed is a key parameter in gas transfer dynamics at air-sea interface and would describe accurately the dynamic state of ocean surface and the physics of gas exchange for steady-state conditions (constant wind speed). However these ideal conditions are rarely verified at sea. Hence, wind speed determines only approximately the real dynamic conditions at the interface.

b) Difficulty in relating the dynamics of the air-water interface in laboratory and small scale experiments to the real conditions at the ocean surface [*Wu*, 1985]. Although hard to quantify, this effect is likely substantial. Hence, large lakes and open-sea experiments should be preferred to parametrize transfer velocity–wind speed relationship at air-sea interface.

c) Role of the bubble mediated gas transfer: in the scaling procedure, the contribution of the bubble mediated transfer can significantly bias the comparison among experiments carried out with gases of very different solubilities, especially for high wind speed regimes [*Asher and Wanninkhof*, 1998a, 1998b].

d) Data processing : The main uncertainties in the calculated transfer velocities are not due to the lack of analytical precision but rather to the various simplifying assumptions that are usually made to deduce transfer velocities from experimental measurements:

- a non-linearity bias affects many published data where average transfer velocities k_{ij} have been computed for average wind speed W_{ij} between successive measurements at times t_i and t_j respectively, according to the standard equation:

$$k_{ij}=h/(t_j-t_i) \ln[(C_i-C_{eq})/(C_j-C_{eq})] \qquad (3)$$

where C_i and C_j are the measured gas concentrations at times t_i and t_j, respectively, C_{eq} is the background concentration and h, the reservoir mean depth. The main drawback of this method is a systematic bias in the determination of the gas transfer velocity–wind speed relationship owing to its nonlinear nature. The offset is a growing function of the wind speed variability and has been shown to amount to 20% [*Jean-Baptiste and Poisson*, 2000] or even more [*Livingstone and Imboden*, 1993].

- the well-mixed reservoir assumption is another usual approximation in tracer experiments. However, even in non-stratified reservoirs, the tracer vertical distribution may not be strictly constant with depth, upper layers being depleted preferentially. These small but significant vertical heterogeneities can have significant consequences on the dispersion of the results when using equation (3) (see figure 12 in *Jean-Baptiste and Poisson* [2000]). Moreover, we see that, in the case of Lake Studer, it leads to a 18% systematic error in the transfer velocity.

- as shown above, the flat bottom reservoir approximation, *i.e.* the use of an average reservoir depth h, instead of a more realistic bathymetry, also leads to further bias.

Substantial errors can also arise from the potential presence of surface films (barrier to diffusion) or even from the calculation of standard wind speeds at 10 m height [*Kwan and Taylor*, 1993; *Upstill-Goddard et al.*, 1993].

CONCLUSION

We have presented new data for the ^{3}He transfer velocity, obtained from two successive evasion experiments in a fresh water lake of the Kerguelen Islands. This study documents the transfer velocity–wind speed relationship in the wind speed range 0-20 m/s. Our results can be described by the power law $k_{3He}=1.45W^{1.5}$, where k_{3He} is the transfer velocity at T=1°C, corresponding to a Schmidt number Sc=310. When scaled to CO_2 at 20°C (Schmidt number Sc=600), the transfer velocities deduced from both time-series are consistent with oceanic $^{14}CO_2$ inventories and are significantly larger than those expected from the *Liss and Merlivat* [1986] parametrization. Finally, our study suggests that a substantial fraction of the scatter among published data could be due to the simplifying assumptions usually made to compute gas transfer velocities from experimental measurements.

Acknowledgements. The Lake Studer project was supported by the Territoires des Terres Australes et Antarctiques Françaises (TAAF).

REFERENCES

Asher W. E. and R. Wanninkhof, The effect of bubble-mediated gas transfer on purposeful dual-gaseous tracer experiments, *J. Geophys. Res.*, 103, 10555-10560, 1998a.

Asher W. E. and R. Wanninkhof, Transient tracers and air-sea gas transfer, *J. Geophys. Res.*, 103, 15939-15958, 1998b.

Broecker W. S., T. H. Peng, G. Mathieu, R. Hesslein and T. Torgersen, Gas exchange rate measurements in natural systems, *Radiocarbon*, 22(3), 676-683, 1980.

Broecker W. S., T. H. Peng, G. Ostlund and M. Stuiver, The distribution of bomb radiocarbon in the ocean, *J. Geophys. Res.*, 90, 6953-6970, 1985.

Cember R., Bomb radiocarbon in the Red Sea: a medium-scale gas exchange experiment, *J. Geophys. Res.*, 94, 2111-2123, 1989.

Clark J. F., P. Schlosser, R. Wanninkhof, H. J. Simpson, W. S. Schuster and D. T. Ho, Gas transfer velocities for SF_6 and ^{3}He in a small pond at low wind speeds, *Geophys. Res. Lett.*, 22, 93-96, 1995.

Jean-Baptiste P., F. Mantisi, A. Dapoigny, M. Stievenard, Design and performance of a mass spectrometric facility for measuring helium isotopes in natural waters and for low-level tritium determination by the ^{3}He ingrowth method. *Appl. Radiat. Isot.*, 43(7), 881-891, 1992.

Jean-Baptiste P. and A. Poisson, Gas transfer experiment on a lake (Kerguelen Islands) using ^{3}He and SF_6, *J. Geophys. Res.*, 105, 1177-1186, 2000.

Kwan J. and P. Taylor, A reassessment of the gas the gas transfer velocity-wind speed relationship from the Siblyback Lake data, *Tellus*, 45B, 296-298, 1993.

Large W. P. and S. Pond, Open ocean momemtum flux measurements in moderate to strong winds, *J. Phys. Oceanogr.*, 11, 324-336, 1981.

Liss P. S., Gas transfer: experiments and geochemical implications, in *Air-sea exchange of gases and particles*, edited by P.S. Liss and W. G. Slinn, pp. 241-298, Reidel, Dordrecht, 1983.

Liss P. S. and L. Merlivat, Air-sea gas exchange rates: introduction and synthesis, in *The role of air-sea exchange in geochemical cycling*, edited by P. Buat-Menard, pp. 113-127, Reidel, Dordrecht, 1986.

Livingstone D. M. and D. M. Imboden, The non-linear influence of wind-speed variability on gas transfer in lakes, *Tellus*, 45B, 275-295, 1993.

Nightingale P. D., G. Malin, C. S. Law, A. J. Watson, P. S. Liss, M. I. Liddicoat, J. Boutin, and R. C. Upstill-Goddard, In situ evaluation of air-sea gas exchange parametrizations using novel conservative and volatile tracers, *Global Biogeochemical Cycles*, 14, 373-387 2000a.

Nightingale P. D., P. S.Liss, P. Schlosser, Measurements of air-sea gas transfer during an open ocean algal bloom, *Geophys. Res. Lett.*, 27, 2117-2120, 2000b.

Smethie W. M., T. T. Takahashi, D. W. Chipman, and J. R. Ledwell, Gas exchange and CO_2 flux in the tropical Atlantic Ocean determined from ^{222}Rn and pCO_2 measurements, *J. Geophys. Res.*, 90, 7005-7022, 1985.

Upstill-Goddard R. C., A. J. Watson and P. S. Liss, A reply to comments by Kwan and Taylor, *Tellus*, 45B, 299-300, 1993.

Wanninkhof R., Relationship between wind speed and gas exchange over the ocean, *J.Geophys. Res.*, 97(C5), 7373-7382, 1992.

Wanninkhof R., W. Asher, R. Weppernig, H. Chen, P. Schlosser, C. Langdon and R. Sambrotto, Gas transfer experiment on Georges Bank using two volatile deliberate tracers, *J. Geophys. Res.*, 98(C11), 20237-20248, 1993.

Wanninkhof R., G. Hitchcock, W. J. Wiseman, G. Vargo, P. B.

Ortner, W. Asher, D. T. Ho, P. Schlosser, M-L. Dickson, R. Masserini, K. Fanning, and J-Z. Zhang, Gas exchange, dispersion, and biological productivity on the west Florida shelf: Results from a Lagrangian tracer study, *Geophys. Res. Lett.*, 24, 1767-1770, 1997.

Watson A. J., R. C. Upstill-Goddard and P. S. Liss, Air-sea gas exchange in rough and stormy seas measured by a dual-tracer technique, *Nature*, 349, 145-147, 1991.

Weiss R. F., Helium isotope effect in solution in water and seawater, *Science*, 168, 247-248, 1970.

Wu J., Parametrization of wind-stress coefficients over water surfaces, *J. Geophys. Res.*, 90(C5), 9069-9072, 1985.

Elise Fourré, LSCE, CEA/Saclay, 91191 - Gif/Yvette cedex, France (email: Fourre@lsce.saclay.cea.fr)

Philippe Jean-Baptiste, LSCE, CEA/Saclay, 91191 - Gif/Yvette cedex, France (email: pjb@lsce.saclay.cea.fr)

Alain Poisson, LPCM, Université Pierre et Marie Curie, 4 place Jussieu, 75252 – Paris cedex 05, France.

Measurement Uncertainty in Gas Exchange Coefficients

J. S. Gulliver and B. Erickson

St. Anthony Falls Laboratory, Department of Civil Engineering, University of Minnesota, Minneapolis

A. J. Zaske

Osmonics, Inc., Minneapolis, Minnesota

K. S. Shimon

Bonestroo, Rosene, Anderlik and Assoc., St. Paul, Minnesota

A technique to measure the gas exchange coefficient (transfer velocity) from a three-dimensional unsteady tracer cloud on large water bodies is described. Arguments based on first principles are presented in support of the technique. For this reason, the measurement uncertainty is apparent and can be quantified analytically. The dual gas exchange measurement technique was tested and results are discussed through computational experiments that simulate a tracer cloud and incorporate random sampling. Measurements of gas exchange coefficient are undertaken in Chequamegon Bay of Lake Superior. The intensive sampling regression technique developed was used in these measurements, and a minimum percent uncertainty was used to determine an optimum analysis period. Results were determined with a 4 to 9 hour sampling period, with no preliminary mixing period.

DETERMINATION OF GAS EXCHANGE COEFFICIENT FROM A THREE-DIMENSIONAL TRACER CLOUD

A primary untested assumption of the dual tracer studies is that the tracers are vertically well-mixed through the length of the experiment. Investigators have gone to extremes to achieve this condition, including waiting a period of two days between dosing with the tracers and sampling, and utilizing large tracer clouds. Still, there is no documentation of whether the vertical mixing is sufficient. In an effort to reduce the mobilization and time required for each field experiment, a new analysis technique that does not include the vertically well-mixed assumption will be developed to analyze data from a fully three-dimensional cloud. This will also allow a reduction in the size of the tracer dose and the sampling time for the analysis.

The general concept of the experiment involves discharging a water slug of two tracers into a water body, similar to *Holley et al.* [1991]. One tracer will be a conservative tracer, with no significant air-water transfer, and the other will be a non-conservative tracer, in that it will have significant gas transfer, and will be labeled the gas tracer. The tracer cloud is then tracked for 8 - 12 hours. Vertical profile samples are taken every 20 - 30 minutes. To compare the two tracer profiles visually, the data points need to be normalized. They are made dimensionless by dividing each concentration by an initial concentration, sampled from the surface immediately following the release of the tracers. These profiles are illustrated in Fig. 1, where C is

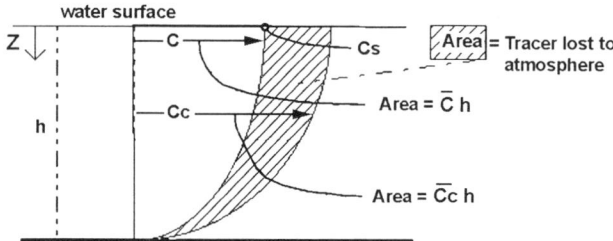

Figure 1. A normalized illustration of the gas and conservative tracer profiles.

the concentration of gas tracer and C_C is the concentration of conservative tracer. From the normalized profile two vertical mean concentrations, $\overline{C}$ and $\overline{C_C}$, can be calculated from the equations

$$\overline{C} = \frac{1}{h}\int_0^h C\,\partial z \qquad \overline{C_C} = \frac{1}{h}\int_0^h C_C\,\partial z \qquad (1)$$

where h is a depth where the flux is negligible and z is the vertical coordinate. Due to horizontal dispersion (primarily due to variations in temporal mean velocities) and turbulent diffusion, both $\overline{C}$ and $\overline{C_C}$ will decrease with time. $\overline{C}$ should decrease faster, as flux into the atmosphere combines with horizontal transport. When determining the dispersion of the gas tracer, it must be understood that the horizontal dispersion of both the normalized gas and conservative tracer concentration is assumed to be equal. Therefore, the temporal gradient in the difference between the two normalized tracer concentration clouds indicates the amount of gas tracer lost from the water body at a given time.

Consider a cylindrical control volume of depth h. The mean concentrations $\overline{C_C}$ and $\overline{C}$ will experience turbulent diffusion in the horizontal plane. In addition, the volatile tracer concentration will be reduced due to surface transfer to the atmosphere. Therefore, similar to *Kilpatrick et al.* [1976] for stream applications and *Watson et al.* [1991] for ocean applications, we will perform a mass balance on the quantity $R = \overline{C}/\overline{C_C}$. The difference is the flux of gas tracer at the surface:

$$F_C = K_L C_S \qquad (2)$$

where C_S is the gas tracer concentration at the surface. This results in a sink term of the quantity R:

$$S_R = K_L C_s / h\overline{C} \qquad (3)$$

If the control cylinder moves with the mean velocity of the tracer cloud, a transport relation for the ratio R (a scalar quantity) is

$$\frac{dR}{dt} = D_x\left(\frac{\partial^2 R}{\partial x^2}\right) + D_y\left(\frac{\partial^2 R}{\partial y^2}\right) - \frac{K_L(C_s)}{h(\overline{C_C})} \qquad (4)$$

where D_x and D_y are horizontal dispersion coefficients, combining the impact of both the vertical velocity profile and turbulent diffusion on horizontal mixing of the cloud. The horizontal dispersive flux terms

$$D_x\left(\frac{\partial^2 R}{\partial x^2}\right) \text{ and } D_y\left(\frac{\partial^2 R}{\partial y^2}\right)$$

should be small compared to the sink rate because the horizontal gradient of R is small. Based upon this assumption, Eq. 4 becomes:

$$\frac{dR}{dt} = -\frac{K_L(C_s)}{h(\overline{C_C})} \qquad (5)$$

Therefore
$$K_L = \frac{-\overline{C_C}h}{C_s}\frac{dR}{dt} \qquad (6)$$

or in dimensionless variables,

$$Sh = \frac{-1}{R_s}\left(\frac{dR}{dT}\right) \qquad (7)$$

where $Sh = K_L h/D$ is a Sherwood number, $R_S = C_S/\overline{C_C}$ is a dimensionless concentration, $T = tD/h^2$ is a dimensionless time, and D is the diffusion coefficient of the tracer gas in water. In theory then, we would need to know $\overline{C}$, $\overline{C_C}$, and C_S, over time and h in order to calculate K_L. Dropping the horizontal flux terms from Eq. 4 is necessary for the analysis, but is an unverified assumption. These two terms, actually, are a major source of uncertainty in the field measurements, as will be discussed later.

DATA ANALYSIS TECHNIQUE

In order to shorten the time required to measure a bulk K_L value with any accuracy, intensive sampling will be employed, with a regression through the data to indicate a temporal and spatial mean surface exchange coefficient. This is similar to the approach used by *Holley et al.* [1991], except that the regression is performed on different vari-

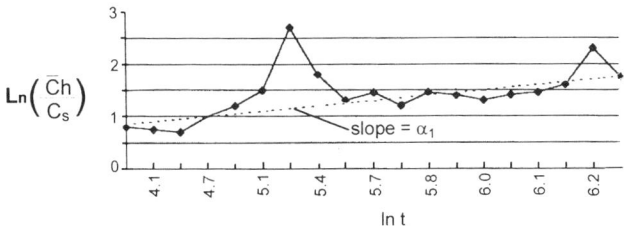

Figure 2. Sample regression of $(\overline{C_Ch}/C_S)$ vs. $\text{Ln}(t)$. The straight line is the log-linear fit of the data.

ables. There are a number of relationships that may be used in the regressions. Our technique uses the following:

1. Regress $\ln(\overline{C_Ch}/C)$ vs. $\ln t$ or $\ln 1/R_S$ vs. $\ln T$ in a linear regression. Then

$$\overline{C_Ch}/C_S = (\beta_1)t^{\alpha_1} \qquad (8)$$

where α_1 and β_1 are fitted constants. A sample for one data set is shown in Figure 2.

2. Regress R vs. $t^{1-\alpha_1}$ (or R vs. $T^{1-\alpha_1}$). Then

$$R = \alpha_2 t^{1-\alpha_1} + \beta_2 \qquad (9)$$

A sample of this regression is shown in Figure 3. Eq. 9 is chosen to result in the proper units for K_L.

Combining Equations 6, 8, and 9 yields:

$$K_L = \beta_1 * t^{\alpha_1} * \left[\frac{\partial}{\partial t}\left(\alpha_2 * t^{(1-\alpha_1)} + \beta_2\right)\right] \qquad (10)$$

or

$$K_L = -\beta_1 * \alpha_2 * (1-\alpha_1) \qquad (11)$$

with the dimensional formulation, β_1 has units of length/time$^{\alpha_1}$ and α_2 has units of time$^{\alpha_1-1}$. Thus K_L has units of length/time.

The precision uncertainty associated with the field sampling is generally much larger than that associated with the analytical technique, which is roughly ± 2% to the 67% confidence interval for the two compounds used as conservative and gas tracers. A technique to determine the precision uncertainty associated with field sampling and incorporated into the mean K_L estimate will therefore be propagated with the first order-second moment analysis [*Abernathy et al.*, 1985].

$$U_{K_L}^2 = \left(U_{\beta_1} * \frac{\partial K_L}{\partial \beta_1}\right)^2 + \left(U_{\alpha_2} * \frac{\partial K_L}{\partial \alpha_2}\right)^2 + \left(U_{\alpha_1} * \frac{\partial K_L}{\partial \alpha_1}\right)^2 \qquad (12)$$

When the partial derivatives are taken from Eq. 11, Eq. 12 becomes:

$$\begin{aligned}U_{K_L}^2 &= \left(\alpha_2 * (1-\alpha_1) * U_{\beta_1}\right)^2 \\ &+ \left(\beta_1 * (1-\alpha_1) * U_{\alpha_2}\right)^2 + \left(\beta_1 * \alpha_2 * U_{\alpha_1}\right)^2\end{aligned} \qquad (13)$$

The variables U_{β_1}, U_{α_1}, and U_{α_2} are the corresponding uncertainty values for each parameter. They are computed to the 67% confidence interval by taking the standard error of each parameter in the regressions, i.e. $\beta_1, \alpha_1, \alpha_2$ and multiplying by their student t-score t_S, i.e. $U_{\beta_1} = t_S * S_{\beta_1}$. Where t_S is the student t-score at the confidence level of interest and S_{β_1} is the corresponding standard error for the parameter β_1.

Thus, the uncertainty in the measurement is a sampling precision uncertainty based on the concept that samples may be affected by edge effects, as will be discussed later. There is also the possibility of sampling bias, which will be addressed in the following section.

NUMERICAL TRACER EXPERIMENTS

The next step in the exploration of this technique was to construct a computer program that would simulate a dual tracer field experiment. This program would then test and provide guidance towards the future application of the technique. A gas exchange coefficient will be entered into the computer code and the numerical experiment will be conducted similar to a field experiment. The tracer cloud will be sampled in a chosen random distribution. The regressions given in Eqs. 8 and 9 will then be run on the numerical tracer data and K_L value will be calculated from Eq. 11 with an uncertainty for the K_L value calculated from Eq. 13. We can then check our calculated results against the entered K_L value.

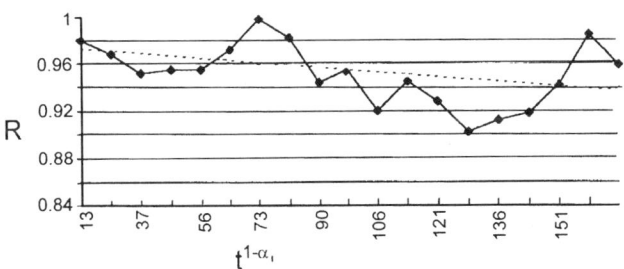

Figure 3. Sample regression of $R = \overline{C}/\overline{C_C}$ vs. t^{1-a_1}.

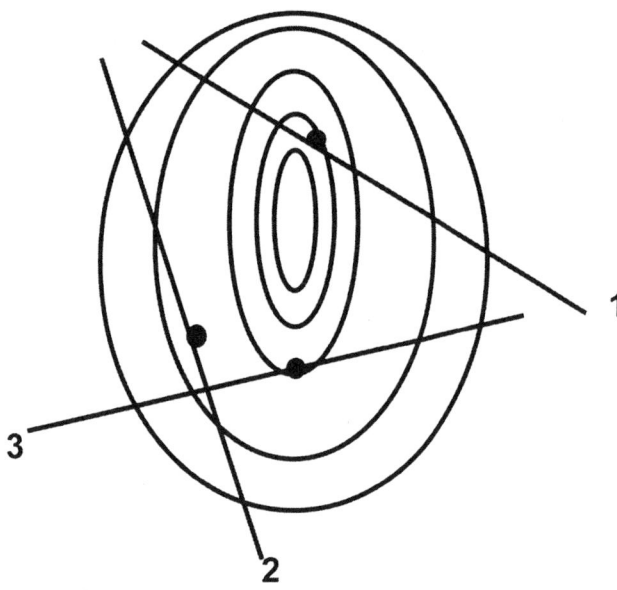

Figure 4. Plan view sketch of dye tracer isoconcentration lines and possible sample locations for three passes, at different times, through the cloud.

In order to have numerical experiments that are representative, certain facts of tracer cloud sampling need to be recognized. The dye cloud is constantly moving and mixing, and therefore it is impossible to be consistent in our sampling location, with respect to the cloud. The field experiments will always attempt to sample at the point of maximum surface concentration, but will probably miss with varying degrees of accuracy, as shown in Fig. 4. Thus, a certain amount of randomness is inherent in this type of experiment because the samples cannot be taken from the same location every time. In the field, the only indicator of the center of the cloud would be the surface concentrations of the dye, which would be measured along a path a illustrated in Fig. 4. It was determined that Monte Carlo sampling would be appropriate to simulate the randomness of field sampling. All of the distributions considered required a mean sampling location and a standard distribution in distance about the mean. Thus, the results of the computational routine could be sampled with a random distribution that we specified.

The weighted mean location and standard deviation of the location were both calculated with respect to the surface concentrations. Then random numbers could be generated about the mean, and constrained by the standard deviation in a normal or uniform distribution. The random numbers were generated on the computer to the following specifications: normal distribution, normal distribution with cutoff at ± 1 sigma, uniform distribution cutoff at ± 1 sigma, uniform distribution with cutoff at ± 2 sigma, and uniform distribution with cutoff at 3 sigma. Using these different scenarios of random sampling, one of the methods should be fairly representative of what an experiment could hope to obtain. A few runs were performed for each case, mass transfer coefficients calculated and then the percent uncertainties were calculated. The expectation was that the average uncertainty should be between 20 - 40% of the calculated value. This was based on preliminary field experiments that we had performed on Lake Waconia, Minnesota [*Shimon*, 1993]. These calculations should also help define a sampling precision for this technique of gas exchange rate measurement. The experiment began with an initial cloud size of 20 m wide and 3 m deep, similar to the Lake Waconia experiments. The water body was 5 m deep and large compared to the tracer cloud size and location in the horizontal dimensions. The wind velocity at 10 m height was 3 m/s, and the gas transfer coefficient was equal to 3 cm/hr. The experiments were conducted for 12-hour periods.

Table 1 shows that the uniform random technique constrained by ± 2 sigma seems to give us the percent uncertainty closest to those from Lake Waconia in the mass transfer calculations. This sampling precision will be an important part of our uncertainty. In fact, it will dominate the other causes of uncertainty. The uniform sampling distribution best represents the tracer sampling instead of a normal distribution because it is difficult to find the peak, as illustrated in Fig. 4.

The computational routine was also used to determine how long the experiment should be run. The average time previously required for the accurate calculation of a gas exchange coefficient, by the techniques described in the literature, is approximately 48 hours. Some experiments, however, had been done in as little as 12 hours. Because this experiment incorporates a new analysis technique, which does not assume a vertically well mixed tracer, it was necessary to estimate how long an experiment should be conducted. Twenty numerical experiments with Monte Carlo samplings of a uniform distribution between ± 2σ were conducted and values of the exchange coefficient and precision uncertainty of the 67 percent confidence interval were calculated for each experiment. The K_L value and

Table 1. Table of calculated percent uncertainties for various sampling scenarios. No sampling was performed outside the cutoff level.

Distribution	Cutoff	Uncertainty
Normal	None	13%
Normal	± 1σ	9%
Uniform	± 1σ	15.9%
Uniform	± 2σ	27.5%
Uniform	± 3σ	51%

precision uncertainty were then calculated for various sampling periods. Table 2 gives the results.

This data gives an indication that a valid estimate for K_L can be calculated after approximately 8 to 10 hours with intensive sampling and our analysis technique. The calculations at 25 and 26 hours are based on sampling done every half-hour for the first 12 hours, and then sampling re-started at the 24th hour and continued every half hour for the next two hours. This is due to the fact that for our experiments, it will not be logistically feasible to sample throughout the night with the equipment and personnel that were available. This data does not conclusively support the "next day sampling" strategy as a significant improvement over a straight 12-hour experiment.

Calculations of mass transfer may also have a significant bias due to the asymmetry of velocity-driven dispersion processes. Multiple runs with the computer program can also assist in the detection, analysis and description of this bias. Table 2 indicates that there may be some bias in our experiment because the mean value of K_L after 20 runs was not identical to that specified at 3 cm/hr into the computational routine. This apparent bias varies consistently from +2% at an experimental period of four hours to -7% at 26 hours. At 12 hours the bias was -3%. This bias sampling uncertainty is small when compared to the precision sampling uncertainty of roughly 20% and greater. Thus, while a bias may be present, its influence is dwarfed by the uncertainty associated with the sampling precision.

COMPARISON OF MEASUREMENT TECHNIQUES

The last use of the computational program is the comparison of the intensive sampling-regression technique to determine K_L with a two-point technique where measurements are made at the beginning and end of a given period. The vertically well-mixed assumption, $C_s \cong \overline{C}$, will also be applied to the two-point technique. Then Eq. 6 becomes

$$K_L = \frac{h(dR/dt)}{R} \quad (14)$$

which can be integrated to give

$$K_L = \frac{h}{\Delta t} \ln\left(\frac{R_i}{R_r}\right) \quad (15)$$

where R_i and R_r are average mixed layer ratio of gas tracer and conservative tracer concentrations at the beginning and end of a time period, respectively. This dual tracer technique is similar to that used in prior tracer experiments. An uncertainty cannot be computed for one day's measurements based on equation 15 because it is only based on two measurements. However, the numerical routine may be used to conduct a number of experiments, and the standard deviation of the results is indicative of the uncertainty (to the 67 percent confidence interval) that can be expected.

Twenty runs of the mass transfer simulation were conducted, and various time intervals were used in each analysis. A standard deviation of the different K_L values was calculated to assist in the comparison of the two techniques. The experiments were started with an initial cloud of area 20x3 meters, a wind speed of $U_{10} = 3$ m/s, and a transfer coefficient of 3 cm/hr. The depth was again 5 m, and the horizontal dimensions of the water body were large, compared to the location and size of the cloud.

Table 3 indicates that the technique designed to consider the three-dimensionality of tracer clouds and incorporate intensive sampling can provide an acceptable value and uncertainty within a shorter measurement period. The mean K_L values are not quite 3.0 cm/hr, indicating some bias. It is unclear what this bias is due to. The positive sign is that the values are well within the uncertainty of the measurements as indicated in table 2. The standard deviations for the intensively sampled 3-D tracer cloud of table 3 are similar to the uncertainty to the 67 percent confidence level in table 2, as expected. Also evident in table 3 is the progression of standard deviation towards smaller values with time of the experiment. The standard deviation of the dual sample technique with the vertically well-mixed assumption is much larger than the new, intensive sampling technique, as would be expected in an actual field experiment.

Based on the computational experiments presented, this 3-D tracer cloud method with intensive sampling should provide an acceptable K_L measurement with shorter experiments than the two-sample technique. In addition, a sampling uncertainty is computed and should provide the means to discard or weigh data more heavily in future analyses.

Table 2. Variations in uncertainty to the 67 percent confidence interval with experimental period. A K_L value of 3 cm/hr was to input to the program with a wind velocity of 3 m/s at U_{10}

Period of the Experiment (hrs)	Mean K_L (cm/hr)	Precision Uncertainty (cm/hr)	Percent Precision Uncertainty (%)
4	3.35	1.29	38.7
6	3.04	0.98	32.3
8	2.73	0.77	28.3
10	2.88	0.67	23.3
12	2.90	0.63	21.5
25 (gap=12)	2.70	0.56	20.2
26 (gap=12)	2.79	0.54	19.4

Table 3. A comparison of K_L calculation with two techniques (Results based on 20 experimental runs)

Time	Intensively Sampled 3-D Tracer Cloud		Dual Samples Vertically Well Mixed Assumption Mean	
	Mean K_L (cm/hr)	Standard Deviation (cm/hr)	K_L (cm/hr)	Standard Deviation (cm/hr)
4	3.35	1.78	3.27	3.35
6	3.04	1.32	3.09	2.44
8	2.73	0.45	3.37	1.78
10	2.88	0.57	3.68	1.69
12	2.9	0.57	3.25	1.77
25	2.78*	0.35	2.89	0.99
26	2.79*	0.34	2.88	1.02

* (gap=12hrs)

FIELD EXPERIMENTS

Four field experiments were conducted on Chequamegun Bay in Lake Superior. The calculation of results for the field experiments began by incorporating all data points that were gathered, analyzed, and calculated. Equations 8 and 9 were fit by least-squares regression to the field data, with the coefficients used in the final K_L calculation (Eq. 6).

Equation 11 was then used to calculate uncertainty, U_{KL}. The best data period for each field experiment was found by iteration where the objective was to minimize U_{KL}/K_L. The data at the beginning and end of a sequence and end of a sequence can have a great impact on the resulting fit. The first couple of data points, collected early in each experiment, also have the potential for being the least accurate measurements. This is due to the lack of size of the cloud and limited mixing of the tracers. Data points were dropped off the beginning and/or end sequentially, then the curve fit process was repeated. These points were then graphed to look at the changes that occurred. Once the optimum period was determined by the minimum U_{KL}/K_L value, K_L and U_{KL} values were recorded for each experimental period. The results are summarized in table 4.

The uncertainty is broken up into three components in Table 4, which represent the uncertainty of each fitted coefficient and can be compared to other field experiments. It can be seen that the source of the higher uncertainty in the September 21 and 22 measurements is the uncertainty associated with the α_2 coefficient.

The results from these experiments are plotted with those of similar, large water body experiments in Fig. 5. The results are below those of Georges Bank and at about the same trend line versus wind speed as the measurements in the North Sea. It should not be surprising, however, that these measurements have different trends with wind velocity. First, the sampling uncertainty of the Georges Bank and North Sea measurements was not evaluated, so the quality of the data cannot be fully assessed. Second, wind velocity is a secondary independent variable that does not directly influence the gas exchange coefficient. Rather, gas exchange coefficient is affected by wind shear, wave breaking, etc. These independent variables are also impacted by wind fetch, wave age, and water velocity. Until the influence of these primary variables is well-characterized, there is little reason to expect all of the data to follow the same trend line.

CONCLUSIONS

Dual tracer studies have been conducted before, and deemed successful. This success typically involved extending the time period of the experiment until a reasonable K_L value was determined. This paper seeks to provide a statistical rationale for the decisions made by investigators. A sampling uncertainty is computed for each experiment that should be included with any experimental data. If this uncertainty is too large for any useful conclusion to be reached, then the field experiment failed to reach its primary goal.

Other investigators have gone to some effort to assure that the water column is vertically well-mixed in order to meet the requirements of data analysis. This is no longer necessary with the three dimensional tracer cloud analysis presented herein. The computational and field experiments have shown that the technique can be relatively accurate, even for a fairly small tracer cloud.

Acknowledgements. This work is the result of research sponsored by the Minnesota Sea Grant College Program supported by the NOAA Office of Sea Grant, Department of Commerce, under Grant No. DOC/NOAA-86AA-D-5G112. The authors would also

Table 4. Field experiment results with dual tracers and intensive sampling

Date	U* (m/s)	U_{10} (m/s)	Time Analyzed (hrs)	Total Experiments (hrs)	K_L (cm/hr)	Uncertainty (cm/hr)
July 15	0.112	3.5	9.0	9.3	2.7	0.6
August 12	0.113	3.5	8.2	9.7	3.5	1.6
September 21	0.056	1.9	7.8	10.2	1.4	4.8
September 22	0.246	6.7	4.3	8.0	12.8	4.6

like to thank the Apostle Islands National Park Service stationed in Bayfield, Wisconsin, for the use of their boat and captain throughout the field experiments.

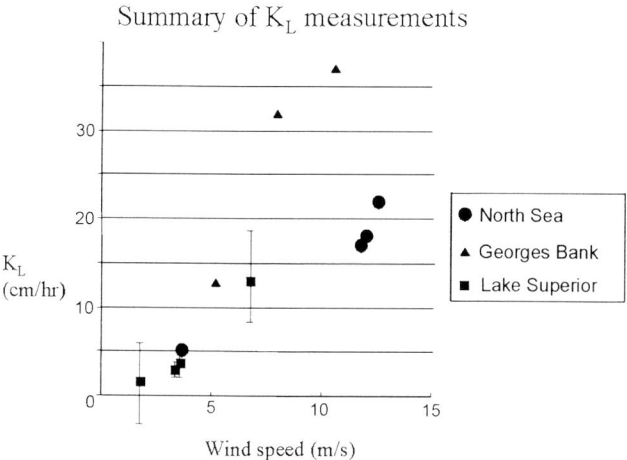

Figure 5. Results of K_L measurements. U_{10} is the extrapolated logarithmic-law velocity that would be present at 10m height if the sea-surface boundary layer extended to 10m. The North Sea data are repoerted in Watson et al., 1991. The Georges Bank datda are reported in Wanninkhof and Asher, 1993.

REFERENCES

Abernathy, R. B., R. P. Benedict, and R. B. Dowdell, "ASME measurement uncertainty," *J. Fluids Engineering*, 107(2), 161-164, 1985

Holley, E. R., C. W. Downer and G. H. Ward, "Tracer gas transfer technique for shallow bays," in *Air Water Mass Transfer*, S. C. Wilhelms and J. S. Gulliver (eds.), ASCE, New York, NY, 234-243, 1991.

Kilpatrick, F. A., R. E. Rathbun, N. Yotsukura, G. W. Parker, and L. L. DeLong, "Determination of stream reaeration coefficients by use of tracers," Book 3, Chapt. A18. *Techniques of Water Resources Investigations of the U.S. Geological Survey*, 1989.

Shimon, K.S., A gas transfer measurement technique for lakes, M.S. Thesis, University of Minnesota, Minneapolis, MN, 146 pp., 1993.

Wanninkhof, R. and W. Asher, "Gas transfer experiment of Georges Bank using two volatile delibrate tracers," *J. of Geophys. Res.*, 98(C11), 20237-20248, 1993.

Watson, A. J., R. C. Upstill-Goddard, and P. S. Liss, "Air-sea gas exchange in rough and stormy seas measured by a dual tracer technique," *Nature*, 349, 145-147, 1991.

J. S. Gulliver and B. Erickson, St. Anthony Falls Laboratory, University of Minnesota, Mississippi River at 3rd Ave. SE, Minneapolis, MN 55414.

A. J. Zaske, Osmonics, Inc., 5951 Clearwater Dr., Minnetonka, MN 55343.

K. S. Shimon, Bonestroo Rosene Anderlok & Assoc., 2335 West Highway 36, St. Paul, MN 55113.

Measurements of Free Surface Turbulence

Joseph J. Orlins

Civil and Environmental Engineering, Rowan University, Glassboro, New Jersey

John S. Gulliver

St. Anthony Falls Laboratory, Department of Civil Engineering, University of Minnesota, Minneapolis, Minnesota

Free-surface turbulence is important to the transfer of heat and mass across an air-liquid interface. In the absence of wind shear and surface waves, turbulence generated below the water surface (as from a streambed) is one of the controlling factors in air-water mass transfer. Research has shown that the temporal fluctuations in the two-dimensional divergence of the surface velocity field are related to the liquid-film mass transfer coefficient. To aid in understanding the relationship between free-surface turbulence and mass transport, an oscillating grid chamber has been used to study turbulence at a water surface. A horizontal grid of square bars was oscillated vertically beneath the water surface. Turbulence generated by the grid propagates up towards the water surface, generating near-two dimensional turbulence right on the free surface. Particle Image Velocimetry (PIV) was used to measure the temporally varying flow pattern on the water surface for three different energy levels in the chamber. Data were recorded over eight different sub-regions of the water surface, and the velocity, vorticity, and two-dimensional divergence were calculated as functions of space and time for each region. The spatial variations in two-dimensional divergence are shown to be similar to those found in flume experiments; this suggests that that the oscillating grid chamber may be used as a direct analogue to open-channel flows for studying interfacial transport phenomena.

INTRODUCTION

Turbulence on the surface of free liquid flows is of importance in the transfer of heat and mass across the liquid-atmospheric interface. Reaeration in streams, mixing in water and wastewater treatment processes, and other natural and engineered processes rely on transfer of mass from one phase to another across such a surface. These transport processes are controlled by both gas- and liquid-film transfer coefficients; for slightly soluble compounds such as oxygen, the transfer rate is dominated by the liquid-side mass transfer coefficient [Jahne and Haussecker, 1998]. The liquid-film coefficient in turn is related to features of the flow, in particular the turbulence structure right at the interface [Gulliver and Tamburrino, 1995]. In many cases (such as open channel flows free from wind shear), the turbulence is generated well below the liquid surface (e.g. at the stream bed or channel bottom) and propagates upwards towards the air-liquid interface.

The present study investigates the turbulence on the free-surface of water in an oscillating grid tank. The tank was constructed for studies of chemical fluxes from sediments to water and from water to the atmosphere. Details of the construction and quantification of bulk turbulence parameters using Laser Doppler Velocimetry, as well as measurements of the oxygen mass transfer coefficient using this device are described elsewhere [*Orlins*, 1999].

The turbulence on the free liquid surface was investigated using Particle Image Velocimetry. In this paper, we will discuss the experimental aspects and data reduction, examine results from the present study, and conclude with a comparison between free-surface turbulence generated with this device and that from other flumes and grid-stirred tanks reported in the literature.

EXPERIMENTAL ASPECTS

Oscillating Grid Chamber

An oscillating grid is often utilized to approach horizontally homogeneous turbulence in a small chamber. The oscillating-grid chamber has been used as a more convenient substitute for flumes and tanks, especially when investigating the transport of potentially toxic substances [*Connolly, et al.*, 1983; *Valsaraj, et al.*, 1997]. The turbulence characteristics in the bulk of the fluid have been well characterized [*e.g., Hopfinger and Toly*, 1976; *Thompson and Turner*, 1975], but the turbulence at the water surface has not previously been measured.

Since there is no mean shear in the oscillating grid mixing chamber, the total kinetic energy (TKE) of the turbulence can be used as a measure of the energy in the system. The TKE can be defined from the r.m.s. velocity fluctuations:

$$\text{TKE} = \tfrac{1}{2}(u'^2 + v'^2 + w'^2) \quad (1)$$

For a grid with square bars, it has been shown [*e.g., Hopfinger and Toly*, 1976; *DeSilva and Fernando*, 1992] that the horizontal (u', v') and vertical (w') r.m.s. fluctuations in the bulk of the fluid can be described by:

$$\begin{aligned} u' = \sqrt{\overline{u^2}} = C_1 M^{0.5} S^{1.5} f z^{-1} \\ w' = \sqrt{\overline{w^2}} = C_2 M^{0.5} S^{1.5} f z^{-1} \end{aligned} \quad (2)$$

where S is the stroke length, f is the oscillation frequency, z is the distance away from a virtual origin, M is the mesh spacing of the grid, and C_1 and C_2 are constants that may depend on the geometric parameters of the grid. Here, u and w are the instantaneous velocity fluctuations ($u = U - \overline{U}$; $w = W - \overline{W}$, where U and W are the instantaneous velocities and the overbars denote the mean values).

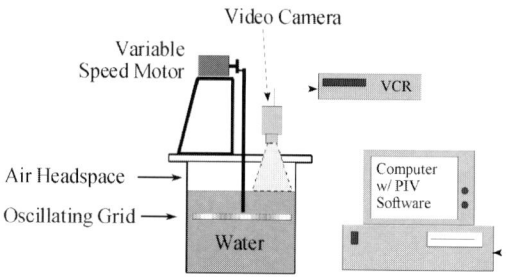

Figure 1. Experimental Setup.

Hopfinger and Toly report values of C_1 and C_2 of approximately 0.25 and 0.27, while DeSilva and Fernando measured values of approximately 0.22 and 0.26, respectively.

For the present work, experiments were conducted in a square chamber 50 cm on a side, with a horizontal 7x7 grid of 12 mm square bars spaced 62.5 mm apart. The grid was connected by a vertical shaft and eccentric drive to a variable-speed DC motor with a programmable speed controller. A stationary sleeve mounted to the lid of the tank surrounded the vertical drive shaft and projected below the water surface, to minimize surface waves caused by the motion of the shaft. The grid stroke length could be varied from 2 to 4 cm, and the vertical oscillation frequency maintained in the range 3 to 7 Hz.

For the tests described here, the water depth was 30.6 cm, the grid centerline elevation was 16.6 cm from the tank bottom, the stroke length was kept constant at 3 cm, and the oscillation frequency was set to 3, 5, or 7 Hz, depending on the test. The corresponding total kinetic energy of turbulence (TKE) calculated using Equation (1) near the water surface could thus be varied from about 0.7 to 4.0 cm^2/sec^2. Normal tap water was used for the experiments.

Particle Image Velocimetry System

The turbulence right on the water surface was measured using Particle Image Velocimetry (PIV), as shown schematically in Figure 1. A commercial Hi-8 video camcorder was used to record the motions of polystyrene particles floating on the water surface for different turbulence conditions in the tank. By recording the particle motions over relatively long times (1-5 minutes), both the spatial and temporal nature of the free-surface turbulence could be investigated.

Image capture system. The camera field of view was set to span approximately 18 cm to obtain the required resolution for capturing the small eddies using the PIV technique. To ascertain the spatial variations in free-surface turbulence within the tank, eight separate sub-areas of the water surface were recorded to provide overlapping coverage of the full width of the tank. Five minutes of the free-surface motions were recorded for each grid oscillation frequency at each of the eight measurement locations. The camera recorded images at a nominal rate of 30 frames per second, resulting in a total of approximately 9000 images that could be digitized for analysis.

Velocity field calculations. The resulting video images were digitized using a computer-based frame-grabber and stored as 8-bit grayscale image files with 640x480 pixels. Image scaling was determined by floating a ruler on the water surface, and determining the number of pixels spanned by a given distance.

The free-surface velocity was calculated from pairs of images using a commercial PIV analysis program (INSIGHT-NT, by TSI, St. Paul, MN) with a cross-correlation algorithm. The spot size for flow field interrogation was 32x32 pixels, with a 50% overlap between analysis spots. This resulted in approximately 26 rows x 38 columns of vectors for each image, depending on the size of the actual image analysis area used. For each operating condition and camera location, velocity fields were calculated for 1800-2100 image pairs (60 to 70 seconds), providing a spatial and temporal record of the free surface motions. Program output consisted of horizontal and vertical pixel displacements for each analysis spot.

Preliminary analysis of the surface flow patterns indicated that the velocities over much of the tank were relatively low, ranging from about 3 to 23 mm/s. In order to distinguish particle motions using the PIV technique, it is best when particles move at least one pixel between the digitized images. In addition, to minimize the number of "lost pairs" between images, the maximum pixel displacement should be less than ¼ of the spot size [*Keane and Adrian*, 1990; 1993]. To achieve these ranges of pixel displacements, the cross-correlation was performed between every third image, with a resulting time differential of 3/30 or 0.1 seconds. Processing was done on all video frames in the sequence, and the resulting 0.1-second displacement fields were interleaved to synthesize a "smoothed" 30 Hz data set. The interleaving was achieved by computing the velocity field between every third frame, without skipping any frames. For example, the first velocity field would be calculated between frames 1 and 4; the second between frames 2 and 5, the third between frames 3 and 6, the fourth between frames 4 and 7, and so on.

Post-processing of velocity files. PIV vector validation was done with a shareware post-processing program. Spurious vectors were removed and vacant vector regions were filled using the program CLEANVEC [*Soloff and Meinhart*, 1999] from the Laboratory for Turbulence and Complex Flow at the University of Illinois at Urbana-Champaign.

The resulting "cleaned" vector files were then converted to engineering units (coordinates → mm, pixel displacements → velocities in mm/s), and the velocity variance (*i.e.*, measurement uncertainty) was calculated for each vector. After the raw pixel displacements were converted to engineering units, the vector fields were then smoothed using a 3x3 (Gaussian) weighting matrix.

The two-dimensional vector flow field on the water surface was used to calculate the vorticity and two-dimensional divergence of the flow field over space and time. The vorticity and two-dimensional divergence of the flow field were calculated using central differences.

Power spectra of velocity and divergence at 9 points in each of the eight imaging regions were calculated from the time series data sets, and then ensemble averaged to create a representative power spectrum for each operating condition.

RESULTS AND DISCUSSION

Qualitative Flow Field Characterization

A sample of the results is shown in Plate 1, which gives typical velocity, vorticity, and 2-D divergence for a grid oscillation frequency of 5 Hz. The corresponding TKE near the free-surface calculated using Equation (1) is approximately 2.0 cm^2/s^2. These measurements are shown for two instants, with a temporal difference of 0.1 seconds.

The velocity field on the water surface was well developed, changing gradually over time and space. Local velocities ranged from near zero to over 20 centimeters per second, depending on the grid oscillation frequency and the location in the tank. For a grid oscillation frequency of 3 Hz, the peak surface velocity was about 8 cm/second. The average magnitude of the surface velocity increased at a grid oscillation frequency of 5 Hz, with regions of "hot spots" or intense rapid surface movement associated with upwelling of fluid from the bulk, as shown in Plate 1a and 1d. At a grid oscillation frequency of 7 Hz, the surface flow patterns were much more active, with strong circulation patterns over some image regions and areas of little surface movement in others.

Eddies form and disappear on the free surface and persist in stable forms for relatively long periods of time. The vorticity field evolves relatively slowly over time (tens of seconds), with spatial scales on the order of 2 cm for the smaller eddies and 10 cm for the larger ones.

There is little variation in vorticity over a time span of 0.1 seconds, as shown in Plate 1b and 1e. However, the vorticity field does undergo substantial temporal changes as the eddies appear, evolve, migrate, and disappear over longer time periods. It appears that eddies first form in the

250 FREE-SURFACE TURBULANCE

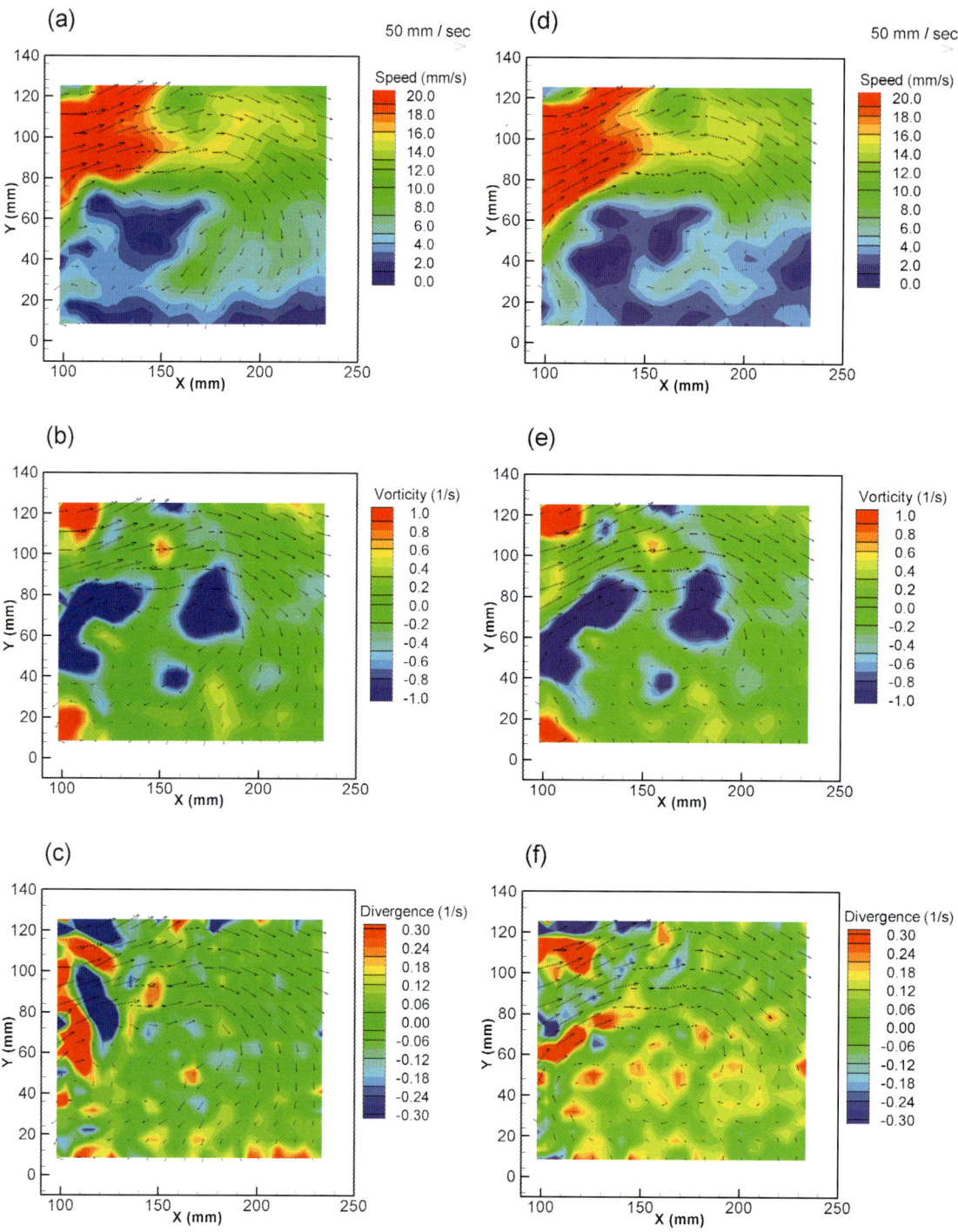

Plate 1. Typical velocity, vorticity, and divergence for grid oscillation frequency of 5 Hz. (a) – (c): time = t; (d) – (f): time = t + 0.1 second.

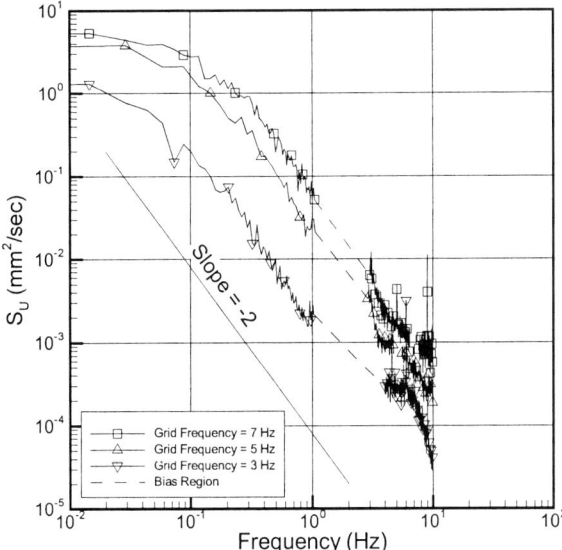

Figure 2. Frequency spectra of horizontal velocity component.

corner regions of the tank for small energy inputs (*i.e.* low oscillation frequencies). As the grid frequency is increased, these vortex structures increase in size and magnitude, and additional regions of high vorticity form near the sides and then near the center of the chamber. At the highest energy level investigated (grid frequency of 7 Hz), the regions of large vorticity magnitude appear evenly distributed over the water surface of the tank.

The two-dimensional divergence (Plate 1c and 1f) has both spatial and temporal scales much smaller than the large-flow features such as vorticity. The length scale of the finer features is on the order of 1 cm. The temporal variations are rapid, with the divergence changing often, at frequencies approaching the image capture rate. The spatial distribution of divergence is qualitatively quite similar to that measured in open channel flows [*Tamburrino*, 1994; *Kumar, et al.*, 1998].

The overall magnitude of the divergence is similar to that found by McKenna and McGillis (2001). In their work, the authors also used a PIV system to measure free-surface turbulence and divergence, but on a much smaller spatial scale. In addition, the variation in surface turbulence and divergence with total energy input to the system (*e.g.* grid oscillation frequency) is not shown. As a result, it is difficult to directly compare those results to the present work.

Spectral Analysis

For each of the eight image analysis regions, time series data were extracted at nine locations from the measured flow field. Frequency spectra of the U-component of veloc-

ity and divergence were computed for each of these 72 locations, and then averaged to create an overall spectrum for each of the three grid oscillation frequencies tested.

The frequency spectra of free-surface velocity are given in Fig. 2. Spectral peaks and noise in a "bias region" at and near the grid oscillation frequencies have been removed for clarity. These curves seem to indicate a slope of –2 on a log-log plot in the frequency decade above that of the peak values. This is so consistent that we believe it may have some relevance for free-surface flows.

Comparisons were made with the velocity spectra in the bulk of the flow, as shown in Figure 3 for a 5 Hz grid oscillation frequency. The spectra of free-surface velocities have a magnitude that is approximately a factor of $10^{3.5}$ lower than the spectra in the bulk flow. This seems to indicate that there is significant dissipation of turbulence as the free surface is approached. In addition, while the bulk flow spectra follow a slope of –5/3 between 1 and 40 Hz, corresponding to the inertial subrange, the spectra on the water surface follows a slope of –2 between 0.1 and 5 Hz.

It is interesting to make a comment regarding the slope of the velocity spectra. In three-dimensional turbulence, a –5/3 slope is representative of the inertial sub-range. However, this is not the case here because close to the free surface, the motion is restricted in the vertical direction and the flow becomes more two-dimensional. For two-dimensional turbulence, the inertial sub-range is characterized by a proportionality to κ^{-3} [*Batchelor*, 1969]. A dependency on $\kappa^{-5/3}$ has been detected in the productive sub-

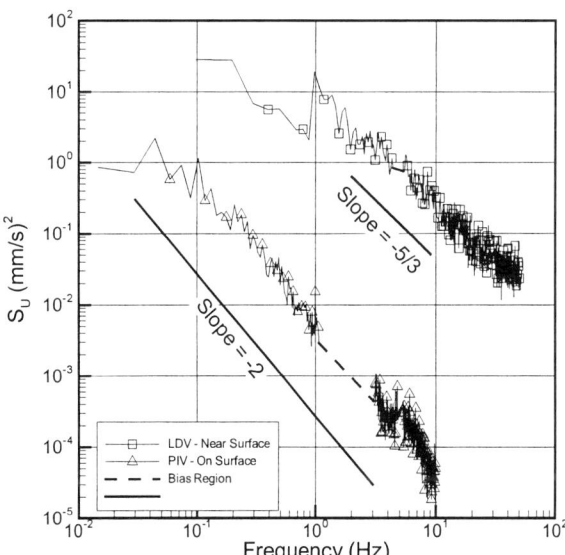

Figure 3. Frequency spectra of horizontal velocity in grid in grid-stirred tank. Top curve: measured with LDV, below water surface. Bottom curve: measured with PIV on water surface.

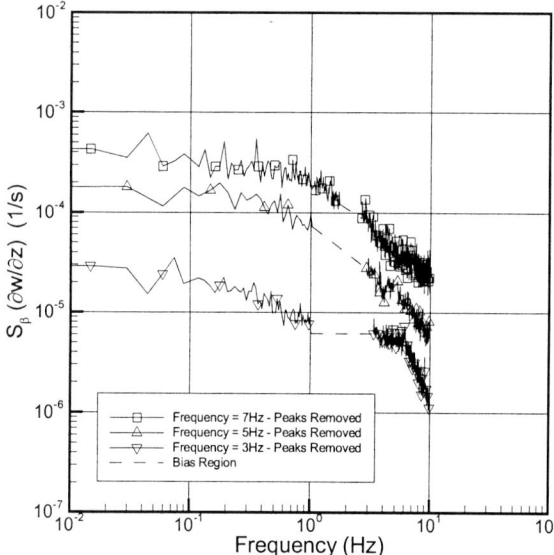

Figure 4. Frequency spectra of 2-D divergence.

range [*Farge*, 1992]. There is, however, little turbulence production at a free-surface with no wind, and we are well below the frequency of the inertial sub-range. Thus, these spectra represent neither the production of turbulence nor the inertial sub-range.

The -3 slope of the inertial subrange for free-surface turbulence seems to occur above a frequency of 5 Hz. There are few studies of free-surface turbulence to compare with these results. Brumley and Jirka (1987), however, used rotating hot-film probes to measure the turbulence below but near the free surface in a grid-stirred tank. In their work, a moving probe was used to measure a turbulence field that was varying in both space and time. Brumley and Jirka calculated spatial power spectra from their velocity measurements. Their results have been normalized and are based on wave number, so it is difficult to make direct comparisons to the present study, but qualitative comparisons can be made.

In Brumley and Jirka's work, the turbulence spectra in the bulk of the flow follow a $-5/3$ slope. As the free surface is approached, a pattern develops in the free-surface velocity spectra with a slope of -3 at the higher frequencies and a slope of -2 at the lower frequencies, similar to what is seen in Fig. 3. Since there is no production of turbulence at the free-surface in these flows, we believe that the -2 slope is related to the stretching of turbulent eddies from 3-dimensional into 2-dimensional forms.

The spectra of 2-D divergence ($\partial w/\partial z$) were also computed, as shown in Fig. 4. The spectra appear to follow a slope of -1 until a low-frequency plateau is reached, similar to the spectra of McCready, *et al.* (1986).

CONCLUSIONS

Measurements made of free-surface turbulence in an oscillating grid chamber were made using Particle Image Velocimetry. The results were compared with prior studies of turbulence below the free surface in oscillating grid tanks. Turbulence intensity increases with increasing grid oscillation frequency, as do surface velocity, vorticity, and two-dimensional divergence. The velocity and calculated vorticity change slowly over time and space, while the divergence fluctuates more rapidly and over a smaller spatial scale. The velocity spectra at the water surface decays in the inertial range more rapidly than in the bulk, which is in agreement with measurements made by Brumley and Jirka (1987) just below the free surface. The predominant slope of the spectra was -2. At higher frequencies, the slope was -3, which corresponds to the dissipation range for free-surface turbulence.

The spatial variability of 2-D divergence in the grid-stirred tank is strikingly similar to that found from Kumar, *et al.*'s (1998) measurements in a laboratory flume, and the spectra of 2-D divergence are similar to those obtained by McCready, *et al.* (1986). Since both heat- and mass-transfer across the air-water interface are related to the spectrum of 2-D divergence, this suggests that the oscillating grid chamber may be appropriate as a direct analogue to open-channel flows for studying interfacial transport phenomena. Identification of the relationship between free-surface turbulence in an open channel flow and surface renewal rate in open-channel flows and stirred chambers will allow more detailed investigations of mass transfer in the future.

Acknowledgments. This research was funded in part by support from the National Science Foundation under grant no. 9522171 and by a Doctoral Dissertation Fellowship from the University of Minnesota. The authors would like to thank Dr. Steve Anderson and Mr. Steven Soloff for their help with various aspects of PIV data processing.

REFERENCES

Batchelor, G.K., Computation of the energy spectrum in homogeneous two-dimensional turbulence, *Physics of Fluids* 12(12), 11,233-239, 1969.

Brumley, B.H., and G.H. Jirka, Near-surface turbulence in a grid-stirred tank, *J. Fluid Mech.* 183, 235-263, 1987.

Connolly, J.P., N.E. Armstrong, and R.W. Miksad, Adsorption of Hydrophobic Pollutants in Estuaries, *J. of Environmental Engineering,* 109(1), 17-35, 1983.

DeSilva, I.P.D. and H.J.S. Fernando, Some aspects on mixing in a stratified turbulent patch, *J. Fluid Mech.* 240, 601-625, 1992.

Farge, M., The continuous wavelet transform of two-dimensional turbulent flows, *Wavelets and Their Applications,* edited by

M.B. Ruskai, G.G. Beylkin, R. Coifman, I. Daubechies, S. Mallat, Y. Meyer, and L. Raphael, pp. 275-302, Jones and Bartlett Publishers, 1992.

Gulliver, J.S., and Tamburrino, A., Turbulent Surface Deformations and Their Relationship to Mass Transfer in an Open-Channel Flow. Air-Water Gas Transfer, Proceedings of 3rd International Symposium on Air-Water Gas Transfer, edited by B. Jahne and E.C. Monahan, pp. 589-600, AEON Verlag & Studio, Hanau, Germany, 1995.

Hopfinger, E.J., & Toly, J.A., Spatially decaying turbulence and its relation to mixing across density interfaces, *J. Fluid Mech.*, 78:1, 155-175, 1976.

Jahne, B.; Haussecker, H., Air-water gas exchange, Annual Review of Fluid Mechanics, 30: 443-468, 1998.

Keane, R.D. & R.J. Adrian, Optimization of particle image velocimeters. Part I: Double pulsed systems, *Meas. Sci. Technol.* 1, 1202-1215, 1990.

Keane, R.D.; Adrian, R.J., Theory of cross-correlation analysis of PIV images, in *Flow Visualization and Image Analysis*, edited by F.T. M Nieuwstadt, pp. 1-25, Kluwer Academic Publishers, 1993.

Kumar, S., R. Gupta, S. Banerjee, An experimental investigation of the characteristics of free-surface turbulence in channel flow, *Physics of Fluids,* 10(2): 437-456, 1998.

McCready, M.A., E. Vassiliadou, and T.J. Hanratty, Computer simulation of turbulent mass transfer at a mobile interface, *AIChE Jour.,* 32(7), 1108-1115, 1986.

McKenna, S.P.; McGillis, W.R. Surface Divergence and Air-Water Gas Transfer. Geophysical Monograph series, edited by M.A. Donelan, W.M. Drennan, E.S. Saltzman and R. Wanninkhof, American Geophysical Union, Washington, D.C., this volume, 2001.

Orlins, J. J., Turbulence and mass transfer with an oscillating-grid, Ph.D. Thesis, University of Minnesota, Minneapolis, MN, 1999.

Soloff, S. M.; Meinhart, C.D.; CLEANVEC: PIV Vector Validation Software. Available from the Laboratory for Turbulence and Complex Flow at the University of Illinois at Urbana-Champaign. URL: http://ltcf.tam.uiuc.edu/

Tamburrino, A., Free-surface kinematics: Measurement and relation to the mass transfer coefficient in open-channel flow, Ph.D. thesis, University of Minnesota, Minneapolis, MN, 1994.

Thompson, S.M., and J.S. Turner, Mixing across an interface due to turbulence generated by an oscillating grid, *J. Fluid Mech.*, 67:2, 349-368, 1975.

Valsaraj, K.T., R. Ravikrishna, J.J. Orlins, J.S. Smith, J.S. Gulliver, D.D. Reible, and L.J. Thibodeaux, Sediment-to-air mass transfer of semi-volatile contaminants due to sediment resuspension in water. *Advances in Environmental Research,* 1(2) 145-156, 1997.

John S. Gulliver, Professor and Department Head, St. Anthony Falls Laboratory, Department of Civil Engineering, University of Minnesota, Mississippi River at 3rd Avenue S.E., Minneapolis, MN 55414 USA. Email: gulli003@tc.umn.edu

Joseph J. Orlins, Assistant Professor, Civil and Environmental Engineering, Rowan Hall, Rowan University, 201 Mullica Hill Road, Glassboro, NJ 08028 USA. Email: orlins@rowan.edu

Gas Transfer Across a Zero-Shear Surface: a Local Approach

Mohamed A. Atmane

Mechanical and Manufacturing Engineering, Trinity College, Dublin, Ireland

Jacques George

Institut de Mécanique des Fluides de Toulouse, France

Dissolved oxygen absorption experiments are carried out at a free surface in an agitated tank. Gas concentration and hydrodynamics are measured in the aqueous boundary layer by LDV and polarographic techniques. Surface renewal models are assessed from the concentration field and the gas transfer velocity. Near surface turbulence is shown to control the gas flux through two hydrodynamic parameters. An attempt to measure the local gas flux is made. In spite of the lack of accuracy, some interesting conclusions are drawn with respect to the scales playing a role in the interfacial gas transfer.

1. INTRODUCTION

Evaluation of the mass flux through the air-water surface is made difficult by the complexity of the interface shape (waves, presence of bubbles, etc.). In this study these difficulties are tentatively overcome by generating a flat shear-free interface. Keeping in mind that the existence of shear, waves or bubbles mainly affects the hydrodynamics by increasing the turbulence level below the surface, a mass transfer model including only the internally generated turbulence parameters is expected. Such a model is likely to be generalised to more complex flows.

The most attractive way to model the mass transfer at a free surface is to consider the surface renewal concept (Danckwerts, 1951). It is based on a space-time scheme in which a volume of liquid flows upwards from the bulk to the surface, captures a given amount of dissolved gas by a diffusion process and returns to the bulk saturated by the gas of interest. This concept is based on the surface renewal frequency f and the derived mass transfer velocity $K_L=(Df)^{1/2}$, where D is the gas molecular diffusivity.

At a wind-sheared surface, the choice of the surface renewal frequency is made obvious by the nature of the flow. The dynamics of the flow allows determination of both length and velocity scales from which the surface renewal frequency is extracted. The aim of the present work is to show that such a frequency still exists in a flow dominated by diffusion. This frequency is related to the measured mass transfer velocity.

In this study, an attempt was also made to assess directly the local mass flux. Such experiments have been carried out by *Chu & Jirka* (1992) in a grid-stirred tank. The difference here is that the velocity is measured in a non-intrusive way (LDV) whereas those authors used the hot-wire technique.

2. EXPERIMENTAL PROCEDURE

Tests are conducted in a jet stirred tank 0.8m high made in Perspex with a square section (0.45mX0.45m). A glass window is fitted on each side of the tank to allow LDV measurements to take place.

A turbulent flow is generated at the bottom of the tank making use of a stainless steel square plate on which are

Table 1. Hydrodynamics conditions and mass transfer results obtained for different tests

Test	Test 1	Test 2	Test 3	Test 4	Test 5	Test 6	Test 7
H_{water} (mm)	410	300	220	280	330	407	407
L (mm)	33	22	14	20	25	32.7	32.7
$k^{1/2} 10^2$ (m/s)	0.64	1.23	1.19	1.62	1.49	1.03	1.55
f (1/s)	0.1939	0.559	0.85	0.81	0.596	0.3149	0.474
z_0 (μm)	495	250	245	366	300	390	317

installed 100 microjets equally spaced (each jet is 0.7mm in diameter). The tank is provided with filtered water through a pump and is connected to a secondary tank which controls the water level. Hence, two parameters control the flow: the water flow rate and the water level.

Hydrodynamics of the water flow inside the tank are investigated using a two component LDV system composed of a laser source (Argon *10W*), the optics kit and the processing units (2 Burst Spectrum Analysers-BSA manufactured by Dantec).

Dissolved oxygen absorption is simulated by creating an oxygen concentration difference in the water layer below the surface. This is made possible by saturating the gas phase with nitrogen. Open circulation between the two tanks ensures that the water in the bulk of the test tank is kept saturated in oxygen. A steady gas transfer process takes then place between the liquid and gas phases.

Local instantaneous oxygen concentration is measured in the aqueous boundary layer using the polarographic technique. A micro-probe, manufactured by PME, with a tip 60 μm in diameter is used.

Velocity and concentration measurements were carried out from the bulk to the immediate vicinity of the free surface (that is a distance of order of 1mm below the interface which was kept flat in the present experiments).

3. RESULTS PRESENTATION AND DISCUSSION

Three sets of experimental results are presented and discussed: the scalar field description, the local mass fluxes and the global mass flux.

DO concentrations are normalised by the concentration difference between the bulk and the surface (C_b and C_S). As shown by *Jähne & Haussecker* (1999) or *Atmane* (1999), the mass flux equation can be solved in order to predict the spatial variation of the concentration as a function of the depth (assuming horizontal homogeneity). Solving this equation with surface renewal model assumptions (which include a surface renewal frequency of the form $f \sim z^a$) one ends up with an exponential solution (with $a=0$ standing for a constant renewal frequency along the same vertical). This exponential function has also been proposed by *Chu & Jirka* (1992) in their experimental work.

For each mean concentration profile, an exponential function is fitted. The normalised concentration profile for a given test condition can be written as:

$$\frac{C - C_b}{C_S - C_b} = e^{-z/z_0}$$

where z_0 is the mass boundary layer thickness.

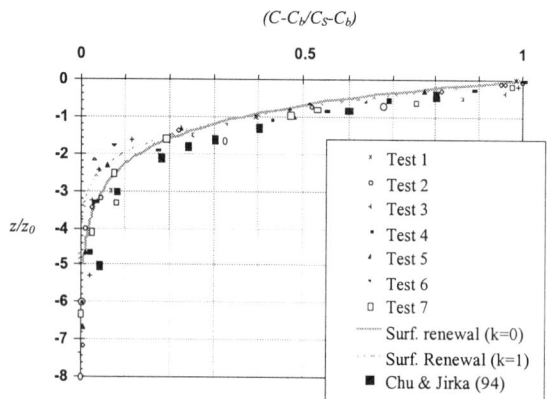

Figure 1. Experimental and theoretical concentration profiles

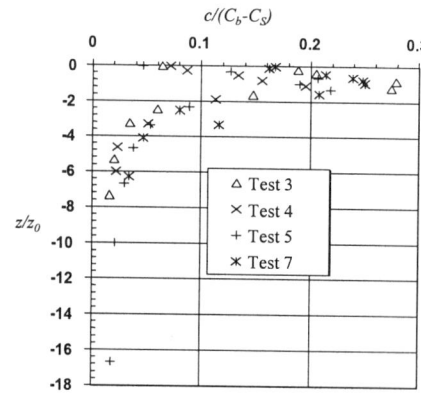

Figure 2. Fluctuating concentration profiles

Experimental hydrodynamics conditions are summarised in table (1) with the mass boundary layer thickness estimated for each test.

Figure (1) depicts the z_0-scaled non-dimensional concentration profiles obtained in our tests together with experimental results of *Chu & Jirka* (1992) and the theoretical profiles resulting from the scalar transport equation.

These profiles show a good agreement with the surface renewal theory. Further evidence on the suitability of this theory will be given later.

The non-dimensional fluctuating concentration profiles are shown on figure (2). Profiles highlight two facts: first, the fluctuations reach a maximum at a depth $z=-z_0$, in any case, where the vertical turbulent velocity fluctuations vanish at the surface, and mass transfer is mainly the fact of molecular diffusion. Moreover, the normalised maximum takes always a value ranging between *0.2* and *0.3* which corresponds to the typical value found in the modelling work of *Prinos et al.* (1995).

Cross correlation $\overline{cw}$ is calculated for each test from the velocity and concentration time varying signals. The cross correlation is normalised by the average flux $J=K_L(C_S-C_b)$.

The non dimensional experimental local fluxes depicted on figure (3) show a large discrepancy and fail even to reproduce the correct trend regardless of the flux values. A better understanding could be achieved through the calculation of the positive (from the water to the air) flux. Therefore, for each depth, the instantaneous positive flux corresponding to a positive couple of fluctuations ($w>0$, $c>0$) or a negative one ($w<0$, $c<0$) is evaluated. This positive flux is then subtracted from the total flux to obtain the 'negative' flux. The negative and positive flux profiles (normalised by the total flux) are shown on figure (4). The striking fact to be noticed is that the separated fluxes are an order of magnitude higher than the total flux and are almost symmetric. Thus, subtracting two large quantities of the same order

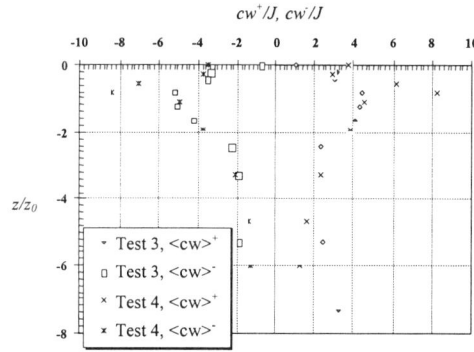

Figure 4. Separated gas flux profiles

results in a small quantity in which the errors are dominant. This explains in part the inaccuracy in predicting the total flux. However, despite the supposed negligible intrusive aspect of the oxygen probe, which is probably the case, another problem concerning this local flux estimation comes from the fact that velocity and concentration measurements do not coincide perfectly in space.

Nevertheless, from a physical standpoint, the positive and negative fluxes express two different transport modes inside the upper water layer. The first mode is assimilated to the surface renewal scheme during which a given volume of water saturated with oxygen rises towards the surface, stays long enough to allow a diffusion process to take place, that is for nitrogen to replace oxygen, and returns to the bulk (positive flux). The negative flux for its part can be explained the following way: the diffusion time (at the surface or in the bulk) is too short for the water volume to exchange dissolved gas. The liquid volume gets back to its initial position without losing (or bringing) a significant amount of oxygen and its contribution to gas transfer is negative. All this scenario is based on the simultaneously recorded velocity and concentration signals. Quantitatively, the explanation holds if we introduce two time scales: a 'residence' time scale controlled by the hydrodynamics and a diffusion time scale. Whenever the diffusion time scale is smaller, the flux is positive, the contrary involves a negative mass flux.

Global oxygen mass transfer through the free surface is usually expressed in terms of the transfer velocity K_L defined as the ratio of the molecular diffusivity to the mass boundary layer thickness:

$$K_L = \frac{D}{z_0}$$

We notice here that z_0 contains all the hydrodynamic characteristics and can be derived experimentally from the exponential form of concentration profiles, it depends on interfacial hydrodynamic characteristics.

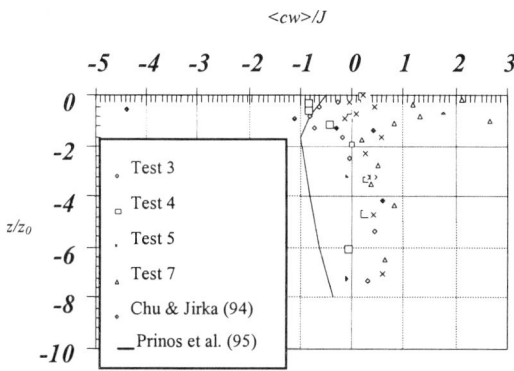

Figure 3. Experimental local gas flux compared to previous works

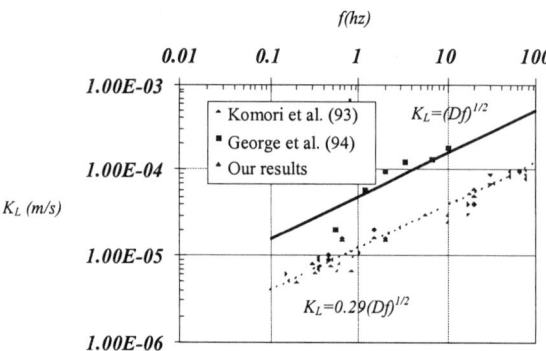

Figure 5. Gas transfer velocity as a function of the renewal frequency

One objective of this study was to separate the effects of both hydrodynamic parameters controlling our experiments on z_0: the square root of the interfacial turbulent kinetic energy (characteristic velocity scale $u_0 = \sqrt{k}$) and the turbulence integral length scale L at the surface. It can be shown that these two scales are totally independent from each other (See *Atmane & George*, 1999 for details on the hydrodynamics). Since k is almost constant between the surface and z_0, the chosen value of k is that of $k(z_0)$.

The characteristic time scale can be written as:

$$T = \frac{1}{f} = \frac{L}{\sqrt{k}}$$

and the frequency f deduced from this time scale is used to assess the surface renewal model applied to our data.

Figure (5) shows the variation of the mass transfer velocity as a function of the experimental renewal frequency. On the same figure are reported *George et al.* (1994) and *Komori et al.* (1989) results. In both previous studies, the mass flux was measured using a gas global balance method and the gas of interest was CO_2. One can notice that experiments by *George et al.* were carried out on this same experimental device whereas *Komori et al.*'s work dealt with a channel flow stressed by a wind. Moreover, the renewal frequency f was computed the same way in experiments carried out in this study and in *George et al.*

In *George et al's* experiments, the interfacial agitation level was larger than in present experiments and, hence, interfacial eddies could develop more freely even in the presence of possible contaminants.

Two conclusions can be derived from figure (6): first, the surface renewal concept designed for flows with convective time scales (mainly sheared surfaces) still holds in the case of an 'isotropic' turbulence. An empirical 'efficiency' factor appears in this figure. It is believed that this factor is closely related to the total area covered by large structures impacting at the free surface over the total area of the free surface. Numerical observations of *Banerjee* (1990) as well as experiments carried out in presence of surfactants *George et al.* (1994) support this point of view.

Second, the mass transfer at a shear-free surface needs more than one dynamic scaling (characteristic velocity) to be well evaluated, it appears to be a multi-scale problem also involving a characteristic length. The role of the distance separating the plane where turbulence is generated from the free surface should be taken into account. In other terms, the turbulence integral length scale, which gives an idea on how turbulence is distributed in wave-number space, plays a major role in the gas flux estimation. The agreement of our results with the surface renewal theory also support the idea that the mass transfer velocity does depend on the Reynolds number through the TKE and the turbulence integral length scale *(Atmane 1999)*. This dependency (figure 7):

$$K_L \propto \sqrt{k} Re^{-1/2}$$

supports the conclusions of *Fortescue and Pearson* (1967), according to whom large eddies control gas transfer; on the contrary the –1/4 dependency which refers to the gas transfer controlled by small eddies (*Lamont and Scott* 1970) is not found to be relevant. In fact, as shown experimentally by *Münsterer et al.* (1995) and *Duke* (1996) intermittent structures impinging from the bulk are capable of removing instantaneously large amounts of dissolved gas from the free surface.

CONCLUSION

Dissolved oxygen fluxes through a flat shear-free air-water surface have been observed. The velocity field has been investigated simultaneously. Better accuracy is needed in the measurement devices to correctly assess the local flux. However, some interesting facts can be highlighted. First, a space-time scheme, in which two different time

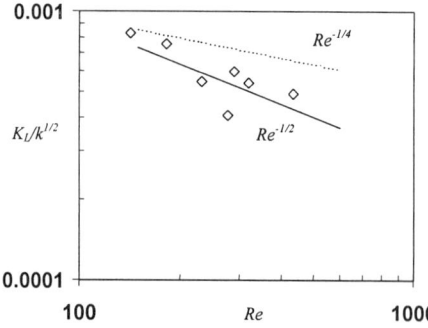

Figure 6. Normalised gas transfer velocity as a function of the Reynolds number

scales govern the local gas flux, can be proposed; the first scale is linked to the hydrodynamics whereas the second one is dictated by the quick diffusion process. Moreover, a good match is found between the mass transfer velocity and the frequency based on turbulence parameters (the turbulence integral length scale and the TKE) through the surface renewal model. Further work is needed in order to investigate the physical meaning of the 'efficiency factor'. A better way of determining its value should be found.

REFERENCES

Atmane, M. A., Approche Locale du transfert de masse interfacial contrôlé par une turbulence de microjets, Ph.D., *INP*-Toulouse, France, 1999.

Atmane, M.A., and J. George, Diffused turbulence distortion by a free-surface. *4th International Symposium on Engineering Turbulence Modeling and Measurement*s, Corsica, France, May 1999.

Banerjee, S., Turbulence structure and transport mechanisms at interfaces. *In 9th Inter. Heat transfer conference*. 1990.

Chu, C. R., and G. H. Jirka. Turbulent gas flux measurements below the air-water interface of a grid-stirred tank. *Int. J. Heat Mass Transfer*, 35, 1957-1196, 1992

Danckwerts, P. V., Significance of liquid-film coefficients in gas absorption. *Ind. eng. Chem.*, 34,1460, 1951

Duke, Steve R., Air-water transfer at wavy interfaces. PhD thesis, University of Illinois at Urbana-Champaign, 1996

Fortescue, G. E. and Pearson, J. R. A., *On gas absorption into a turbulent liquid*, Chem. Eng. Sci., 22, 1163, 1967.

George, J., and F. Minel, and M. Grisenti,. Physical and hydrodynamical parameters controling gas-liquid mass transfer. *Int. J. Heat Mass Transfer*, 37,11,1569-1578, 1994

Jähne, B., and H. Haussecker,. Air-water gas exchange. *Annu. Review of Fluid Mech*, 1998

Komori, S., and Y. Murakami, and H. Ueda,. The relationship between surface-renewal and bursting motions in an open-channel flow. *J. Fluid Mech.*, 203, 103-123, 1989.

Lamont, J. C. and Scott, D. S., *An eddy cell model of mass transfer into the surface of a turbulent liquid*, AIChE Journal, 16, 513, 1970.

Muensterer T., Mayer H.J., Jähne B., Dual-tracer measurements of concentration profiles in the aqueous boundary layer, In B. Jähne, Editor. Gas transfer at water surfaces, 1995.

Prinos, P., and M. A. Atmane, and J. George,. Gas flux measurements and modeling below an air-water interface. In B. Jähne, Editor. Gas transfer at water surfaces, 1995.

Atmosphere-Ocean Gas Exchange Due to Bubbles Generated by Wind Wave Breaking

Roman S. Bortkovskii

Main Geophysical Observatory, St. Petersburg, Russia

The problem of gas transfer by bubbles is far from having a final solution. Previous models described upper ocean aeration due to bubbles generated by whitecapping (Bortkovskii, 1998). In the new version the bubble transfer of the main gaseous constituents of air is considered. The bubble size distribution in time-depth space is found from field data. Winds during the measurements reached up to 20 m/s. Within foam patches, the bubble gas transfer through the interface is determined using the calculated flux of gas from bubbles into water. Taking into account the relative sea surface foam coverage and foam life-time the surface-averaged gas fluxes are found. Changes of dissolved gas content in the upper ocean layer during duration-limited gales are considered. A drastic increase of air-sea gas exchange due to bubble transfer is predicted in high winds.

1. INTRODUCTION

The diffusive gas flux through the air water interface is expressed by the known relationship: $Q_d = V_L \Delta C$, where Q_d is the gas flux and ΔC is the difference between gas solubility and gas concentration in a subsurface water layer. The gas transfer rate V_L is normally assumed to depend mainly on wind velocity.

However it was shown that at winds exceeding 10-12 m/s the gas transfer by bubbles, generated by wind-wave breaking, becomes the dominant mechanism [*Memery and Merlivat, 1985; Thorpe, 1982; Woolf and Thorpe, 1991; Leifer et al., 1995*]. Furthermore ΔC, the gas concentration difference in the subsurface water layer, becomes vanishingly small at strong winds [*Memery and Merlivat, 1985; Oost et al., 1995*]. So, under these conditions the diffusive gas transfer ceases, while the gas flux from bubbles continues as the interior bubble pressure is increased due to Laplace effect. Thus the mechanism of gas transfer changes: the only effective mechanism of air-sea gas exchange at strong winds is bubble transfer.

The above-mentioned results on bubble gas transfer modeling are important, but some unresolved questions remain. Available data on bubble field structure are obviously insufficient. The phase of bubble submergence wasn't considered in previous estimates. It was shown [*Bortkovskii, 1997*] that water temperature significantly influences near-surface processes but in the recent models this influence has not been completely considered.

2. FIELD DATA AND THEIR PROCESSING

Bubble sizes and concentrations were determined using remotely controlled cameras floating on the actual sea surface and 10 cm beneath it [*Bezzabotnov et al., 1986*]. Data were obtained at winds ranging from 10 to 20 m/s. The water temperature was close to 15°C. Photographs were taken every second, starting from the instant of a whitecap's appearance at the point where the float was drifting, and continuing up to the time when the foam disappeared. Bubble concentrations were found for bubble

Table 1. Rise speed W and coefficient K_2 for O_2 at various r and at water temperatures 0 and 30 °C

r, cm	0.01	0.05	0.09
		0° C	
W, cm/s	1.2	11.0	23.0
K_2, cm/s	0.22	0.29	0.55
		30° C	
W, cm/s	2.3	16.1	34.0
K_2, cm/s	0.45	0.54	1.00

radii from 0.01 to 0.2 with radius increment 0.02 cm. The minimum size was limited by the camera resolution. In spite of the limitation, the results are compatible with those obtained by higher resolution techniques [*Medwin and Breitz, 1989; de Leeuw and Cohen, 1995*].

It was assumed that bubbles, entrained in the water beneath a moving whitecap, begin to rise just behind the rear of the whitecap, at the instant $t=0$. It is possible to relate time t, at which bubbles of radius r_{10} were fixed at depth $z=10$ cm, with maximal depth z_M reached by them at $t=0$:

$$z_M(r_{10}) = t\overline{W}(r_{10}, z_M) + 10 \quad (1)$$

Here $\overline{W}(r_{10}, z_M)$ is mean bubble rise velocity, found taking into account the change in bubble radius during the rise. By this means, vertical profiles of bubble concentration size-spectra referred to the instant $t=0$ were found.

The data show that even under a whitecap the bubble void fraction at depth $z=10$ cm doesn't exceed 0.1 at winds up to 20 m/s. This means that the single bubble approach can be applied to the problem of multiple bubbles discussed here.

3. SINGLE BUBBLE DYNAMICS AND MASS EXCHANGE

For a single bubble, the system of equations takes the form:

$$dz/dt = -W + w \quad (2)$$

$$dr/dt = [3RT(dm/dt)/(4\pi r^2) - r\rho_w g(dz/dt]/(3p - 2\sigma/r) \quad (3)$$

$$dm_i/dt = -K_i s_i r^2 (pm_i/m - p_i) \quad (4)\text{-}(7)$$

Here w is the turbulent vertical velocity component of water motion, R is the universal gas constant, T is the water temperature, K, m is the bubble mass, ρ_w is the water density, g=980 cm/s², K_i, s_I, m_i and p_i are respectively the exchange coefficient, solubility, mass and partial pressure of i-th gas constituent with $i=1...4$, i.e., of nitrogen, oxygen, argon and CO_2 respectively. The pressure in a bubble p is the sum of the atmospheric pressure at $z=0$, the hydrostatic term $\rho_w gz$ and the Laplace term $2\sigma/r$, where σ is the water surface tension.

The influence of turbulent water velocity fluctuations was found to be insignificant [*Memery and Merlivat, 1985; Bortkovskii, 1998*], as are the Langmuir circulation effects, so for rising bubbles we take $w=0$.

To roughly evaluate the contribution to the gas flux from a bubble while submerged, the downward water velocity w beneath whitecaps was estimated. At the maximal depth reached by bubbles the condition $w=-W(r)$ was taken. As the w estimates agree satisfactorily with results by *Longuet-Higgins and Turner, 1974* and *Thorpe and Hall, 1983*, they were input in (2) for submerging bubbles. Numerical experiments show that the bubble gas transfer is visibly less in cold water than in warm water. Furthermore, the bubble concentration decreases with depth more rapidly in cold water than in warm. The effect of temperature on basic variables is seen from Table 1.

4. GAS TRANSFER BY BUBBLE POPULATION

Numerical solutions of the system of equations (2)-(7) allow the calculation of gas influxes into thin layers from bubbles of all sizes starting to rise at various levels z_0:

$$G_{ji} = \int_{r_m}^{r_M} \int_{z_M}^{z_j} \frac{\delta m_{ji}(r, z_0)}{\delta z} F(r, z_0) dr\, dz_0 \quad (8)$$

Here r_M and r_m are the greatest and the smallest bubble radii, z_M is the maximal depth reached by bubbles, i.e. maximal z_0, z_j is the middle depth of j-th layer, δm_{ji} is the mass increment of i-th gas contained in a bubble referred to j-th layer, δz is the layer thickness, and $F(r, z_0)$ is the bubble concentration spectrum. Gas fluxes Q_i through the interface, found by integration of influxes G_{ji} over z, are referred to the life-time of the foam patch, t_f.

Expressions for influxes from sinking bubbles differ from (8) as all these bubbles start to move from the interface, but the contribution to gas transfer from bubbles at

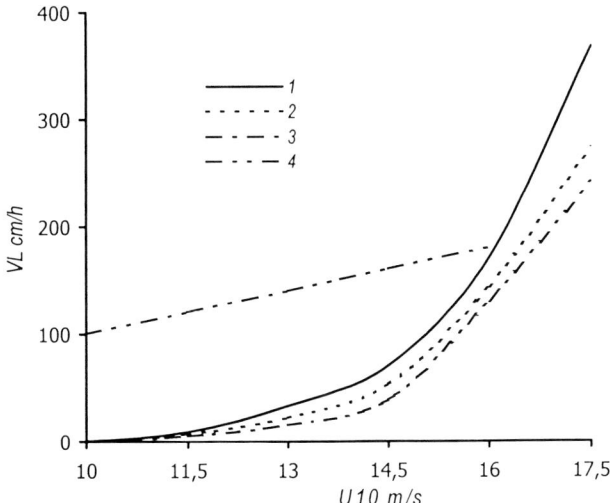

Figure 1. Oxygen transfer rate versus wind speed. Lines 1, 2, 3 represent bubble transfer at water relative saturation 0.95, 0.90 and 0.85 respectively. Line 4 shows field determinations for CO_2 by *Donelan* and *Drennan*, 1995

this stage of life is relatively small. The surface averaged gas flux transferred by bubbles is found to be

$$\overline{Q_i} = Q_i S_f / t_f \qquad (9)$$

where S_f is the relative sea surface coverage by foam. The values S_f and t_f were found as functions of wind speed and of water temperature (see Bortkovskii, 1998)

5. DISCUSSION

Influxes and fluxes for various degrees of water saturation S are calculated. The oxygen flux dependence on S is slightly nonlinear and fluxes are positive, i.e. from air to sea, in the area of supersaturation, $S>1$. Similar results are obtained for CO_2 bubble transfers, but positive fluxes are limited by smaller supersaturation values than those for O_2.

The oxygen bubble transfer velocity at relative saturation $S=0.95$ and wind speeds of around 15 m/s is close to the V_L values obtained in field experiments [*Donelan* and *Drennan, 1995*], (Figure 1), but it is much higher than known diffusive model evaluations. It is clear that field experiments give the total gas flux, i.e. the sum of all transfer mechanisms. When S is close to unity, the diffusive gas transfer ceases, but the bubble transfer continues, so bubble V_L has a discontinuity here. The dependence of bubble V_L on S is conditioned by the above-noted nonlinearity of the bubble gas flux dependence on S.

Changes of dissolved gas content in the upper ocean due to a jump from moderate to strong winds were considered on the basis of a numerical model [*D'Alessio et al., 1998*]. The quasistationary vertical profiles of the current velocity components, of the turbulent velocity scale and of the water temperature, calculated first for wind speed 7 m/s, were taken as initial values. Initial profiles of O_2 content were taken from data by *Levitus* [*1994*]; the near surface part of the profiles was interpolated on a basis of modeling [*Prinos et al., 1995*]. It is found that just after the start of a gale the saturation in the upper water layer grows quickly, but then the rate of increase of S goes down. Due to the growth, both the diffusive and bubble gas fluxes decrease with time, but after 10 hours of gale action the bubble flux is reduced by only 20%, whereas the diffusive flux is decreased 22 times. So ratio of these fluxes grows with time, and ratio of the bubble flux to the pre-gale diffusive flux remains almost unchanged (Figure 2). Thus, the previous results [*Bortkovskii, 1998; 1999*] are confirmed: due to bubble transfer, the air-sea gas exchange during a storm is increased by a factor of 10.

6. CONCLUSIONS

The field data by *Bezzabotnov et al.* [*1986*], obtained at winds up to 20 m/s, are used as a basis to estimate the bubble gas transfer contribution to air-sea gas exchange.

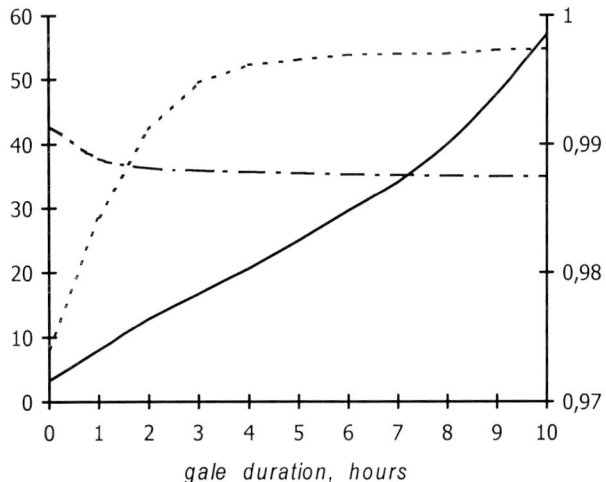

Figure 2. Dependencies of O_2 exchange parameters on duration of gale action. The solid line is the ratio of the bubble flux to the simultaneous diffusive flux, the dash-dotted line is the ratio of bubble flux to the referred to the prestorm ($U_{10}=7$ m/s) diffusive flux (left scale). The dotted line represents water saturation S at $z=90$ cm (right scale)

A model including bubble movement and mass balance is developed, and the gas transfer by bubble clouds is estimated. Peculiarities of CO_2 transfer are noted.

A drastic increase of air-sea gas exchange during storms is shown.

Acknowledgments. The work was supported by the Russian Foundation for Basic Investigation, Grant 97-05-64738, and by INTAS Grant, Project 96-2089.

REFERENCES

Bezzabotnov, V.S., R.S. Bortkovskii and D.F. Timanovski, On the structure of the two-phase medium generated by wind-wave breaking. *Izvestia Akad. Nauk SSSR, Phys. Atmos. Ocean, 6(11)*, 1186-1193, 1986 (in Russian)

Bortkovskii, R.S., Water temperature effects on the ocean surface state and on processes in near-surface air and water layers. *Izvestia Rus. Akad. Nauk, Phys. Atmos. Ocean, 33(2)*, 266-273, 1997 (in Russian)

Bortkovskii, R.S., Aeration of the ocean upper layer due to bubbles generated in whitecaps, in *Remote Sens.Pacif. Ocean by Satellites*, Ed. R.A.Brown, USA, Seattle, 354-362, 1998

Bortkovskii R.S. Mechanisms of gas transfer through sea surface and their contributions to atmosphere -ocean gas exchange, in *Contemporary Investig. Main Geophys. Observatory*, Russia, St. Petersburg, 70-83, 1999 (in Russian)

D'Alessio, S.J.D., K.Abdella and N.A.McFarlane, A new second-order turbulence closure scheme for modeling the oceanic mixed layer. *J. Phys. Ocean., 28, 8*, 1624-1641, 1998

de Leeuw, G, and L.H.Cohen Air-Sea Gas Transfer, Bubble size distribution in coastal seas, in *Air-Sea Gas Transfer, Select. Papers* of 3d Intern. Symp., Heidelberg, 325-336, 1995

Donelan, M.A. and W.M.Drennan, Direct field measurements of the flux of carbon dioxide. In: *Air-Sea Gas Transfer, Select. Papers* of 3d Intern. Symp., Heidelberg, 677-684, 1995

Leifer, I.F., W.A.Asher and P.J.Farley, A validation study of bubble mediated air-sea gas transfer modeling for trace gases, in *Air-Sea Gas Transfer Select. Papers* of 3d Intern. Symp., Heidelberg, 269-283, 1995

Levitus, S., and T.P.Boyer, World Ocean Atlas 1994, vol.2, 186 pp., Washington, D.C., 1994

Longuet-Higgins, M.S., and J.S.Turner, An «entraining plume» model of a spilling breakers, *J. Fluid Mech., 63, 1*, 1-20, 1974

Medwin, H., and N.D.Breitz, Ambient and transient bubble spectral densities in quiescent seas and under spilling breakers, *J. Geophys. Res., 94, 9C*, 12751-12759, 1989

Memery, L., and L.Merlivat, Modeling of gas flux through bubbles at the air-water interface, *Tellus, 37B, 4-5*, 272-285, 1985

Oost, W.A., W.Kohsiek, G.de Leeuw et al., On the discrepancies between CO_2 flux measurements methods, in *Air-Water Gas Transfer, Select. Pap.* 3d Intern. Symp., Heidelberg, 723-733, 1995

Prinos, P., M.Atmane, and J.George, Gas flux measurements and modelling below an air-water interface, in *Air-Water Gas Transfer, Select. Pap.* 3d Intern. Symp., Heidelberg, 49-58, 1995

Thorpe, S.A., On the clouds of bubbles formed by breaking wind-waves in deep water, and their role in air-sea gas transfer, *Phil. Trans. Roy. Soc. Lond., A304, 1483*, 155-251, 1982

Thorpe, S.A., and A.J.Hall, The characteristics of breaking waves, bubble clouds, and near-surface currents observed using side-scan sonar, *Continent. Shelf Res., 1, 4*, 353-384, 1983

Woolf, D.K., and S.A.Thorpe, Bubbles and the air-sea exchange of gases in near-saturation conditions, *J. Mar. Res., 49*, 435-466,1991

Dr. Roman S. Bortkovskii, Voeykov Main Geophysical Observatory, 7 Karbyshev str., 194021 St.Petersburg, Russia.
E-mail: rsb@main mgo rssi.ru

The Effect of Bubbles on Air-Water Oxygen Transfer in the Breaker Zone

Shohachi Kakuno[1], Douglas B. Moog[2], Tetsuya Tatekawa[3], Kenji Takemura[4], and Tatsuya Yamagishi[5]

The effect of bubbles entrained in the breaker zone on air-water oxygen transfer is examined. First, the area of bubbles entrained by breakers generated on a sloping bottom in a wave tank is analyzed using a color image sensor which can count the pixel number of a specific color in a frame. It was found that the time-averaged pixel number over a wave period has a strong relationship to the energy dissipation rate per unit mass of the breaker. The time-averaged pixel number is then incorporated with some modification into an equation proposed by Eckenfelder for the calculation of the mass transfer coefficient from bubble surfaces in an aeration tank. The coefficient resulting from the modified equation shows a strong relationship between the mass transfer coefficient and the dissipation rate.

1. INTRODUCTION

The ocean is expected to be a major sink for carbon dioxide as its atmospheric concentration increases. In addition, oxygen that is entrained from air into water, in a process known as "reaeration", is indispensable to aquatic creatures and plays a vital role in the aquatic environment of the coastal zone. Moreover, water bodies exchange toxic compounds with the atmosphere.

The present paper considers air-water oxygen transfer in the breaker zone, with special attention to the effect of entrained bubbles. To consider air-water oxygen transfer caused by breakers, *Hosoi et al.* (1986), and *Moutzouris and Daniil* (1994) experimented with breakers generated on a uniformly-sloping bottom in a wave tank. *Kakuno et al.* (1995, 1998), however, pointed out that their experimental method should be modified, and made experiments to reconsider reaeration through the surfaces of breakers with a method in which the aeration volume was isolated by a vertical thin film made of vinyl (see Fig.1). The thin film, loosely stretched and easily movable to some extent to confirm little wave reflection, was placed to prevent diffusion or convection of dissolved oxygen (DO) from the breaker zone to the offshore side in the tank. They demonstrated the validity of the method, showing that the mass transfer coefficient k_L* had a strong relationship to the square of the energy dissipation rate per unit mass, ε, as shown in Fig. 2 where ε was calculated from the model by *Nadaoka et al.*(1986) or *Okayasu et al.*(1990). A method to calculate ε from the models is described in detail in *Kakuno et al.* (1998).

The mass transfer coefficient, in general, has a relationship to the reaeration coefficient, k_2, as

$$k_2 = k_L(A/V) \qquad (1)$$

so that once k_2, which can be obtained from experiments, is known, k_L may be evaluated from Eq.(1), where A is the aeration area and V is the aeration volume. For the breakers on a sloping bottom, as shown in Fig.1(a), they used the volume which is restricted by the thin film as V, and considered that the A should be n times A_p, the horizontally projected area of the breaker zone (the area of the still water surface between the breaking line and the shore line):

---

[1]Dept. Civil Engrg., Osaka City Univ, Sugimoto, Sumiyoshi-Ku, Osaka, 558-8585, Japan
[2]Dept. Geological Science, Case Western Reserve Univ., Cleveland, Ohio, USA
[3]Hitachi Ship Building Co. Ltd., Osaka, Japan.
[4]Ministry of Transport, Kobe, Japan
[5]Graduate Student, Osaka City Univ., Osaka, Japan

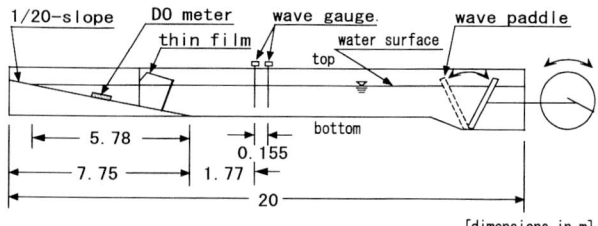

(a) Wave tank and experimental setup

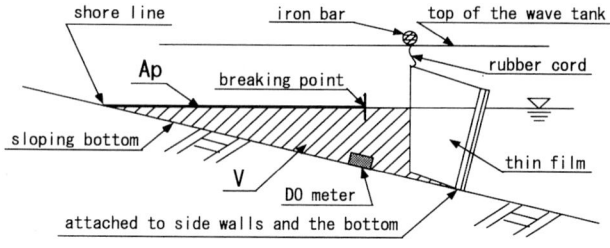

(b) Aeration volume isolated by a thin film and aeration area

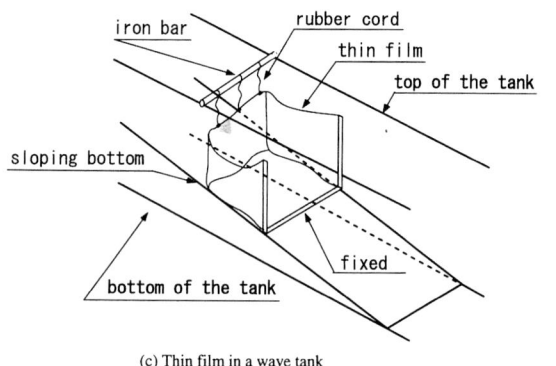

(c) Thin film in a wave tank

Figure 1. Experimental setup to measure dissolved oxygen (DO) in breakers in a wave tank.

$$k_2 = k_L^*(A/V) = k_L^*(nA_p/V) \quad (2)$$

We denote here k_L^* as the mass transfer coefficient which is evaluated with introducing $A=nA_p$. The constant $n(=A/A_p)$ is considered to be the ratio of the surface area through which gas transfer occurs, A, to the projected area, A_p, and obviously, it may be considered to be greater than 1 in the case of breakers that have complex water surfaces and many bubbles. They assumed that n may be expressed as

$$n = \alpha / B_t \quad (3)$$

where α is a constant, and B_t is the Breaker Type Index after *Galvin* (1968):

$$B_t = H_b / gT^2 \tan\theta \quad (4)$$

where H_b is the breaker height, g is the gravitational constant, T is the wave period, and θ is the angle of the bottom slope. They obtained the relationship between the mass transfer coefficient and the energy dissipation rate per unit mass shown in Fig.2 from Eq.(2) through (4) with k_2 obtained from experiments, regardless of bottom slope and the type of breaking, spilling or plunging. However, as is well known, if the small eddy model is adopted, k_L^* should be proportional to $\varepsilon^{1/4}$. Hence, it was concluded that the difference must be caused by entrained bubbles, and a study to clarify their effect was subsequently performed.

2. MEASUREMENTS OF ENTRAINED BUBBLES IN THE BREAKER ZONE

Studies of breakers on a sloping bottom using image analysis have been performed by *Watanabe et al.* (1995), *Miyamoto et al.* (1998), *Yamada et al.* (1999), *Chang and Liu* (1999), and *Mutsuda* (2000). The main focus of the studies has been on breaker kinematics and dynamics. Only Mutsuda analyzed the bubble region by processing images from a high-speed video camera. However, a systematic analysis was not carried out in his study. The current study relates oxygen absorption to a variable indicating the density of entrained bubbles. This variable is the proportion of the area falling within any bubble in a vertical projection of the breaker zone - specifically, the number of fixed-size pixels corresponding to any bubble in a digital image taken through the side wall.

2.1. Experimental Method

A series of experiments to examine the vertically projected area of bubbles entrained by breakers on a sloping bottom in a wave tank were performed. Experiments were

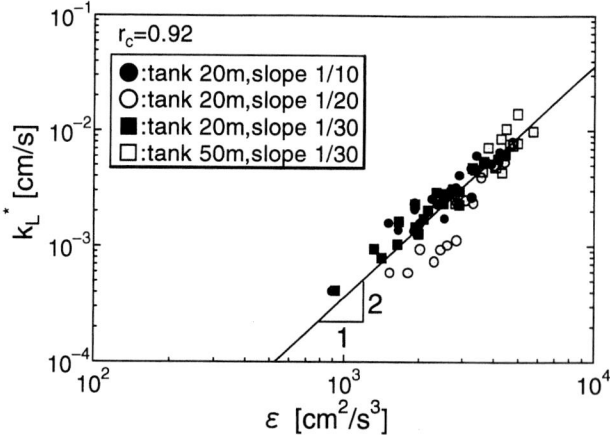

Figure 2. k_L^* as a function of ε

carried out using a two-dimensional wave tank 20m long, 50cm wide, and 50cm high, with a wooden 1/15-slope at one end, as shown in Fig. 3. The water used in the experiment was fresh water. Breakers generated on the sloping bottom were spilling and plunging. The bubble area installed in the tank. Also, to prevent thinning of the water is comprised of many bubbles of different sizes. It was optically recorded through the side glass wall of the tank by a color image sensor - that is, a special video camera that can count the number of pixels of a specific color in a frame. The water in the tank was colored blue with dye, while the other side wall of the tank was colored red. The number of pixels having other colors, which corresponds to the bubble area, was counted (Plate 1). In order to avoid oblique images in the color image sensor, a partition was color by diffusion, a thin film similar to the one used in the experiment to measure the dissolved oxygen was loosely stretched off the breaking point. Measurements in a frame were carried out after the wave crest proceeded by about 2 cm. When it almost reached the edge of a frame, the color image sensor was moved toward the shore by 10 cm to 20 cm, and the same operation was repeated at that point. The distance between the color image sensor and the glass wall of the tank was kept constant at 50cm. This procedure was applied from the breaking point to near the shore line, and repeated three times to obtain an averaged value. The total number of runs was 19, with a wave period, T, from 0.89 s to 1.32 s, deep water wave height, H_0, from 2.3 cm to 10.2 cm, and water depth, h, from 31 cm to 36 cm, so that the deep water wave steepness, H_0/L_0, from 0.008 to 0.057. In Table 1, these values are shown with breaking water depth, h_b, and the breaker type. Q_a shown in Table 1 will be defined and explained later.

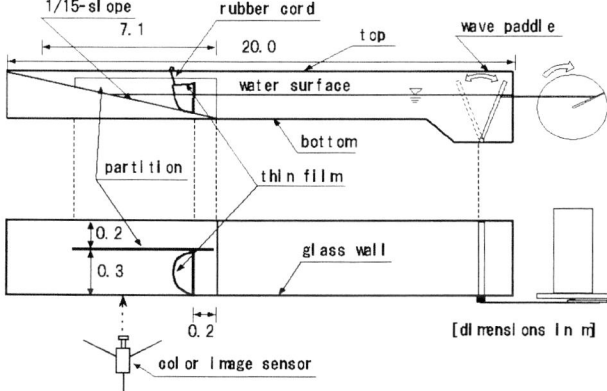

Figure 3. Experimental setup to measure bubble volumetric (areal) property

2.2. Experimental Result

2.2.1. Change of bubble area with progress of breaker

Fig. 4 (a) through (d) show examples of 2 sets (4 cases) of plunging breakers and spilling breakers of almost equal height. The measurement was not carried out in the vicinity of the shore line (S-LINE), since the form of the wave crest became obscure as it approached the shore line. However, since bubbles were recognized even near the shore line, an interpolated, straight broken line is drawn in the figures. It is seen that the pixel number, corresponding to the bubble area, increases rapidly just after passing the breaking point (B.P.) with the development of breaking, and decreases after a peak. The spilling case is characterized by a gradual increase starting just after the breaking point, while the plunging case exhibits a sharp peak a little after the breaking point.

2.2.2. Bubble area per unit time.

In order to obtain the average pixel number over one wave period, the pixel number shown in Fig. 4 was integrated from the breaking point to the shore line, and divided by the wave period T. This value, designated Q_a, is shown in Table 1, and in Fig. 5 as a function of the energy dissipation rate per unit mass, ε, which was obtained from the model *of Nadaoka et al.*(1986) and *Okayasu et al.*(1990), which gave the highest correlation to the mass transfer coefficient k_L^* in the previous study (*Kakuno et al.* 1998). As shown in the figure, Q_a has a strong correlation correlation to the cube of ε regardless of the breaker type, spilling or plunging; therefore, it may be concluded that bubble area entrained by breakers is also controlled by the cube of ε, similar to the relationship between k_L^* and ε.

3. THE EFFECT OF BUBBLE AREA ON AIR WATER OXYGEN TRANSFER IN THE BREAKER ZONE

It was concluded that time-averaged bubble area is strongly controlled by the energy dissipation rate per unit mass. The next stage should be to consider the relationship between the bubble area and the mass transfer coefficient k_L^*. Although some literature has addressed mass transfer at the bubble surface, progress has been limited by its complexity. For example, *Kawase and Moo-Young* (1992) concluded that k_L at the surface of bubble is constant regardless of bubble diameter and flow condition. On the other hand, *Eckenfelder* (1959) proposed an expression which depends on the bubble diameter and bubble volume rate for evaluation of the mass transfer coefficient from bubbles in an aeration tank:

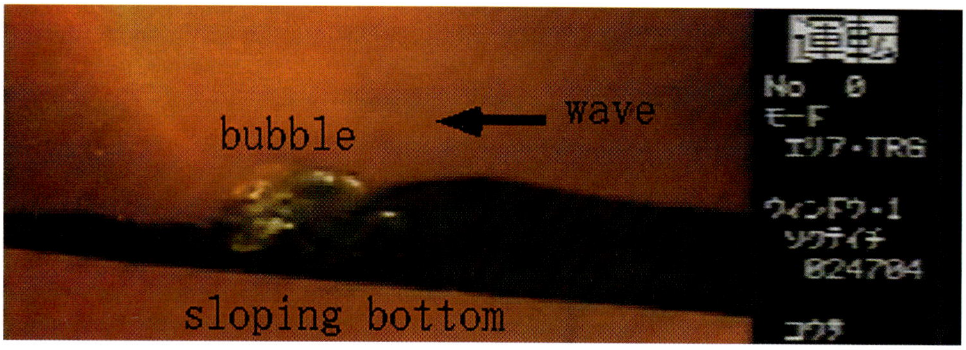

Plate 1. A shot by a color image sensor

Table 1. Experimental conditions and bubble area per unit time obtained

RUN	T[s]	H_0[cm]	h[cm]	H_0/L_0	H_b[cm]	h_b[cm]	Q_a[1/s]	Type
1	1.03	5.5	34.0	0.033	7.0	7.4	387,473	plunging
2	0.89	7.0	34.0	0.057	7.7	9.5	622,694	spilling
3	0.89	5.0	34.0	0.040	6.5	6.6	254,115	plunging
4	0.89	4.4	34.0	0.036	5.5	5.9	157,764	plunging
5	1.05	4.8	34.0	0.028	6.7	6.6	173,298	plunging
6	1.05	8.5	34.0	0.049	9.8	11.5	770,094	spilling
7	1.29	2.8	33.0	0.011	4.6	4.5	62,322	plunging
8	1.03	10.2	36.0	0.061	11.2	13.8	1,215,203	spilling
9	1.03	8.2	33.0	0.049	9.5	11.1	918,502	spilling
10	1.00	4.6	33.0	0.029	6.0	6.4	78,417	plunging
11	1.00	3.9	33.0	0.025	5.3	5.5	67,577	plunging
12	1.32	2.3	33.0	0.008	4.1	4.0	58,119	plunging
13	1.23	3.0	31.0	0.013	4.8	4.5	38,402	plunging
14	1.23	7.7	34.0	0.033	9.7	10.5	460,273	plunging
15	1.16	10.1	33.0	0.048	11.7	13.4	1,039,840	spilling
16	1.16	9.0	33.0	0.043	10.8	11.9	811,581	spilling
17	1.16	9.2	33.0	0.044	10.7	12.4	1,143,129	spilling
18	1.14	7.0	33.0	0.034	9.0	9.7	772,319	plunging
19	1.04	7.9	33.0	0.047	9.1	10.6	970,529	spilling

$$k_L = \frac{Cl^{2/3} Q_a'}{AD_a S_c^{1/2}} \qquad (5)$$

where C is a constant, l is tank liquid depth, Q_a' is the bubble volume rate, or air flow, A is the aeration area, D_a is the bubble mean diameter, and S_c is the Schmidt number. That is, the mass transfer coefficient at the bubble surface is supposed to be proportional to the bubble volume rate and 2/3 power of bubble migration length, or tank liquid depth, while inversely proportional to the bubble mean diameter. This formula was adopted in the present study. C, S_c and D_a are fixed as constants for the present. Although the bubble size should have a spectrum clearly, and therefore; D_a should be determined taking the bubble size spectrum into consideration, it is fixed as a constant because of scant knowledge about the spectrum. Then, we assume that the bubble migration length is proportional to the breaker height, H_b, and that the aeration area, A, is expressed as nAp with $n=1/B_t$, as in the previous study. In addition, Q_a' is replaced by the time-averaged pixel number corresponding to bubble area per unit time, Q_a, measured in the present study. Fig. 6 shows the mass transfer coefficient obtained from Eq.(5) as a function of ε. The gradient of the straight line in the figure is 1:3. As shown in the figure, the mass transfer coefficient obtained by substituting the time-averaged pixel number corresponding to the entrained bubble area has a strong relationship with ε, which can also be seen in Fig. 2. Therefore, it may be concluded that mass

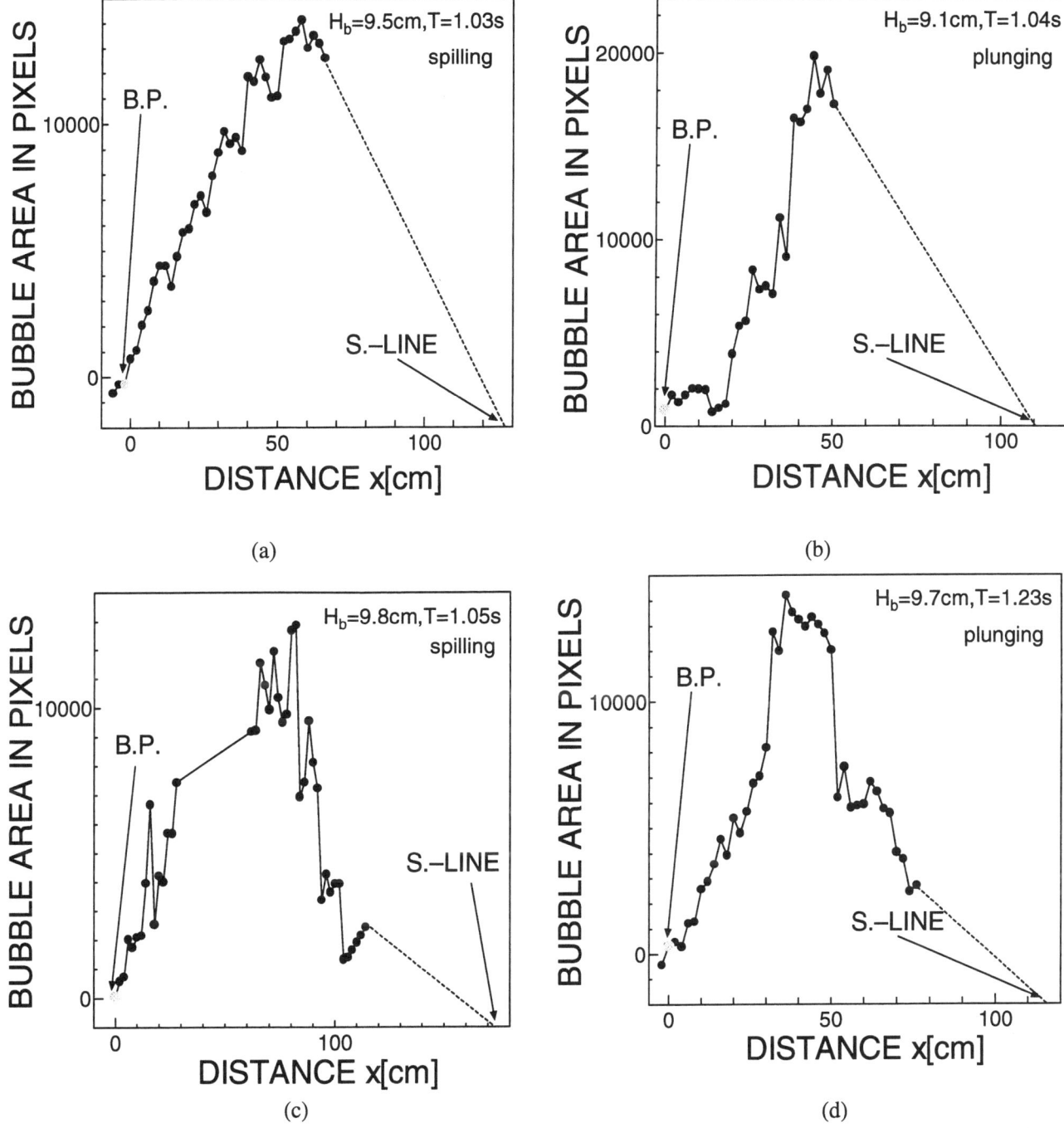

Figure 4. Change of bubble area with progress of breaker

transfer in the breaker zone is strongly controlled by the entrained bubble volume.

4. CONCLUSIONS

Entrained bubbles in the breaker zone, generated on a slope in a wave tank, were analyzed using images from a color image sensor. The results and conclusions may be summarized as follows:

1) The time-averaged pixel number, corresponding to the vertically-projected bubble area per unit time, is strongly controlled by the energy dissipation rate per unit mass of the breaker, and is proportional to its cube.

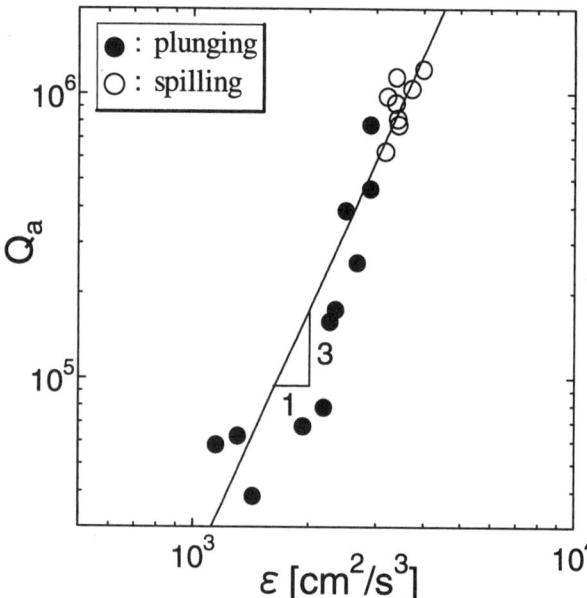

Figure 5. Bubble area per unit time as a function of ε

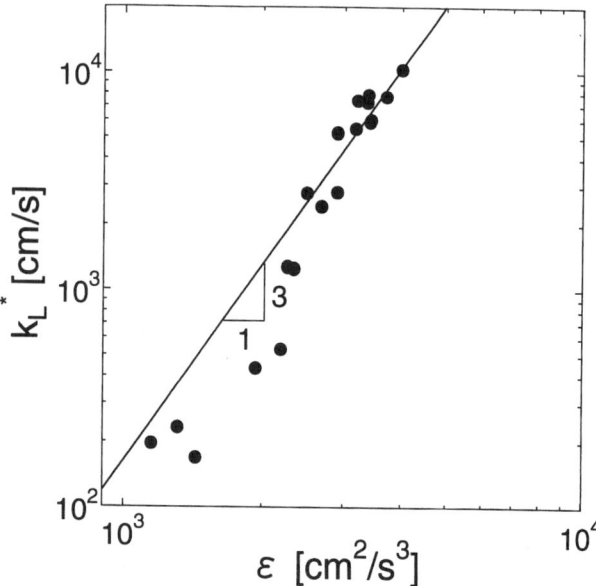

Figure 6. Mass transfer coefficient as a function of ε

2) The equation of Eckenfelder for calculating the mass transfer coefficient from bubble surfaces in an aeration tank was modified for the breaker zone.

3) The equation gives a mass transfer coefficient proportional approximately to the cube of the energy dissipation rate per unit mass in the breaker.

4) This result agrees with the previous study by the authors; therefore, it may be concluded that the mass transfer in the breaker is determined by entrained bubble volume.

5) More research should be done, preferably using salt water, to examine the effects of other bubble properties such as bubble diameter, bubble population, and bubble penetration depth, etc., and to clarify the mechanism of bubble entrainment.

Acknowledgement. We thank the reviewers for their valuable comments and suggestions on the earlier version of this paper. Shohachi Kakuno is grateful for the support by the Grant-in-Aid for Scientific Research (No. 10450185), the Ministry of Education, Science and Culture, Japanese Government.

REFERENCES

Chang, K.-A. and P. L.-F. Liu, Experimental investigation of turbulence generated by breaking waves in water of intermediate depth, *Phys. Fluids*, 11, 11, 3390-3400, 1999.

Eckenfelder Jr., W.W., Absorption of oxygen from air bubbles in water, *J. Sanitary Engrg. Div.*, Proc. ASCE, 85, SA-4, 89-99, 1959.

Galvin, C. J. Jr., Breaker type classification on three laboratory beaches, *J. Geoph. Res.*, 73, 12, 3651-3659, 1968.

Hosoi, M. and H. Murakami, Effect of breaking waves on dissolved oxygen and organic matter, *Proc. 20th Intnl. Conf. On Coastal Engrg.*, 2498-2512, 1968.

Moutzouris, C.I. and E.I. Daniil, Water oxygenation in the vicinity of coastal structures due to wave breaking, *Proc. 24th Intnl. Conf. On Coastal Engrg.*, 3167-3177, 1994.

Kakuno, S., M. Saitoh, Y. Nakata, and K. Oda, The air-water oxygen transfer coefficients with waves determined by using a modified method, *Air-Water Gas Transfer*, B. Jahne and E.C. Monahan, eds., AEON Verlag, 577-587, 1995.

Kakuno, S., D. B. Moog, T. Tatekawa, N. Shintani, and T. Shigematsu, Air-water oxygen transfer in the breaker zone based on energy dissipation., *Proc. Coastal Engrg.*, JSCE, 45, 66-70, 1998, (in Japanese).

Kawase, Y. and M. Moo-Young, Correlations for liquid-phase mass transfer coefficients in bubble column reactors with Newtonian and non-Newtonian fluids, *Can. J. Chem. Engrg.*, 70, 48-54, 1992.

Miyamoto, T., M. Nagao, S. Arai, and M. Kamioka, Simultaneous measurements of the surface profile and particle velocity distribution by an image analysis, *Proc. JSCE*, 53, 264-265, 1998, (in Japanese).

Mutsuda, H., A development of a numerical scheme with high accuracy for air-water mixed phase fluid flow due to large wave breaking, *PhD dissertation thesis for Gifu Univ.*, 2000, (in Japanese).

Nadaoka, K., and F. Hirose, Modeling of water particle dispersion under breaking waves in the surf zone, *Proc. 33rd Japanese Conf. On Coastal Engrg.*, 26-30, 1986, (in Japanese).

Okayasu, A., A. Watanabe, and M. Isobe, Modeling of energy transfer and undertow in the surf zone, *Proc. 22nd Intnl. Conf. On Coastal Engrg.*, 123-135, 1990.

Watanabe, Y. and H. Saeki, A consideration on kinematics in the breaker zone by image analysis, *Proc. Coastal Engrg.*, JSCE, 42, 116-120, 1995, (in Japanese).

Yamada, H., K. Takigawa, and K. Takayama, A relation among internal property of breakers, breaker type, and surface profile - image analysis, *Proc. Coastal Engrg.*, JSCE, 46, 135-139, 1999, (in Japanese).

Bubble Size Distributions on the North Atlantic and North Sea

Gerrit de Leeuw and Leo H. Cohen

TNO Physics and Electronics Laboratory, The Hague, The Netherlands

Bubble size distributions were measured at open sea with optical bubble measuring systems (BMS) deployed from buoys at depths from 0.4 to 1.5 m. The BMS measures the bubbles in a small sample volume that is monitored with a video camera. The images are analyzed to obtain bubble size distributions in the diameter range from 30 to 1000 μm. The BMS was deployed in the North Sea from the research tower Meetpost Noordwijk and from ships in the North Atlantic, yielding a validated data set of circa 300 bubble spectra obtained during seven cruises in different seasons. The measurements were made in a variety of wind speeds, water temperatures and atmospheric stability's. Bubble concentrations vary by 2 orders of magnitude, the spectra peak at diameters of 50-80 μm, and the slope varies from −1.8 to −5. Often a second peak is observed at 200-300 μm. The largest variations were observed between different experiments, while during a single experiment the spectra showed less variation, indicating seasonal influences. Wind speed dependencies were observed during single experiments, while in other experiments such effect was not statistically significant. The measurements were all made with the same technique, and the results are similar to data presented in the literature obtained with different techniques.

1. INTRODUCTION

Bubbles in the ocean are important to a variety of processes with applications in the fields of acoustic propagation in the water, electro-optical propagation in the atmosphere, corrosion, meteorology, climate, etc. Air-sea gas transfer is directly enhanced by bubbles [*Woolf*, 1993; 1997], and bubble-mediated air-sea gas transfer is the most significant mechanism associated with breaking waves [*De Leeuw et al.*, 2001]. Bubbles are the most important source of sea spray aerosol at moderate wind speeds. The size and number of sea spray droplets produced by a single bubble depends strongly on the bubble size (e.g., *Blanchard* [1983]). The sea spray source function is a major issue in climate models (e.g., *Gong et al.* [1997]). The possible effect of bubbles on aerosol deposition has been indicated [*Williams*, 1982], although experiments are not conclusive as regards the existence of such effect [*Larsen et al.*, 1995].

Both for gas transfer and for sea spray aerosol production the bubble size distribution is a key parameter. However, the little information available from the literature shows that the bubble spectra are very variable both as regards the concentrations and the spectral shape. Both parameters obviously change with environmental conditions, but a general parameterisation is not available.

In this contribution, bubble size distributions are presented that were measured with optical bubble measuring systems (BMS) in the North Atlantic and in the North Sea. The large variability between various bubble spectra is confirmed, even for measurements at the same location and with the same system. Effects of

Gas Transfer at Water Surfaces
Geophysical Monograph 127
Copyright 2002 by the American Geophysical Union

Table 1. Overview of deployments of the TNO-FEL bubble measuring system

Experiment	Location		period	water depth (m)	BMS depth (m)	wind speed (m s^{-1})	water temp. (°C)	air temp. (°C)	bubble spectra #
OMEX	Belgica	N. Atlantic	Mar 95	>100	0.38	7-14	11.3-12.2	8.8-11	14
OMEX	Valdivia	N. Atlantic	Jul 95	>100	0.4	3-12	15.7-19.8	16.8-19.3	29
BUBBLES II	Inisheer	N. Atlantic	Aug/Sep 95	70	1.5	1-11	17-20	12-20	185
ASGASEX/ MAPTIP	MPN	North Sea	Sep/Oct 93	18	0.5, 1	3-17	15.3-17.2	10.9-19.1	23
ASGAMAGE-A	MPN	North Sea	May/June 96	18	0.5	1-11	9.3-12	8.7-10	23
ASGAMAGE-B	MPN	North Sea	Oct 96	18	0.5	5-13	14.3-14.7	7.1-16.4	10
ANICE	MPN	North Sea	Aug 99	18	0.5	5-13	18.6-19.2	17.5-19.4	14

meteorological parameters and wave height are discussed, but the influence of oceanographic parameters such as supersaturation or biological activity are not accounted for in the present analysis.

2. BUBBLE MEASUREMENTS

Bubble size distributions were measured using two optical bubble measuring systems. Most measurements were made with the BMS described in *De Leeuw and Cohen* [1994; 1995]. The system was developed to measure the size of single bubbles in sea water, in the equivalent diameter range, D, from 30 to 1000 µm, from which the bubble size distribution was obtained. The BMS uses a diode laser to illuminate the sample volume (2.0x0.29x0.19 cm^3) that is imaged by a telescope on a ccd camera. The length of the sample volume, 2.0 cm, is determined by the focal length of the optical system, and is mechanically limited by conical tubes in which windows and lenses are mounted. The conical shape has been chosen to reduce the creation of turbulence near the sample volume. The camera signal is fed into a dedicated processing board for on-line analysis of the size and shape of objects in the sample volume. The images are also recorded on S-VHS video tape as a back-up. The BMS was calibrated with simulated bubbles, i.e. circles drawn on paper which were reduced photographically to a known size. By using several simulated bubble sizes a calibration curve could be obtained. Based on this exercise, the smallest bubble size that can reliably be measured with the BMS has been determined as 30 µm.

This system has been modified into a smaller and lighter unit, the Mini-BMS, that is described in detail in *Leifer et al.* [2001], see also *De Leeuw and Leifer* [2001]. The optical configuration is exactly the same.

For both systems automatic data processing routines have been developed to derive the bubble size distributions. The routines discriminate between bubbles and other objects based on several criteria described in *Leifer et al.* [2001].

The results from the Mini-BMS compare favourably with those from a Large-BMS [*Leifer et al.*, 2001]. The Large-BMS is also an optical bubble measuring system, but with a different optical design, using a larger field of view and measuring much larger bubbles than the Mini-BMS, using different processing algorithms.

The BMS has been used for deployments from ships in the North Atlantic and from several buoy systems in the North Atlantic and in the North Sea. An overview of the deployments and the experimental conditions is presented in Table 1. The measurements in the North Atlantic were made during OMEX cruises between Oostende (Belgium) and Lisbon (Portugal) in March 1995, and during a cruise between Reykjavik (Iceland) and Hamburg (Germany) in July 1995. The North Atlantic buoy measurements were made with the BMS mounted on a toroid anchored in deep water (70 m), at 10 km SW of the island Inisheer (Aran Islands, Ireland), during a period of 5 weeks in August/September 1995. The measurements in the North Sea were all undertaken from Meetpost Noordwijk (MPN), a tower in the North Sea in water of about 18 m deep, during four different campaigns between 1992 and 1999.

Commonly, the BMS is manually deployed and most of the data obtained are discontinuous time series. In contrast, the buoy measurements off Inisheer were made with an automatic recording system, using a TV transmitter and a receiver on the island. This yielded an almost continuous time series of bubble size distributions, at 3-hourly intervals. All size distributions were averaged over 15-20 minutes to allow for reasonable statistics on the bubble size distributions.

3. RESULTS AND DISCUSSION

Results from the deployments before early 1995 were presented in *De Leeuw and Cohen* [1994; 1995]. In summary, the ASGASEX/MAPTIP experiments showed a clear wind speed dependence, as well as a dependence on wind direction (fetch). The latter obscured the effect of wind speed on the bubble concentrations, and only trends could be analysed. The data set was too small to make a sub-division according to wind direction. The bubble distributions peaked at diameters of 50-60 μm, and the concentrations for larger bubbles decreased with D^S, with values for S between -3 and -4. The OMEX-Belgica data, obtained in the North Atlantic, showed a strong wind speed dependence. The oceanic size distributions were observed to peak at 50 – 70 μm and fall off with values of S varying between -2 and -3.

3.1 North Atlantic Data

The OMEX Belgica data, averaged over wind speed intervals of 7-8 m s^{-1} (low wind speed) and 11-13 m s^{-1} (high wind speed), are reproduced in Figure 1 together with data from the OMEX Valdivia cruise and the Galway deployment, each averaged over similar wind speed intervals (see Table 2). The wind speed intervals are not equal because selections had to be made for data availability. The spectra presented in Figure 1 show a very large variability with differences in peak concentrations, for

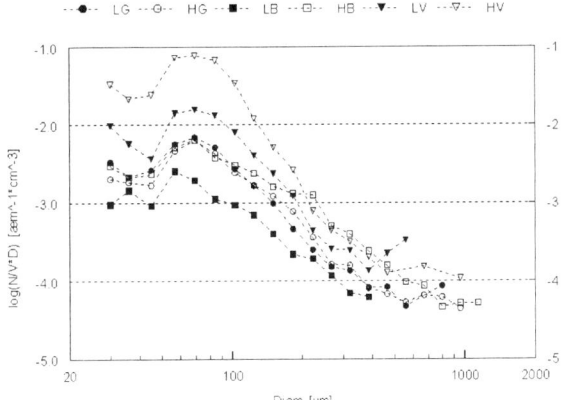

Figure 1. Bubble spectra measured in the North Atlantic, averaged over low and high wind speed intervals. Key: L= low wind speed, H=high wind speed, G=Galway (Inisheer deployment), B=Belgica, V=Valdivia.

either low or high wind speed, of one order of magnitude, and spectral shape varying from –2 to –4.2 (see Table 2). For larger bubbles, the concentrations are much closer. Intuitively this might be thought of as an artifact due to the small sample volume. However, the comparison with the Large-BMS, having a much larger sample volume, showed an excellent agreement in overlapping size ranges between 300 and 1000 μm diameter.

The bubble concentrations measured from the Belgica appear the lowest, in both the low and the high wind speed interval, for bubbles smaller than about 100 μm. In contrast,

Table 2. Low and high wind averaged bubble spectral characteristics

ID	Experiment	Wind speed interval (m s^{-1}) Low	High	spectral peak P1 (μm)	slope S1	spectral peak P2 (μm)	slope S2
North Atlantic:							
LB	OMEX-Belgica	7-8		60	-2		
HB			11-13	70	-2	200	-2.5
LV	OMEX-Valdivia	6-8		70	-3.5		
HV			10-12	70	-4.2		-2.4
LG	BUBBLES II	5-6		70	-3		
HG			9-11	70	-2.6		-1
North Sea:							
LAS	ASGASEX	6-8		60	-3.3	250	-5.2
HAS			11-13	70	-4.7	350	-4.7
LAA	ASGAMAGE-A	6-7		55	-5	350	N/A
HAA			10-11	55	-4.7	300	-3.5
LAB	ASGAMAGE-B	6-8		50	-2.2		
HAB			11-13	55	-1.8	350	-5.2
LAN	ANICE	8-9		60	-2.0		
HAN			12-13	70	-2.3		

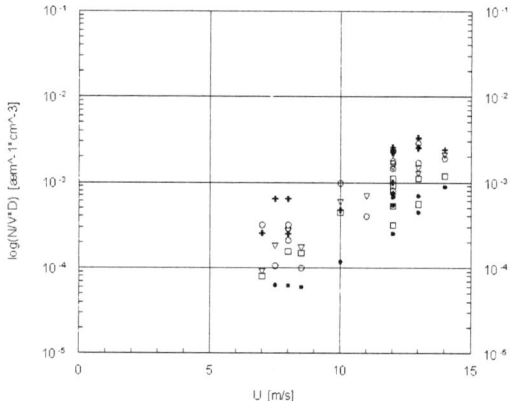

Figure 2. Effect of wind speed on the bubble concentrations during the OMEX-Belgica cruise between Oostende (Belgium) and Lisbon (Portugal) in March 1995, for bubbles of 150 (+), 183 (o), 222 (v), 267 ( ) and 321 (•) μm.

for the larger bubbles sizes, D > 200 μm, the high wind concentrations appear highest. This may indicate that different mechanisms are involved in the generation of small and large bubbles.

The high wind speed concentrations of bubbles larger than about 200 μm measured on the Belgica and the Valdivia are quite similar, and a factor of 2-3 higher than those measured near Inisheer. The former two data sets were both measured on a small buoy, which allowed for very shallow deployment, at a depth of about 0.4 m. The large toroid used for the Inisheer deployment required that the BMS was mounted at a depth of 1.5 m below the water surface. The sampling depth, together with the lower wind speeds encountered near Inisheer, may explain the smaller concentrations in the latter data set. The depth to which the bubble plume is entrained depends on various parameters such as wave height, swell, wind speed, and transport in the water column such as turbulence and Langmuir circulation. In addition, large bubbles rise faster and therefore have a shorter residence time than smaller bubbles. Hence, the average concentrations of small bubbles are expected to be higher than those of larger ones.

The bubble spectra presented here are averages over 15-20 minutes, and hence present the 'background' distributions rather than those in the bubble plumes. Laboratory experiments on the evolution of bubble plumes generated by breaking waves show the change in bubble spectra during the plume life time, and the transition to the background distribution in the senescence phase [*Leifer and De Leeuw*, 2001; *De Leeuw and Leifer*, 2001]. Obviously, the concentrations of the larger bubbles decrease fastest. While the bubble populations inside the plumes are shallow, the average bubble distribution decreases with S=-3.1. The spectral features of the oceanic bubble distributions are further discussed in section 3.3.

Remarkably, in the small size range, the low wind Belgica bubble concentrations are similar to those observed in high winds in the Inisheer data set. Furthermore, the Inisheer low and high wind data are very similar. No explanation for these observations is readily available, other that during the Inisheer deployment the wind speeds were rather low, up to 11 m s^{-1}, as compared to 14 m s^{-1} during the Belgica cruise. Also, a statiscally meaningful wind speed dependence has not been observed in the Inisheer data, although a relation between concentrations and wind speed has been observed in the time series. In cases of rising wind speeds over extended periods, also the bubble concentrations increased, similar to observations during the 1993 ASGASEX experiments on the North Sea [*De Leeuw and Cohen*, 1994]. In the latter case the observations could be related to fetch effects.

The effect of wind speed on the bubble concentrations during the two OMEX cruises is shown in Figures 2 and 3. For all sizes, the concentrations clearly increase with wind speed. As discussed in *De Leeuw and Cohen* [1995], the wind speed dependence varies with bubble size: the wind speed dependence of the larger bubbles is similar to that observed by *Wu* [1981], whereas the concentrations of the smaller bubbles follow the wind speed relation derived by *Walsh and Mulhearn* [1987].

The bubble concentrations measured during the OMEX-Valdivia cruise on the North Atlantic do not show a clear wind speed dependence (Figure 3) although for the higher wind speeds occurring in this data set the concentrations for all sizes are certainly higher than at low wind speeds. The range of wind speeds for which data are available from this cruise is smaller than for the Belgica cruise, and in general

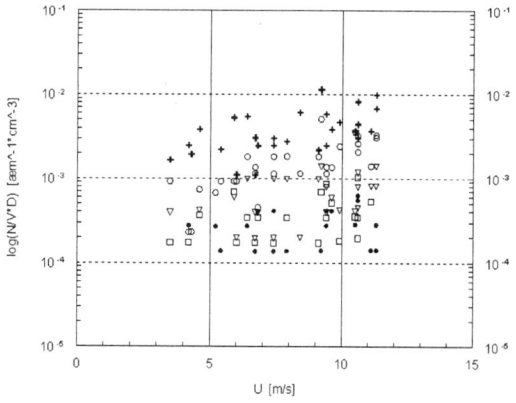

Figure 3. Effect of wind speed on the bubble concentrations during the OMEX-Valdivia cruise between Reykjavik (Iceland) and Hamburg (Germany) in July 1995, see further Figure 2.

the wind speeds were lower. Hence, also smaller bubble concentrations would be expected if wind speed were the only parameter determining them. Figure 3 shows that significant concentrations were observed also at the lower wind speeds encountered. The data indicate an increase of the bubble concentrations starting at wind speeds of 6-7 m s^{-1}, but Figure 3 is certainly not conclusive in this respect. The origin of bubbles at lower wind speed is not clear. Possibly biological activity plays a role and the observed bubbles are produced by micro-organisms in the water.

Atmospheric stability has been identified as a factor in whitecap ratio [*Monahan and O'Muircheartaigh*, 1986]. Therefore, also an effect is expected on the bubble concentrations. However, because the wind speed is expected to be the governing parameter, the effect of wind speed has to be removed before any other relation can be investigated. This requires a large data set, which is available only for the 1995 BUBBLES II experiment off Inisheer. Figure 4 shows the bubble size distributions averaged for various ASTD's in the low and high wind speeds intervals, where ASTD is the air-sea temperature difference in °C which is taken as a measure for atmospheric stability. The data in Figure 4 show a trend in the concentrations which generally increase for atmospheric stability changing from stable/neutral (0>ASTD>-1) to increasingly unstable (ASTD < -1), except for the largest ASTD (-3 to –4 in Figure 4).

Finally, the effect of water temperature has also been identified by *Monahan and O'Muircheartaigh* [1986] as a factor influencing whitecap ratio. For bubbles the water

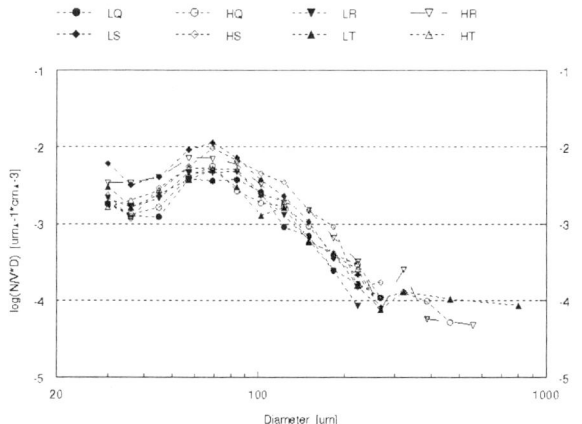

Figure 4. Bubble size distributions averaged for low and high wind speed intervals, for four different air-sea temperature difference (ASTD's) ranges. Data for the BUBBLES II experiment off Inisheer: L = low wind speed (5-6 m s^{-1}); H = high wind speed (7-8 m s^{-1}); Q: 0<ASTD<-1; .R: -2<ASTD<-1; S: -3<ASTD<-2; T: ASTD<-4.

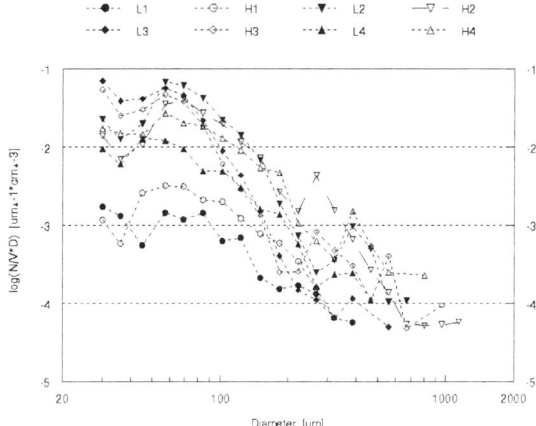

Figure 5. Bubble spectra measured in the North Sea, averaged over low and high wind speed intervals. Key: L = low wind speed, H = High wind speed; 1 = ANICE; 2 = ASGASEX; 3 = ASGAMAGE-A; 4 = ASGAMAGE-B.

temperature should affect the bubble size because of thermal expansion. In the present data set, water temperatures ranged from 11-12 °C for the Belgica data, to almost 20 °C for the Inisheer data. However, the bubble concentrations measured on the Valdivia, with water temperatures of 15.7-19.8 °C, were much higher than in the other two data sets.

Apparently also other factors play a role. Biological effects were already indicated. Saturation is expected to be also of major influence, but no data are available to support any conclusion to explain differences observed between the three cruises.

3.2 North Sea Data

The bubble size distributions available from the North Sea experiments are compiled in Figure 5. As in Figure 1, also here the data are binned in low and high wind speed intervals (see table 2). However, as in the North Atlantic data, wind speed is not the governing factor and large differences are observed between different data sets. The low and high wind speed data for each data set are much closer than the overall variation between different deployments. For instance, the ANICE concentrations for the smaller bubbles are an order of magnitude lower than for the other data sets. For the larger bubbles, on the other hand, the concentrations for all data sets are much closer, as was also observed in the North Atlantic data in Figure 1.

The differences between the different data sets seem to depend on the season rather than on meteorological or oceanic parameters. It might be argued that in the North Sea riverine outflow may play a role, but since similar variations are observed in the North Atlantic near the shelf, coastal effects on the overall variations in the size distributions are not likely the governing factor. As indicated in *De Leeuw and Cohen* [1994], coastal effects appear to cause variations in the concentrations because of different fetches.

As for the North Atlantic, no data are available to support conclusions as regards seasonal effects. However, it may be expected that the algae cycle does affect the sea water saturation.

3.3 Spectral Shape

The spectral shape parameters of the bubble size distributions are presented in Table 2. The spectra peak at equivalent diameters between 50 and 70 µm, and often a second peak is observed at diameters varying from 200 to 350 µm. Bubble spectra presented in the literature usually assume that the concentrations decrease with size as a power law distribution. As shown in Table 2, the values of S vary between -1.8 and -5.

These spectral shape characteristics are in good agreement with data presented in the literature, as well as with laboratory observations (see above). Bubble size distributions were reported to peak at 68 µm [*Walsh and Mulhearn*, 1987], 60-70 µm [*Kolovayev*, 1976], 40-50 µm [*Johnson and Cooke*, 1979] and 30-93 µm [*Baldy and Bourguel*, 1987]. Values for the power law S reported in the literature range between –3 and –7. (Note that data in the literature are presented as $N(D)=CD^n$, i.e. the number of bubbles per cm^3, whereas here the number of bubbles per size increment per cm^3, dN/dD is plotted; this implies that S = n-1). *Medwin and Breitz* [1989] report an average value of n=-3.7, *Walsh and Mulhearn* [1987] give an average value of n=-5 (with data ranging between -3.9 and -7), *Medwin* [1970] reported n=-3, and *Kolovayev* [1976] found n=-4.5. *Johnson and Cooke* [1976] determined a slope n=-5.5 and *Wu* [1988] reported n=-5.

4. CONCLUSIONS

Bubble spectra were presented for a wide variety of conditions, measured in the North Atlantic and in the North Sea. All data were obtained with similar instruments, using the same physical principles, both as regards the measurements and the processing. Nevertheless large differences are observed, with orders of magnitude in concentrations, and spectral slopes varying between –1.8 and –5. A similar range of values was reported in the literature, but obtained with different instrumentation using either acoustical or optical methods. Comparison between these methods shows different results for the smaller particles, with higher concentrations measured with acoustical techniques. However, the present optical BMS was calibrated both as regards size and number of bubbles, and the decrease of the concentrations of bubbles smaller than the spectral peak at 50-70 µm is a real phenomenon.

The differences between the results from the various deployments are ascribed to seasonal variations, but no data are available to support this conclusion. However, similar variations are observed between different locations in the North Atlantic during different seasons, and at the same location in the North Sea throughout the years.

Acknowledgments. The data presented here were obtained during a large number of projects supported by the European Commission EC DG XII, contracts MAS2-CT93-0056, MAS3-CT96-0056, MAS3-CT95-0044, ENV4-CT97-0594, the Netherlands Ministry of Defence, assignments A92KM614 and A95KM785, and TNO-FEL internal funding.

REFERENCES

Baldy, S. and M. Bourguel, Bubbles between the wave trough and wave crest levels, *J. Geophys. Res.* 92, 2919-2929, 1987.

Blanchard, D.C., The production, distribution, and bacterial enrichment of the sea-salt aerosol, in *Air-sea exchange of gases and particles*, edited by P.S. Liss and W.G.N. Slinn, Reidel, pp. 407-454, 1983.

De Leeuw, G., and L.H. Cohen, Measurements of oceanic bubble size distributions, in *OCEANS94, Proc. Vol. II*, 694-699, 1994.

De Leeuw, G., and L.H. Cohen, Bubble size distributions in coastal seas, in *Air-Water Gas Transfer*, edited by B. Jähne and E.C. Monahan, AEON Verlag & Studio, Hanau (GE), pp. 325-336, 1995.

De Leeuw, G., and I. Leifer, Bubbles outside the plume during the LUMINY wind wave experiment, in *Gas Transfer and water Surfaces*, edited by M.A. Donelan, W.M. Drennan, E.S. Salzman, and R. Wanninkhof, pp. xxx-xxx, AGU, this volume, 2001.

De Leeuw, G., G.J. Kunz, G. Caulliez, D.K. Woolf, P. Bowyer, I. Leifer, P. Nightingale, M. Liddicoat, T.S. Rhee, M.O. Andreae, S.E. Larsen, F.A. Hansen and S. Lund, LUMINY - An overview, in *Gas Transfer and water Surfaces*, edited by M.A. Donelan, W.M. Drennan, E.S. Salzman, and R. Wanninkhof, pp. xxx-xxx, AGU, this volume, 2001.

Gong, S.L., L.A. Barrie, and J.-P. Blanchet, Modeling sea-spray aerosols in the atmosphere. 1. Model development. *Journal of Geophysical Research*, 102, 3805-3818, 1997.

Johnson, B.D. and R.C. Cooke, Bubble populations and spectra in coastal waters: A photographic approach, *J. Geophys. Res.*, 84, 3761-3766, 1979.

Kolovayev, P.A., Investigation of the concentration and statistical size distribution of wind-produced bubbles in the near-surface ocean layer. *Oceanology* 15, 659-661, 1976.

Larsen, S.E., J.B. Edson, P. Hummelshoj, N.O. Jensen, G. de

Leeuw and P.G. Mestayer, Dry deposition of particles to the ocean surfaces, *OPHELIA* 42, 193-204, 1995.

Leifer, I. and G. de Leeuw, Bubble measurements in breaking-wave generated during the LUMINY wind-wave experiment, *in Gas Transfer and water Surfaces*, edited by M.A. Donelan, W.M. Drennan, E.S. Salzman, and R. Wanninkhof, pp. xxx-xxx, AGU, this volume, 2001.

Leifer, I., G. de Leeuw and L.H. Cohen, Optical measurement of bubbles: system design and application, Submitted for publication in *J. Atm and Ocean. Tech.*, 2001

Medwin, H., In-situ acoustical measurements of bubble populations in coastal ocean waters, *J. Geophys. Res.* 75, 559-611, 1970.

Medwin, H., and N.D. Breitz, Ambient and transient bubble spectral densities in quiescent seas and under spilling breakers, *J. Geophys. Res.*, 94, 12751-12759, 1989.

Monahan, E.C. and I.G. O'Muircheartaigh, Whitecaps and the passive remote sensing of the ocean surface, *Int. J. Remote Sensing* 7, 627-642, 1986.

Walsh, A.L. and P.J. Mulhearn, Photographic measurements of bubble populations from breaking wind waves at sea. *J. Geophys. Res.* 92, 14553-14565, 1987.

Williams, R.M., A model for the dry deposition of particles to natural water surfaces. *Atmos. Env.* 16, 1933-1938, 1982.

Woolf, D.K., Bubbles and the air-sea transfer velocity of gases, *Atmosphere-Ocean* 31, pp. 517-540, 1993.

Woolf, D.K., Bubbles and their role in gas exchange, *in The Sea Surface and Global Change*, editors P. S. Liss and R. A. Duce, pp. 173-205, Cambridge University Press, 1997.

Wu, J., Bubble populations and spectra in near surface ocean; summary and review of field experiments. *J. Geophys. Res.* 86, 457-463, 1981.

Wu, J., Bubbles in the near-surface ocean - a general description. *J. Geophys. Res.* 93, 587-590, 1988.

G. de Leeuw and L. H. Cohen, TNO Physics and Electronics Laboratory, P.O. Box 96864, 2509 JG The Hague, The Netherlands (e-mail: deleeuw@fel.tno.nl)

Measurements of Large Bubbles in Open-Ocean Whitecaps

Dale Stokes and Grant Deane

Marine Physical Laboratory, Scripps Institution of Oceanography, La Jolla, California

Svein Vagle and David Farmer

Institute of Ocean Science, Sidney, British Columbia, Canada

There is theoretical evidence that the enhanced gas transfer observed during storms is due to bubble-mediated transport. Calculations done by Keeling [1993] suggest that the dissolution of highly soluble gasses such as CO_2 is particularly sensitive to the formation of large (greater than 0.3mm radius) bubbles within whitecaps. Estimates of the numbers and sizes of bubbles within whitecaps are difficult to obtain, and the lack of this data is currently a limiting factor in model development. A new optical instrument that can image bubbles in high void fraction plumes has been deployed in a series of open-ocean experiments. The numbers and sizes of bubbles within 30 cm of the surface and directly beneath breaking waves have been estimated. Size distribution estimates within a 3.8 cubic centimeter volume are made every 40 milliseconds for a few seconds during a breaking event. The largest bubbles imaged were 3 mm in radius. Void fractions between 1 and 10 percent persisting for a second or so during plume formation were observed. Initial estimates of bubble-mediated gas transfer based on these new data are presented.

INTRODUCTION

Breaking waves in the open ocean create dense plumes of bubbles. These plumes can extend 0.5m or more below the surface and have 0.1 or greater void fractions of air in the first second or so after wave breaking. The bubbles within plumes have a wide range of radii, extending from 10's of microns to centimeters, and are thought to be responsible for the enhanced gas transfer velocity across the air-sea interface observed at high wind speeds. Bubble-mediated gas transport has been the subject of a number of theoretical studies [eg. *Merlivat* and *Memery*, 1983; *Jähne et al.*, 1984; *Keeling*, 1993] and is sensitive to the numbers and sizes of bubbles within plumes.

One of the motivations for the present study is the importance of bubble size spectra in plumes to air-sea gas transport, and the difficulty in estimating bubble spectra within the acoustically and optically dense plumes that form immediately after wave breaking. The measurements reported here follow on from the work of Bezzabotnov *et al.* [1986] who made what appear to be the first optical measurements of bubbles in open-ocean whitecaps (also see *Monahan* and *Zietlow*, 1969, *Cipriano* and *Blanchard*, 1981, *Medwin* and *Daniel*, 1990, Loewen *et al.*, 1995, Leighton et al., 1996 and *Deane*, 1997).

The size distributions reported here were obtained with an optical instrument that operates within the interior of plumes [*Stokes* and *Deane*, 1999]. The instrument has also been used in the laboratory to compare its performance with existing laboratory data sets.

The field deployment of the bubble sizing instrument is described in section 3 and representative bubble size distributions are presented in section 4. The implications of the observed open-ocean distributions for air-sea gas exchange are discussed in section 5, followed by the concluding remarks in section 6.

DEPLOYMENT OVERVIEW

Open ocean deployments were conducted from the research platform FLIP during experiments off the coast of Southern California (034° 11.022' N, 122° 24.171' W). The wind speed, significant wave height, wave period and water temperature respectively were (7-10) m/s, (2.5-3.5) m, (9-12) s and 13.2 degrees Celsius. The bubble counting instrument, BubbleCam [see *Stokes* and *Deane*, 1999], was mounted on a surface-tracking frame along with an additional wide-angle underwater video camera (for visually estimating the location of the subsurface sensors relative to the bubble plumes), pressure sensor (for continuous measurement of BubbleCam depth), and a hydrophone (for simultaneous acoustic recordings). The surface-tracking frame was tethered beneath one of FLIP's booms approximately 10 meters away from the hull with data and power cables connected to topside control systems for recording and real-time monitoring of the instruments. Deployed in this manner, the BubbleCam imaging volume (3.8 cubic centimeters directly in front of the instrument) was positioned approximately 30 cm beneath the air/sea interface, facing into the wind and breaking waves. Details of the instrument calibrations and measured flow distortions can be found in Stokes and Deane [1999]. Bubble size distribution estimates were made every 40 ms following a wave breaking event over the surface-tracking frame. Bubble size distribution measurements were supplemented with meteorological data (wind speed, direction, air temperature, humidity) from instruments mounted on the ship superstructure.

OBSERVATIONS

The rapid injection of air and the resulting high void fractions within the bubble plumes during their formation can be seen in Plate 1. The depth time series was derived from a pressure sensor mounted on the instrument, and was not corrected for flow speed. In this event, bubbles as large as 1.6 mm in radius were generated in the first half-second of plume formation, and void fractions as great as 1.6% were measured more than 40 cm beneath the ocean surface (see Deane [1997] for a discussion of the void fraction calculations]. Bubbles as large as 3 mm in radius and void fractions as high as 10% were recorded during other whitecap events. Peaks in both the void fraction and bubble density data suggest a great deal of heterogeneity within the forming plume.

The bubble size distribution calculated from a 0.5 s average (during which peak void fractions were observed) of 11 breaking events is shown in Plate 2 (open red circles). The distribution from Bezzabotnov *et al.* [1986] has been scaled for a void fraction of 0.0175, corresponding to a wind speed of 8.5m/s (see their figure 6), in the middle of the range of wind speeds observed during the BubbleCam deployment. The Deane and Stokes size distribution is similar to those previously measured within alpha plumes in the surf zone [Deane, 1997; Deane and Stokes, 1998; Leighton et al., 1996]. The distributions are also similar to previous laboratory flume studies for bubbles greater than about 600 microns in radius. However, for smaller bubbles, the present field measurements suggest densities up to a factor of ten higher than those previously observed.

GAS FLUX MODELING

The open ocean bubble size distributions have been included in a model of bubble plume mediated gas flux based on the work by Thorpe [1982]. Exchange between the plume bubbles and the seawater was controlled by the partial pressure of gas inside the bubbles and in solution (Henry's Law), a wind varying transfer velocity, and the gas solubility. Bubble volume then varied as a result of dissolution and hydrostatic pressure [*Thorpe*, 1982]. The model did not include bubble coalescence or the effects of surfactants on gas exchange across the bubble air / water interface. The initial conditions assumed that all gases were 100% saturated in seawater and the plume injection had a vertical injection velocity of 0.4 m/s (Figure 1). Turbulence was superimposed on the downward plume advection with a diffusivity of K=0.002 m²/s [*Woolf* and *Thorpe*, 1991]. Plume injection continued until the bubbles reached a depth of 0.5 m (estimated from underwater video observations), at which point both turbulent and downward flow were stopped and the bubble distribution was allowed to evolve under the influence of buoyancy and dissolution only. Due to the high initial downward velocity, the size distribution predicted at 30-40cm where the measurements were made were essentially the same as the distribution at the surface. Bubble rise velocities were estimated from Keeling [1993] for bubbles smaller than 100 μm radius and from figure 2.10 in Fan and Tsuchiya [1990] for larger bubbles. The model was repeated for wind speeds between 0 and 30 m/s, using the estimates of increasing bubble con-

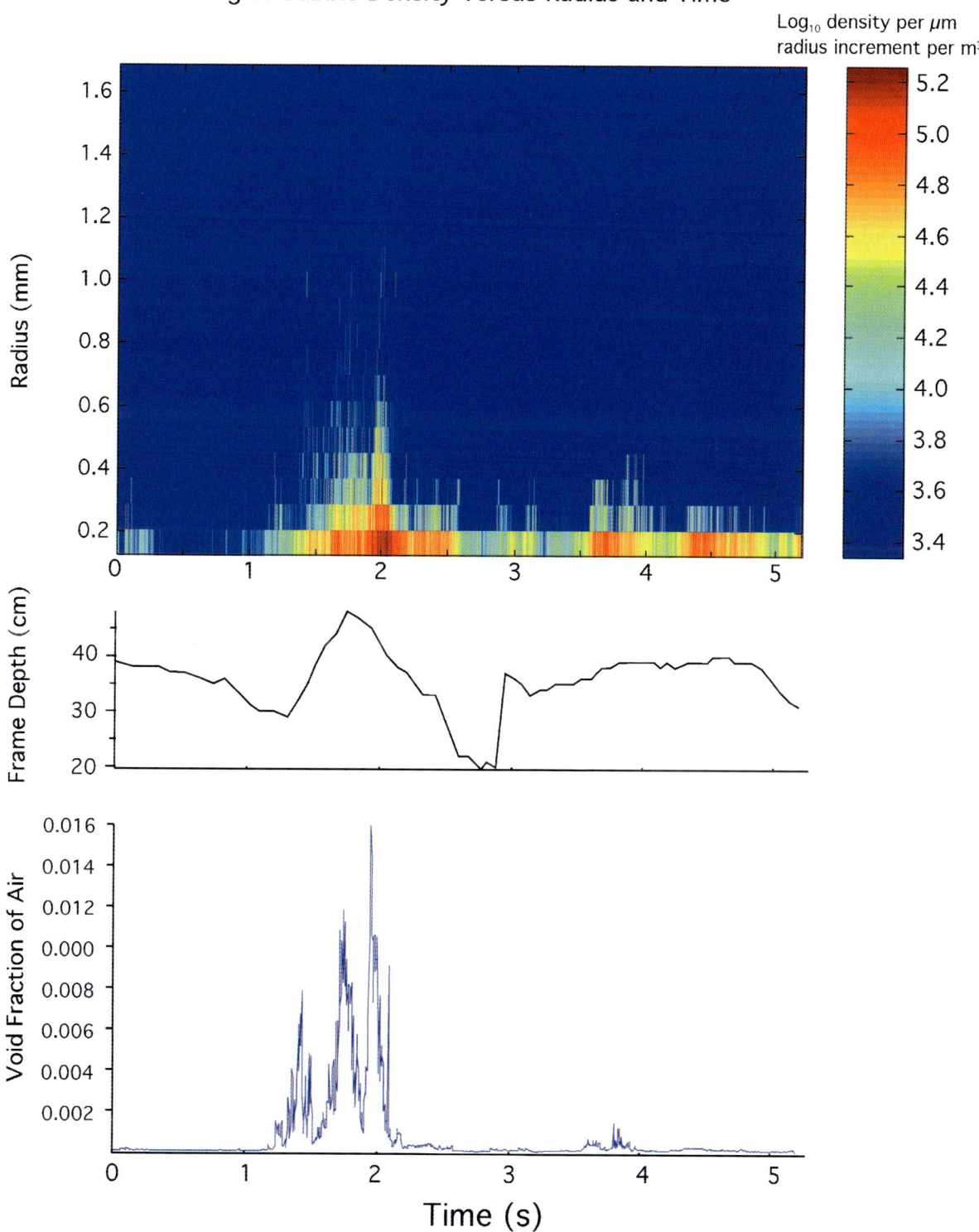

Plate 1. Plate shows in descending order, a contour map of Log bubble density (number of bubbles per cubic meter per μm radius increment) versus bubble radius and time, the frame depth (in cm) versus time and the void fraction of air versus time. The radii bin centers are centered on multiples of 80 micrometers, starting at 160 micrometers.

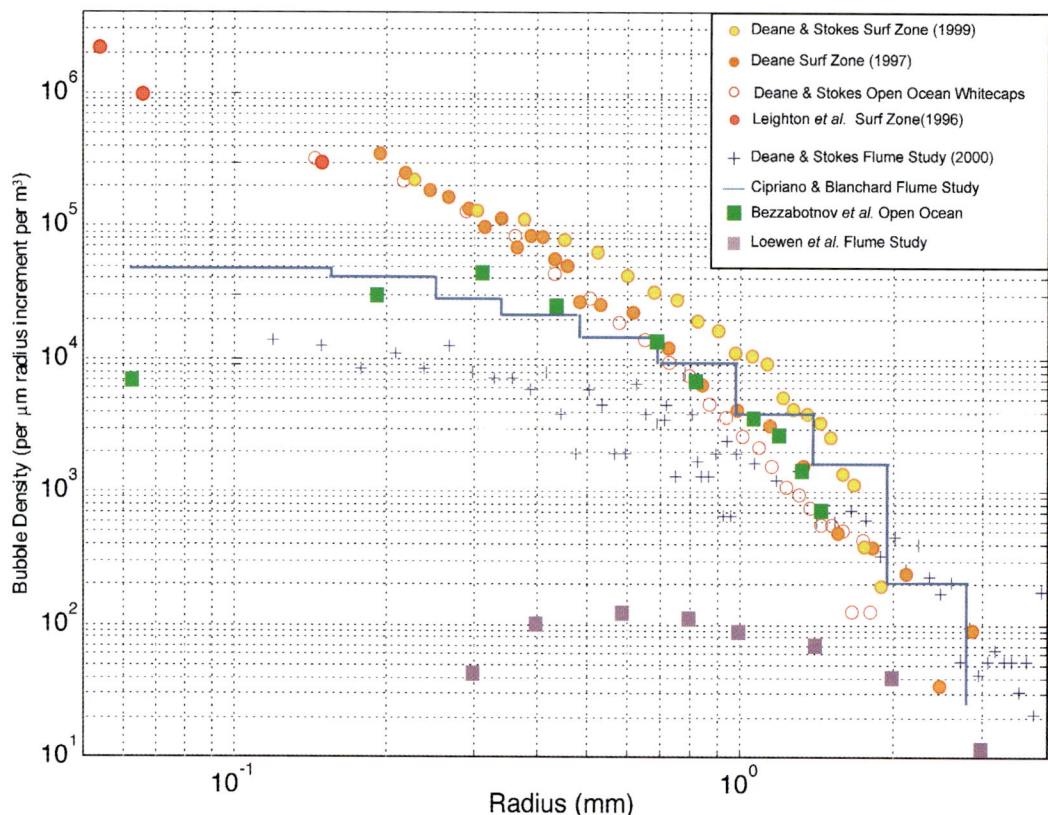

Plate 2. Collected bubble size distributions in salt water from both laboratory flume and open-ocean experiments. The range resolution of the Deane and Stokes data is 80 micrometers, and the smallest bubble radius that can be reliably detected is 160 micrometers.

centration and whitecap coverage with increasing surface wind speed described by Monahan and O'Muircheartaigh [1980].

The results of the modeling for CO2 are shown in Figure 2. Included are the results using plumes seeded with bubble size distributions previously measured by Cipriano and Blanchard [1981] and Bezzabotnov et al. [1986] as well as the recent open ocean measurements. As expected, bubble mediated gas flux increases with increasing wind speed. In addition, increasing the numbers of smaller bubbles to the levels suggested by the present study enhances the gas flux by a factor of 2 to 4.

CONCLUSIONS

One of the most significant findings of this study is the determination that the size distribution of bubbles beneath whitecaps in moderate seas (wind speed 7-10 m/s) is significantly different from that observed in the less energetic breaking events studied in laboratory flumes. Bubbles less than 300 micrometers in radius are at least 10 times more numerous in open ocean whitecaps, with the differences between laboratory and field distributions increasing with decreasing bubble radius. The decreasing numbers of bubbles less than 300 micrometers in radius observed by Bezzabotnov et al. is not consistent with the present study, or the measurements of Leighton et al. in the surf zone, and is

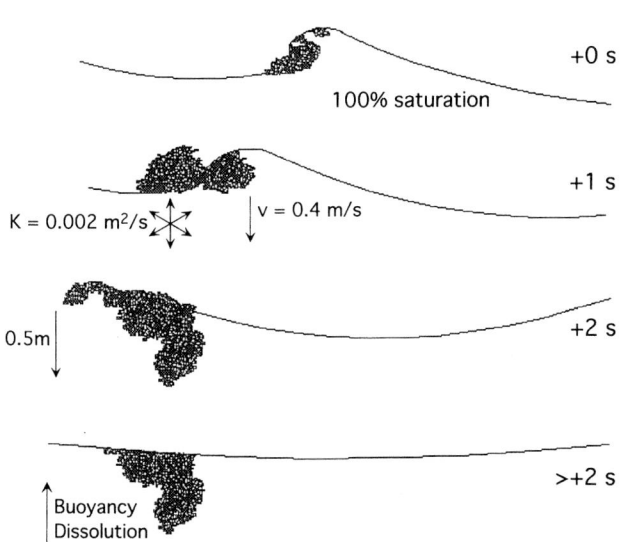

Figure 1. Cartoon of an open ocean white cap showing gas saturation, turbulent diffusivity, initial vertical plume velocity and plume injection depth for the bubble gas dissolution model.

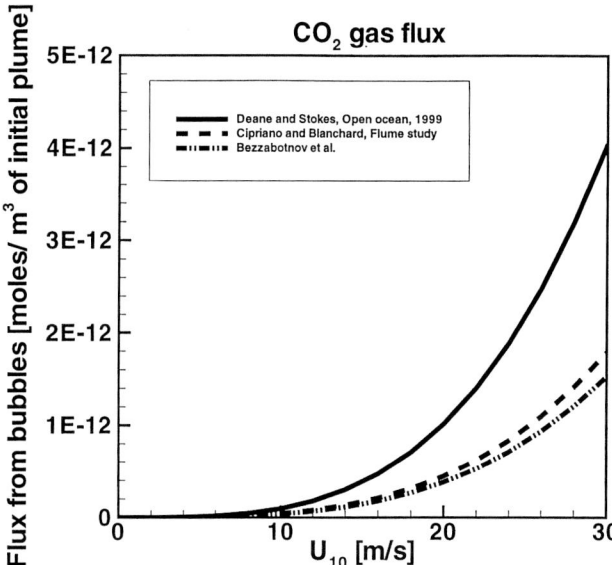

Figure 2. Plot of CO2 gas flux injected by a bubble plume versus windspeed. The curves are based on the model described in the text and are shown for the Deane and Stokes [this study], Cipriano and Blanchard [1981] and Monahan and Zeitlow [1969] bubble size distributions.

possibly due to limitations in the optical system they used.

The differences between the oceanic and laboratory bubble plumes has important implications for air-sea gas flux calculations. For example, model calculations based on flume data lead to the conclusion that large bubbles (radius greater than 300 microns) are important for the transport of highly soluble gasses like CO2 [Keeling, 1993]. However, smaller bubbles are considerably more numerous in the open ocean than expected from flume studies, leading to the conclusion that CO2 transport is actually dominated by bubbles smaller than 300 microns in radius and the soluble gas flux may be 2 to 4 times greater than earlier estimates, with the difference increasing with higher wind speed.

These preliminary results emphasize the importance of collecting data under field conditions. Although open ocean data is difficult to obtain, the differences between field and laboratory data are significant. It is worth noting the bubble size distributions observed beneath breaking surf are similar to those found in the open ocean study, despite the quite different routes to wave breaking in the two environments. The similarity of the bubble size spectra between the open ocean plumes and those in the surf zone suggest that the surf zone might be used as a convenient natural laboratory for some plume studies.

The current data sets are limited because they were measured at only a single depth within the plume. Furthermore the depth is different between the flume and ocean plumes and this must be considered when making direct comparisons between them. Future studies need to measure the bubble size distribution as a function of depth through the plume, quantify the plume volume and characterize both these properties as a function of energy dissipated by wave breaking.

Acknowledgments. These experiments were funded by the physical oceanography program of the National Science Foundation and the underwater acoustics program of the Office of Naval Research.

REFERENCES

Bezzabotnov, V. S. Bortkovskii, R. S. and D. F. Timanovskii, On the structure of the two-phase medium generated at wind-wave breaking, *Izvestiya Akademii Nauk SSSR, Fizika Atmosfery I Okeana*, 22: 1186-1193, 1986.

Cipriano, R. J. and D. C. Blanchard, Bubble and aerosol spectra produced by a laboratory 'breaking wave,' *J. Geophys. Res.*, 86: 8085-8092, 1981.

Deane, G. B., Sound generation and air entrainment by breaking waves in the surf zone, *J. Acoust. Soc. Amer.* 102: 2671-2689, 1997

Deane, G. B. and M.D. Stokes, Air entrainment processes and bubble size distributions in the surf zone, *J. Phys. Oceanography.* 29: 1393-1403, 1999.

Fan, L-S., and K. Tsuchiya, *Bubble wake dynamics in liquids and liquid-solid suspensions*, p. 363. Butterworth-Heinmann, Stoneham, MA., 1990.

Jähne B., Wais, T. and M. Barbaras, A new optical measuring device: A simple model for bubble contribution to gas exchange, in *Gas Transfer at Water Surfaces*, edited by W. Brutsaiert and B.H, Jirka, pp. 237-246, D. Reidel, Hingham, MA, 1985.

Keeling, R. F., On the role of large bubbles in air-sea gas exchange and supersaturation in the ocean. *Journal of Marine Research*, 51: 237-271, 1993.

Leighton, T. G., A. D. Phelps and D. G. Ramble, Acoustic bubble sizing: from the laboratory to surf zone trials, *Acoustics Bull.*, May/June pp. 5-9, 1996.

Loewen M. R., O'Dor, M. A., and M. G. Skafel, Laboratory measurements of bubble size distributions beneath breaking waves, in *Air-Sea Gas Transfer, Third International symposium on Air-Water Gas Transfer*, edited by B. Jähne and E. C. Monahan, pp. 337-349, AEON Verlag & Studio, Hanau, Germany, 1995.

Medwin, H. and A. C. Daniel, Jr., Acoustical measurements of bubble production by spilling breakers, *J. Acoust. Soc. Am.*, 88, 408-412, 1990.

Merlivat, L. and L. Memery, Gas exchange across an air-water interface: experimental results and modeling of bubble contribution to transfer, *J. Geophys Res.*, 88:707-724, 1983.

Monahan, E. C. and I. O'Muircheartaigh, Optimal power-law description of oceanic whitecap coverage dependence on wind speed, *J. Phys. Oceanogr.*, 10; 2094-2099, 1980.

Monahan, E. C. and C. R. Zeitlow, Laboratory comparisons of fresh-water and salt-water whitecaps, *J. Geophys. Res.*, 74: 6961-6966, 1969.

Thorpe, S. A., On the clouds of bubbles formed by breaking wind-waves in deep water, and their role in air-sea gas transfer, *Phil. Trans. Soc. Lond.*, A304, 155-210, 1982.

Stokes, M. D. and G. B. Deane, A new optical instrument for the study of bubbles at high void fractions within breaking waves, *IEEE Journal of Oceanic Engineering* 24: 300-311, 1999.

Woolf, D. K. and S. A. Thorpe, Bubbles and air-sea exchange of gases in near-saturation conditions, *J. Marine Research*, 49, 435-466, 1991.

M. Dale Stokes, Marine Physical Laboratory, Mail Code 0238, Scripps Institution of Oceanography. 8820 Shellback Way. La Jolla, CA 92037-0238

Grant B. Deane, Marine Physical Laboratory, Mail Code 0238, Scripps Institution of Oceanography. 8820 Shellback Way. La Jolla, CA 92037-0238

Svein Vagle, Institute of Ocean Sciences, 9860 West Saanich Road, Sidney, British Columbia, V8L 4B2, Canada

David Farmer, Institute of Ocean Sciences, 9860 West Saanich Road, Sidney, British Columbia, V8L 4B2, Canada

The Effects of Bubbles on Mass Transfer Across the Breaking Air-Water Interface

Satoru Komori and Ryuta Misumi

Department of Mechanical Engineering, Kyoto University, Kyoto, Japan

The effects of bubbles entrained into water by wave breaking on the mass transfer in wind-driven turbulence were experimentally and numerically investigated. The bubble entrainment rate and relative velocity between a bubble and liquid were measured using both a phase Doppler anemometer(PDA) and a laser Doppler velocimeter(LDV). The results show that the bubble entrainment rate increases with wind speed and the relative velocity in wind-driven turbulence is smaller than the terminal velocity based on the Stokes' law. Further, the mass transfer velocity for a bubble was estimated by means of a direct numerical simulation(DNS) over a wide range of the particle Reynolds number Re_p. The DNS results show that the mass transfer velocity for a bubble depends on Re_p but not on the ambient flow structure. It is also found that the net contribution of the bubbles to the total mass transfer across the breaking air-water interface is at most 7% in wind-driven turbulence.

1. INTRODUCTION

Recently the accurate estimation of the mass transfer rate has attracted a special interest in the global-warming problem related to the exchange of carbon dioxide between atmosphere and ocean. The mass transfer velocity of CO_2 increases with increasing wind speed or wind shear in the free-stream wind speed range of $U<5$m/s, where capillary like waves are present. Then, the transfer velocity tends to level off in the high speed range(5m/s$\leq U \leq 12$m/s), where wind waves are rapidly developing. For higher wind speeds of $U>12$m/s, mass transfer velocity increases rapidly again [*Komori et al.*, 1993, 1995]. For reference, the transfer velocity k_L of CO_2 for salt water with the same NaCl concentration of 3.5wt% as sea water, measured in a wind-wave tank [*Komori et al.*, 1999], is plotted in Figure 1 against the free-stream wind speed U, together with field measurements against the wind speed U_{10} at 10m elevation from the air-water interface [*Cember*, 1989; *Watson et al.*, 1991; *Broecker et al.*, 1985]. The best-fitting curve for the laboratory data in Figure 1 is given by

$$k_L = -2.353 \times 10^{-4} + 1.285 \times 10^{-4} \times U \\ -1.873 \times 10^{-5} \times U^2 + 1.140 \times 10^{-6} \times U^3 \\ -2.226 \times 10^{-8} \times U^4. \quad (1)$$

It is impossible to quantitatively compare the laboratory measurements with the field measurements, as long as a similarity parameter which can well correlate the mass transfer velocity is not found from a fluid-mechanical point of view. However, the behavior of the laboratory data of k_L against wind speed resembles that of the field measurements as shown by a dot-dash-line in Figure 1:

$$k_L = -1.187 \times 10^{-4} + 5.245 \times 10^{-5} \times U_{10} \\ -5.237 \times 10^{-6} \times U_{10}^2 + 1.746 \times 10^{-7} \times U_{10}^3. \quad (2)$$

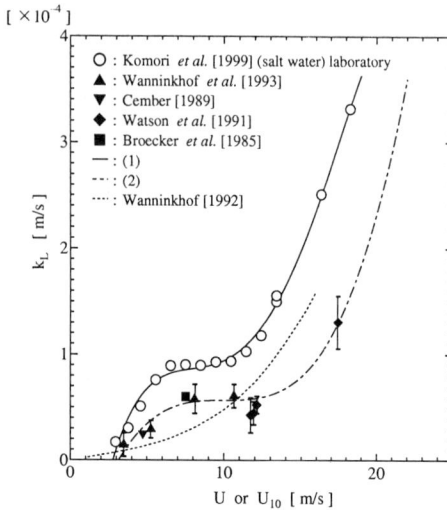

Figure 1. Distributions of mass transfer velocity of CO_2 against free-stream wind speed U or 10m wind speed U_{10}.

This behavior of k_L was physically explained by relating k_L with the frequency of the appearance of surface-renewal eddies [Komori et al., 1993, 1995, 1999]. Recent field and laboratory measurements [Schimpf et al., 2001; Gulliver and Zaske, 2001] also show similar trends to our laboratory measurements. Despite the facts, it is quite curious why most of the ocean chemists roughly trace the field data by using monotonously increasing curves of Wanninkhof [1992] and Liss and Merlivat [1986]. It should be emphasized that the field data of k_L in the wind speed range of 6m/s< U_{10} <12m/s should be treated not as the scattering data but as the data showing a level distribution against U_{10}. Of course, we should not mix the field data of k_L measured in sea and lakes in the low wind speed range without wave breaking, since k_L for sea water is 50% smaller than k_L for fresh water under non-breaking surface conditions in the low wind speed range [Komori et al., 1995, 1999].

Especially for high wind speeds of U>12m/s, wind waves are intensively broken in a wind-wave tank and many bubbles and spray droplets are entrained into water and air flows, respectively. Some experiments have shown that the enhanced air-sea gas exchange may be caused by the bubble entrainment associated with breaking waves [Farmer et al., 1993; Merlivat and Memery, 1983]. Thus, it is generally said that the entrained bubbles play an important role in various small-scales of air-sea interaction processes, especially for the exchange of climate relevant trace gases such as carbon dioxide and methane gas in the high wind speed region. However the detailed mass transfer mechanism across sheared air-water interfaces with entrained bubbles has not been clarified. On the other hand, from an industrial point of view, an averaged void fraction (the volume fraction of air bubbles in the total volume) in industrial gas absorption apparatus, e.g. a bubble column, is greatly larger than that beneath the breaking air-sea interface. This suggests that the bubbles entrained by breaking waves may not offer an additional route for gas exchange, that is, the presence of bubbles may not lead to significant enhancement of the mass transfer velocity.

A number of previous studies have acquired bubble measurements in laboratory and field experiments, and they have proposed some models of the bubble mediated mass transfer [Merlivat and Memery, 1983; Woolf and Thorpe, 1991; and Keeling, 1993]. These studies have tried to explain the mass transfer from bubbles in wind-induced turbulence by using an orbital motion flow model or other flow models, but the explanations were rarely satisfactory because of lack of direct measurements of bubble properties required in the models. Microbubbles can persist in the water for a long time because they have small rise velocities due to buoyancy and are easily advected by currents. Bubble mass transfer depends on the bubble size and ambient liquid flow conditions. It is, therefore, essential to measure the bubble size and the relative velocity between a bubble and ambient liquid. By means of the direct measurements of the bubbles size and relative velocity, it becomes possible to accurately estimate the mass transfer velocity due to the entrained bubbles.

The purpose of this study is to investigate both bubble motions in wind-induced turbulence experimentally and mass transfer by the bubbles numerically, and then to clarify the effects of the entrained bubbles on the mass transfer across the breaking air-water interface. To investigate the contribution of bubbles to the mass transfer, the bubble entrainment rate into water by wave breaking and the relative velocity between a bubble and liquid were measured using both a phase Doppler anemometer(PDA) and a laser Doppler velocimeter(LDV). The mass transfer velocity for a bubble was estimated by means of direct numerical simulation(DNS) over the wide range of the particle Reynolds number.

2. MEASUREMENTS OF PROPERTIES OF BUBBLES ENTRAINED BY WAVE BREAKING

2.1. Experiments

The experimental apparatus and measurement system used here are shown in Figure 2. The experimental

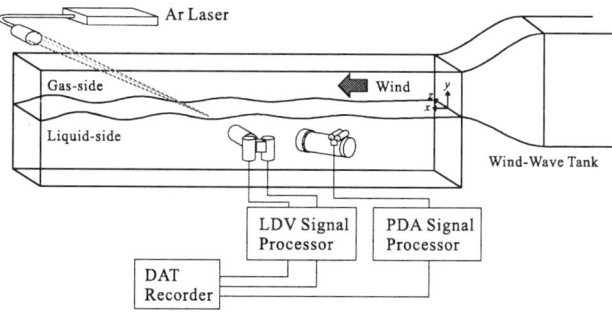

Figure 2. Schematic diagram of experimental apparatus.

apparatus was a wind-wave tank, and it had a glass test section of 7m long, 0.3m wide and 0.8m high. The water depth in the tank was 0.5m and the vertical height of the air flow above the air-water interface was 0.3m. Nonlinear three-dimensional waves under various conditions were driven in the wind-wave tank by winds with free-stream velocity of $U=$2-20m/s. Especially for high free-stream velocity of $U>$12m/s, wind waves were intensively broken and many bubbles and spray droplets were entrained into water and air flows. The details of the apparatus are described in *Komori et al.*[1993, 1995].

Instantaneous streamwise and vertical bubble velocities and ambient fluid velocities were simultaneously measured at $x=$5m from the entrance of the test section($x=$0) by using of a two-colour argon-ion laser phase Doppler anemometer PDA(DANTEC 58N-80 SYSTEM) and a two-colour laser-Doppler velocimeter LDV(DANTEC 55X SYSTEM) with a forward scattering mode, respectively. The measurement volume made from the laser beams was the ellipsoid of $0.25 \times 0.25 \times 1.00$mm, and small ceramic particles with the 1μm diameter which was sufficiently small to distinguish from a bubble were used for seeding the liquid flow. The bubble diameter was also measured using the PDA. The sampling interval and sample size were 0.00025s and 1,200,000, respectively. The measured velocity signals were transmitted into a digital recorder(SONY PC-208A) and were processed statistically by a digital computer for calculating the relative velocity between a bubble and ambient liquid. In addition, bubble motions were visualized by shooting the reflected high-power argon ion laser sheet(LEXEL model 95) and the density of bubble was counted from the visualized flames recorded by a high speed video system(NAC HSV-400). The measurements were carried out for free-stream wind speeds of 9~20m/s and water depths of $y=-3\sim-25$cm.

2.2 Experimental Results and Discussion

Figure 3 shows the bubble size spectra, N_{norm}, normalized with bubble density at the free-stream wind speed of $U=$18.3m/s. The bubble size spectra have approximately the same shape for different water depths of $y=-3.0\sim-10.0$cm. Here the value of $y=-10$cm corresponds to the value of about 3 times the wave height. The distributions of N_{norm} have peak values at about 50μm diameter and decrease linearly with increasing bubble diameter:

$$N_{\text{norm}} = \exp(11.3) \times d^{-3.66}. \quad (3)$$

Similar spectra were obtained for different wind velocities of $U=$9.4~18.3m/s. The bubble size spectra in Figure 3 are in good agreement with the laboratory measurements [*Geißler and Jähne*, 1995].

Figure 4 shows the bubble density, N, against the water depths. The bubble density decreases with increasing the water depth. The bubble density at $U=$18.3m/s where waves are intensively broken is about 10 times larger than that at $U=$9.4m/s where the wave breaking is weak. Table 1 shows the variations of the bubble void fraction, ψ, and the area of the air-water interface occupied by bubbles in unit volume, A_b, against the wind speeds. The void fraction of bubbles in wind-induced turbulence is less than 0.01% even at wind speed $U=$18.3m/s with intensive breaking waves, and these values are quite small compared to values of $\psi=$2~40% and $A_b=$1.0~6.5m^2/m^3 in industrial bubble columns which

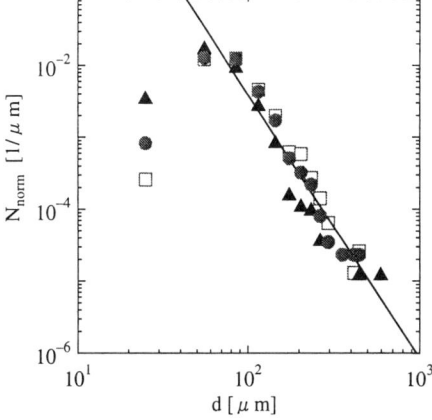

Figure 3. Bubble size spectra normalized with bubble density; ●: $y = -3$ cm, □: $y = -5$ cm, ▲: $y = -10$ cm. The solid line indicates (1).

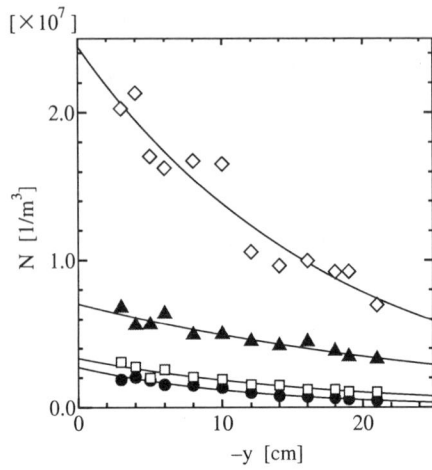

Figure 4. Vertical distributions of bubble density; $\Diamond$: $U = 18.3$ m/s, $\blacktriangle$: $U = 14.8$ m/s, $\square$: $U = 11.5$ m/s, $\bullet$: $U = 9.4$ m/s.

Table 1. Void fraction ψ and bubble interface area A_b.

U [m/s]	ψ [%]	A_b [m^2/m^3]
9.4	9.652×10^{-5}	7.178×10^{-3}
11.5	1.479×10^{-4}	1.100×10^{-3}
14.8	3.960×10^{-4}	2.945×10^{-2}
18.3	1.109×10^{-3}	8.246×10^{-2}

are designed to get effective gas absorption. Accordingly it is not expected that the bubbles entrained in wind-induced turbulence play a significant role on the mass transfer across the breaking air-water interface. The area of the free surface cannot exactly be estimated because of three-dimensionality of breaking waves. If the two-dimensionality of the waves is assumed, the ratio of the bubble interface to the free surface is estimated to be about 8% at U=18.3m/s.

Figure 5 shows the probability density of the vertical relative velocity between a bubble and liquid, $|v_f - v_p|$, and the probability density of the terminal velocity based on the Stokes' law, $U_t(= \frac{gd^2}{12\nu})$. It is found that relative velocity in wind-induced turbulence $|v_f - v_p|$ is smaller than the Stokes' terminal velocity U_t.

3. DNS FOR ESTIMATION OF MASS TRANSFER BY A BUBBLE

3.1 Direct Numerical Simulation(DNS)

The flow geometry and coordinate system for computations are shown in Figure 6. The imposed flow is a uniform non-shear flow around a spherical bubble. The governing equations for an incompressible flow in cylindrical coordinates are given by:

$$\boldsymbol{\nabla} \cdot \boldsymbol{V} = 0, \quad (4)$$

$$\frac{\partial U}{\partial t} + (\boldsymbol{V}\cdot\boldsymbol{\nabla})U = -\frac{\partial p}{\partial x} + \frac{1}{Re_p}\boldsymbol{\nabla}^2 U, \quad (5)$$

$$\frac{\partial V}{\partial t} + (\boldsymbol{V}\cdot\boldsymbol{\nabla})V - \frac{V^2}{r}$$
$$= -\frac{\partial p}{\partial r} + \frac{1}{Re_p}\left(\boldsymbol{\nabla}^2 V - \frac{V}{r^2} - \frac{2}{r^2}\frac{\partial V}{\partial \theta}\right), \quad (6)$$

$$\frac{\partial W}{\partial t} + (\boldsymbol{V}\cdot\boldsymbol{\nabla})W + \frac{VW}{r}$$
$$= -\frac{1}{r}\frac{\partial p}{\partial \theta} + \frac{1}{Re_p}\left(\boldsymbol{\nabla}^2 W - \frac{W}{r^2} + \frac{2}{r^2}\frac{\partial W}{\partial \theta}\right), (7)$$

where all quantities are normalized by the bubble diameter d and mean velocity of fluid on the streamline

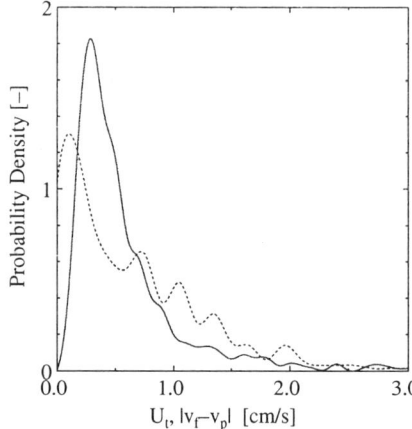

Figure 5. Comparison of probability density of the relative velocity $|v_f - v_p|$(a dotted line) and that of the terminal velocity U_t, at U=18.3m/s and y=−5.0cm.

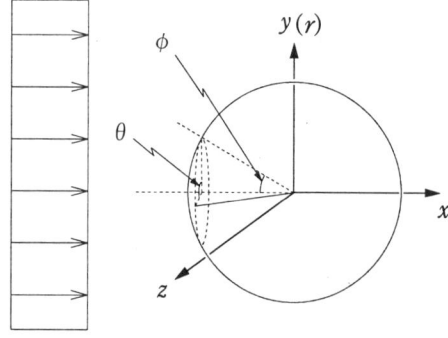

Figure 6. Coordinate system for a spherical bubble.

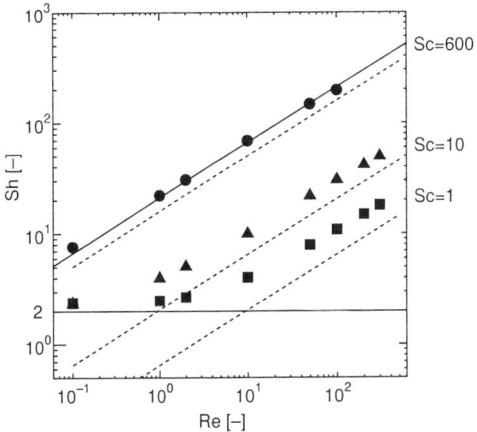

Figure 7. Variations of Sh for Re_p.

through the centre of a sphere U_c. The dimensionless parameters appearing in the governing equations are the particle Reynolds number $Re_p(=\frac{dU_c}{\nu})$ and the Schmidt number $Sc(=\frac{\nu}{D_L})$. Here, ν and D_L are the kinematic viscosity for liquid and the molecular diffusivity of mass, respectively. The free slip condition was used as the boundary condition on the surface of a bubble. The time variations of gas concentration inside a bubble and bubble size were neglected, and the concentration field inside a bubble was assumed to be homogeneous.

The three-dimensional Navier-Stokes(N-S) equations and mass transport equation were directly solved using a finite difference scheme based on the maker and cell (MAC) method. The numerical procedure used here was essentially the same as that used by *Kurose and Komori* [1999] except the boundary conditions and addition of the mass transport equation. The computations were performed for particle Reynolds numbers of $1 \leq Re_p \leq 300$ and Schmidt numbers of $1 \leq Sc \leq 600$. The dimensionless mass transfer rate from a bubble surface was obtained as the Sherwood number Sh by integrating the concentration gradient:

$$Sh = \frac{k_{Lb} \cdot d}{D_L} = -\frac{d}{A}\int_A \frac{\partial C}{\partial \eta} dA. \quad (8)$$

3.2 Numerical Results and Discussion

Figure 7 shows the variations of Sh against Re_p for $Sc=1$, 10 and 600. The solid line shows the relation (9) which was obtained from DNS for a CO_2 bubble with $Sc=600$ and dotted lines show the analytical solutions based on the Stokes' law.

$$Sh = 0.86\sqrt{Re_p \cdot Sc} \quad for \quad Sc=600 \quad (9)$$

The DNS data satisfy the relation of $Sh \propto \sqrt{Re_p \cdot Sc}$ for the wide range of Re_p for $Sc=600$. For other tracer gases such as SF_6, the coefficient in (9) should be determined by performing the DNS for an appropriate Sc of the tracer gas. For small Schmidt numbers of $Sc=1$ and 10, the values of Sh are higher than the Stokes' law solution. The reason is due to that the Stokes' analysis is not appropriate for low Re_p and low Sc because of the assumption of the infinite Peclet number $Pe(=Re_p \cdot Sc)$.

4. CONTRIBUTION OF BUBBLES TO MASS TRANSFER VELOCITY

Mass transfer velocity for a CO_2 bubble is given by (9) from the DNS results. Therefore, the mass transfer rate by all bubbles k_{Lbt} can be estimated by substituting the experimental data of the bubble diameter d (Figure 3), the relative velocity $|v_f - v_p|$ (Figure 5) and the bubble density N (Figure 4) and by summing up (9) for all bubbles. Figure 8 shows the comparison of k_{Lbt} with the total mass transfer velocity for the air-water interface k_{LS}, directly measured through the CO_2 desorption experiment by *Komori and Shimada* [1995]. The mass transfer mediated by bubbles is little in the free-stream velocity range of $U \leq 12$m/s, and is about 7% even at $U=18$m/s where wind-waves are intensively broken. This results support our previous study [*Komori et al.*,

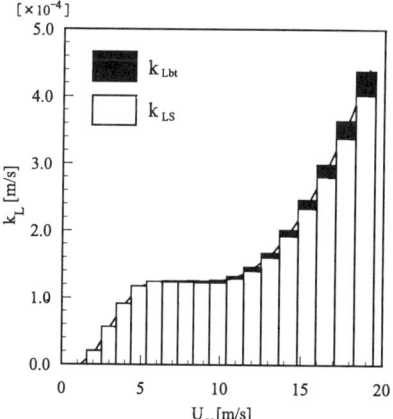

Figure 8. Contribution of bubbles to the mass transfer across the air-water interface.

1999] that the rapid increase of k_L in the high wind speed region with breaking waves is due to both the increase of the surface renewal frequency of ripple-like eddies and the disappearance of surface contamination of very tiny surface-active impurities by wave-breaking.

5. CONCLUSIONS

We have investigated the effects of bubbles on the mass transfer across the breaking air-water interface. The results are summarized as follows:

(a) The bubble size spectra near the air-water interface normalized by the bubble density have peak values at about $50\mu m$ diameter independent of the water depth. The void fraction of the bubbles entrained by breaking waves has extremely small values less than 0.01%.

(b) The relative velocity between a bubble and fluid is smaller than the terminal velocity based on the Stokes' law.

(c) The contribution of the entrained bubbles to the total mass transfer across the breaking air-water interface is only 7% even at $U=18m/s$ where wind-waves are intensively broken.

Acknowledgements. This research was supported by the Japanese Ministry of Education, Science and Culture through Grants-in-Aid(Nos.11450077 and 00107484). The direct numerical simulations were carried out by using a super computer NEC SX-4 of the Center for Global Environment Research, National Institute for Environmental Studies, Environmental Agency of Japan.

REFERENCES

Broecker, W.S., T.H. Peng, G. Ostlund and M. Stuiver, The distribution of bomb radiocarbon in the ocean, *J. Geophys. Res.*, 90, 6953-6970, 1985.

Cember, R., Bomb radiocarbon in the Red Sea : A medium-scale gas exchange experiment, *J. Geophys. Res.*, 94, 2111-2123, 1989.

Farmer, D., C.L. McNeil and B.D. Johnson, Evidence for the importance of bubbles in increasing air-sea gas flux, *Nature*, 361, 620-623, 1993.

Geißler, P. and B. Jähne, Measurements of bubble size distributions with an optical technique based on depth from focus, *in Air-Water Gas Transfer*, edited by B. Jähne and E.C. Monahan, pp.351-362, Aeon Verlag, Hanau, Germany, 1995.

Gulliver, J.S. and A.J. Zaske, Computational experiments to quantify measurements uncertainty in gas exchange coefficient, *in Gas Transfer at Water Surfaces*, edited by M.A. Donelan, W.M. Drennan, E.S. Saltzman and R. Wanninkhof, pp.xxx-xxx, AGU, this volume, 2001.

Keeling, R., On the role of large bubbles in air-sea gas exchange and supersaturation in the ocean, *J. Marine Res.* 51, 237-271, 1993.

Komori, S., R. Nagaosa and Y. Murakami, Turbulence structure and mass transfer across a sheared air-water interface in wind-driven turbulence, *J. Fluid Mech.*, 249, 161-183, 1993.

Komori, S. and T. Shimada, Gas transfer across a wind-driven air-water interface and the effects of sea water on CO_2 transfer, *in Air-Water Gas Transfer*, edited by B. Jähne and E.C. Monahan, pp.553-569, Aeon Verlag, Hanau, Germany, 1995.

Komori, S., T. Shimada and R. Misumi, Turbulence structure and mass transfer at a wind-driven air-water interface, *in Wind-over-Wave, Couplings: Perspective and Prospects*, edited by S.G. Sajjadi, N.H. Thomas and J.C.R. Hunt, pp.273-285, Oxford University Press, Oxford, 1999.

Kurose, R. and S. Komori, Drag and lift forces on a rotating sphere in a linear shear flow, *J. Fluid Mech.*, 384, 183-198, 1999.

Liss, P.S. and L. Merlivat, Air-sea gas exchange rates : introduction and synthesis, *in The Role of Air-Sea Exchange in Geochemical Cycling* (ed. P.Buat-Menard), pp.113-127, 1986.

Levich, V.G., *Physicochemical Hydrodynamics*, Prentice - Hall, 1962.

Merlivat, L. and L. Memery, Gas exchange across an air-water interface: experimental results and modeling of bubble contribution to transfer, *J. Geophys. Res.*, 88, 707-724, 1983.

Schimpf, U., H. Haußecker and B. Jähne, On the investigations of air-water gas exchange and surface micro turbulence using thermography, *in Gas Transfer at Water Surfaces*, edited by M.A. Donelan, W.M. Drennan, E.S. Saltzman and R. Wanninkhof, pp.xxx-xxx, AGU, this volume, 2001.

Wanninkhof, R., Relationship Between Wind Speed and Gas Exchange Over the Ocean, *J. Geophys. Res.*, 97, 7373-7382, 1992.

Wanninkhof, R., W.E. Asher, R. Weppernig, H. Chen, P. Schlosser, C. Langdon and R. Sambrotto, Gas transfer experiment on Georges Bank using two volatile deliberate tracers, *J. Geophys. Res.*, 98, 20237-20248, 1993.

Watson, A.J., R.C. Upstill-Goddard and P.S. Liss, Air-sea gas exchange in rough and stormy seas measured by a dual-tracer technique, *Nature*, 349, 145-147, 1991.

Woolf, D.K. and S.A. Thorpe, Bubbles and the air-sea exchange of gases in near-saturation conditions, *J. Marine Res.*, 49, 435-466, 1991.

S. Komori and R. Misumi, Department of Mechanical Engineering, Kyoto University, Kyoto 606-8501, Japan. (e-mail: komori@mech.kyoto-u.ac.jp)

LUMINY – An Overview

G. de Leeuw[1], G.J. Kunz[1], G. Caulliez[2], D.K. Woolf[3], P. Bowyer[4], I. Leifer[1,4,8], P. Nightingale[5], M. Liddicoat[5], T.S. Rhee[6,9], M.O. Andreae[6], S.E. Larsen[7], F.Aa Hansen[7] and S. Lund[7]

Experiments were undertaken in the Large Air Sea Interaction Simulation Tunnel of IRPHE-IOA, Laboratoire de LUMINY, in Marseille, France, aimed at improving our understanding of the effects of breaking waves on gas transfer, and providing parameterisations for the transfer velocities. Detailed studies were made of breaking wave phenomena, bubbles and turbulence in water and air, and exchange rates of gases with a variety of physical properties (CO_2, CH_4, N_2O, DMS, CH_3Br, ^4He and SF_6). A simple scaling of air-water transfer velocities with friction velocity and Schmidt number breaks down at high wind speeds. A solubility-dependent enhancement of transfer velocity by bubbles can explain only part of the behaviour. An "interfacial resistance" model can explain much of the outstanding behaviour at high wind speeds. Bubble-mediated transfer, surface disruption by turbulence and surfacing bubbles, and interfacial resistance, are all identified as significant to air-sea gas exchange at high wind and sea states.

1. INTRODUCTION

The LUMINY project, designed around laboratory experiments in the Large Air Sea Interaction Simulation Tunnel of the Institut de Recherche sur les Phénomènes Hors Equilibre, Laboratoire Interactions Océan-Atmosphère Luminy (IRPHE-IOA), in Marseille, France, (the project is named after the location in LUMINY) was aimed at improving our understanding of the effects of breaking waves on air-sea gas transfer, and providing parameterisations for the transfer velocities. The main effort was the simulation of breaking waves and to determine their respective effects on air-water gas transfer. Experimental methods addressed three key areas: 1) waves and turbulence; 2) bubbles; and 3) gas transfer. Theoretical and numerical modelling studies of gas transfer mechanisms were an essential part of the study.

2. METHODOLOGY

During two periods, 10 weeks in total, experiments were undertaken in which whitecaps were simulated with bubbles generated by aeration devices (a submerged grid of porous ceramic tubes) and by waves generated by wind and/or a wave maker. Bubble spectra were measured with optical techniques [*Leifer et al.*, 2001]. Measurements of the wave characteristics and the turbulent structure in the air-section were made to quantify breaking wave phenomena and momentum transfer at the air-water interface

---

[1]TNO Physics and Electronics Laboratory, The Hague, The Netherlands

[2]Laboratoire Interactions Océan-Atmosphère, Institut de Recherche sur les Phénomènes Hors Equilibre, Marseille, France

[3]James Rennell Division, Southampton Oceanography Centre, Southampton, United Kingdom

[4]Department of Oceanography, National University of Ireland Galway, Galway, Ireland

[5]Plymouth Marine Laboratory, Plymouth, United Kingdom

[6]Biogeochemistry Dept., Max Planck Institut for Chemistry, Mainz, Germany

[7]Risø National Laboratory, Department of Wind Energy and Atmospheric Physics, Denmark

[8]current address: Chemical Engineering Department, University of California, Santa Barbara, Santa Barbara, California, USA

[9]also at: Dept of Oceanography, Texas A&M University, USA

[*Caulliez*, 2001]. Gases with a variety of physical properties (CO_2, CH_4, N_2O, DMS, CH_3Br, ^4He and SF_6) were introduced into the tunnel and gas concentrations were measured in the water and in the air in order to determine the exchange rates in a variety of conditions. The air-water exchange of "total gas" was also measured. Individual processes were simulated in isolation in order to determine the separate effects of waves, bubbles, and wind stress on the exchange coefficients.

2.1 Retrieval of Gas Transfer Rates

The gas exchange rates were determined from the variation of the concentrations in the water and in the air using three different methods:
1. An analytical model was developed at TNO that describes the exchange of gas between the water and air sections of the tunnel and the leakage from the air section to the laboratory [*Kunz*, 1998]. After determination of the tunnel leakage rate with this model, it was applied to CO_2 data. The model describes the concentrations well, even though chemical buffering was not accounted for.
2. A four-reservoir model was developed that not only takes into account the leakage of the tunnel head space to the laboratory, but also exchange between the laboratory and the atmosphere. In addition, to account for the chemical buffering in the water phase, data on the pH and total inorganic carbon (TIC) are used to determine the CO_2 exchange coefficients.
3. The third model considers only the concentrations of the gas in the water phase, and its time evolution as determined by the concentrations in the water and in the air. Gas is exchanged across the surface of each submerged bubble and at different rates over the surface of the tank, but the measurements provide only the integrated effect over the whole tank. When the aeration devices are fed with air from the head space of the tank, the "mass budget" for the dissolved gas in the tank can be written as (see *Woolf* [2001]):

$$V\, dC_w/dt = A\, \{K_b\, [(1+\delta)C_a/H - C_w] + K_o\, [C_a/H - C_w]\} \quad (1)$$

in the absence of any net source in the tank, where V is the volume of the tank, A is the surface area of the tank, C is the concentration in the head space, H is the Henry's law constant of the gas, δ is a small fraction associated with compression of the bubbles, C_w is the concentration of the dissolved gas, dC_w/dt is the rate of change of C_w, K_b is the "integral" gas transfer velocity for exchange across the surface of the bubbles, and K_o is the "integral" gas transfer velocity for exchange directly across the air-water interface. Time series of C_a and C_w were used to generate sets of values $\{C_a, C_w, dC_w/dt\}$ from which the unknowns δ, K_o and K_b can be estimated.

Similar experiments were conducted with the single difference that the aeration devices were not fed from the head space, but from air outside the laboratory (essentially background atmospheric levels). In this case, the mass budget equation for dissolved gas may be written:

$$V\, dC_w/dt = A\, \{K_b\, [(1+\delta)C_{atm}/H - C_w] + K_o\, [C_a/H - C_w]\} \quad (2)$$

where C_{atm} is the atmospheric concentration of the gas. As long as C_{atm} and C_a are quite different it is relatively easy to discriminate the influence of bubble-mediated exchange, K_b, and direct exchange, K_o. The effect of switching the air source is illustrated in Figure 1.

3. RESULTS

3.1 Turbulence and Wave Breaking

A strong effort was made to characterise wave breaking and wind stress [*Caulliez*, 2001]. The wave breaking conditions can be described by one parameter only, namely the breaking rate R_d, and this parameter is directly related to the friction velocity in air u_{*a}, irrespective the wave conditions at the surface. This somewhat surprising result suggests that at high wind speeds in the tunnel the increase in the turbulent momentum flux throughout the air surface boundary layer is to a large extent controlled by the increase in occurrence and intensity of breaking phenomena.

3.2 Bubbles

Bubble size distributions were determined with different optical techniques [*Leifer et al.*, 2001]. The bubble measuring system was used, among others, to monitor characteristics of bubble plumes resulting from waves breaking at fetches from 17.5 m to 25 m. Results are presented and discussed in *Leifer and De Leeuw* [2001] and *De Leeuw and Leifer* [2001]. Enormous variety was observed from plume to plume in regards to bubble density, plume physical dimensions, the intensity and extent of both structured and unstructured turbulence, and induced fluid flows leading to diverse evolutionary paths.

3.3 Gas Exchange

Experiments with aeration devices. Several experiments were conducted using the aeration devices together with various combinations of wind and waves. The relatively

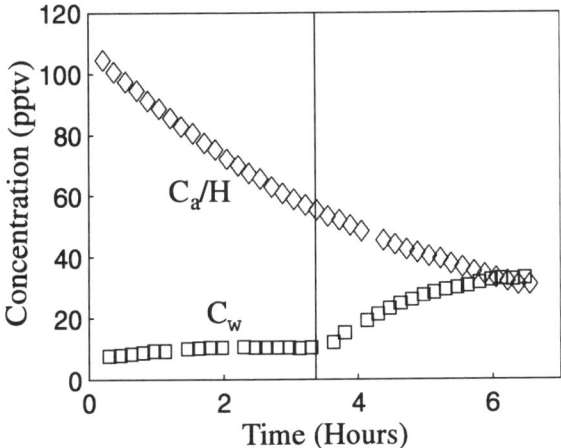

Figure 1. Concentrations of SF_6 in water and in air, measured at a wind speed of 2.5 m/s. The lhs panel shows data obtained with bubbles aerated with air from outside the tunnel air space, in the rhs panel the air is aerated from the tunnel head space.

controlled characteristics of the aeration devices allow for comparison of experimentally determined transfer coefficients with theoretical expectation. This is a vital part of developing a reliable parameterisation, since then confidence can be gained in the predictions of gas transfer coefficients based on other observations of bubbles (produced by breaking waves in the laboratory or the field). In addition, in these experiments the contributions to total transfer of exchange across the surface of the bubbles and exchange across the main water surface can be separated.

Pairs of invasion experiments with the aeration devices fed sequentially from outside of the laboratory and from the headspace, cf. Figure 1, were conducted for four different combinations of wind, waves and bubble flow rate. Separate polynomials were fitted to C_w for each experiment. The coefficients, K_b and K_o, were estimated for both experiments in the pair using eqs. 2 and 1. The data from the first of the pair establishes the "balance" between bubble stripping and direct transfer, but often leaves considerable uncertainty in the absolute value of each. The second of the pair only determines the total coefficient, $K_T = K_o + K_b$ (and if wished an approximate value of "δ"), but in combination with the first experiment usually improves the accuracy of the individual coefficients.

These analyses were applied to air, CH_3Br, He, SF_6 and N_2O. The contributions of "direct" transfer (directly across the planar water surface) and "bubble-mediated" transfer (*via* exchange across the surface of bubbles) were successfully separated. The bubble-mediated transfer is very high for the very poorly soluble gases, but lower for relatively soluble gases. For example in one experiment K_b is estimated at 60 cm h^{-1} for SF_6, but only 1-2 cm h^{-1} for CH_3Br.

Wind and wave experiments. Transfer coefficients were determined using method 3 described in Section 2.1. Only one coefficient is computed, $K_T = K_o + K_b$, compared to two coefficients in the aeration experiments. In comparing transfer velocities of different gases, the orthodox approach is to "convert" the raw values to an "equivalent" value for a Schmidt number Sc of 600:

$$K_{600} = (Sc/600)^{-0.5} \, K \qquad (3)$$

This transform is usually considered appropriate for direct transfer in moderate or high wind speeds and evidence of a breakdown is usually interpreted as evidence of bubble-mediated transfer (e.g., *Wanninkhof et al.* [1993]).

Transfer velocities have been calculated for five of the gases. The results are presented in Figure 2 as function of the friction velocity in the water u_{*w}:

$$u_{*w} = u_{*a} \, (\rho_a/\rho_w)^{0.5}, \qquad (4)$$

The results can be separated into three groups with a model description of the apparent behaviour, cf. *Woolf* [2001] for a discussion of models in energetic conditions. The simplest model is for transfer velocity to be proportional to friction velocity in water u_{*w}:

$$K = A \, (Sc/600)^{-0.5} \, u_{*w} \qquad (5)$$

where A is a non-dimensional constant. The behaviour of N_2O is relatively well described by this model with A=0.006.

The behaviour of He and air can be described better by including the influence of breaking waves:

$$K \, (Sc/600)^{0.5} = A \, u_{*w} + B \, R_d \qquad (6)$$

The behaviour of He is described reasonably well with A=0.006 and B=0.01, for air the value of B is slightly higher.

The transfer velocities of SF_6 and CH_3Br are suppressed at high wind stresses compared to the other gases. This behaviour can be simulated by a model

$$K = K_o \, K_i / (K_o + K_i) \qquad (7)$$

where $\quad K_o = A \, (Sc/600)^{-0.5} \, u_{*w}$
$\qquad\quad K_i = 0.3$ mm/s
and $\quad\;\; A = 0.01$

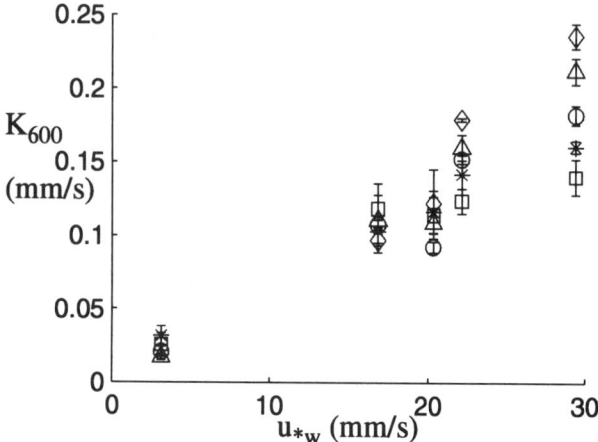

Figure 2. Transfer velocity of five gases plotted against friction velocity. Experiments have been grouped into five subsets and the mean and standard error of transfer velocity and the mean of the friction velocity for each subset are plotted for air (◊), He (Δ), N_2O (o), SF_6 (*) and CH_3Br (!).

The additional term, K_i, is interpreted as the inverse of a fixed interfacial resistance associated with processes at the actual air-sea interface. It is proposed that these processes become rate-limiting at sufficiently high stirring rates for all gases. The interfacial resistance depends on (largely obscure) molecular properties of the gas and is relatively large for SF_6 and CH_3Br. It is proposed that breaking wave mechanisms enhance the transfer of all gases at high wind speeds, while an interfacial resistance limits the maximum transfer of all gases. These two influences counteract such that they may effectively cancel out (as is the case for nitrous oxide in these experiments) or one or other may dominate. Review of other laboratory studies has affirmed the existence of a significant interfacial resistance to the transfer of SF_6 and some other gases [*Woolf*, 2001].

4. DISCUSSION AND CONCLUSIONS

Typical values of flow rate for the aeration devices (~200 litres/min equivalent to ≈ 0.032 litres m^{-2} s^{-1}) compare to a crude estimate of the "flushing rate" of the upper ocean of ≈ 0.125 litres/(m^2 second), when the whitecap coverage is 2% (at a wind speed of approximately 12m/s; [*Woolf*, 1993]). Thus the strong influence of bubble-mediated exchange for some gases and the moderate influence of surface disruption found experimentally appears to be associated with a quite moderate flux of bubbles. This implies that breaking wave mechanisms are very significant to air-sea gas exchange. Note that this conclusion is dependent on an uncertain estimate of bubble fluxes in oceanic conditions. The results also imply that any enhancement in air-sea exchange at high wind speeds should be particularly pronounced for gases such as SF_6 and He.

Bubble-mediated transfer appears to be the most significant mechanism associated with breaking waves, though for relatively soluble gases, disruption by surfacing bubble plumes and the turbulence generated by breaking will be relatively important.

Acknowledgements. The LUMINY project was supported by the European Commission EC DG XII, contract ENV4-CT95-0080, and by the participants' institutes. The contribution of TNO-FEL was supported by the Netherlands Ministry of Defence, assignment A95KM786. The PML contribution was supported by the UK National Environment Research Council and UEA by UK D.O.E. Grant number EPG/1/1/1278. The authors would like to thank IRPHE-IOA for the extensive laboratory support provided.

REFERENCES

Caulliez, G., Statistics of the geometric properties of breaking wind waves observed in laboratory, Geophysical Monograph (this volume), edited by M.A. Donelan, W.M. Drennan, E.S. Salzman, and R. Wanninkhof, American Geophysical Union, Washington, DC, 2001.

De Leeuw, G., and I. Leifer, Bubbles outside the plume during the LUMINY wind-wave experiment, Geophysical Monograph (this volume), edited by M.A. Donelan, W.M. Drennan, E.S. Salzman, and R. Wanninkhof, American Geophysical Union, Washington, DC, 2001.

Kunz, G.J., CO2 gas exchange during the LUMINY Main Experiment. *TNO Physics and Electronics Laboratory*, Report. 1998.

Leifer, I. and G. de Leeuw, Bubble measurements in breaking-wave generated during the LUMINY wind-wave experiment, Geophysical Monograph (this volume), edited by M.A. Donelan, W.M. Drennan, E.S. Salzman, and R. Wanninkhof, American Geophysical Union, Washington, DC, 2001.

Leifer, I., G. de Leeuw and L.H. Cohen, Optical measurement of bubbles: system design and application, *submitted to J. Atm and Ocean. Tech.*, 2001.

Rhee, T. S., *The process of air-water gas exchange and its application*, Ph.D. thesis, Texas A&M University, College Station, Texas, 2000.

Wanninkhof, R., W. Asher, R. Wepperning, H. Chen, P. Schlosser, C. Langdon and R. Sambrotto, Gas transfer experiment on Georges Bank using two volatile deliberate tracers. *J. Geophys. Res.*, 98, pp. 20237-20248, 1993.

Woolf, D.K., Bubbles and the air-sea transfer velocity of gases, *Atmosphere-Ocean* 31, pp. 517-540, 1993.

Woolf, D.K., Gas transfer in energetic conditions, Geophysical Monograph (this volume), edited by M.A. Donelan, W.M. Drennan, E.S. Salzman, and R. Wanninkhof, American Geophysical Union, Washington, DC, 2001.

G. de Leeuw, TNO Physics and Electronics Laboratory, P.O. Box 96864, 2509 JG The Hague, The Netherlands (e-mail: deleeuw@fel.tno.nl)

Bubbles Outside the Plume During the LUMINY Wind-Wave Experiment

Gerrit de Leeuw

TNO Physics and Electronics Laboratory, The Hague, The Netherlands

Ira Leifer

Chemical Engineering Department, University of California, Santa Barbara, California

Since many bubble-mediated processes are size dependent, it is often necessary to characterize the bubble distribution over the full size spectrum. For example, in regards to bubble-mediated gas transfer, small bubbles are important for insoluble gases like helium, while large bubbles are important for soluble gases, like dimethyl sulfide. In order to measure bubbles over the full size spectra, two complementary Bubble Measuring Systems (BMS) were deployed in the LUMINY wind-wave experiment. The TNO Mini-BMS measured smaller bubbles over the radius (r) range $15 < r < 500$ µm and had a well defined measurement volume. The NUIG Large-BMS measured bubbles over the range $200 < r < 5000$ µm and was designed to examine bubbles in bubble plumes without affecting the plume motions. The Large-BMS provided time resolved distributions, appropriate for the highly transient bubble plumes where large ($r > 500$ µm) bubbles are located. The Mini-BMS time averaged the bubble distributions, which is appropriate for the largely background bubble populations. The two systems were intercompared during a series of bubble frit experiments, and produced good agreement. The distribution was observed to have a power law dependency S in agreement with most other observations and Senescence observations by the Large-BMS. Characteristics of the bubble concentrations are discussed in terms of wind speed, fetch and additional paddle waves.

1. INTRODUCTION

The influence of breaking waves on air-sea gas transfer has been investigated in the LUMINY project [*De Leeuw et al.*, 2001]. One of the main conclusions from the LUMINY project is that bubble-mediated air-sea gas transfer appears to be the most significant mechanism associated with breaking waves. Within the limits of the LUMINY analysis, it appears reasonable to calculate bubble-mediated transfer from measured bubble distributions.

This conclusion was obtained from experiments in the Large Air Sea Interaction Simulation Tunnel of the Institut de Recherche sur les Phénomènes Hors Equilibre, Laboratoire Interactions Océan-Atmosphère Luminy (IRPHE-IOA), in Marseille, France. Two experiments were undertaken. The LUMINY Pilot experiment in September 1996 was conducted as a test experiment to try

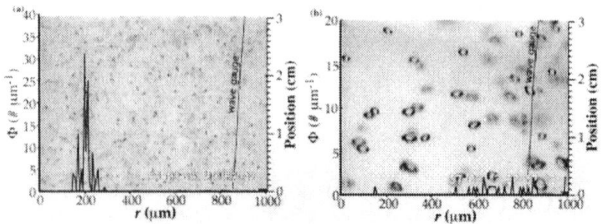

Figure 1. Population distributions and sample images for bubble streams from (a) working and (b) broken bubble frits. Scale is shown in upper right of image with large line equivalent to 1 cm, smaller dimensions equivalent to 1 mm.

experimental procedures and techniques that were used in the Main experiment, 15 February - 4 April, 1997. Measurements were made of the turbulent structure in the air-section to quantify breaking wave phenomena and turbulence in water and air [*Caulliez et al.*, 1999; *Caulliez*, 2001]. Gases with a variety of physical properties (CO_2, CH_4, N_2O, DMS, CH_3Br, ^4He and SF_6) were introduced into the tunnel and gas concentrations were measured in the water and in the air in order to determine the exchange rates in a variety of conditions. The air-water exchange of "total gas" was also measured. Individual processes were simulated in isolation in order to determine the separate effects of waves, bubbles, and wind stress on the exchange coefficients. The effects of bubbles were investigated through simulation of whitecaps with bubbles generated by aeration devices (a submerged grid of porous ceramic tubes) and by waves generated by wind and/or a wave maker. Often bubbles were generated using both techniques simultaneously. Bubble spectra were measured with optical techniques described briefly below.

2. EXPERIMENTS

Aeration devices mounted on grids were submerged over the length of the large air sea interaction tunnel at a depth of about 0.5 m below the water surface. They were fed with air from the tunnel air section over the water, thus introducing air with the same concentration of the study gases that would have been introduced when the air would have been entrained by breaking waves. For a number of paired experiments [*De Leeuw et al.*, 2001] the aerators were fed with atmospheric air free of any of the study gases.

Gas transfer studies were undertaken at three reference wind speeds: 2.5, 10 and 13 m s^{-1}. The two higher wind speeds generated breaking waves. The wave breaking process was further modified by adding a surface modulation using a paddle in the water at the tunnel entrance, simulating a swell. The waves generated by these methods are referred to as 'wind waves' and as 'amplified paddle waves'. Bubbles entrained in the water during the wave breaking process were strongly variable and based on the bubble plume characteristics 7 plume categories could be identified [*Leifer and De Leeuw*, 2001]. To further enhance bubble-mediated gas transfer, often also the aeration devices were used in combination with wind and paddle waves.

These various combinations of aeration intensity, wind velocity and swell gave rise to different bubble plumes that were studied using two optical bubble measuring systems (BMS). The Large-BMS and the Mini-BMS are described in detail in *Leifer et al.* [2001]. In brief, the TNO Mini-BMS consists of a light source illuminating a well-defined sample volume that is monitored with two cameras. The overview camera served to monitor part of the plume and the interaction of the plume with the Mini-BMS. No effect was found. Bubbles inside the sample volume, in the range between 15 and 500 μm equivalent spherical radius, r, were monitored through a telescope onto a ccd camera. Images are recorded on video tape for later processing using automated image processing software. The Mini-BMS was size calibrated. The Large-BMS consisted of multiple video cameras allowing for simultaneous observations at multiple scales and locations. The system was designed to provide non-invasive measurements of time-resolved bubble distributions, fluid velocities, and bubble plume structural details. The Large-BMS observed bubbles for 150 μm < r < 5000 μm. These observations were analysed in conjunction with the TNO Mini-BMS. Good agreement was found for the overlapping ranges in calibration experiments, for the background LUMINY bubble distribution and for the LUMINY aeration devices. Both sytems were mounted on a moveable chariot for observations from 15 to 30 m fetch. The Mini-BMS was usually mounted at a fixed depth of 0.07 m below the mean water surface, the Large-BMS could measure bubbles at a range of depths.

Bubble size distributions were measured at fixed positions along the accessible tunnel fetch. As far as possible, the same positions were maintained for the various different experimental types to allow for intercomparison of the bubble size distributions and plume characteristics at each position in the different conditions. In the highest wind speeds combined with paddle waves the carriage was sometimes pushed by the wind and the waves, thus providing a scan over a certain distance. In addition, scans were made on purpose to sample the mean bubble size distributions along the tunnel length. This is important for the gas exchange experiments because the gas transfer rate retrieved from the measured water and air concentrations results from all processes occurring along the length of the tunnel, and thus represents an integration over the whole tunnel.

3. RESULTS

3.1 Aeration Devices

Images of bubble plumes released by the aeration devices, recorded with the Large-BMS, are shown in Figure 1. The sizes of the bubbles produced by the two devices are very different. The population distribution shown in Figure 1a, is very narrow with a width of about 20 μm peaking around 205 μm radius. The width of the bubble distributions represents variations due to varying bubble size, bubble oscillations, and size change with depth of field. Although single bubbles of this size do not oscillate in a stagnant fluid, in the dense bubble stream produced by the bubble frits, interactions between bubbles caused shape oscillations. In contrast, bubbles from the second aerator type (Figure 1b) are significantly larger, and the distribution is broad with a peak at about 750 μm. Although bubbles of this size normally exhibit large shape change variations, the broadness of the distribution is significantly larger than observed from single bubbles (Leifer et al., 2000a).

The bubble distributions shown in Figure 1 may represent natural variability between aeration devices. The strong variations occurring between different devices are illustrated in Figure 2. The bubble size distributions in Figure 2 were measured with the Mini-BMS at fetches between 20 and 30 m, in a wind speed of 10 m s^{-1}. The line represents a continuous scan made in the same conditions, immediately following the first six measurements. The individual size distributions are strongly bi-modal. The shape of the distributions of the smallest bubbles are similar and they peak at a radius of about 35 μm. Concentrations

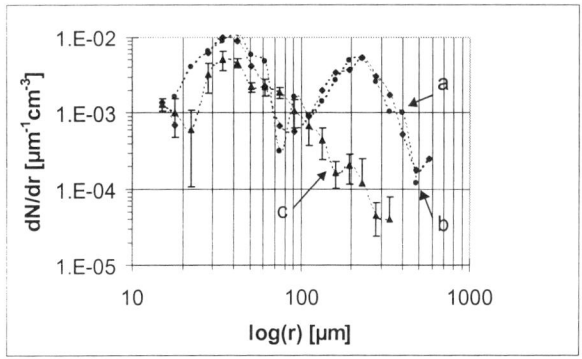

Figure 3. Bubble size distributions resulting from scans over fetches from 15 to 30 m:
a) u=2.5 m s^{-1}, paddle waves, 2 bubbler units (◆),
b) u=10m s^{-1}, no paddle waves, 2 bubbler units (●)
c) u=13 m s^{-1}, paddle waves, no bubbler units (▲)
Uncertainties are shown as error bars on distribution c.

vary by a factor of 2-3. However, for radii larger than 70-80 μm the distributions start to diverge both as regards the shape and the concentrations. The peaks of the second mode are observed to vary between 130 and 230 μm. The peak concentrations vary by about two orders of magnitude. The concentration variations in either mode are not related to fetch.

The scan represented by the solid line in Figure 2 was obtained by slowly moving the carriage at constant speed (0.5 m s^{-1}) from 20 to 30 m. The bubble distribution is an average over many individual aeration devices and is therefore thought to be representative for the mean bubble size distribution in the whole tunnel, for the chosen condition.

Other examples of scans are shown in Figure 3, for different situations as regards bubble intensity, wind speed and wave generation by the paddle. Curve c is for a wind speed of 13 m s^{-1}, and bubble generation only from breaking wind waves. The concentrations fall of with r$^{-2.5}$. Distributions a and b were all measured in the presence of artificial bubble generation using the aerators. The most striking change is the addition of the second mode of bubbles peaking around 200 μm, while the spectral shape of the first mode is more or less retained. But note that also in the wind-wave generated bubble size distribution a weak but nevertheless clear large bubble mode is present that peaks at 200 μm. The concentrations in the small mode increase by about a factor of 2. The concentrations in both modes are similar to those for the scan shown in Figure 2.

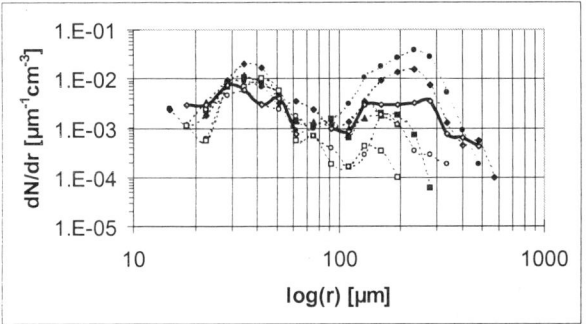

Figure 2. Bubble size distributions measured at fetches of 20 (◆), 22 (■), 24 (▲), 26 (●), 28 (□) and 30 (o) m; the line is a continuous scan. Wind speed 10 m s^{-1}, no paddle.

The similarity of the concentrations of the smaller bubbles produced by breaking wind waves and by the bubblers, and the large difference in the concentrations of

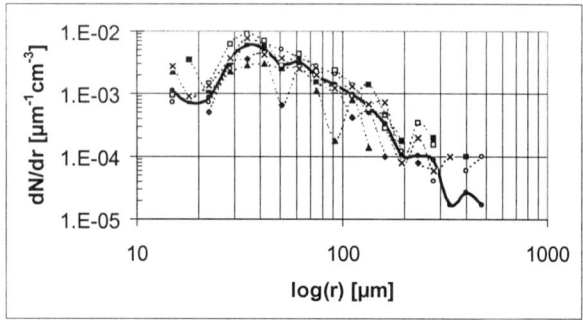

Figure 4. Bubble size distributions measured in a wind speed of 13 m s^{-1} and paddles, at fetches of 17.8 (■), 20 (◆), 22.7 (▲), 25 (o), 27.5 (□) and 30 m (x). The line is the average of all six distributions.

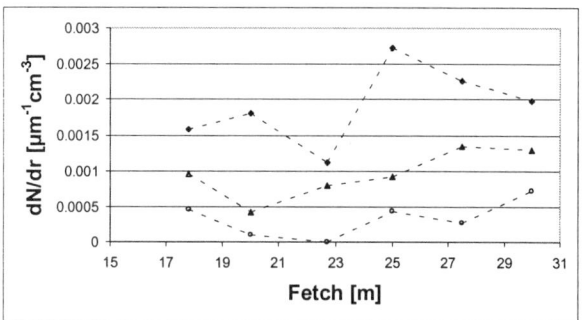

Figure 5. Variation with fetch of bubble concentrations measured in a wind speed of 13 m s^{-1} and mechanical wave generation with the paddle wave. Bubble radii: 75 (◆), 111 (▲) and 160 (o) μm.

the larger bubbles, together with the observation of very narrow distributions of bubbles produced by the aerators, suggest that the smaller bubbles observed in the absence of breaking wind waves are produced by a secondary generation mechanism: bubbles produced by large bursting bubbles. Observations of this process were described in [*Leifer et al.*, 2000b]. As observed by *Spiel* [1998], when a bubble larger than 1200 μm radius bursts, the film cap pulls very rapidly back, generating aerosols. The rapidly rolling back cap continues downward after intersecting with the surface, forming bubbles.

It has been suggested that the observation of small bubbles could in fact be biological organisms rather than bubbles. However, the image processing program uses a routine that discriminates between bubbles and other organisms. Based on many visual observations it is assumed that bubbles are spheroidal whereas other objects may not be. Further, when a bubble is exactly in focus, it appears as a dark object with a small white spot in the middle. A bubble that is slightly out of focus is completely dark. In contrast, most other objects like algae are partially transparent. In the Mini-BMS algorithm a transparency criterion was set to accept only objects with a black/white ratio larger than 90%.

3.2 Breaking Waves

In this section only bubble size distributions will be discussed that were obtained with the aeration devices switched off. In this situation bubbles are only created by breaking waves. At the short fetches in the tunnel, the wave field is not fully developed and the wave breaking rate is expected to change strongly with fetch. Measurements show a change of the wave breaking rate with increasing fetch over the fetches of interest (15-30 m) [*Caulliez*, 1999]. This leads to the expectation that also the bubble spectra will change with fetch. Indeed, a large number of different bubble plume types have been observed, and the characteristics change with fetch [*Leifer and De Leeuw*, 2001].

Amplified paddle waves. Figure 4 shows bubble size distributions measured with a wind speed of 13 m s^{-1} in the presence of a paddle wave. The average distribution, shown as the line in Figure 4, falls off with r$^{-2.5}$, i.e. the same slope as scan c in Figure 3 that was made in the same conditions.

Looking at the individual bubble spectra in Figure 4, the expected fetch dependence is not very clear and the distributions appear interspersed. However, plotting the concentrations as a function of fetch, shown in Figure 5 for three bubble sizes, shows a trend of an initial decrease for the smaller fetches, followed by an increase for fetches larger than about 22 m. The trend is steepest for the smallest bubbles (i.e. longest life time) and slowest for the largest bubbles (i.e. diffuse easiest).

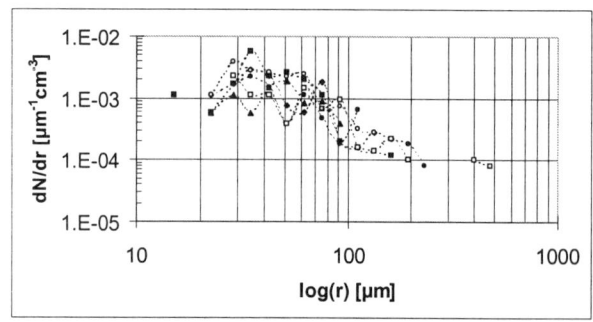

Figure 6. Bubble size distributions measured in a wind speed of 10 m s^{-1} and paddles, at fetches of 17.8 (■), 20 (◆), 22.7 (▲), 25 (o), 27.5 (□) and 30 m (x). The line is the average of all six distributions.

The variation of the bubble concentration with fetch has similarity to the trend in the wave breaking rate for wind amplified paddle waves reported by *Caulliez* [1999] for 10 m s^{-1} wind speed. The breaking rate decreased from 17 m to 21 m fetch, slightly increased to 26 m and then decreased again in the next 2 m.

Figures 6 and 7 show examples of bubble size distributions created by amplified paddle waves at a wind speed of 10 m s^{-1}. The bubble concentrations are significantly lower than those shown in Figure 4 (by a factor of 2-3) and the spectral peaks are not as well defined as in the higher wind speed case. For most spectra a peak is observed between 30 and 40 μm. However, on average the spectra fall off with r$^{-1.6}$ and for radii of 200 μm the concentrations are similar to those at 13 m s^{-1}. On the other hand, this is about the largest size that is observed with statistical significance by the Mini-BMS.

The fetch dependence for this situation is shown in Figure 7. For the smallest bubbles of 61 μm the concentrations increase with fetch, in contrast to the observation at the higher wind speed, which behave similar to the 75 μm bubbles in Figure 5. For larger bubbles no significant trend is observed with fetch.

Wind waves. For wind waves the breaking rate reported by *Caulliez* [1999] for a wind speed of 13 m s^{-1} decreased gradually between 15 and 26 m. The bubble size distributions and fetch dependence for this situation are presented in Figures 8 and 9. The bubble spectra peak at slightly smaller radii of circa 35 μm, as opposed to about 40 μm when the paddle waves are on. The concentrations are seen to fall off somewhat faster, with r$^{-3.1}$ (determined from a fit to the average distribution).

Also the variation with fetch is influenced by the paddle waves. The initial decrease observed in the presence of paddle waves for all sizes now only occurs for the larger ones, and continuous also for longer fetches. This is in agreement with observed wave breaking rate behaviour.

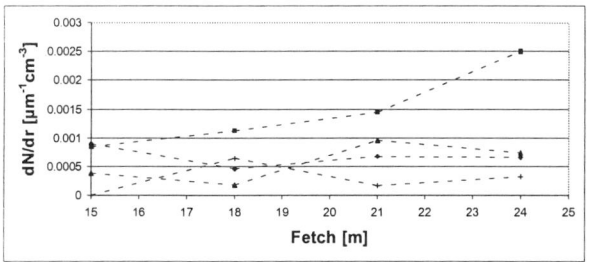

Figure 7. Variation with fetch of bubble concentrations measured in amplified paddle waves at a wind speed of 10 m s^{-1}. Bubble radii: 61 (■), 75 (♦),111 (▲) and 133 (x) μm.

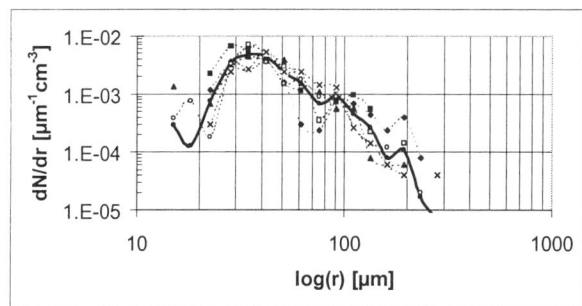

Figure 8. Bubble size distributions measured in a wind speed of 13 m s^{-1} (no paddles), at fetches of 16.5 (■), 18 (♦), 21.5 (▲), 24 (o), 26.5 (□) and 29 m (x). The line is the average over all six distributions.

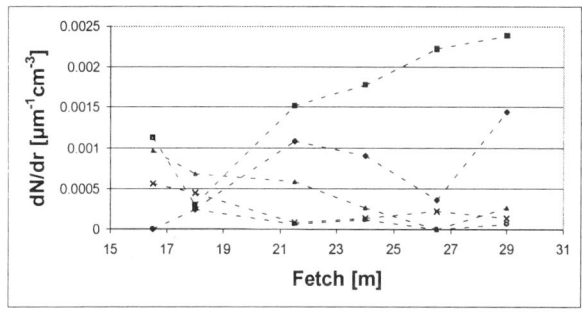

Figure 9. Variation with fetch of bubble concentrations measured in a wind speed of 13 m s^{-1} (no paddles). Bubble radii: 61 (■), 75 (♦),111 (▲) 133 (x) and 160 (o) μm.

The concentrations of the smaller bubbles, however, show a significantly different behaviour. In the presence of paddle waves, the 61 μm bubbles, although having higher concentrations, followed the same trend as the 75 μm bubbles, whereas in Figure 9 they continue to increase. The concentrations of the 75 μm bubbles, in contrast to the previous case, now initially increase and then decrease.

4. DISCUSSION AND CONCLUSIONS

Material presented in this manuscript provides input for calculations of bubble-mediated transfer, indicated as the most significant mechanism associated with breaking waves. Quantitative information is now available to test models and bubble parameterisations (e.g., *Woolf* [1993; 1997]).

Presented bubble size distributions show significant differences between the various situations encountered during the LUMINY experiment. Bi-modal distributions were generated by the aeration devices, which, in spite of

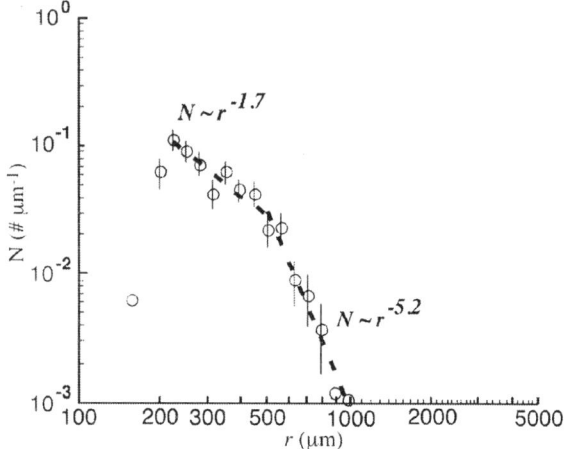

Figure 10. Senescence phase of a Narrow Shallow Diffuse bubble plume from a breaking wave for 0.52 to 0.66 s after formation. Plume population (total number of bubbles in plume), N, as a function of radius, r, was fit with a least squares linear regression analysis over the range shown. Smaller r population is Senescence, larger r population is plume.

their inherent different behaviour, create an average distribution similar to that generated by breaking waves, with a second mode of large bubbles superimposed. The generation mechanism of both distributions is anticipated to be quite different, as suggested by independent measurement showing the creation of small bubbles by bursting larger ones. The 'normal' aeration devices appear to generate bubble spectra that peak around 200 μm, whereas a number of other devices were encountered that generated much larger bubbles

The bi-modal character of the bubble size distributions, as opposed to the more regular distributions created by breaking waves, may allow for the separation of the effect of large bubbles from other effects [*Woolf*, 2001; *De Leeuw et al.*, 2001] and the quantification of bubble mediated gas transfer [*Woolf*, 1993; 1997].

Significant differences were observed between bubble spectra generated by waves breaking at different wind velocities and by amplified paddle waves, both as regards the concentrations and spectral shape and the fetch dependence. These differences allow for detailed studies of gas transfer mechanisms [*De Leeuw et al.*, 2001]. Results were used to derive new parameterisations for gas transfer in terms of the friction velocity that takes into account the molecular properties of the gas in question.

A parallel study of bubble plumes generated by breaking waves reveals the existence of a variety of different plume types which can be classified using several criteria [*Leifer and De Leeuw*, 2001]. The life cycle of the bubble plume can be divided into four phases, formation, injection, rise and senescence. The senescence phase is the plume remnants, i.e. the background population which the Mini-BMS studied. The bubble populations change significantly from one phase to another, because of the different bubble transport properties, see *Leifer and De Leeuw* [2001] for examples. The spectral shape evolves and because of the difference in rise time with bubble size, the slope of the populations becomes steeper. For example, the slope is −1.6 in the radius range 200-500 μm, for the population in the senescence phase of an narrow shallow diffuse plume, see Figure 10, quite similar to the slope observed for the spectra of amplified paddle waves at 10 m s^{-1} (Figure 6).

Acknowledgements. The LUMINY project was supported by the European Commission EC DG XII, contract ENV4-CT95-0080, The contribution of TNO-FEL was supported by the Netherlands Ministry of Defence, assignment A95KM786. The authors would like to thank IRPHE-IOA for the extensive laboratory support provided.

REFERENCES

Caulliez, G. LUMINY final report, in *Breaking Waves and Air-Sea Gas Transfer (LUMINY), Contract ENV4-CT95-0080. Final report: 1 February 1996- 31 January 1999*, edited by G. de Leeuw, TNO Physics and Electronics Laboratory, Report FEL-99-C122, 1999.

Caulliez, G., Statistics of the geometric properties of breaking wind waves observed in laboratory, in *Gas Transfer and water Surfaces*, edited by M.A. Donelan, W.M. Drennan, E.S. Salzman, and R. Wanninkhof, pp. xxx-xxx, AGU, this volume, 2001.

Caulliez, G., L. Jaouen, S. Larsen, F.A. Hansen, S. Lund, G. de Leeuw, D.K. Woolf, P. Bowyer, I. Leifer, G. Kunz, P.D. Nightingale, T.-S. Rhee, M.I. Liddicoat, J.M. Baker, S. Rapsomanikis, S. Hassoun, and L.H. Cohen, Wind and wave characteristics observed during the LUMINY gas transfer experiments, in *Greenhouse gases and their impacts on the climate system: the status of research in Europe*, edited by R. Valentini and C. Brüning, pp. 108-112, European Commission, Report EUR 19085 EN, 1999.

De Leeuw, G., G.J. Kunz, G. Caulliez, D.K. Woolf, P. Bowyer, I. Leifer, P. Nightingale, M. Liddicoat, T.S. Rhee, M.O. Andreae, S.E. Larsen, F.A. Hansen and S. Lund, LUMINY - An overview, in *Gas Transfer and water Surfaces*, edited by M.A. Donelan, W.M. Drennan, E.S. Salzman, and R. Wanninkhof, pp. xxx-xxx, AGU, this volume, 2001.

Leifer I.S., R.K. Patro, and P. Bowyer, A study on the temperature variation of rise velocity for large clean bubbles. *J. Atm. and Ocean. Tech.* 17(10), 1392-1402. 2000a.

Leifer, I., G. de Leeuw and L.H. Cohen, Secondary bubble production from breaking waves: the bubble burst mechanism, accepted for publication in *Geophys. Res. Letters*, 2000b.

Leifer, I. and G. de Leeuw, Bubble measurements in breaking-wave generated during the LUMINY wind-wave experiment, in *Gas Transfer and water Surfaces*, edited by M.A. Donelan, W.M. Drennan, E.S. Salzman, and R. Wanninkhof, pp. xxx-xxx, AGU, this volume, 2001.

Leifer, I., G. de Leeuw and L.H. Cohen, Optical measurement of bubbles: system design and application, submitted for publication in *J. Atm and Ocean. Tech.*, 2001

Spiel, D., On the births of film drops from bubbles bursting on seawater surfaces, *J. Geophys. Res.* 103(C11), pp. 24,907-24918, 1998.

Woolf, D.K., Bubbles and the air-sea transfer velocity of gases, *Atmosphere-Ocean* 31, pp. 517-540, 1993.

Woolf, D.K., Bubbles and their role in gas exchange, in *The Sea Surface and Global Change*, edited by P. S. Liss and R. A. Duce, pp. 173-205, Cambridge University Press, 1997.

Woolf, D.K., Gas transfer in energetic conditions, in *Gas Transfer and water Surfaces*, edited by M.A. Donelan, W.M. Drennan, E.S. Salzman, and R. Wanninkhof, pp. xxx-xxx, AGU, this volume, 2001.

G. de Leeuw, TNO Physics and Electronics Laboratory, P.O. Box 96864, 2509 JG The Hague, The Netherlands (e-mail: deleeuw@fel.tno.nl).

I. Leifer, Chemical Engineering Department, University of California, Santa Barbara, Santa Barbara, California, 93106-5080, USA (email: ira.leifer@bubbleology.com).

Bubble Measurements in Breaking-Wave Generated Bubble Plumes During the LUMINY Wind-Wave Experiment

Ira Leifer

Chemical Engineering Department, University of California, Santa Barbara, Santa Barbara, California

Gerrit de Leeuw

TNO Physics and Electronics Laboratory, The Hague, The Netherlands

Affordable, high quality video equipment has allowed the development of bubble measurement systems (BMS). A multi-camera BMS was developed and deployed to observe wind-wave generated bubble plumes in the IRPHE wind-wave tunnel during the LUMINY wind-wave experiment. The BMS visualized plume dynamics and large ($200 > r > 5000$) bubble populations within the bubble plumes. Large variability was observed between bubble plumes in terms of density, dynamics, size, and penetration depth. To avoid losing information through averaging, a plume-type classification scheme was developed and plume type probabilities measured at different fetches. Populations of different plume types were analyzed and were multimodal.

1. INTRODUCTION

Bubble plumes from breaking waves are either dominant or very important to geophysical processes as diverse as air-sea gas transfer [*Liss et al.*, 1997], marine aerosol formation [*Resch and Afeti*, 1992], and surface microlayer enrichment [*Blanchard*, 1989]. A Bubble Measurement System (BMS) studied bubble plumes during the LUMINY wind-wave experiment in the Large Air Sea Interaction Simulation Tunnel of the Institut de Recherche sur les Phénomènes Hors Equilibre, Laboratoire Interactions Océan-Atmosphère Luminy (IRPHE-IOA), Marseille, France. LUMINY studied the importance of waves and bubbles to air-water gas transfer (described in *De Leeuw et al.*, [2001]).

2. EXPERIMENTAL SET-UP

The BMS is shown schematically in Figure 1 and described in detail along with analysis techniques in *Leifer et al.* [2000a]. Briefly, multiple video cameras allowed simultaneous imaging at multiple scales and locations. The system provided non-invasive measurements of time-resolved bubble distributions, fluid velocities, and plume structure. The overview camera was repositionable to observe from many angles. Bubbles were observed over the equivalent spherical radius, r, range 100 to 5000 µm, where 5000 µm was the largest bubble observed. These observations were analyzed in conjunction with a Mini-BMS deployed by TNO Physics and Electronics Laboratory (TNO-FEL), The Hague, The Netherlands. The Mini-BMS observed bubbles over the range $15 < r < 500$ µm. Good agreement was found for the overlapping ranges both in calibration experiments, and for the background LUMINY bubble distribution. Both systems were on a chariot moveable from 17.5 to 25 m fetch.

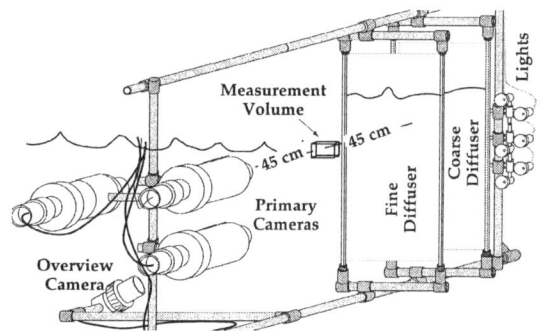

Figure 1. Bubble measurement system set-up.

Movies were digitized and analyzed by computer or hand. Hand analysis, in conjunction with bubble trajectory determination was required in the dense bubble plumes to analyze partially and completely obscured bubbles. Due to the optical set-up (wide angle, non-telecentric) required to image bubble plumes, uncertainty of the bubble and measurement volume sizes was introduced. Based upon the change in scale, for the wide field of view, FOV, camera any single bubble had an error of circa ±25%. However, the error in the distribution was less since a bubble appears larger when closer and smaller when further. Further size uncertainty arises from irregular bubble shape oscillations. In a stagnant fluid, oscillations begin at $r \sim 700$ μm [*Patro et al.*, 2001]. Complex shape oscillations cause circa 10-15% variation in r for 2000-μm bubbles, decreasing slightly for larger r [*Leifer et al.*, 2000]. This problem is worse in the highly turbulent flow of a bubble plume.

A comparison of the distribution for the wide and narrow FOV cameras when looking at the same plume (Figure 8) shows very good agreement. Errors at the small radius cutoff are necessarily larger, the divergence of the two curves shows that the smallest bubble reliably identifiably was 4-5 pixels (200-250 μm) in diameter for the wide FOV camera. (The bubble halo was used to determine r and if in-focus). For most bubble plumes, the depth of field was sufficient to identify all bubbles in the plume.

Bubble plume distributions are generally described by a power law dependency [*Johnson and Cooke*, 1979; *Medwin and Breitz*, 1989],

$$\Phi(r, t) \sim r^{-S(t)} \qquad (1)$$

where Φ is the bubble population (# μm^{-1}), r is the equivalent spherical radius, t is time, and S is the power law exponent and may vary with t. Values of S were calculated by least-squares, linear-regression analysis of the log of both sides of (1). Populations in this paper are scaled by the ratio of plume size with the measurement volume (since internally, plumes were well mixed) as determined by the overview camera. The populations were calculated by histogramming the time series of r in logarithmically spaced bins normalized to # μm^{-1}.

3. RESULTS

Bubble plumes are highly transient and localized phenomena. As such, spatial and temporal averaging greatly underestimates the number of bubbles formed - the injection, or initial distribution, particularly for large bubbles. A classification scheme was developed based on bubble plume horizontal extent, w, penetration, z_p, and bubble density, summarized in Table 1. The normalized probability of each plume type at each fetch was analyzed for 13 m s^{-1} wind speed and paddle waves and is shown in Figure 2 along with the total probability. Both BDDi and BDDe plumes were very rare (one each during 1500 waves), and were merged into the BSDi and NDDe plume categories, respectively.

The total probability distribution shows that major plumes are an order of magnitude less probable than minor plumes. Major plumes were most likely, and micro plumes least likely at 22.5 m fetch, where wave breaking was at a peak [*De Leeuw et al.*, 2001].

4. BUBBLE PLUMES

4.1. Bubble Plume Life Cycle

A bubble plume's life can be divided into four phases, Formation, Injection, Rise, and Senescence. Formation was observed to occur during the first 0.1 s or less. During

Table 1. Bubble plume classification criteria.

Type	Symbol	Criteria
Broad	B	$w > 30$ cm
Narrow	N	$w < 30$ cm
Shallow	S	$z_p < 30$ cm
Deep	D	$z_p > 30$ cm
Dense	De	for $t < 1$ s, bubbles obscure background
Diffuse	Di	for $t < 1$ s, background not obscured.
Major		Dense, Broad and Shallow or Narrow and Deep
Minor		Neither Major nor Micro
Micro		Narrow, Shallow, Diffuse, Small[a]
None		Wave has no bubble plume

[a]Additional Micro criteria: $w < 10$ cm, $z_p < 10$ cm, $\Phi < 100$.

Injection the plume rapidly descended, initially at a circa 30° angle, then tilting towards vertical. In general the distribution did not evolve significantly, at least until maximum penetration. The Injection population is important for bubble model initialization [*Leifer*, 1995]. Injection ends at maximum penetration and was followed by the Rise phase, when the mass of bubbles rises towards the surface. The Rise and Injection phases lasted roughly the same time. Finally, the Senescence phase consisted of the smaller (mostly $r < 200$ µm) plume remnants after the main plume surfaces. Senescence corresponds to typical bubble observations, i.e., the background distribution.

4.2. Micro Bubble Plumes

While micro plumes were the most common plume type, they contained few bubbles. Micro plumes generally formed in the trough rather than at the crest. A micro plume is shown in Figure 3a during Injection, approximately 0.02 s after formation. The plume penetrated 5 cm and persisted for 0.36 s. Injection and average Φ for micro plumes are shown in Figure 3b. Error bars are 1 standard deviation. The Injection Φ is fairly flat for $r < 650$ µm, and decreased sharply for larger r. The small peak at $r \sim 650$ µm is statistically significant.

4.3. Minor Bubble Plumes

Minor plumes were generally shallow and diffuse. Of the four Minor plume types, NSDi were overall dominant, although NDDi dominated at fetches where major plumes were most common (20-22.5 m). NSDi plumes were most common at the earliest (least developed waves) fetch, 17.5 m. BSDi plumes were always significant, but never dominant. Only NSDe plumes were insignificant, presumably

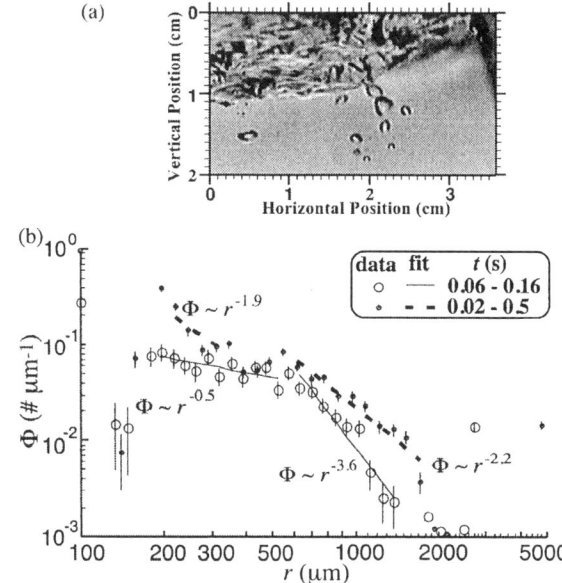

Figure 3. Micro plume (a) image during Injection and (b) population distributions, Φ, for Injection ($t = 0.06$-0.16 s) and average ($t = 0.02$ - 0.50 s), where t is time. Legend on figure.

because the energy required to create sufficient bubbles to form a dense bubble cloud is generally sufficient to disperse the plumes (i.e., plume is dense for short time).

4.3.1. Minor Plumes: NSDi. Images of a typical NSDi plume during Injection ($t = 0.06$-0.16 s) are shown in Figure 4. This plume only penetrated about 5 cm, and had a lifetime of circa 0.5 s. NSDi plumes were also "richer" in small bubbles relative to dense and major plumes, as can be seen by the steepness of Φ for $r < 500$ µm.

The populations for five NSDi plumes were combined and the t, r-resolved and t-averaged distributions are shown in Figure 5. In early Injection Φ increased as bubbles formed and were advected into the FOV, and then remained constant (horizontal contours). During the Rise phase Φ decreased as indicated by the downward sloping contour of all but the smallest ($r < 250$ µm) bubbles. Smaller bubbles rose slower and fewer were lost at the surface. The contours are nearly parallel indicating the plumes were well mixed. Converging contours would indicate z_p decreased with increasing r as larger bubbles left the FOV sooner. Note rise velocity, V_B, decreases for $r > 700$ [*Patro et al.*, 2001]. Senescence begins after most of the larger bubbles surfaced, indicated by the nearly vertical contours at $t = 0.5$ s. NSDi plumes contained distinct small and large ($r > 700$ µm) bubble populations, with $S = -2.6$, -3.6, respectively.

4.3.2. Minor Plumes: NSDe. Although NSDe plumes were the least common minor plume, particularly at the

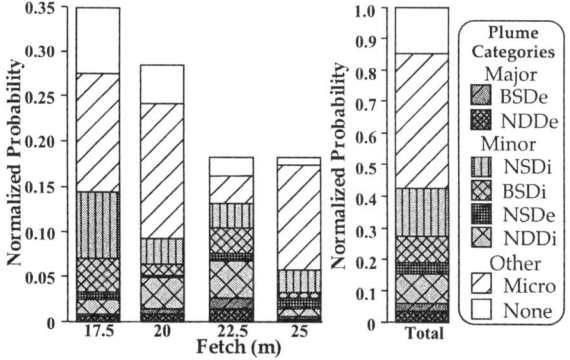

Figure 2. Bubble plume category probability with fetch and total. Classification scheme described in text.

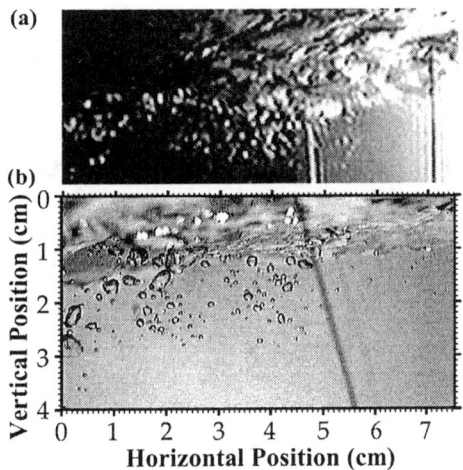

Figure 4. Minor NSDi plume, Injection phase image from (a) overview and (b) primary camera.

shortest fetch, 17.5 m, (earliest wave development), they had the highest large bubble population. A NSDe plume is shown in Figure 6a. The interface can be seen due to the upwards angle required to image near the crest. The plume remained very dense throughout its 1.0 s lifetime and z_P was circa 10 cm. NSDe plumes were relatively "poor" in small bubble populations as can be seen in Figure 6b for the Injection Φ. Three peaks at 250, 900, and 2000 μm are visible, although the 900-μm peak is poorly resolved. Both large and small populations decreased with S similar for the 250-μm, and 2000-μm populations. The Injection phase lasted circa 0.5 s. Some size segregation with z occurred as the Rise phase ended almost simultaneously for bubbles larger than $r = 1200$ μm ($t = 0.7$ s). This cannot be explained by varying V_B since V_B decreases for $r > 700$ μm [*Patro et al.*, 2001] and must be due to segregation during Injection.

4.3.3. Other Minor Plumes. Three other minor plume types were observed, BSDi, BDDi, and NDDi. All these diffuse plumes were quite similar to NSDi plumes, with many features in common. Specifically, bi-modal with a second peak circa 700 to 900 μm, similar S for both small and large distributions, and a similar relative importance of small to large bubbles. Specifically, the small r cut-off peak at 200 μm was typically an order of magnitude greater than the large bubble peak. This peak also appeared in dense plumes, although it was shifted to slightly larger r.

4.4. Major Bubble Plumes

Significantly larger plumes were also observed. Overview images of the major plume types are shown in Figure 7 at maximum penetration. Images are all the same scale. Major plumes persisted longer than other plumes, from 1 to 1.5 s.

4.4.1. Major Bubble Plumes: BSDe. BSDe plumes were most common at 22.5 m fetch where wave breaking was most intense, and were more sensitive to fetch than other major plumes. Additionally, the probability of BSDe plumes relative to NDDe plumes was greatest at this fetch and least at the earliest fetch, suggesting that BSDe plumes were more energetic than NDDe plumes. The several BSDe plumes analyzed showed three clear peaks, one at circa 500 μm, and one at circa 1800 μm, as well as the small radius limit. The larger peaks were shallow during Injection ($-2.8 < S < -3.1$).

4.4.2. Major Bubble Plumes: NDDe and BDDe. The most common major plume types were NDDe and BDDe, and based on plume dynamics and total large bubble population, the most energetic. The t-averaged Φ of a BDDe and NDDe plume (from Figure 7) are shown in

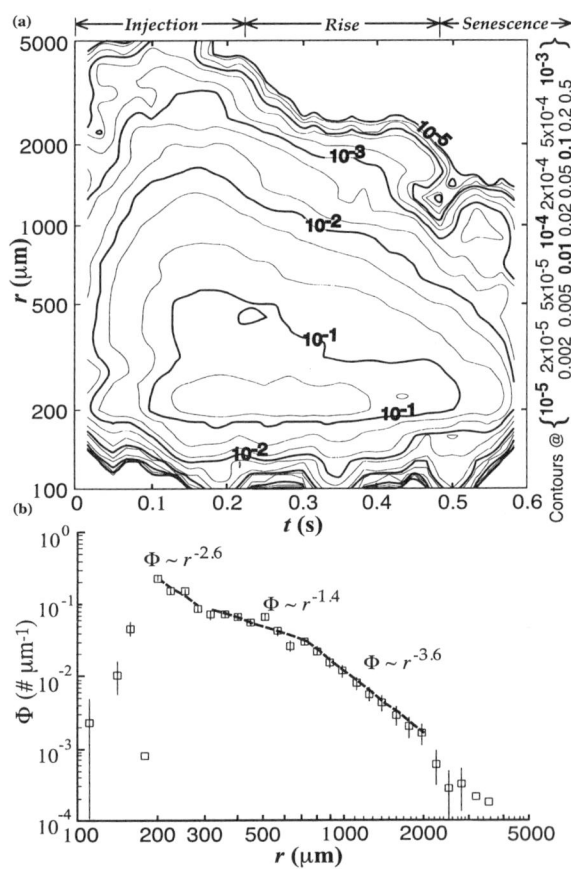

Figure 5. Total minor NSDi population, Φ, (a) and time average (b). Contour levels and life phases on figure.

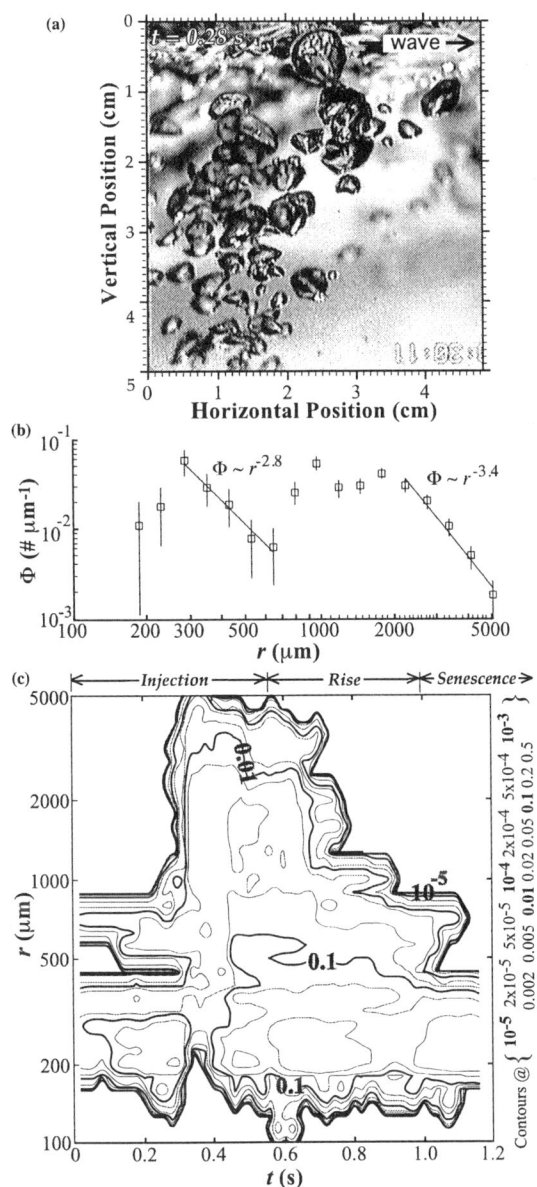

Figure 6. Image of NSDe during Injection at time, $t = 0.28$ s (a), Injection population, (b) and t, time, radius, r, contour plot of total NSDe plume population, Φ (c). Contours on figure.

Figure 8. A narrow and wide FOV camera both observed the same BDDe plume, and as can be seen, the agreement between the two distributions is good. The small r peak for the wide FOV is at the BMS size limit. The distributions shown are different from other dense plumes in that they are bi-modal, with widely separated peaks (near 1500 μm). Not all NDDe plume showed this type of distribution, about half showed a tri-modal distribution with peaks similar to the NSDe and BSDe plumes.

4.5. Background Population

Breaking waves were the primary bubble source. Thus, the background population is generated by the smaller bubbles that diffuse out of the bubble plumes. Background bubble observations by the TNO Mini-BMS (presented in *De Leeuw and Leifer* [2001]) showed agreement with Senescence bubble populations observed by the Large-BMS for the overlapping r ranges. Specifically, S was -1.6 to -3.1 for both systems for $r > 100$.

5. BUBBLE CLEANLINESS

A key parameter to bubble hydrodynamics [*Leifer et al.*, 2000a], and probably bubble formation is surface cleanliness. It has been observed that large ($r > 700$ μm) bubbles rising in contaminated water remain clean for many seconds [*Patro et al.*, 2001]. Bubbles from an NSDe plume were observed surfacing and bursting immediately (<20 ms), indicating cleanliness. A possible explanation is that the rapid forward surface motions associated with microscale breaking and surface renewal [*Jessup et al.*, 1997] sweep away the microlayer and its contaminants during plume formation.

6. DISCUSSION AND CONCLUSIONS

The most novel result from this research is that bubble plumes are bi- or tri-modal. Visual inspection shows this is not an experimental artifact, e.g., in Figure 6a there are clearly many very large bubbles and small bubbles with few intermediate. Published distributions (e.g., *Asher et al.*, [1997]; *Haines and Johnson*, [1995]) are monomodal with S typically between -3 and -4 for background Φ [*Johnson and Cooke* 1979; *Medwin and Breitz* 1989] and $S \sim -2$ for initial distributions [*Haines and Johnson*, 1995; *Leifer*, 1995] or near the wave crest [*Baldy and Bourguel*, 1987]. However, these plumes averaged together many plumes of diverse types and the background thereby eliminating the multiple peaks. Also, few distributions including large bubbles ($r > 500$) have been published, and thus few span a range necessary to see the $r \sim 700 - 900$ μm peak, much less the third peak at circa $r \sim 1900 - 2200$ μm. From the point of view of bubble hydrodynamics, these are two important transitional points. At 700 μm, $V_B(r)$ decreases as bubbles begin to oscillate and buoyancy forces begin to dominate over surface tension (i.e., the bubble surface can oscillate). In the second range, the oscillations become complex, and $V_B(r)$ increases again [*Leifer et al.*, 2000a].

The plume probability distribution with fetch (Figure 2) is useful for understanding the relationship between plume

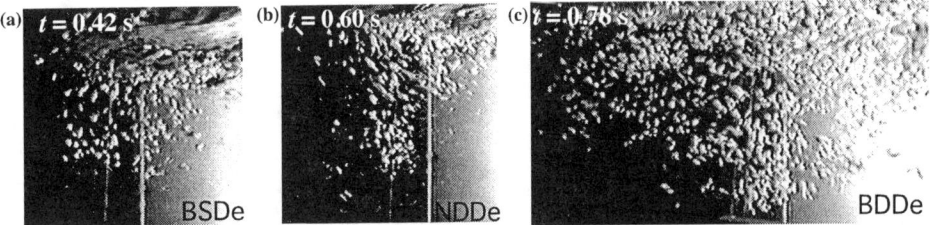

Figure 7. Overview images for the three major plume types, BSDe, NDDe, and BDDe.

types and wave development and dissipation. (Note: plume energy equals wave energy loss). For example, although narrow minor plumes were slightly more likely to be shallow versus deep, at 17.5 m fetch (i.e., least developed waves) these plumes were primarily shallow, but at greater fetches (20 m, 22.5 m), the more energetic deep narrow minor plumes dominated. Similarly, the significantly larger (energetic) major plumes also increased. At the final fetch (25 m) where the waves were presumably regenerating after breaking, narrow minor plumes are once again primarily shallow. Thus shallow minor plumes are likely less energetic (e.g., dissipative) than deep minor plumes, and dense plumes require more energy than diffuse plumes to form. Also suggested is that deep plumes are less energetic than broad plumes despite comparable bubble populations. Movie analysis suggested that broad plume formation required interaction between two narrow plumes during formation. This may explain why broad plume were rarer than narrow plumes.

Bubble plumes contain significantly enhanced bubble concentrations relative to background bubble populations including very large bubbles (r<5000 μm). Despite significantly smaller populations, minor plumes are not negligible compared to major plumes. By analyzing different plume types separately, considerable insight into bubble dynamics and behavior can be gained. In conjunction with plume probabilities, total bubble populations can be determined.

Acknowledgments. The LUMINY project was supported by the European Commission EC DG XII, contract ENV4-CT95-0080, The contribution of TNO-FEL was supported by the Netherlands Ministry of Defense, assignment A95KM786. The authors would like to thank IRPHE-IOA for the extensive laboratory support provided.

REFERENCES

Asher, W.E., L.M. Karle, and B.J. Higgins, On the difference between bubble-mediated air-water transfer in freshwater and seawater. *J. Mar Res.* 55(5), 1-34, 1997.

Baldy S. and M. Bourguel, Bubbles between the wave trough and wave crest levels. *J. Geophys. Res.* 92C, 2919-2929, 1987.

Blanchard, D.C., The ejection of drops from the sea and their enrichment with bacteria and other materials: A review. *Estuaries* 12(3), 127-137, 1989.

De Leeuw, G., G.J. Kunz, G. Caulliez, L. Jaouen, S. Badulin, D.K. Woolf, P. Bowyer, I. Leifer, P. Nightingale, M. Liddicoat, T.S. Rhee, M.O. Andreae, S.E. Larsen, F.Aa Hansen, and S. Lund, LUMINY - An Overview, Geophysical Monograph (this volume), edited by M.A. Donelan, W.M. Drennan E.S. Saltzman, and R. Wanninkhof, American Geophysical Union, 2001.

De Leeuw, G., and I. Leifer, Bubbles outside the bubble plume during the LUMINY wind-wave experiment, Geophysical Monograph (this volume), edited by M.A. Donelan, W.M. Drennan E.S. Saltzman, and R. Wanninkhof, American Geophysical Union, 2001.

Haines, M.A and B.D. Johnson, Injected bubble populations in seawater and fresh water measured by a photographic method *J. Geophys. Res.* 100(C4), 7057-7068, 1995.

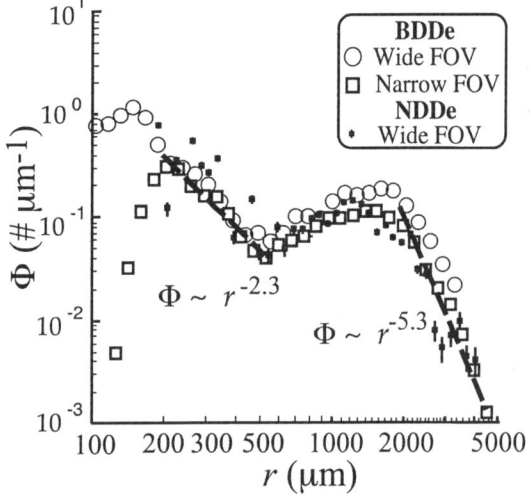

Figure 8. Time average populations, Φ, for NDDe and BDDe plumes. Data for BDDe plumes is for two different cameras. Data key on figure.

Jessup A.T., C.J. Zappa, and H. Yeh, Defining and quantifying microscale wave breaking with infrared imagery. *J. Geophys. Res.* 102(C10), 23145-23153, 1997.

Johnson, B.D, and R.C. Cooke, Bubble populations and spectra in coastal waters: a photographic approach, *J. Geophys. Res.* 92(C2), 3761-3766, 1979.

Leifer, I.S., 1995. A validation study of bubble mediated air-sea gas transfer modeling. Ph.D. Thesis, Georgia Institute of Technology, Atlanta, Georgia, USA.

Leifer I.S., R.K. Patro, and P. Bowyer, A study on the temperature variation of rise velocity for large clean bubbles. *J. Atm. and Ocean. Tech.* 17, 1392-1402, 2000a.

Leifer, I., G. De Leeuw, and L.H. Cohen. Optical measurement of bubbles: System Design and Application. *J. Atm. and Ocean. Tech.* Submitted. 2000b.

Liss, P.S., A.J. Watson, E.J. Bock, B. Jähne, W.E. Asher, N.M. Frew, L. Hasse, G.M. Korenowski, L. Merlivat, L.F. Phillips, P.Schluessel, and D.K. Wolf, Physical processes Asher, W.E., L.M. Karle, and B.J. Higgins, On the difference between bubble-mediated air-water transfer in freshwater and seawater. *J. Mar Res.,* 55(5), 1-34, 1997.

Medwin, H. and N. D. Breitz, Ambient and transient bubble spectral densities in quiescent seas and under spilling breakers. *J. Geophys. Res.*, 94C, 12571-12759, 1989.

Patro, R.K., I. Leifer , and P. Bowyer. Better bubble process modeling. Geophysical Monograph (this volume), edited by M.A. Donelan, W.M. Drennan E.S. Saltzman, and R. Wanninkhof, American Geophysical Union, 2001.

Resch, F., and G. Afeti, Submicron film drop production by bubbles in seawater. *J. Geophys. Res.* 97., 3679-3683, 1992.

I. Leifer, Chemical Engineering Department, University of California, Santa Barbara, Santa Barbara, California, 93106-5080, USA (email: ira.leifer@bubbleology.com).

G. de Leeuw, TNO Physics and Electronics Laboratory, P.O. Box 96864, 2509 JG The Hague, The Netherlands (e-mail: deleeuw@fel.tno.nl).

An Experimental Study of Bubble Mediated Gas Exchange for a Single Bubble

Nobuhito Mori, Masahiro Imamura, and Ryosuke Yamamoto

Central Research Institute of Electric Power Industry(CRIEPI), Chiba, Japan

An experimental study of bubble mediated gas exchange for a single bubble was performed. Medium sized air bubbles were generated into a water column by a computer-controlled electromagnetic valve. Temporal variations of gas concentration were analyzed by a gas chromatography with a head space method. The total amount of gas exchange for a single bubble with several radii are compared with the experimental results and theory.

1. INTRODUCTION

Accurate estimates of gas transfer near the air-sea interface are very important for environmental mechanisms of the earth. A wind dependent gas transfer model by Liss and Merlivat(1986) is widely used. The gas transfer velocity of the Liss and Merlivat model increases rapidly when the wind speed exceeds 13m/s. This rapid increase of gas transfer velocity over 13m/s wind speed are explained by several reasons such as enhancements of wind and breaking wave induced turbulence, breaking wave induced air bubbles, and sea spray. Merlivat and Memery(1983) and Woolf(1997) discuss bubble mediated gas exchange. Merlivat and Memery(1985) formulated contribution of bubbles mediated gas transfer and they concluded that the bubble mediated gas transfer has an important role for air-sea interaction processes. Monahan(1991) also studied bulk relationships between whitecap coverage and gas transfer enhancement. These previous studies show that bubble mediated gas transfer is an important factor to explain enhancement of gas transfer velocity over 13m/s wind speed.

Estimating the total amount of bubble mediated gas exchange, is difficult. For example, the following subjects underlie this study: relationships between whitecap coverage and wind speed, vertical bubble cloud distributions, radii distributions of bubble clouds, and the behavior of rising bubbles. Particularly, the gas exchange due to medium sized bubble is difficult to estimate because it's halfway between fully dissolved(small sized bubble) and insoluble(large sized bubble) as pointed out by Keeling(1993).

The final goal of this study is to measure highly accurate gas transfer bubble clouds under several conditions. In this study an instrument is built for generating medium sized single bubbles. The amount of exchanged gas for a rising medium sized single bubble with several radii is measured. Finally, it is compared with the experimental results and a theory.

2. EXPERIMENTAL SETUP

2.1. A Single Bubble Generation

The experiments were conducted in a rectangular vertical column (Figure 1). Gas tanks were connected at the bottom of the rectangular column (W 0.1m × D 0.1m × H1.0m) by way of second buffer. Thin metal needles with several radii were used for generating medium sized single bubbles at the outlet of gas in the column. A micro radius glass needle can generate micro size bubbles but it wasn't used in here. The gas

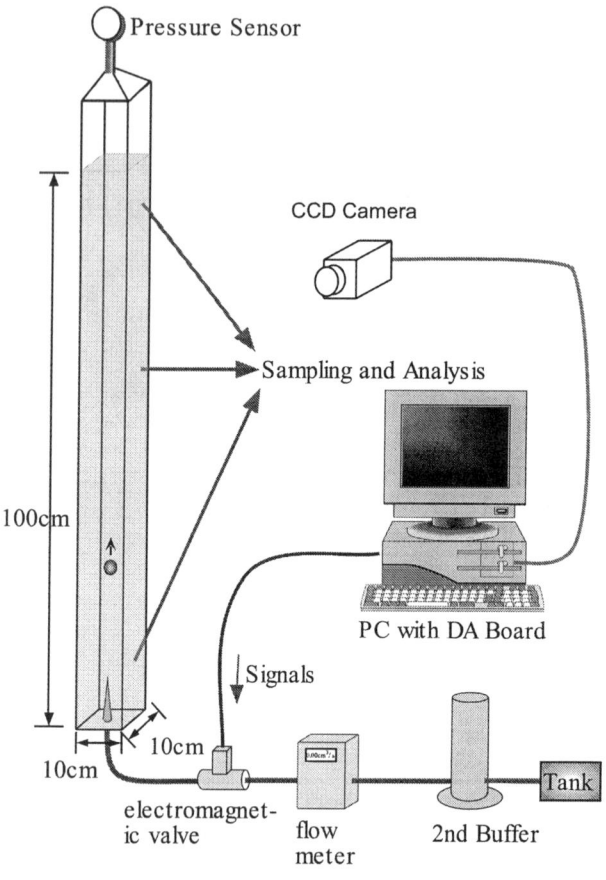

Figure 1. An illustration of the experimental setup.

flow was controlled by a computer controlled electromagnetic valve. Single bubbles were generated in the column with different radii, discharge periods and total number of bubbles. The rising bubble was observed by a CCD camera with a shutter speed of 1/400s and the shape/radius of the bubble was calculated by the sequence of CCD images.

Pure water and 302ppm N_2O gas were used for this experiment. The water temperature in the column can be changed from 0 to 40C, but the air and water temperature were fixed as 15C during this experiment. The discharge period between the individual bubbles was fixed at 1s to avoid circular flow in the column and duration of the experiments was conducted over 10000s. The radii of the measured bubbles ranged from 1 to 5mm.

2.2. Method of Chemical Analysis

The water was sampled at constant intervals into glass vial bottles at several vertical points in the column. The samples were immediately poisoned with 5ml of 30 % formaldehyde and analyzed within 12 hours of collection. The concentration of N_2O was analyzed using a head space method. The collected samples for N_2O analysis were set on a auto sampler. About 5ml of head space gas, (N_2) was introduced into the glass vial with a syringe. Subsequently, in order to equilibrate the gas and liquid, the samples were kept in thermostat water bath($40 \pm 0.5C$) for at least 2 hours. The N_2O in the head space gas was withdrawn by a gas syringe and was subsequently injected into a gas chromatograph equipped with a $^{63}N_i$ electron capture detector. The temporal variations of gas concentration were measured repeatedly by this procedure. Finally, the total amounts of gas exchange for a single bubble were calculated from the measured N_2O concentration over time.

3. RESULTS AND DISCUSSION

Figure 2 shows the temporal variations of gas concentration in the column for 2mm radius bubbles. The filled circles in the figure indicates experimental data and the lines are linear and quadric regression curves. The bubbles were discharged from the outlet of the column exactly every second, therefore it is possible to convert the unit from 1s to one bubble in this case. The input gas concentration was relatively low and the Ostwald coefficient of N_2O is large. Therefore the temporal evolution of gas concentration in the water increases very slowly in time and does not reach saturated conditions until the end of the experiment. The gas concentration increased linearly in the beginning and then increases as quadric curve as shown in Figure 2. The total amount of gas exchanged and gas transfer velocities for the single rising bubble can be calculated by the re-

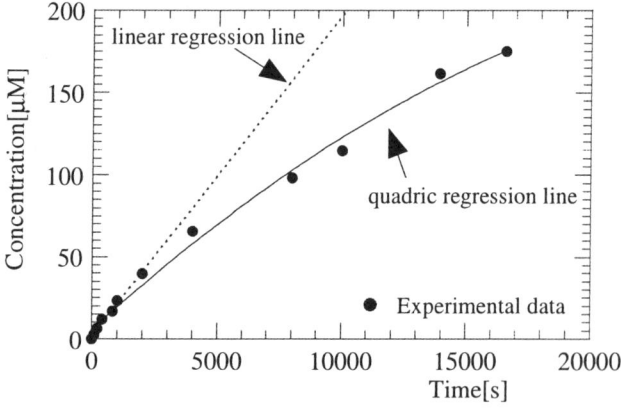

Figure 2. Temporal evolution of gas concentration in the column for a 2mm radius bubble.

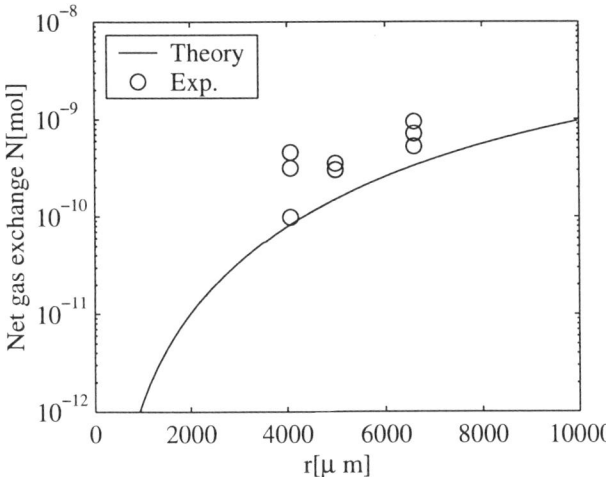

Figure 3. Comparison of total amount of gas exchange N for a single bubble between the experimental results and theory.

gression curves. A linear approximation is adopted for estimation of the total amount of gas exchange in this case, because it is easily compared with theory. The parameterization of total amount of gas exchanged and gas transfer velocity for water nearly saturated with gas is the topic for a future study.

Figure 3 shows a comparison of experimental data and theory for the total amount of gas exchange for the medium sized single bubble. The circles in the figure indicate the experimental results and solid line indicates the theory developed by Keeling(1993):

$$N(r,z) = \frac{P_g V_0}{RT}\left\{1 - \exp\left[\frac{z_s(2H_0 - z_s)}{2H_0 H_{eq}}\right]\right\}$$
$$+ \frac{P_l V_0}{RT}\sqrt{\frac{\pi H_0}{2H_{eq}}}\exp\left(\frac{H_0}{H_{eq}}\right)$$
$$\times \left[Erf\left(\frac{H_0}{\sqrt{2H_0 H_{eq}}}\right) - Erf\left(\frac{H_0 - z_s}{\sqrt{2H_0 H_{eq}}}\right)\right] \quad (1)$$

where H_0 and H_{eq} are defined by

$$H_0 = \frac{P_0}{\rho g}, \quad (2)$$

$$H_{eq} = \frac{V_0}{RT}\frac{U}{kr^2 S}. \quad (3)$$

and P_g and P_l are the partial pressures of gas in the bubble and the water, z_s is the initial position of the bubble, T the water temperature. The solubility and diffusion coefficient for N_2O gas at $15C$ are 7.784 atm/cm^3 and 1.80×10^{-5}cm^2/s, respectively, and $H_{eq} = 0.25$cm. The rise velocity of the bubbles are 30cm/s. The total amount of gas exchange of a single bubble from the bottom to top of the column N monotonically increases in the theory. There is quantitative difference between the experiments and the theory, but the experimental result shows qualitative agreement with the theory.

4. CONCLUSION

The gas exchange for a single medium sized bubble was obtained experimentally. The experimental results show that the gas exchange for the medium sized bubble is radius dependent. The data shows qualitative agreement with Keeling's theory but there is quantitative difference between them. The parameterization of total amount of gas exchange and gas transfer velocity for a gas near saturation, and micro size bubble experiments will be investigated in future study.

REFERENCES

Keeling, R., On the role of large bubbles in air-sea gas exchange and super saturation in the ocean. *J. Marine Res.* 51, 237–271, 1993.

Liss, P. and L. Merlivat, Air-sea gas exchange: introduction and synthesis., In P. Buat-Menard (Ed.), *The role of air-sea exchange in geochemical cycling*, Reidel, Dordrecht. 113-127, 1986.

Merlivat, L. and L. Memery, Gas exchange across an air-water interface: experimental results and modeling of bubble contribution to transfer. *J. Geophys. Res.* 88, 707–724, 1983.

Merlivat, L. and L. Memery, Contribution of bubbles to gas transfer across an air-water interface, W.Brustaiert and B.Jirka(Ed.), In *Gas Transfer at Water Surface*, Reidel, 247-253", 1985.

Monahan, E.C., Enhancement of air-gas exchange by oceanic whitecapping, In *Air-Water Mass Transfer*, 2nd International Symposium on Gas Transfer at Water Surfaces, ASCE, 608-617, 1991.

Woolf, D., *Bubbles and their role in gas exchange*, NY: Cambridge Univ. Press, 173–205, 1997.

Nobuhito Mori, Department of Hydraulics, Central Research Institute of Electric Power Industry, 1646 Abiko, Abiko, Chiba 2701194, Japan.
(email:mori@criepi.denken.or.jp)

Masahiro Imamura, Department of Environmental Science, Central Research Institute of Electric Power Industry, 1646 Abiko, Abiko, Chiba 2701194, Japan.
(email:mima@criepi.denken.or.jp)

Ryosuke Yamamoto, Department of Hydraulics, Central Research Institute of Electric Power Industry, 1646 Abiko, Abiko, Chiba 2701194, Japan.
(email:r-yama@criepi.denken.or.jp)

Better Bubble Process Modeling: Improved Bubble Hydrodynamics Parameterization

Ranjan Patro

Department of Physics and Physical Oceanography, Memorial University of Newfoundland, St. John's, Canada.

Ira Leifer

Chemical Engineering Department, University of California, Santa Barbara, California, USA.

Peter Bowyer

Department. of Oceanography, National University of Ireland, Galway, Ireland.

Proper modeling of bubble-mediated processes requires both good observations and parameterizations. Although one of the most important bubble parameterizations is the rise velocity, V_B, published studies of V_B in natural waters (i.e., sea water, marsh water, lake water) are largely unavailable; most studies are for "clean" distilled water. Also poorly studied is the effect of temperature, T on V_B. An examination of V_B in seawater showed that for bubbles with radius, $r > 700$ µm, V_B was not significantly different from the value for distilled water. Analysis of V_B with depth, showed a decrease in V_B as the bubble rose over a distance, suggesting bubble contamination can take a significant time (1 m or more), and this time increases with increasing r. Also, hydrodynamic contamination was fastest in marsh and lake waters, where the water was collected close to sediments. Experiments to measure V_B over the range $0 < T < 40°C$ showed that for non-oscillating bubbles, $V_B(T)$ increases with T; while for larger bubbles, $V_B(T)$ decreases with T due to oscillations. A three-part parameterization of $V_B(r,T)$ with transitions at $Re = 1, 540$, and the onset of oscillations (itself T dependent) was developed.

1. INTRODUCTION

The bubble rise velocity, V_B, is of both academic and practical interest. Investigation of bubble behavior is important to fluid dynamics and mass transfer [*Tsuchiya et al.*, 1997], oceanic noise [*Medwin and Breitz*, 1989], aerosol generation [*Monahan*, 1986], and chemical and industrial applications [*Clift et al.*, 1978]. Several parameters affect V_B including size, temperature, T, and the presence of surface active materials, or surfactants. Although V_B is well characterized for water at 20°C and other liquids, the relationships between V_B and T and for V_B in natural waters remains unquantified. This research investigated V_B for bubbles in various natural waters and over a range of T.

2. EXPERIMENTAL SET-UP AND PROCEDURE

The experimental studies were performed in a Plexiglas tank 12 cm square by 60 cm tall. Bubbles over the range $360 < r < 4500$ µm, where r is the equivalent spherical radius, were generated from a regulated air flow through drawn capillary tubes inserted in a rubber stopper in one tank wall, 5 cm from the bottom. Two video cameras were used to simultaneously observe r and V_B. Images were recorded for later digitization and analysis. Analysis was by routines written in NIH Image (developed at the U.S. National Institutes of Health and available on the Internet at http://rsb.info.nih.gov/nih-image/) and MatLab (The MathWorks, Nantick, MA). The x, y position of each bubble, its major and minor axes (determined by a best fit ellipse), and the time were calculated for each frame. Further data analysis including outlier removal and compensation for hydrostatic changes were performed in MatLab. A detailed description of the set-up, procedure, and analysis methodology, is provided in *Leifer et al.* [2000]. V_B was measured at 20°C for various natural waters collected in polyethylene containers from locations in County Galway, Ireland at a depth of 10 cm, and then filtered.

3. RESULTS

3.1. Observation of Rise Velocity

A comparison of observed V_B at $T = 20°C$ for distilled, i.e., clean, and natural waters as well as from other researchers is shown in Figure 1. Also shown is the developed V_B parameterization presented below for clean (this work) and dirty (immobilized interface) bubbles from *Clift et al.* [1978]. For clean bubbles, V_B increases with r until the onset of oscillation after which V_B decreases.

In contrast, the dirty V_B parameterization increases monotonically with r. Figure 1 also shows experimentally determined V_B for natural waters at room temperature (19 - 21°C). The first water analyzed was seawater, and as shown in Figure 1 (crosses) V_B for bubbles in seawater was no more 1-2 cm s^{-1} lower than that for distilled water and significantly greater than V_B dirty. Hypothesizing that perhaps this was a salinity effect, fresh lake water was analyzed and produced similar "clean" results. However, saltmarsh water showed a significant decrease from the clean V_B parameterization over a wide range of r, although larger bubbles ($r > 2000$ μm) had V_B close to that of clean water. Canal water was also tested and V_B was found similar to the saltmarsh water.

The clean behavior of bubbles in seawater is surprising given the prevalence of surfactants in seawater [*Liss et al.*, 1997]. It was hypothesized that these bubbles had not achieved equilibrium and thus had accumulated insufficient surfactants to affect V_B. If true, V_B should decrease as a bubble rises, and in fact, this was the case. Figure 2 shows $V_B(z)$ segregated into 1-cm depth bins and averaged, where z is height above the capillary tube, and z increases towards the surface. Error bars are indicated by the length of the vertical lines determined by the standard deviation for all bubbles in each depth bin. Also shown the least-squares, linear-regression fit (i.e. $V_B = q\,z + C$, where q is slope and C is the initial velocity) to each data set. The

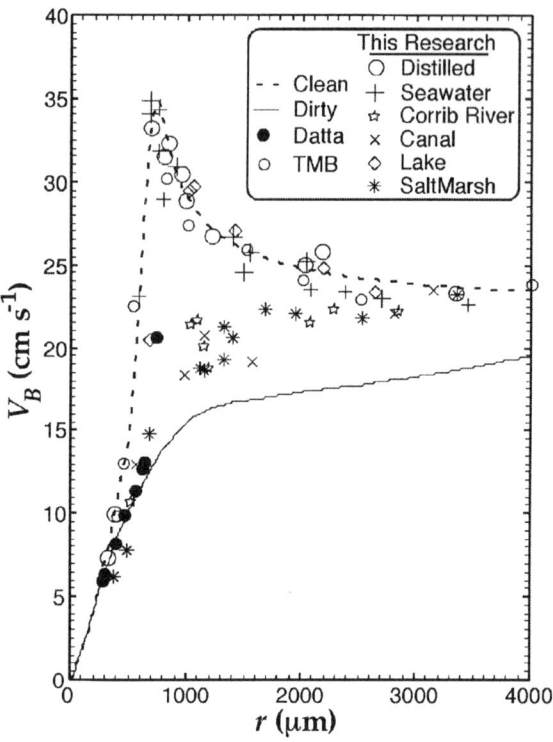

Figure 1. Rise velocity, V_B, as a function of radius, r, at 20°C from observations, other researchers, dirty parameterization from *Clift et al.*, [1978], and clean parameterization given by (1) and (3). Data key on figure. Datta - *Datta et al.* [1950]; TMB - *Haberman and Morton* [1953]. Collected waters were from the vicinity of Galway, Ireland (53° 17' N; 9° 3.6' W).

correlation coefficient, R^2, q, and C are shown in Table 1. As can be seen in Figures 2a and 2b, smaller bubbles ($r < 500$ μm) become immobilized after rising less than 1 cm and hence there was no deceleration of V_B with z. As a result values of R^2 are small.

In contrast, for intermediate ($600 < r < 2000$ μm) bubbles, V_B decreased as they rose. The high R^2 show the decrease in V_B is significant with z. This suggests that the bubbles become progressively dirtier as they rise and accumulate surfactants, and thus a greater percentage of the surface becomes immobilized. For very large bubbles ($r > 2500$ μm), the fluid flow around the bubble is much faster, bubble motions are more complex, and there is a decreased sensitivity to surfactants and low R^2 is small. Note that these size limits are approximates.

3.2. Temperature Dependence of V_B

Although bubble processes of geophysical interest occur in waters spanning a wide range of T, the effect of T on V_B

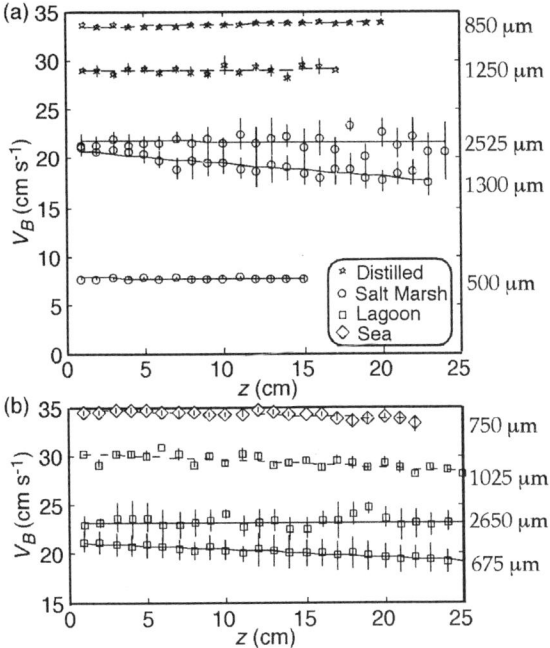

Figure 2. Rise velocity, V_B, versus height, z, above release depth, for bubbles of various sizes and waters. The legend for water type is on Figure 2a, bubble radii marked on figure.

has not been systematically studied. A series of experiments to measure $V_B(T)$ over the T range 0 - 40°C was conducted in distilled water. It was observed that V_B increased with T for smaller bubbles and decreased for larger bubbles. Although for non-oscillating bubbles $V_B(T)$ could be explained by changes in viscosity and density, this is not true for oscillating bubbles. Furthermore, there was a strong relationship between V_B and ζ, the oscillation parameter, defined as V_B/V_x where V_x is the horizontal velocity. Figure 3 shows $V_B(T)$ and ζ for 375, 1000, and 2100 μm bubbles as well as observed *Miyagi* [1927].

The decrease in $V_B(T)$ for the oscillating bubbles suggests that energy from the buoyant rise is transformed into horizontal motions (i.e., trajectory oscillations) and shape oscillations. For the 1000-μm bubbles shown in Figure 3b, V_B decreased with T, while ζ increased from 19% at 3°C to a maximum of 50% at 36°C.

For significantly larger 2100-μm bubbles, V_B also decreased with T although less strongly. The decrease in the T dependency of V_B with increasing r is clearly shown by a comparison of the 2100-μm bubble results (Figure 3c) with the 1000-μm bubble results (Figure 3b). The decrease in the T dependency of V_B with increasing r is clearly mirrored in the T dependency of ζ. Changes in T cause much less change in ζ for 2100-μm bubbles (Figure 3c) than for smaller (i.e. 1000-μm) bubbles (see Figure 3b). Oscillations for the 2100-μm bubbles increased from the lowest observed temperature, 4°C to 31°C at which point they remained constant, although V_B continued to decrease with T.

At $T = 20°C$, oscillations begin at approximately $r = 700$ μm, or $Re \sim 450$ [*Clift et al.*, 1978] and V_B decreases with r; however, at $Re \sim 1000$ ($r \sim 2000$ μm), path oscillations decrease, while bubble shape oscillations, especially higher modal oscillations, increase, and V_B no longer decreases with r.

4. DISCUSSION

In the natural environment, entirely clean surfaces are even less likely due to the presence of carbohydrates, proteins, fatty acids, [*Liss et al.*, 1997] and other organic and inorganic substances (note, salt is an ionic surfactant). The magnitude of the surfactant effect depends upon the surfactant kinetics and equilibrium surface concentration, C_s, as well as T and r. The observed size dependency of the surfactant effect in natural waters is in agreement with the trend for industrial surfactants where V_B decreased more for smaller bubbles [*Okazaki*, 1964].

In contaminated water, stress from the bubble's motion convects surfactants towards the downstream hemisphere, creating a gradient of C_s. This locally reduces the surface tension, σ, resulting in a tangential force towards regions of higher σ. Local surface viscosity is reduced causing decreased interfacial mobility. This interfacial retardation is called the Marangoni effect. Even for a contaminated interface, in the absence of gradients in σ, there is no Marangoni effect. Therefore, surfactants do not always cause a Marangoni effect. For example this occurs when surface diffusion is much faster than surface convection [*Quintana*, 1992]. Surfactant molecules diffuse from the bulk fluid to the new interface created at the upstream pole where $C_s < \alpha_s C_b$, where α_s is the surfactant surface solubility and C_b is bulk concentration. The surfactant is con-

Table 1. Coefficients for $V_B(z)$ regression analysis fit and its correlation coefficients of the fit.

r(μm)	C	q	R^2
489	7.82	-0.009	0.69
677	21.46	-0.078	0.96
747	34.86	-0.047	0.83
1016	30.41	-0.74	0.79
1313	20.97	-0.139	0.91
2529	21.86	-0.008	0.08
2645	23.36	-0.0002	0.0003

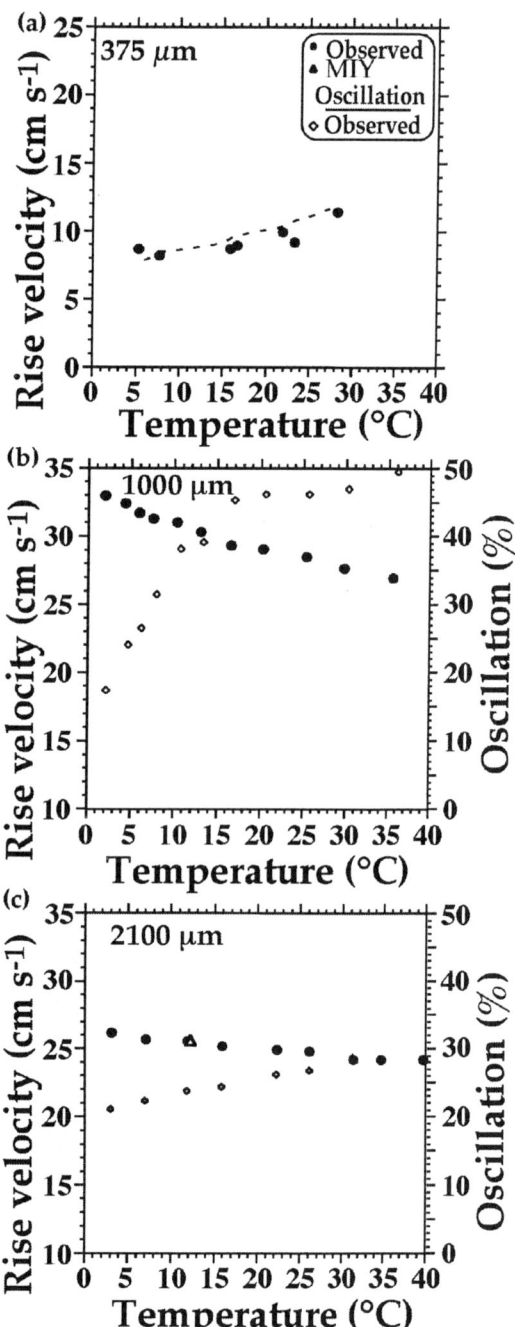

Figure 3. Observed temperature variation of bubble rise velocity for bubbles with radii (a) 375 μm, (b) 1000 μm, and (c) 2100 μm. Also shown oscillation parameter in (%) where it applicable. MIY-*Miyagi* [1927].

vected towards the downstream pole where it accumulates until $C_s > \alpha_s C_b$ and then desorbs and diffuses into the bulk fluid. The surfactant also diffuses against the surface convection towards the upstream hemisphere. Gradients in C_s are strongest in the downstream hemisphere, thus interface mobility is greater in the upstream hemisphere.

An analytic solution for the surfactant surface distribution and its effect on V_B, the Stagnant Cap Model, SCM, was developed by *Sadhal and Johnson* [1983], and is shown schematically in Figure 4. In the model, all the surface tension gradients occur across an immobilized cap in the downstream hemisphere. As a result, V_B is largely unaffected for increasing C_s until a stagnant cap angle of 30 to 45° is reached, at this point V_B decreases very rapidly for small increases in C_s. Once the cap extends above the equator, further growth once again has minimal effect on V_B. The SCM has been experimentally verified with industrial surfactants [*Duineveld*, 1995].

Based on the SCM, the simplest explanation for the "clean" behavior of bubbles in seawater is that they have not accumulated sufficient surfactant during their rise for the stagnant cap to extend greater than 30°. Since smaller bubbles have less surface area and lower convective forces, they accumulate surfactants more rapidly than larger bubbles. Thus V_B decreases more rapidly for smaller bubbles as they rise. Very small bubbles achieved a fully developed stagnant cap after a few centimeters rise, and thereafter $V_B(z)$ was constant with z (Figure 2).

Ionic surfactants form a double layer at the interface which affects the bubble hydrodynamics [*Borwanker and Wasan*, 1988], and the Marangoni effect also becomes dependent upon ionic strength. This was experimentally verified by *Fdihla and Duineveld* [1996] for the ionic surfactant sodium dodecyl sulfate. Unlike non-ionic surfactants, the transition between clean and dirty was not rapid, but spread over a wide concentration range. The lack of a rapid transition in Figure 2 is far more suggestive of the effect of ionic surfactants than the SCM prediction. And of course saltwater contains ionic surfactants such as NaCl.

In natural waters, the lowest V_B was for shallow waters, i.e., waters collected close to sediments. This suggests that surfactants from decomposition are much stronger than those from phytoplankton production. To test this hypothesis seawater was allowed to "age" in a glass tank in the sun and V_B was measured for 1000-μm bubbles every few days. No change was observed until the second week; when the water became the slightly yellow tinged, indicating plankton mortality, and V_B significantly decreased. It is likely that at this point algae began to decompose which may have been cause cell lysis, or bacterial exudation.

Although the experimental bubbles were produced from capillary tubes, there is strong reason to believe the results are applicable to oceanic bubbles. Wind-wave generated bubbles in bubble plumes were observed during the LUMINY experiment described in *De Leeuw et al.*, [2001] to burst immediately upon surfacing [*Leifer and De Leeuw*,

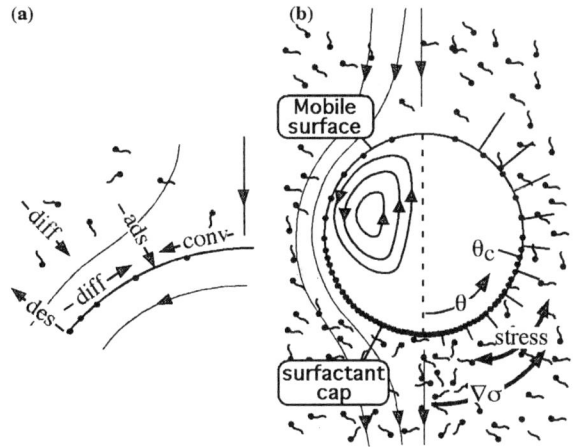

Figure 4. Schematic of Stagnant Cap Model for showing (a) the transport processes affecting surfactants on a bubble, and (b), the surface tension variation, σ, (radial line length) with zenith angle, θ. Key: ads- adsorption, des - desorption, diff - diffusion, conv - convection.

2001] a strong indication of bubble cleanliness [MacIntyre, 1972]. Presumably wave breaking and the bursting of surfacing bubbles sweeps away the surface microlayer [Leifer and De Leeuw, 2001]. Observations of the oscillations of oily bubbles rising from a natural hydrocarbon seep strongly indicated that larger ($r > 1500$ μm) bubbles do not become contaminated even after rising 70 m [Leifer, unpublished].

5. $V_B(T)$ PARMETERIZATION

For spherical bubbles in laminar flow (i.e. $Re < 1$) the Navier-Stoke's equation can be solved analytically yielding the Hadamard-Rybczynski's solution for V_B for both mobile and immobile interfaces [Levich, 1962]. For larger, clean bubbles though, V_B diverges from the solution. Thus a power law modification of this equation was used to parameterized $V_B(r,T)$ for non-oscillating bubbles,

$$V_B = c\frac{1}{3}gr^d v^n \quad (1)$$

Table 2. Coefficients for Eqn. (1), the V_B parameterization for clean non-oscillating bubbles.

Re	r (μm)	c	d	n
<1	<60	0.666	2.00	-1.00
1-150	60-500	0.139	1.372	-0.64
150-420	550-660	11.713	2.851	-0.64
420-470	660-700	0.156	1.263	-0.64
470-540	700-850	0.021	0.511	-0.64

Table 3. Coefficients for Eqn. (3), the V_B parameterization for clean oscillating bubbles.

H	K	r_C	V_{Bm}	m_1	m_2
-4.792×10^{-4}	0.733	0.0584	22.16	-0.849	-0.815

where c, d, and n are coefficients given in Table 2. Further details of the derivation can be found in Leifer et al. [2000].

For clean oscillating bubbles, V_B decreased with both increasing T and r, and a completely different analytic parameterization was developed. The onset of oscillation is not a simple function of Re but involves the shape since at different T, the onset occurs at different Re [Leifer et al., 2000]. The r for the onset of oscillation varies with T according to the relation

$$r_P = 1086 - 16.05 T_P \quad (2)$$

where r_P and T_P are peak r and T. Outside the observed range (7 - 26.8°C), (2) is presumably unreliable since, for example, it predicts $r_P = 0$ μm for $T_P = 78°C$, i.e., a zero radius bubble oscillates.

For clean oscillating bubbles, the following empirical parameterization was developed from observation and is shown in Figure 1. It is in good agreement with Clift et al. [1978] at 20°C and is applicable for oscillating bubbles for $0 < T < 30°C$ and $r_P < r < 4000$ μm, and is,

$$V_B = \{V_{Bm} + H(r - r_c)^{m_1} e^{\{K(r-r_c)^{m_2} T\}} \quad (3)$$

where H, K, m_1 and m_2 are coefficients, r_C is a critical radius below which the parameterization suggests bubbles do not oscillate for any T, and V_{Bm} is the minimum V_B for oscillating bubbles. Since (3) is analytical rather than empirical, the coefficients H and K do not have physical meaning. The coefficients are given in Table 3. A comparison between the parameterization (predicted) and observed values of V_B is shown in Figure 5. The correlation coefficient was 0.9455. The derivation of (3) can be found [Leifer et al., 2000].

6. CONCLUSIONS

A study of bubble hydrodynamics for many fresh and salt waters, including seawater, showed that bubbles larger than approximately 600-μm radius behaved as though in distilled water. Bubbles in saltmarsh water and canal water, though, were found to have V_B values lower than those for clean bubbles, indicating surfactants had at least partially immobilized the bubble interface. Based on the data presented in this paper, a reasonable approach to modeling shallow bubbles in seawater is to simulate small bubbles as dirty, and large bubbles as clean with a transition at circa 600 μm. However, the transition radius increases the longer the bubbles are in the water column.

An empirical parameterization of $V_B(r,T)$ for oscillatory and non-oscillatory bubbles that correctly incorporates the effect of T was presented. The parameterization is applicable to bubbles in clean freshwater. Additionally the param-

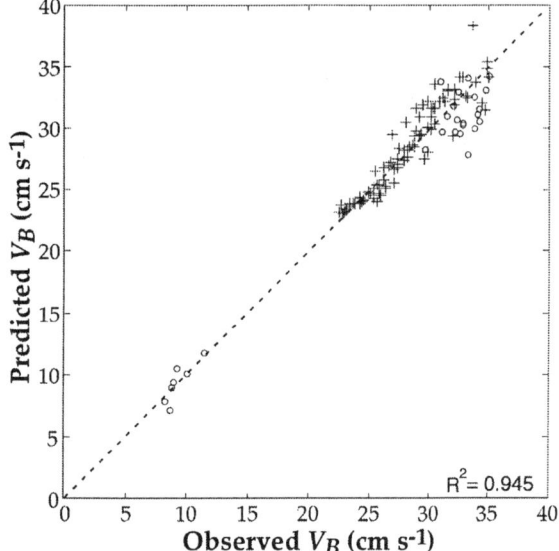

Figure 5. Comparison of parameterized and observed rise velocity, V_B, for oscillating bubbles (+). Also shown is predicted and observed V_B for non-oscillating 375 μm bubbles (o). The correlation coefficient, R^2, was calculated only for oscillating bubbles.

eterization should not be applied for $T > 40°C$ or to other liquids. Since, V_B in seawater for large (i.e. oscillating) bubbles indicates that these bubbles are clean, it is reasonable to apply (3) to seawater. However, experimental verification is clearly required. Further investigation of the effect of surfactants on bubbles in natural waters is also clearly needed.

Finally, V_B is far more sensitive to the effect of surfactants than T. Since bubble gas transfer velocity, k_{Bub}, is a function of V_B, underestimates of V_B cause k_{Bub} to also be underestimated.

REFERENCES

Borwanker, R.P. and D.T. Wasan, Equilibrium and dynamics of absorption of surfactants at the fluid/fluid interfaces. *Chem. Eng. Sci.*, 43, 1323-1337, 1988.

Clift, R., J. R. Grace, and M. E. Weber, *Bubbles, Drops, and Particles*, Academic Press, New York, 1978.

Datta, R.L., D.H. Napier and D.M. Newitt, The properties and behavior of gas bubbles formed at a circular orifice. *Trans. Inst. Chem. Eng.*, 28, 14-26, 1950.

De Leeuw, G., G.J. Kunz, G. Caulliez, L. Jaouen, S. Badulin, D.K. Woolf, P. Bowyer, I. Leifer, P. Nightingale, M. Liddicoat, T.S. Rhee, M.O. Andreae, S.E. Larsen, F.Aa Hansen, and S. Lund, LUMINY - An Overview, *This volume*, 2001.

Duineveld, P.C., The rise velocity and shape of bubbles in pure water at high Reynolds number. *J. Fluid Mech.*, 292, 325-332, 1995.

Fdhila, R.B., and P.C. Duineveld, The effect of surfactants on the rise of a spherical bubble at high Reynolds and Peclet numbers. *Phys. Fluids*, 8, 310-32, 1996.

Haberman W.L., and R.K. Morton, An experimental investigation of the drag and shape of air bubbles rising in various liquids. *The David W. Taylor Model Basin.*, 55, Navy Dept., Washington 7 D.C, 1953.

Jamialahmadi, M., C. Branch, and H. Müller-steinhagen, Terminal bubble rise velocity in liquids. *Trans. IChemE.*, 72A, 119-122, 1994.

Leifer I., and De Leeuw, Bubble measurements in breaking-wave generated bubble plumes during the LUMINY wind-wave experiment, *This Volume*, 2001.

Leifer I., R.K. Patro, and P. Bowyer, A study on the temperature variation of rise velocity for large clean bubbles. *J. Atm. Ocean. Tech.*, 17(10), 1392–1402, 2000.

Levich, V. G., *Physico-Chemical Hydrodynamics*, Prentice-Hall, Englewood Cliffs, N.J., 1962.

Liss, P.S., A.J. Watson, E.J. Bock, B. Jähne, W.E. Asher, N.M. Frew, L. Hasse, G.M. Korenowski, L. Merlivat, L.F. Phillips, P.Schluessel, and D.K. Wolf, Physical processes in the microlayer and the air-sea exchange of trace gases. In *The Sea Surface and Global Change*, edited by P.S. Liss, and R.A. Duce, pp. 1-33, Cambridge University Press, Cambridge, UK, 1997.

MacIntyre, F., Flow patterns in breaking bubbles. *J. Geophys. Res.*, 77, 5211- 5228, 1972.

Medwin, H. and N..D. Breitz, Ambient and transient bubble spectral densities in quiescent seas and under spilling breakers. *J. Geophys. Res.*, 94C, 12, 571-12, 759, 1989.

Miyagi-Kôgakuhakusi, O., The motion of an air bubble in rising water. *Tohoku Imp. Univ. Tech. Reports*, V-VI, 135-171, 1927.

Monahan, E.C., The ocean as a source for atmospheric particles. In *The Role of Air-Sea Exchange in Geochemical Cycling*, edited by P. Buat-Menard, pp. 129-163, D. Reidel Publishing Company, Hingham, Massachussetts, 1986.

Okazaki, S., The velocity of ascending air bubbles in aqueous solutions of a surface active substance and the life of the bubble on the same solution. *Bull. Chem. Soc. Jpn.*, 37, 144-150, 1964.

Quintana, G.C., The effect of surfactants on flow and mass transport to drops and bubbles. In *Transport Processes in Bubbles, Drops, and Particles* edited by R.P. Chhabra and D. De Kee, pp. 87-113, Hemisphere Pub. Corp., New York, 1992.

Sadhal S., and R.E. Johnson, Stoke's flow past bubbles and drops partially coated with thin films. *J. Fluid Mech.*, 126, 237-250, 1983.

Tsuchiya, K., H. Mikasa, and T. Saito, Absorption dynamics of CO_2 bubbles in a pressurized liquid flowing downward and its simulation in seawater. *Chem. Eng. Sci.*, 52, 4119-4126, 1997.

R. Patro, Physics Department, Memorial University of Newfoundland, St. John's, Canada, A1B3X7, (email: ranjan@physics.mun.ca).

I. Leifer, Chemical Engineering Department, University of California, Santa Barbara, Santa Barbara, California, 93106-5080, USA (email: ira.leifer@bubbleology.com).

P. Bowyer, National University of Ireland, Galway, Ireland, (email: peter.bowyer@nuigalway.ie).

Development and Testing of an Eddy Accumulator

W. John Cooper and Mehran Alaee

National Water Research Institute, Environment Canada, Burlington, Ontario, Canada

Mark Donelan

Rosenstiel School of Marine and Atmospheric Science, University of Miami, Miami, Florida

Environmental contaminants are in general of such low concentration that pre-concentration and extraction techniques are required to measure them. In order to estimate their fluxes over the air/water interface an eddy accumulator is required to integrate samples over a period of time and produce a sufficiently concentrated sample to measure the flux in the environment. A proportional sampling eddy accumulator was designed, developed and tested using the Gas Transfer Flume facility in the Hydraulics Laboratory of the National Water Research Institute. Verification of the accumulator was accomplished through comparison to simultaneous eddy correlation measurements conducted with carbon dioxide gas and an infrared absorption CO_2 analyzer. The correlation and accumulation methods agree within 5% of the total concentration of CO_2 for all test runs, and the absolute mean error was 11 ppm compared to correlation at concentration differences between 24 to 117 ppm CO_2. These results have verified the design. The prototype could be further refined for accumulation measurements in the field to determine the flux of toxic contaminants.

1. INTRODUCTION

Vapour exchange across the air-water interface is one of three (along with wet and dry precipitation) major methods of transfer of atmospheric contaminants to a body of water. In order to accurately estimate net fluxes of contaminants across the air/water interface, information on all three methods of exchange should be known. Wet and dry deposition can be determined by direct collection and analysis of precipitation but vapour exchange must, in general, be estimated at present.

Fluxes of contaminants in this state across the air/water interface are governed by Henry's Law and the mass transfer velocity. The mass transfer velocity of contaminants can be determined experimentally in laboratory conditions using a gas transfer flume; however, validating the results of such modeling is difficult due to the other environmental factors present in nature which affect the vapour state transfer of the chemical.

Eddy correlation is a method of direct flux measurement that provides in-situ data for compounds whose concentrations can be determined in real-time such as CO2 (*Donelan & Drennan*, 1995). The flux G_n of quantity C, including water vapour, may be expressed by

$$G_n = \overline{wc} \mp WC / \rho$$

where (ρ) is the air density (assumed constant) and w is the vertical velocity; the upper case letter denotes the mean value and the lower case, the fluctuations about the mean.

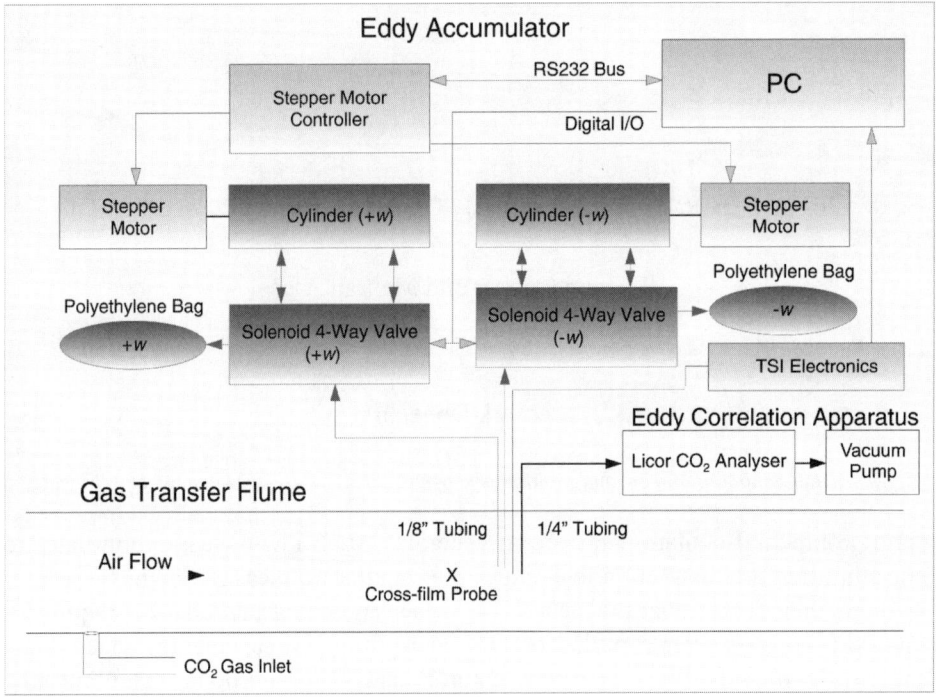

Figure 1. Eddy Accumulator Block Diagram

The second term on the right vanishes because the mean vertical velocity is taken to be zero.

Unfortunately many compounds of environmental interest are of such low concentration in nature that real-time analysis is difficult using today's technologies without pre-concentration and extraction techniques; thus the need for an eddy accumulator to integrate samples and obtain sufficient concentrations for subsequent analysis. The main thrust of this work is to proportionally sample according to the magnitude and direction of the vertical component of the wind velocity, bypassing the need for a fast sensor to measure flux in real-time. The vertical velocity is used to determine the conditional sampling by triggering one of either of the two identical samplers in keeping with the direction of vertical velocity. The amount sampled is kept proportional to the magnitude of the vertical velocity and stored in the appropriate (positive or negative) reservoir. Therefore, the positive and negative reservoirs contain air with contaminant concentrations, respectively related to:

$$\overline{w^+ (C + c)} \text{ and } \overline{w^- (C + c)}$$

When these two quantities are subtracted, we may estimate the flux wc (*Kraus & Businger*, 1994), suitable corrections having been made for density effects (*Webb et al.*, 1980).

2. TECHNICAL APPROACH

The accumulator itself is composed of two identical systems which are alternately triggered to sample air based on the direction of the vertical wind velocity (Figure 1). Wind velocity measurements are made in the gas transfer flume using a X-film hot wire probe. For each sampling interval (200 msec) the magnitude and direction of the vertical velocity is calculated using an iterative routine to process the anemometer cross-film voltages read by the computer. Once the magnitude and direction of the vertical velocity is known, the appropriate sampler is triggered and the amount sampled is directly proportional to the magnitude of w. Each sampler consists of a pneumatic bi-directional cylinder and a four-way solenoid valve which simultaneously extracts gas from the flume using a 1/8" diameter tube and pumps the previously drawn sample into a polyethylene bag. The pistons of the cylinders are fixed in place; stepper motors are used to precisely move the cylinder housings back and forth to draw the current sample and pump the prior sample in a reciprocal fashion. A preliminary run prior to each accumulation was done to characterise the flow and 3 standard deviations of w were used to set the maximum sampling volume for any sample. This unit is a proportional sampling accumulator in that the sample volume in each time interval is directly proportional to the vertical wind velocity. The accumulator was operated at a sampling frequency of 5 Hz and this rate was used for both the correlation and accumulation methods.

Eddy correlation was done by drawing the flume air through a CO2 analyser with a high volume vacuum pump and measuring the flux of CO2 throughout the accumulation period. The CO2 analyser used throughout these tests was the Licor LI-6262 non-dispersive infrared (NDIR) gas analyser. The instantaneous CO2 concentration multiplied by the corresponding scaled w at each data point was then calculated and summed for both directions to yield the expected concentration in each bag. Following the accumulation period, the bags were then alternately pumped out by their respective cylinders through the CO2 analyser to determine the mean bag concentration. The typical volume collected in each bag was approximately 500 cc.

In order to quickly and repeatedly test and verify the accumulator operation, the Gas Transfer Flume was operated as a wind tunnel without water and pure CO_2 was directed into the centreline at the bottom of the flume upstream at a distance of 9.3 m from the accumulator inlet tubes. The upward diffusion of the plume of CO_2 provided a vertical flux at the location of the X-film and CO_2 inlet tubes. The incorporation of a flow controller on the CO_2 gas allowed large, controlled and repeatable fluxes to be obtained within the flume without the delays associated with spiking and maintaining a high water concentration. Thus individual accumulator runs could be accomplished in less than one hour with a large CO_2 flux yielding sufficient bag concentrations for analysis. The sensitivity of the Licor analyser to pressure also caused calibration concerns. Calibration of the analyser is done with gas bottles of a known CO_2 concentration. These gases have to be presented to the Licor unit at the same pressure as that encountered during measurement. Correlation calibration was done by supplying the bottled gases under vacuum; accumulator calibration was done by pumping the bottled gases through the unit using an accumulator cylinder.

3. RESULTS AND DISCUSSION

A total of eighteen runs were done to evaluate the accumulator. These runs were done over a range of six wind speeds(u) from nominally 1 m/sec to 5 m/sec with three different levels of flux generated at each wind speed. These levels of flux produced typical concentrations in the bags ranging from 600 to 1000 ppm following the usual accumulation period of 45 minutes. Some runs were repeated as subtle changes in the X-film anemometer caused an imbalance of the accumulator - one cylinder would significantly oversample and lead to errors. Typically on a good run the total volume of the samples for each cylinder would be within 10 percent of each other.

The results of the eighteen runs are shown in Figure 2 with the concentration difference in ppm by both methods vs. wind speed at three different flux levels indicated on the vertical axis. Figure 2 shows the individual runs as bar graph pairs: the lower darker bar of each pair is the calculated concentration difference between the bags as predicted by eddy correlation; the upper lighter bar of each pair is the measured concentration difference between the bags following accumulation of the samples. Accumulator error can be expressed as the difference between the two methods with the calculated eddy correlation difference taken to be the true value. The absolute mean error is 11 ppm and the standard deviation of that error is 7 ppm. The error bars in Figure 2 represent ± two standard deviations of the mean error and 15 of the 18 runs presented are within that tolerance, with only one run outside three sigma. Figure 3 presents the accumulator error as a percentage of the difference between the two methods divided by the CO2 concentration difference predicted by eddy correlation. These results correspond to an average absolute percentage error of 19% compared to eddy correlation with a standard deviation of 14% as expressed in terms of the (error / calculated correlation difference) x 100% for each run.

4. CONCLUSIONS

These results are acceptable for the determination of fluxes of trace gases in the environment. The relative uncertainty of chemical measurements of trace gases can be several times the anticipated error of an accumulator of this type. Further design and development of a more robust field unit using a sonic anemometer to measure wind velocity could be undertaken in the future. The stability and reliability of the stepper motor design feature should provide precise and consistent sampling over the longer sampling periods necessary in a field environment. Further testing of the unit with air/water boundary layer conditions and on a fixed field platform would be necessary to fully evaluate the accumulator.

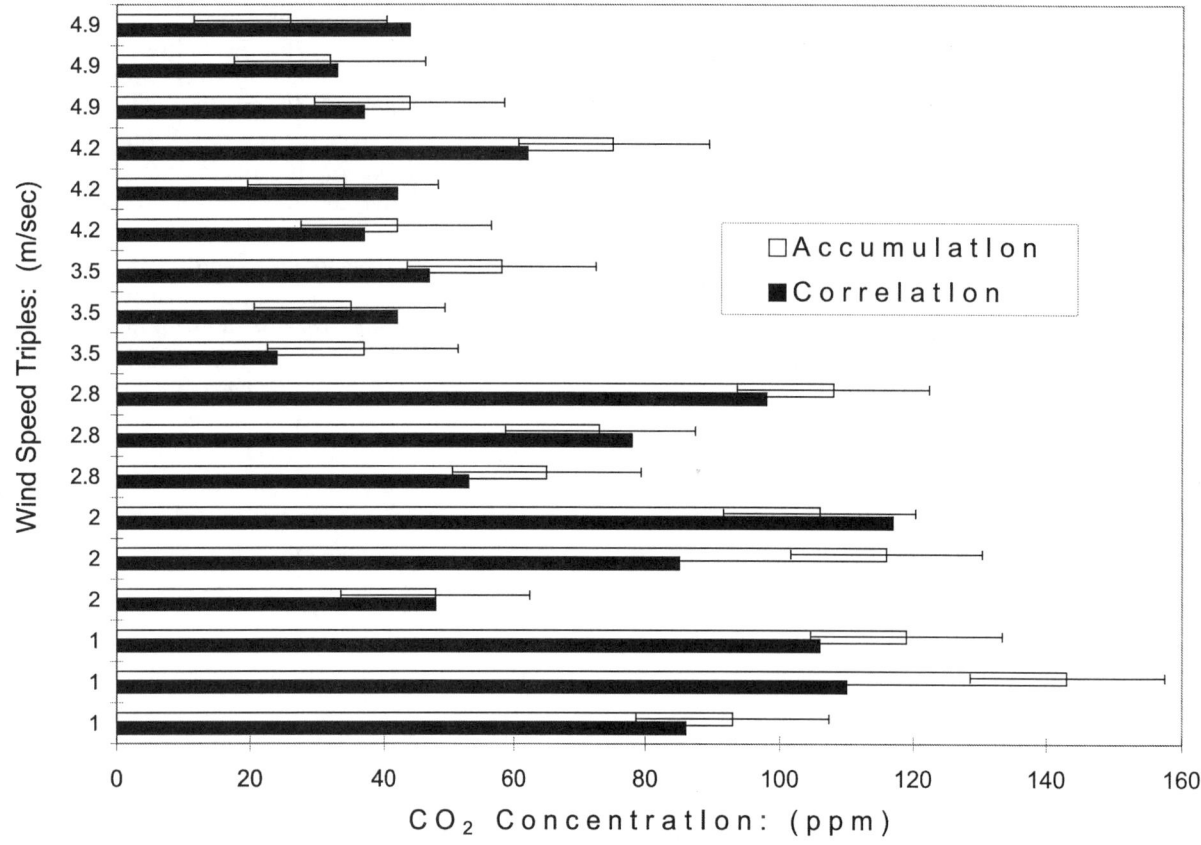

Figure 2. Eddy Correlation Calculated Difference vs. Eddy Accumulator Measured Difference

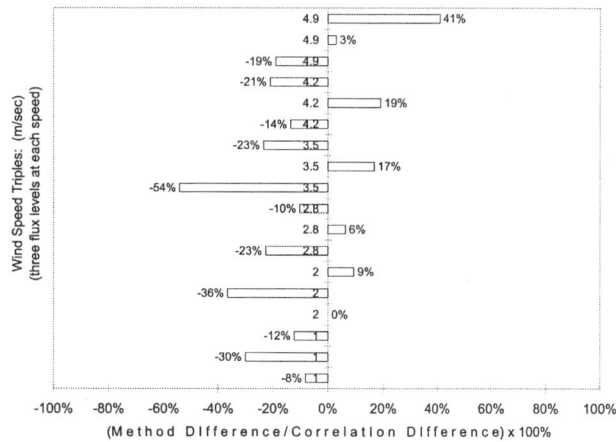

Figure 3. Accumulator Error: Difference in Methods / Correlation Difference

Acknowledgments. The authors wish to thank the NWRI Engineering staff and machine shop for their support, in particular Niels Madsen, P.Eng. for the mechanical design of the accumulator. The authors gratefully acknowledge Environment Canada for their financial support of this study.

REFERENCES

Donelan, M. A. and W. M. Drennan, Direct field measurements of the flux of carbon dioxide, in *Air-water gas transfer*, edited by B. Jahne and E. C. Monahan, pp. 677-683, AEON Verlag & Studio, 1995

Kraus, E. B. and J. A. Businger, *Atmosphere-ocean interaction*, 350 pp., Oxford University Press, New York, ISBN 0-19-506618-9, 1994

Webb, E. K., Pearmen, G. I. And Leuning, R., Correction of flux measurements for density effects due to heat and water vapour transfer, *Quart. J. R. Meteorol. Soc.,* 106, 85-100, 1980.

M. Alaee, NWRI/AEPB, Environment Canada, 867 Lakeshore Road, Burlington, Ontario, Canada. L7R 4A6.

W. J. Cooper, NWRI/AEMR, Environment Canada, 867 Lakeshore Road, Burlington, Ontario, Canada. L7R 4A6

M. A. Donelan, 4600 Rickenbacker Causeway, Miami, Florida.

A Multi-year Time Series of Global Gas Transfer Velocity from the TOPEX Dual Frequency, Normalized Radar Backscatter Algorithm

David M. Glover, Nelson M. Frew, and Scott J. McCue

Woods Hole Oceanographic Institution, Woods Hole, Massachusetts

Erik J. Bock

University of Heidelberg, Heidelberg, Germany

The feasibility of calculating gas transfer velocity directly from a remotely sensed measure of the sea surface roughness presents an unique opportunity. We present here the six year time series (1993–1998) of TOPEX data processed into gas transfer velocity and examine the variability of these results in space and time. The seasonal and interannual variability of the regional patterns yield insight into the sensitivity of the altimeter-based gas transfer velocity to phenomena such as ENSO. We compare the results of this time series to a similar time series created through the application of more traditional wind speed-gas transfer velocity parameterizations to the wind speed estimates made by the National Center for Environmental Prediction reanalysis project for the same period. Our approach to deriving transfer velocity is subject to parameters calibrated against field and laboratory measurements. The intercomparison would be greatly enhanced by *in situ* measurements contemporaneous with the altimetry and the newly launched QuikScat scatterometry mission. These *in situ* measurements are one of our best constraints on the remote observations: wind stress *vs.* surface roughness. Examples are given of how, even at a limited number of sites around the globe, these *in situ* measurements may improve the remotely sensed time series of air-sea gas transfer velocity.

1. INTRODUCTION

For certain global geochemical applications our understanding of the space and time distribution of the gas transfer velocity (k) is inadequate. *Wallace* [1995] points out that an uncertainty of a factor of two in our estimation of k when combined with systematic errors in the measurement of the partial pressure difference of CO_2 across the air-sea interface can lead to unacceptable error in our ability to calculate the marine flux of this gas. When applied to calculating the global flux of "excess" (anthropogenic) CO_2 the problem becomes amplified. *Gruber et al.* [1998], when examining the flux of $\delta^{13}C$ across the air-sea interface at the Bermuda Atlantic Time-series Site (BATS), also found that their calculations were particularly sensitive to uncertainties in the gas transfer velocities. Both of these studies approached the calculation of k in the community standard fashion by estimating the transfer velocity from a knowledge of the wind speed and sea surface temperature (SST). However, there is accumulating evidence that the transfer velocity is sensitive to environmental conditions other than wind speed and SST, in particular, the presence of surfactants [*Frew et al.*, 1999], boundary layer instability [*Erickson*, 1993], and wave fetch [*Wanninkhof*, 1992]. What is needed, then, is a direct measurement of the surface roughness expressed by the small gravity-capillary wave portion (the gas-exchange-active portion) of the surface wave spectrum.

Through their statistical analysis of sun glitter from the sea surface *Cox and Munk* [1954] produce an isotropic, Gaussian, geometric optics model of backscatter. This model and its application to microwave backscatter is reviewed by *Jackson et al.* [1992] where they show, for nadir looking active microwave sensors, the relationship between the mean square slope of the surface waves and the normalized backscatter is straightforward. The power of the specular reflection captured in the radar backscatter is inversely related to the mean square slope and the mean square slope has been linearly related to the gas transfer velocity by a number of workers [*Jähne et al.*, 1987; *Hara et al.*, 1995; *Bock et al.* 1999]. Combining these two relationships is a logical outgrowth.

The remainder of this paper is divided into three more sections. In section 2 we will discuss the basis for our algorithm that derives gas transfer velocity from altimeter normalized radar backscatter. In the third section we describe the results of applying this algorithm to six years of TOPEX data and compare these results to results obtained by applying the *Wanninkhof* [1992] (hereafter referred to as W92) and *Liss and Merlivat* [1986] (hereafter referred to as LM86) wind speed based algorithms to the same time period of National Center for Environmental Prediction (NCEP) reanalysis wind data [*Kalnay et al.*, 1996]. In the final section we discuss these results, problems with the current altimeter algorithm, and directions for future research.

2. ALGORITHM

The details of the development, calibration, and verification of the algorithm that computes gas transfer velocity from normalized radar backscatter is given in *Frew et al.* [2000]. For purposes of discussion we will give a brief description of the algorithm here.

From the work of *Cox and Munk* [1954] we have an expression for backscatter from the geometrical optics (GO) model assuming the surface wave field is best approximated by an isotropic, Gaussian surface (as given in *Jackson et al.* [1992]):

$$\sigma^\circ_{GOG} = \rho_g \left(\frac{\sec^4 \theta}{\langle m_g^2 \rangle} \right) e^{\left(\frac{-\tan^2 \theta}{\langle m_g^2 \rangle} \right)}, \quad (1)$$

where ρ_g is the effective reflectivity, $\langle m_g^2 \rangle$ is an effective mean square slope, and θ is the illuminating pulse incidence angle. For instruments probing only in the nadir direction, Equation 1 becomes:

$$\langle m_n^2 \rangle = \frac{\rho'_n}{\sigma^\circ(0^\circ)}, \quad (2)$$

where the subscript n is now used to indicate nadir and the meaning of the ρ'_n has changed to reflect the approximation of the original azimuthally integrated cross-section by the returned power measured at nadir.

We exploit the dual frequency nature of the TOPEX altimeter by making use of the differential scattering (Ku vs. C bands) isolating a portion of the small gravity-capillary wave spectra related to gas exchange. Since the effective wavelength of surface features sampled is approximately three times the incident wavelength, the combination of the Ku (2.1 cm) and C (5.5 cm) bands resolves to examining the gravity-capillary wave spectra in the 6.3 to 16.5 cm region. *Bock et al.* [1999] have shown that gas exchange is linearly related to the mean square slope of waves with a cutoff wavelength of approximately 16.5 cm. We effectively remove the contribution of longer waves to the mean square slope by differencing the two bands in the following manner:

$$\langle \tilde{m}_n^2 \rangle = \frac{\rho'_n(Ku)}{\sigma^\circ_{Ku}} - \frac{\rho'_n(C)}{(\sigma^\circ_C + \alpha)}, \quad (3)$$

where we refer to $\langle \tilde{m}_n^2 \rangle$ as the partial mean square slope and α is a small correction factor applied to the C-band backscatter to account for the lack of an absolute calibration as reported in *Chapron et al.* [1995].

The linear relationship between mean square slope and gas transfer velocity allows us to calculate k directly from Equation 3 by including a slope and intercept, empirically determined from laboratory wind-wave tank experiments. In this fashion we arrive at the following equation describing our algorithm:

$$k = 7000 \left(\frac{0.38}{\sigma^\circ_{Ku}} - \frac{0.48}{(\sigma^\circ_C + 0.5)} \right) \left(\frac{Sc}{660} \right)^{-0.5}, \quad (4)$$

where we have optimized ρ'_n to field slope data from a range of values given in *Jackson et al.* [1992] and α to fit laboratory data. Note: in this version (3.0) of our algorithm the best fit to the calibrating laboratory data was had when the intercept was constrained to be zero. We report transfer velocities calibrated to a Schmidt number (Sc) of 660 at 20°C for CO_2.

The remainder of the data processing was carried out as follows. The once per second TOPEX data was extracted from the merged geophysical data records (MGDR), generation B, CD-ROMs and standard corrections applied. Data records were rejected based on the quality flags described in Benada [1997]. Additional corrections were made to the C-band $\sigma°$ by removing the atmosphere attenuation overcorrection using vapor-induced path delay and brightness temperature parameters from the MGDR [Benada, 1997]. We then applied the rain flag criterion described by Tournadre and Morland [1997] to eliminate 1 Hz records that were affected by rain events. The data were then binned into monthly bins and regridded onto a 2.5° × 2.5° grid. The data were further filtered by applying an ice mask derived from the monthly SSM/I ice data. The once per second records were block averaged to yield monthly means. Records with a Ku-band $\sigma°$ greater than 16 dB or less than 0 dB were removed before averaging as recommended by Benada [1997].

3. RESULTS

In what follows we present the results of applying the above algorithm to the six year record of TOPEX data starting with cycle 11 (Jan. 1993) and concluding at the end of cycle 232 (Dec. 1998); 1 cycle equals a 10 day exact repeat orbit period. In addition, we process the same six years of wind speed data from the NCEP reanalysis project as a comparison. Because the backscatter-transfer velocity algorithm is under development we adhere to comparison of patterns rather than absolute values. Although we do not expect the algorithm to change dramatically there will likely be quantitative changes but no qualitative changes. We expect the patterns displayed in this analysis to persist and comparison to more traditional methods for estimating gas transfer velocity can be instructive.

3.1. A Six Year Time Series

The results of the six year time series of backscatter-derived transfer velocity are shown as zonal averages in Plate 1. The overall pattern of seasonal variation is clearly seen in Plate 1, with the maximum transfer velocities in each hemisphere's corresponding wintertime. Additionally, there is an anti-correlated period of low to very low transfer velocities along the equator. At

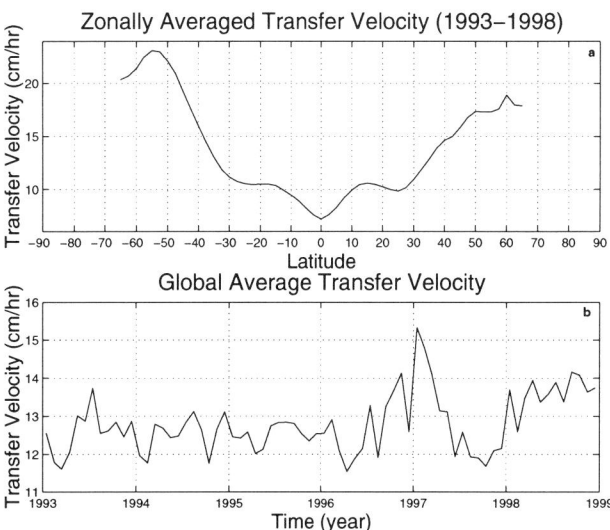

Figure 1. In a) the average zonal transfer velocities over the time series 1993-1998 and b) the global monthly average transfer velocities. Note that in b) both hemispheres are averaged together smoothing out any seasonal cycle.

mid-latitudes (20°–40°N) there is a period of low transfer velocities developing each year in the summertime. A similarly low austral summertime low, zonally averaged, transfer velocity does not appear in any year in the 20°–40°S zone. This could be explained by the greater fetch at these latitudes when compared to the northern hemisphere. The one exception is 1998, which does not develop the very low summertime zonally averaged transfer velocities at northern mid-latitudes. Additionally, early 1997 has the highest zonally averaged transfer velocities in the northern subpolar region. The extremely low zonal averages along the equator are interrupted during two periods: late winter–early spring in 1997 (El Niño) and in late autumn–early winter in 1998. The causal factor for the weakening of the equatorial low seen in 1998 is not readily obvious, although it appears to be preceded by a similar weakening of the mid-latitude lows farther north.

Collapsing Plate 1 into a time series climatology yields the results shown in Figure 1a. Averaged over the six years, the pattern of higher transfer velocities poleward displays an asymmetry, highest between 50°–60°S. This is not surprising since at those latitudes the fetch is greatest, to the north land intervenes and to the south the seasonal ice-pack covers the air-sea interface. In Figure 1b we have processed the data to produce a time series of the monthly global average transfer velocities. Treating both hemispheres together smooths out the seasonal pattern mentioned in Plate 1. However,

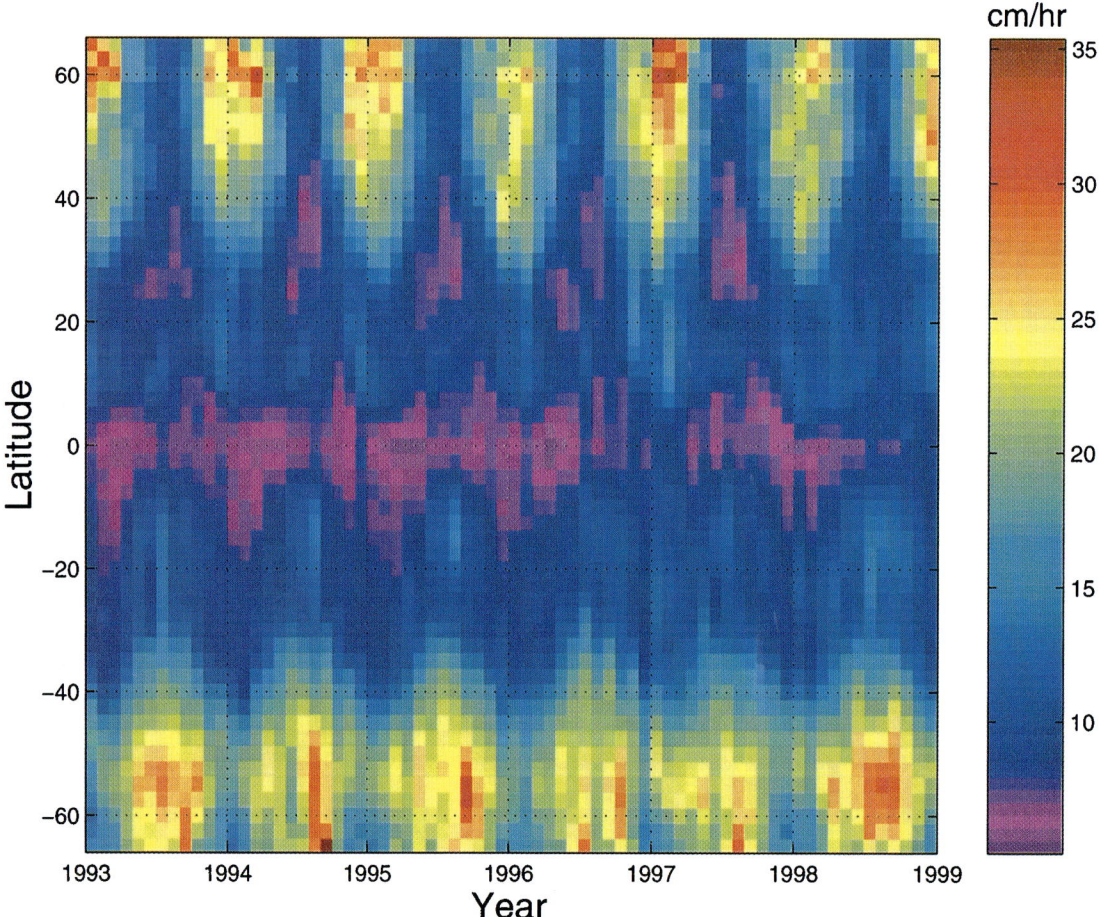

Plate 1. The monthly zonal average time-series (1993–1998) of the TOPEX-derived gas transfer velocity. The minimum value (the darkest purple) is approximately 5 cm/hr.

starting in mid-1996 there is a distinct increase in the global average transfer velocity peaking in Jan. 1997, approximately two to three months before the generally recognized beginning of the last El Niño. After decreasing later in 1997, the transfer velocities in 1998 return to a value higher than the average obtained from the 1993–1995 period.

3.2. Comparison to Wind Derived Transfer Velocities

To evaluate the patterns seen in the altimeter-derived transfer velocities, we obtained the six-hourly wind vector data from the NCEP reanalysis project *Kalnay et al.* [1996] for the same period. The wind vector data was combined into wind speeds and regridded onto the same $2.5° \times 2.5°$ grid used with the TOPEX-derived k's. We then applied these wind speeds to the wind speed-transfer velocity relationships of W92 and LM86. The results of this application were then binned to produce monthly estimates of transfer velocity for comparison.

The comparison of the three differently derived transfer velocities over the six years, globally, is done by examining the time series at the nine Joint Global Ocean Flux Study (JGOFS) time series sites. These stations are enumerated in the caption of Figure 2.

A very interesting result of this exercise is the fact that over most the globe the TOPEX algorithm produces k's that are more in line with results from the LM86 relationship than the W92 relationship. At the BATS, NABE, and ARAB sites an annual cycle of transfer velocity is clearly seen in all three relationships, it is less obvious at the STNP, PFR, and KERFIX sites. At HOT there does not seem to be an annual cycle apparent although all three relationships are close in value as well as pattern. At the EQPAC site the TOPEX relationship seems to go back and forth between the W92 and LM86 relationships and gives a more constant long term pattern. The long term pattern of low k's at the beginning of the time series (1993-1994), arching to higher values in the middle (1995-1996), and then returning to lower values in the later years (1997-1998) seen at the EQPAC site does not seem to be apparent in the TOPEX results. It should be noted that the values of k at EQPAC were so low that a different ordinate had to be used in Figure 2d to see these differences.

Immediately apparent in Figure 2 is the fact that, for the most part, the k's from W92 are larger than either the k's from LM86 or the TOPEX algorithm. This is not unexpected with respect to LM86 because W92 is a quadratic function of wind speed while LM86 varies only linearly. However it is of interest to note the places and times the TOPEX algorithm and W92

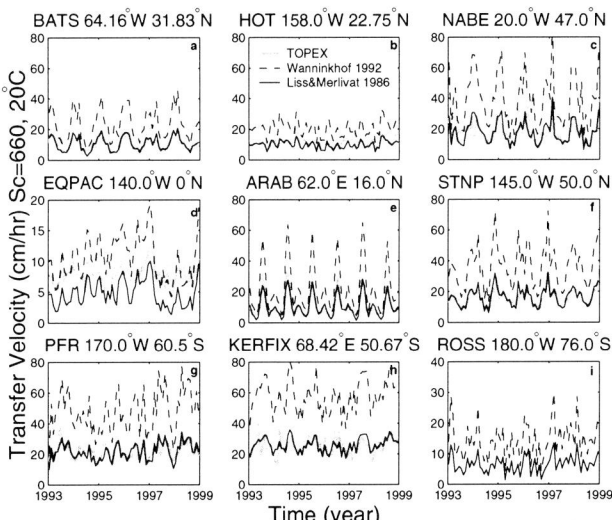

Figure 2. The six year time-series of gas transfer velocity at the nine JGOFS time-series sites around the globe. Gas transfer velocities derived from the TOPEX algorithm are shown as a heavy gray line, from *Wanninkhof* [1992] as a dashed line, and from *Liss and Merlivat* [1986] as a thin line. Note, the inclination of the TOPEX/Poseidon satellite is $66.03°$ and as such there is no coverage for the ROSS site. The sites are Bermuda Atlantic Time-series Site (BATS, $64.16°$W, $31.83°$N), Hawaiian Ocean Time-series (HOT, $158.0°$W, $22.75°$N), North Atlantic Bloom Experiment (NABE, $20.0°$W, $47.0°$N), Equatorial Pacific (EQPAC, $140.0°$W, $0.0°$N), Arabian Sea (ARAB, $62.0°$E, $16.0°$N), Subarctic North Pacific at station P (STNP, $145.0°$W, $50.0°$N), Antarctic Polar Front Region (PFR, $170.0°$W, $60.5°$S), Kerfix time series (KERFIX, $68.42°$E, $50.67°$S), and the Ross Sea site (ROSS, $180.0°$W, $76.0°$S).

do agree in their time series pattern. At the ARAB site the W92 and TOPEX relationship seem to respond more strongly to the bimodal nature of the monsoon season than does the LM86 relationship. In the late winter of 1997 in the EQPAC region a general overall decrease in gas transfer velocity (due to the onset of the latest El Niño) is apparent in all three relationships, but the decrease in TOPEX is not as sharp as W92 or LM86.

4. DISCUSSION

Accurate knowledge of gas transfer velocity is important for many geochemical applications. In low to moderate sea-state conditions the space-time patterns of our algorithm provide an alternate view of global gas transfer velocities. We find an average gas transfer velocity of 13 cm/hr over the six year time series, which is at variance with the 22 cm/hr reported in W92. The average transfer velocity from GEOSECS and TTO (under admittedly fair weather conditions) is found to range from approximately 10 to 13 cm/hr [*Peng et al.*, 1979; *Smethie et al.*, 1985]. Apparently the backscatter-derived transfer velocities are sensitive to factors other than wind speed alone (fetch, surfactants, *etc.*), but there remain issues to be resolved.

The problematic issues with the TOPEX algorithm are two fold: ground truth calibration and space-time coverage. We are addressing the first one with vicarious calibration using field work underway and the second problem with either overlap coverage with Jason-1 or by leveraging altimetry experience to scatterometry with its better space-time coverage. The drawback of using vicarious *in situ* calibration is that the data rate is very low. Ocean going research vessels that happen to be in the right place at the right time, making the appropriate measurements, are few and far between. The prospects for improving the space-time coverage with either another altimeter or by expanding our research to include scatterometry (or both) are, however, excellent. Comparisons between the TOPEX σ°_{Ku} and SeaWinds σ°_{Ku} are currently underway. Because the scattering mechanisms are different (specular *vs.* Bragg) the geometry of the directional modulation seen in the scatterometer presents an algebraic obstacle to overcome. The differences in scattering mechanisms remains a research problem to be resolved.

An additional issue involves breaking waves and their consequent amplification of air-sea gas exchange (*i.e.* k should increase with breaking waves). Currently our algorithm poorly constrains this process, however, some breaking wave roughness is present in σ° at wind speeds above ~ 7 m/s [*Chaperon et al.*, 1995]. The option of creating a hybrid algorithm using data from other sensors is being explored with the expectation that white capping (breaking waves) can be parameterized with a combination of passive optical and microwave sensors. This hybrid algorithm should be particularly useful during high wind events [*Wanninkhof and McGillis*, 1999].

Of the future work needed, the following seems to us to be tantamount: a concerted effort to tie ground observations to satellite observations (both altimetry and scatterometry) and improvement of space-time coverage. For coordinated ground-satellite observations properly instrumented observation towers, located either near mid-ocean island chains or coastal environments, may obviate many of the problems mentioned above in our discussion of vicarious *in situ* calibration. Although multiple, dual-frequency altimeters in the same orbit (TOPEX and Jason-1) will help improve the temporal coverage, the spatial coverage will still be limited by the exact repeat orbit. Extension of this technique to wider scanning instruments, such as SeaWinds, must be undertaken.

Finally, at cycle 236 (Feb. 1999), the TOPEX team switched to the alternate ("B-side") altimeter and at this point our time series must stop. Until we can obtain a longer time series and *in situ* field data to calibrate our algorithm, the backscatter returns from the "B-side" are not verifiable with respect to gas transfer velocity.

Acknowledgments. We would like to thank the Jet Propulsion Laboratory for making the TOPEX/Poseidon MGDR data available to us via CD-ROM. We extend our thanks to the National Snow and Ice Data Center for making the SSM/I ice data available to us. We also thank Scott Doney for his comments on this manuscript and access to the NCEP reanalysis project wind velocities. This work has been supported by NASA grant no. NAGW-2431 and JPL contract no. 961425. (Woods Hole Oceanographic Institution contribution no. 10338).

REFERENCES

Benada, J.R., *Merged GDR (TOPEX/Poseidon) Generation B User's Handbook, Version 2.0 D-11007*, 124 pp., Physical Oceanography Distributed Active Archive Center (PO.DAAC), Jet Propulsion Laboratory, California Institute of Technology, 1997.

Bock, E.J., T. Hara, N.M. Frew, and W.R. McGillis, Relationship between air-sea gas transfer and short wind waves, *J. Geophys. Res.*, *104*(C11), 25821–25831, 1999.

Chapron, B., K. Katsaros, T. Elfouhaily, and D. Vandemark, A note on relationships between sea surface roughness and altimeter backscatter, in *Air-Water Gas Transfer, Selected Papers from the Third International Symposium on Air-Water Gas Transfer July 24–27, 1995, Heidelberg University*, edited by B. Jähne and E.C. Monahan, pp 869–878, AEON Verlag & Studio, Hanau, 1995.

Cox, C. and W. Munk, Statistics of the sea surface derived from sun glitter, *J. Mar. Res.*, *13*(2), 198–227, 1954.

Erickson III, D.J., A stability dependent theory for air-sea gas exchange, *J. Geophys. Res.*, *98*(C5), 8471-8488 1993.

Frew, N.M., E.J. Bock, R.K. Nelson, W.R. McGillis, J.B. Edson, and T. Hara, Spatial variations in surface microlayer surfactants and their role in modulating air-sea exchange, in *13th Conference on Boundary Layers and Turbulence*, 10–15 January 1999, pp. 421–424, American Meteorological Society, Boston, MA, 1999.

Frew, N.M., D.M. Glover, E.J. Bock, S.J. McCue, W.R. McGillis, and R. J. Healy, Estimation of global CO_2 transfer velocity fields using TOPEX altimeter backscatter, *In preparation*.

Gruber, N., C.D. Keeling, and T.F. Stocker, Carbon-13 constraints on the seasonal inorganic carbon budget at the BATS site in the northwestern Sargasso Sea, *Deep-Sea Res. I*, 673–717, 1998.

Hara, T., E.J. Bock, N.M. Frew, and W.R. McGillis, Relationship between air-sea gas transfer velocity and surface roughness, in: *Air-Water Gas Transfer, Selected Papers from the Third International Symposium on Air-*

Water Gas Transfer July 24–27, 1995, Heidelberg University, edited by B. Jähne and E.C. Monahan, pp 611–616, AEON Verlag & Studio, Hanau, 1995.

Jackson, F.C., W.T. Walton, D.E. Hines, B.A. Walter, and C.Y. Peng, Sea surface mean square slope from K_u-band backscatter date, *J. Geophys. Res.*, 97(C7), 11,411–11,427, 1992.

Jähne, B., K.O. Münnich, R. Bösinger, A. Dutzi, W. Huber and P. Libner, On the parameters influencing air-water gas exchange. *J. Geophys. Res.*, 92(C2), 1937–1949, 1987.

Kalnay, E., M. Kanamitsu, R. Kistler, W. Collins, D. Deaven, L. Gandin, M. Iredell, S. Saha, G. White, J. Woollen, Y. Zhu, M. Chelliah, W. Ebisuzaki, W. Higgins, J. Janowiak, K.C. Mo, C. Ropelewski, A. Leetma, R. Reynolds, and R. Jenne, The NCEP/NCAR 40-year reanalysis project. *Bull. Amer. Meteor. Soc.*, 77, 437-471, 1996.

Liss, P. and L. Merlivat, Air-sea gas exchange rates: Introduction and synthesis, in *The Role of Air-Sea Exchange in Geochemical Cycling*, edited by P. Buat-Ménard, pp. 113–127, NATO Advanced Science Institutes Series C, D. Reidel, Boston, MA, 1986.

Peng, T.-H., W.S. Broecker, G.G. Mathieu, and Y.-H. Li, Radon evasion rates in the Atlantic and Pacific Oceans as determined during the GEOSECS program, *J. Geophys. Res.*, 84, 2471–2486, 1979.

Smethie, W.M. Jr., T. Takahshi, D. Chipman, and J.R. Ledwell, Gas exchange and CO_2 flux in the Tropical Atlantic Ocean determined from ^{222}Rn and pCO_2 measurements. *J. Geophys. Res.*, 90, 7005–7022, 1985.

Tournadre, J. and J.C. Morland, The effect of rain on TOPEX Poseidon altimeter data: A new rain flag based on Ku and C band backscatter coefficients, *IEEE Trans. Geosci. Remote Sens.*, 35, 1117–1135, 1997.

Wallace, D.W.R., Monitoring global ocean carbon inventories, *OOSDP Background Report Number 5*, 54 pp, Ocean Observing System Development Panel, Texas A&M University, College Station, TX, 1995.

Wanninkhof, R., Relationship between wind speed and gas exchange over the ocean, *J. Geophys. Res.*, 97(C5), 7373–7382, 1992.

Wanninkhof, R. and W.R. McGillis, A cubic relationship between air-sea CO_2 exchange and wind speed, *Geophys. Res. Lett.*, 26(13), 1889–1892, 1999.

E. Bock, Interdisciplinary Center for Scientific Computing University of Heidelberg, Im Neuenheimer Feld 368, 69120 Heidelberg, Germany. (e-mail: erik@iwr.uni-heidelberg.de)

N. Frew, D. Glover, and S. McCue, Dept. Marine Chemistry and Geochemistry, Mail Stop 25, Woods Hole Oceanographic Institution, Woods Hole, MA 02543. (e-mail: nfrew@whoi.edu; dglover@whoi.edu; smccue@whoi.edu)

A Global, High Resolution, Satellite-Based Model of Air-Sea Isoprene Flux

David J. Erickson III and Jose L. Hernandez

Computer Science and Mathematics Division, Oak Ridge National Laboratory, Oak Ridge, Tennessee

A procedure is described where satellite data from different sensors are merged to compute global air-sea isoprene flux estimates. Observational relationships based on cruise data are used to constrain a global satellite based model of ocean to atmosphere isoprene flux. The strong relationship between surface ocean isoprene concentration and chlorophyll concentration is used to estimate the surface ocean concentration of isoprene on a monthly basis at $2° \times 2.5°$ resolution. Monthly mean NASA SeaWiFS chlorophyll estimates are used to drive the isoprene concentration distributions. The global computed range of isoprene in the surface ocean is 1 – 100 pmol l^{-1}. 4-D assimilated surface meteorological variables from the Data Assimilation Office (DAO) at NASA/GSFC are used to compute the global isoprene transfer velocity field. The range in ocean to atmosphere flux is 0.1-200 ug C m^{-2} d^{-1}. The global integrated flux of isoprene from the ocean to the atmosphere is 0.085 Tg C yr^{-1} with an error estimate of at least 100%. This estimate is a factor of 3 - 10 lower than previous estimates, most likely due to an under representation of the oceanic gyre regions in previous global extrapolations. This procedure will be used in the future when co-located in time and space SeaWiFs data and DAO assimilated meteorological fields are available. Since the atmospheric residence time of isoprene is on the order of hours, the ocean source of isoprene is likely to be critical in determining marine boundary layer O_3, OH and general oxidizing capacity in remote marine regions.

1. INTRODUCTION

The flux of isoprene from the ocean to the atmosphere has been studied experimentally for the past 35 years (Wilson et al., 1970; Swinnerton and Lamontagne, 1974; Donahue and Prinn, 1990, 1993; Broadgate, 1995). Surface ocean waters are usually supersaturated with respect to the overlying atmosphere, leading to a net flux from the ocean into the marine boundary layer. The globally integrated ocean source of isoprene to the atmosphere is several orders of magnitude less than continental sources. However, since the atmospheric residence time of isoprene is on the order of hours, the continental source of isoprene makes an impact on atmospheric chemistry mainly over continents and near-shore coastal regions. This means that the ocean source of isoprene is the dominant isoprene source over much of the Earth's marine regions.

Here, we construct an estimate of the ocean source of isoprene to the atmosphere using remotely sensed ocean color data from SeaWiFS and 4-D assimilated surface meteorological variables. The SeaWiFS ocean color data is coupled with the observed relationship presented by Broadgate et al. (1997) between surface ocean chlorophyll concentration and surface ocean isoprene concentration. Broadgate et al. (1987) obtained water samples from the North Sea and the Southern Ocean and found a strong correlation between chlorophyll and isoprene concentration. This relationship was found to be robust in both hemispheres. This relationship may stem from the possibility that increased chlorophyll is also associated with increased bacterial production of isoprene (Kuzma et al., 1995) and isoprene production has also been seen in phytoplankton cultures (Moore et al., 1994).

The observed close correlation between oceanic chlorophyll concentration and isoprene concentration may stem from the degradation products of the phytol side chain of the chlorophyll molecule. This is basically a hydrophobic terpenoid side chain esterified to a propionic acid substitute in ring IV of the chlorophyll molecule. This chain has the potential to degrade into 5-carbon molecules. The chlorophyll molecule also has a fused cyclopentanone ring along with four pyrrole rings. Each of these rings has unsaturated carbon bonds and the structure of isoprene could conceivably result from the breakdown of these rings. In summary, the chlorophyll molecule is a probable direct precursor to isoprene in the surface ocean and results in a much tighter relationship between these two molecules than for other compounds such as DMS, CO, CO_2 and OCS. The observed relationships allow several estimates of the global surface ocean distribution of isoprene to be computed. Here, the main relationship derived by Broadgate et al. (1997) is used, and ongoing work is addressing the refinements of this relationship and/or other relationships. These calculations are compared to existing observational data sets. Global transfer velocity estimates are created using NASA/GSFC Data Assimilation Office (DAO) 4-D assimilated data products. The meteorological data are from 1990. Once the global surface ocean isoprene concentration and global transfer velocity fields are computed, a global, monthly map of isoprene flux from ocean to atmosphere is prepared. Ongoing work includes implementing this ocean source of isoprene into a global, chemical transport model (CTM) of the atmosphere so as to quantify the impact of ocean isoprene fluxes on lower atmosphere chemistry.

2. MODEL INPUT AND SATELLITE DATA DESCRIPTION

The SeaWiFS Transfer Radiometer (SXR) instrument was built for the Sea-viewing Wide Field-of-view Sensor (SeaWiFS). This instrument has a unique optical design that allows 6 independent optical paths. It was launched in September of 1997 and there have been several post-launch 'ground truthing' exercises to examine the validity of the data. A description of the results of these activities and ongoing and planed analyses are located on the SeaWiFS Project homepage (http://seawifs.gsfc.nasa.gov/SEAWIFS.html). In this work, the monthly mean surface ocean chlorophyll concentration in ug l^{-1}, for 1998, was obtained and re-gridded on the conventional DAO grid system of 2° latitude by 2.5° longitude.

The procedure described here is a precursor to a calculation where the DAO meteorological data and SeaWiFS ocean color data are co-located in time and space. Since the SeaWifs sensor was launched in 1997 and the DAO assimilated GEOS-1 data products are only available until 1995; this work will outline the procedure that will be used over the next 5 years or so. So as to create a global ocean to atmosphere isoprene flux calculation, the global surface wind speed and sea surface temperature were obtained for 1990. In this study, we used monthly mean values for 1990 and in the future we will examine the influence of using higher frequency (i.e. 6–12 hour) meteorological variables in estimating trace gas flux. We also note that Wanninkhof and McGillis (1999) provide different formulations of the wind speed dependence of the transfer velocity for use with wind fields of differing temporal averaging characteristics. The data obtained are from the Data Assimilation Office (DAO) and a 4-D assimilated data. These data are satellite based observations that are fed into a global model of atmospheric motions and processes. This results in a comprehensive and dynamically consistent dataset, which is specific for the period of time that the satellites were obtaining data. A detailed description of the assimilation methods and a discussion of the positive and negative attributes of the derived products are available at the DAO homepage (http://dao.gsfc.nasa.gov).

3. MODEL CALCULATIONS

The flux of isoprene from the ocean to the atmosphere is computed via equation 1:

$$F_{isoprene} = k_w [C_{so}] \quad (1)$$

Where k_w is transfer velocity and $[C_{so}]$ is surface ocean isoprene concentration. We present methods to compute k_w in units of cm hr^{-1} and $[C_{so}]$ in units of pmol l^{-1}. In following sections $F_{isoprene}$ has units of ug m^{-2} d^{-1}.

Broadgate et al. (1997) found a very strong relationship between surface ocean isoprene concentration and simultaneously measured chlorophyll concentration for both North Sea and southern ocean experiments. The relationship used is $[C_{so}] = 6.43$ [chlorophyl] + 1.2. Note that the y intercept is critical for very low chlorophyll concentrations since these are the conditions that exist in the large gyre regions that make up a large proportion of oceanic regions. Several other relationships are being examined (W. Broadgate, personal communication), including some that take take surface radiation fluxes into account. Here, the equation above is the only published relationship available, and ongoing and future research will consider a range of relationships. The error estimate on this first global calculation is at least 100%.

The transfer velocity calculations were done via equation 2 (Wanninkhof, 1992):

$$K_w = 0.31(U^2)(Sc/660)^{-1/2} \quad (2)$$

Where K_w is transfer velocity (cm hr^{-1}), U is wind speed (m s^{-1}), and Sc is Schmidt number computed as equation 3:

$$Sc = 335.6 \, (MW)^{1/2}(1-0.066T+0.002043T^2-0.000026T^3) \quad (3)$$

T is sea surface temperature (°C) and MW is molecular weight, which for isoprene is 68 g mol^{-1}. The K_w values were computed for each month on the DAO grid system (2° latitude by 2.5° longitude).

4. GLOBAL SURFACE OCEAN ISOPRENE CONCENTRATION DISTRIBUTIONS

Using the observation based relationship of Broadgate et al., 1997, the global surface ocean isoprene concentration was computed and the results for four 3 month averages are presented in Plate 1a-d. The gyre regions are characterized by isoprene values of 0.2-5 pmol l^{-1}. Maximum values occur in the coastal and mid-latitude regions of the oceans with values ranging from 5-80 pmol l^{-1}. Overall, the concentrations reflect the high chlorophyll coastal regions and mid-high latitude biologically active regions in summer months. The North Sea is characterized by values between 20 and 80 pmol l^{-1}, in close agreement with the measurements of Broadgate et al. (1997). Note that the observations in Broadgate et al. (1997) were from 1992-1994, and the SeaWifs data presented here is for 1998, so that interannual variability may play some role in any differences between the model and measurements. Note the prominent large surface area of the gyre regions with isoprene concentrations of 1-3 pmol l^{-1}. These regions play a large role in the global integration of the fluxes. An important aspect of this calculation is the spatially heterogeneous character of the predicted isoprene field. This is critical when estimating the impact of oceanic isoprene concentrations on marine boundary layer chemistry.

The most difficult aspect of this modeling exercise is to assign error estimates on the computed values of surface ocean isoprene concentration. The range of surface ocean isoprene observations in the Broadgate et al. (1997) work was 0.09 pmol l^{-1} to 100 pmol l^{-1}. In the Bonsang et al. (1992) work the range was 2 pmol l^{-1} - 110 pmol l^{-1}. It is noted that the higher values in both works were in regions of high productivity and summer season. The computed values in Plate 1 vary from ~1 pmol l^{-1} to roughly 50 pmol l^{-1}. The spatial distributions are consistent with the observational data and with the seasonality observed. To be sure, the paucity of additional observations to validate and evaluate the calculations here makes an assignment of an error estimate very difficult. On the other hand, it is unlikely that a global high-resolution data base of surface ocean isoprene measurements will become available over the next many decades. So, the use of satellite data to estimate the concentrations and fluxes of many trace gases, including isoprene, is the main point of this research. Isoprene was selected due to its surface ocean concentrations being strongly correlated with chlorophyll, a quantity that is available from remote sensing platforms. The error in the transfer velocity estimates is at least a factor of 2. The error in the surface ocean isoprene concentration is also at least a factor of 2. Therefore, the error in the regional and global isoprene fluxes estimates is at least a factor of 5.

5. GLOBAL TRANSFER VELOCITY DISTRIBUTIONS

The global transfer velocity is computed for each month. Maximum values range from 20-55 cm hr^{-1} in January and reflect the mean surface turbulence as assimilated by the GEOS-1 satellite system. The trade winds are much more prominent in these calculations than in previous General Circulation Model (GCM) based global transfer velocity estimates (Erickson, 1989, Erickson, 1993). We also note the strong signature of the monsoonal winds in July. Overall, the general features of the transfer velocity field are in agreement with other global estimates of this quantity. Due to length constraints, we do not include figures of this quantity.

6. GLOBAL OCEAN TO ATMOSPHERE ISOPRENE FLUX

The global flux of isoprene to the atmosphere varies from 0.1-2 ug m^{-2} d^{-2} in gyre regions up to 2-30 ug m^{-2} d^{-1} in coastal and high wind, high chlorophyll regions. The values integrate to 0.085 Tg C yr^{-1}. These values should be

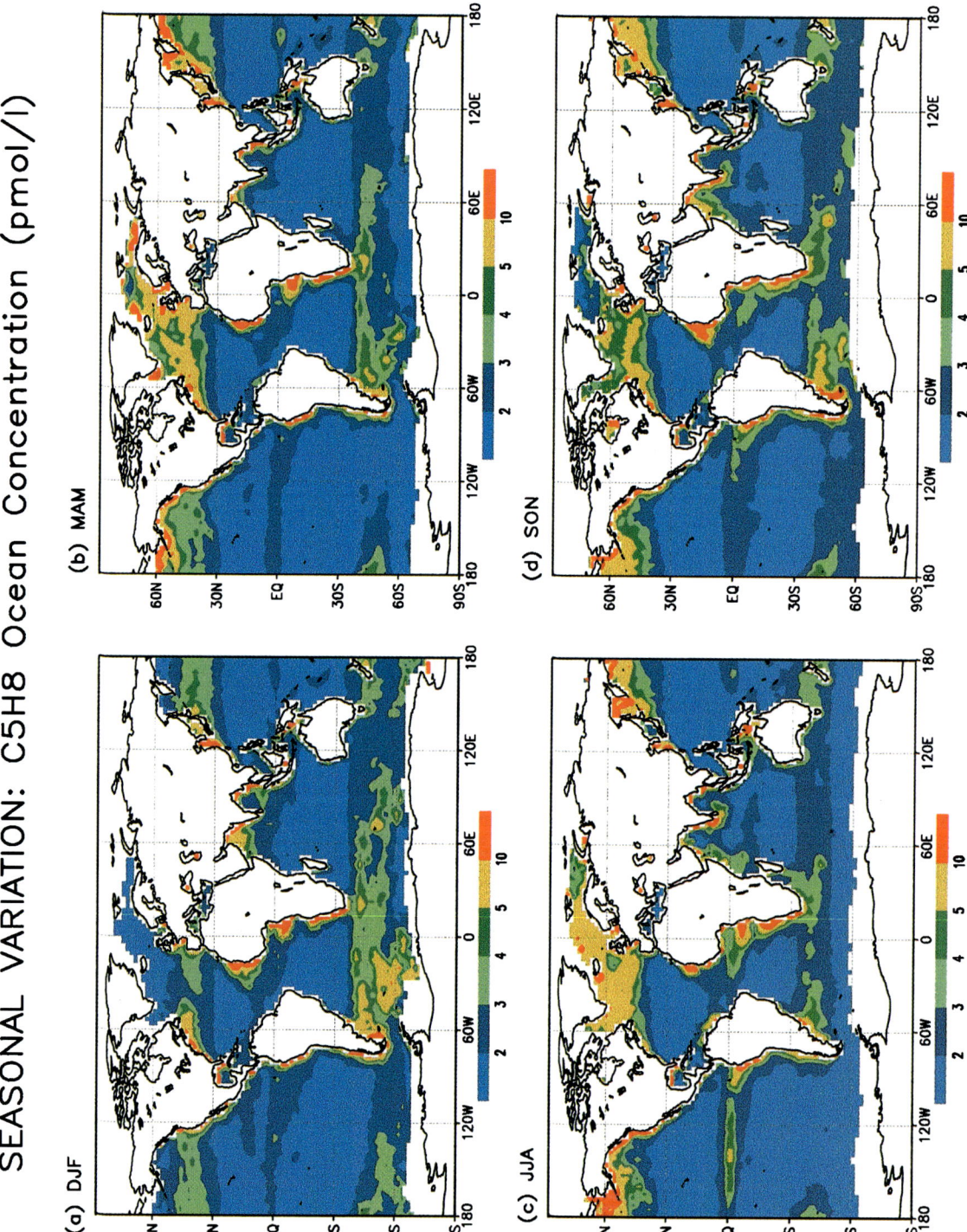

Plate 1 a,b,c,d. Global surface ocean isoprene concentration computed for the 4 seasons. The units are pmol l^{-1}. Note the spatial heterogeneity of the isoprene concentrations. There is a clear seasonality in the mid-high latitudes of each hemisphere.

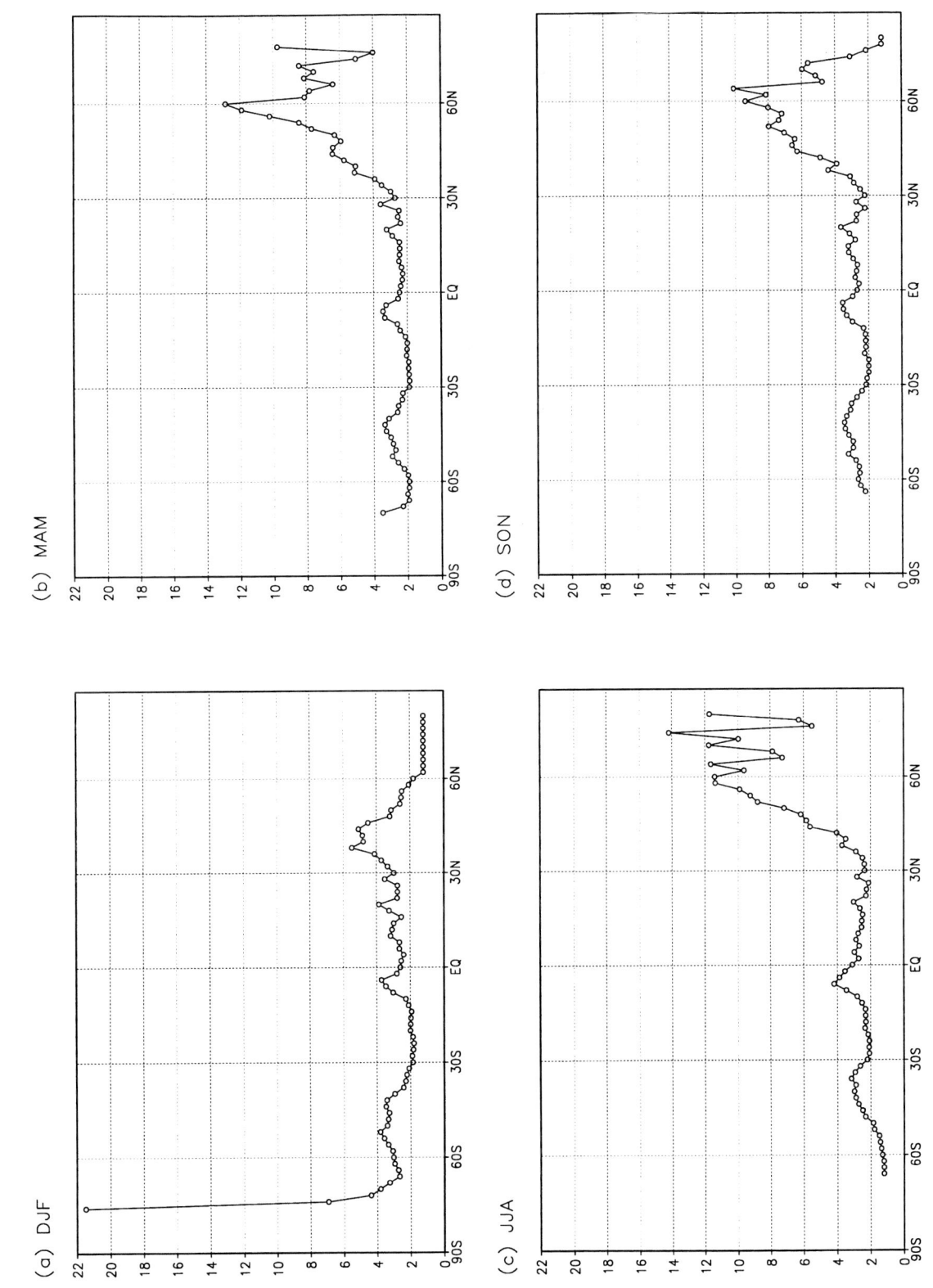

Figures 1 a,b,c,d. The zonal average of the global plots (Plate 1)

338 GLOBAL HIGH RESOLUTION MODEL OF AIR-SEA ISOPRENE FLUX

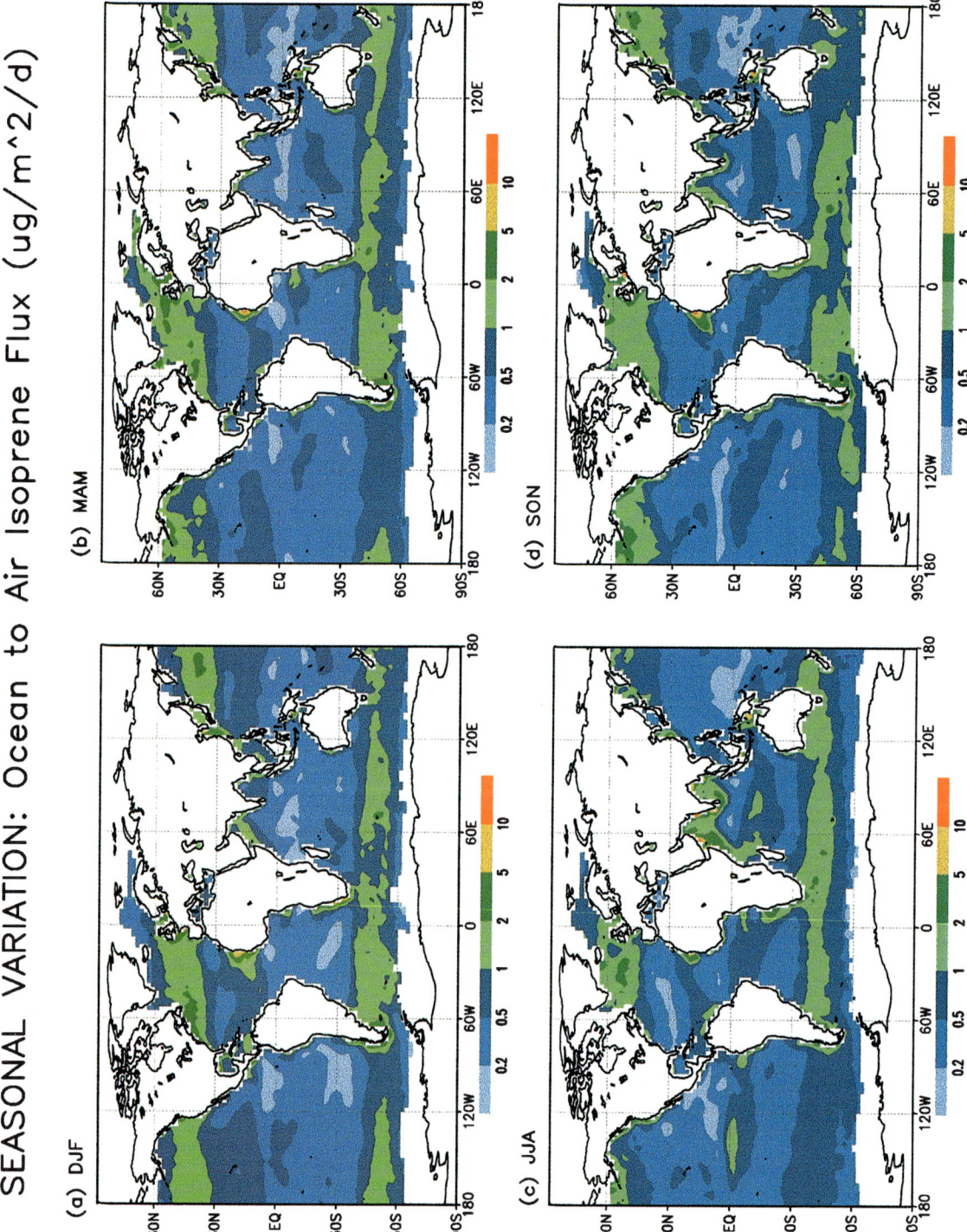

Plate 2 a,b,c,d. Global ocean to atmosphere flux of isoprene. Units of ug m-2 day-1. The spatial structure is related to the distributions of surface ocean isoprene and the structure of the global transfer velocity field.

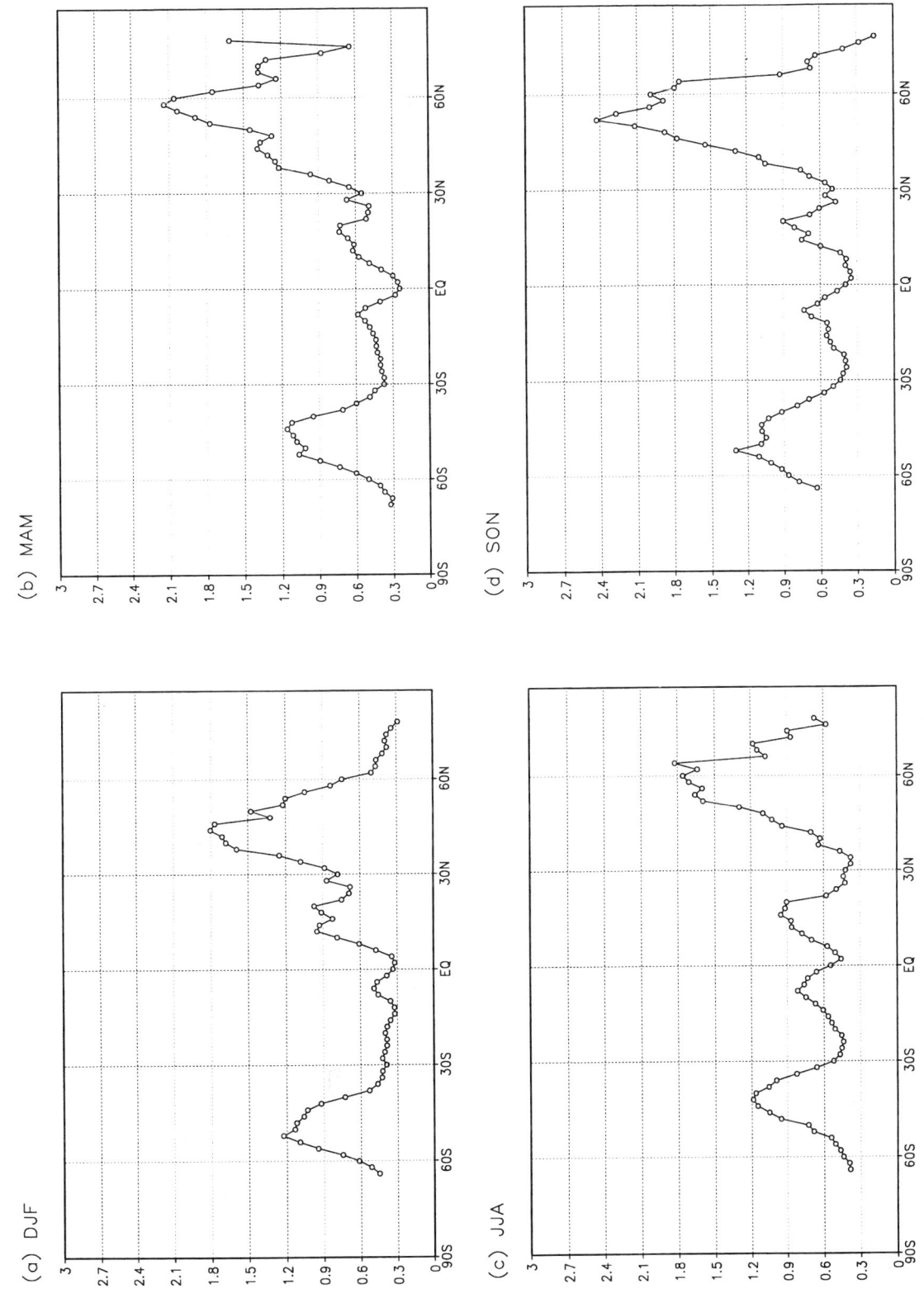

Figures 2 a,b,c,d. The zonal average of the global plots (Plate 2)

Table 1

Global ocean to atmosphere flux of isoprene (C_5H_8) as a function of month

Month	Flux g (x10^{10}) C month^{-1}
January	0.71
February	0.68
March	0.69
April	0.67
May	0.68
June	0.71
July	0.70
August	0.75
September	0.72
October	0.76
November	0.75
December	0.73
Total annual flux	0.085 Tg C yr^{-1}

considered a lower limit due to the gyre regions having very low surface ocean isoprene values and they may be low. Also, these calculations are only as good as the 1998 SeaWiFS data, and there may be some evidence that the SeaWiFS data may be low in gyre regions. The reader is referred to the SeaWiFS web page above to assess this possibility. It is anticipated that other relationships between ocean chlorophyll and isoprene may result in higher global integrals. Since the goal of this work is to create a global ocean source of isoprene for use in Chemical Transport Models (CTM's), it should be noted that there is a significant error estimate associated with this flux field, as discussed above. The main point is that the flux values range over 3 orders of magnitude and the spatial distribution of these fluxes will be critical in determining the impact of oceanic isoprene on atmospheric chemistry.

7. CONCLUSIONS

The computational framework and germane data sets have been merged to create global satellite based air sea trace gas fluxes, using isoprene as an example here. The fluxes obtained integrate to 0.085 Tg C yr^{-1} as isoprene, with an error estimate of at least 100%. This result may be much lower than the actual flux, and more observational data analysis is required to decrease the estimated error. The flux values show a 3 order of magnitude range depending on season and location. Table 1 shows the monthly variability in the global fluxes, showing a clear signal of the southern hemisphere summer ocean source. The seasonal variability at specific locations is consistent with that seen by Broadgate et al. (1997). The goal of ongoing work is to refine these flux estimates and identify when and where the ocean source of isoprene is influencing atmospheric chemistry. Several different relationships between surface ocean isoprene, and other non-methane hydrocarbons, and chlorophyll concentrations are being investigated. In addition, there may be a relationship with the surface radiation field that impacts isoprene production that should be taken into account.

There are several areas where these calculations can be improved. The transfer velocity estimates, as for all trace gases, need to be better evaluated. Using satellite measurements to directly estimate transfer velocities, instead of a parameterization, may better constrain this term in the flux equation. As the quantitative quality of the SeaWiFS data is further validated, the estimation of surface ocean trace gas concentrations will improve. The flux calculations presented here could be implemented in an atmospheric chemistry Chemical Transport Model (CTM). This would allow the quantitative assessment of the impact of the ocean source of isoprene on atmospheric boundary layer reactive chemistry. The expectation is that in oceanic regions far from continental sources the impact of the ocean source of isoprene, and other non-methane hydrocarbons, would be appreciable. The isoprene ocean source described here is available from the authors via ftp.

Acknowledgments. DJE and JLH would like to acknowledge the Laboratory for Atmospheres at NASA Goddard Space Flight Center for support through USRA. We would like to thank S. Steenrod for assistance in the flux calculations, M. Bosilovich for help in obtaining the DAO data for 1990 and W. Gregg for supplying the SeaWifs data.

REFERENCES

Bonsang, B., M. Kanakidou, G. Lambert and P. Monfray, The marine source of C2-C6 aliphatic hydrocarbons, *J. Atmos. Chem.*, 6, 3-20, 1988.

Bonsang, B., C. Polle and G. Lambert, Evidence for the marine production of isoprene, *Geophys. Res. Lett.*, 19, 1129-1132, 1992.

Broadgate, W. J., Non-methane hydrocarbons in the marine environment, PhD Thesis, 229 pp., University of East Anglia, Norwich, UK, 1995.

Broadgate, W.J, P. S. Liss and S. A. Penkett, Seasonal emissions of isoprene and other reactive hydrocarbon gases from the ocean, *Geophys. Res. Lett.*, 24, 2675-2678, 1997.

Donahue, N. M. and R. G. Prinn, Non-methane hydrocarbons in the remote marine boundary layer, *J. Geophys. Res.*, 95, 18387-18411, 1990.

Donahue, N. M. and R. G. Prinn, In Situ non-methane hydrocarbon measurements on SAGA 3, *J. Geophys. Res.*, 98, 16915-16932, 1993.

Erickson, D. J. III, Variations in the global air-sea transfer velocity fields of CO_2, *Global Biogeochem. Cycles*, 3, 37-41, 1989.

Erickson, D. J. III, A stability dependent theory for air-sea gas exchange, *J. Geophys. Res.*, 98, 8471-8488, 1993.

Kuzma, J., M. Nemeck-Marshall, W. H. Pollock and R. Fall, Bacteria produce the volatile hydrocarbon isoprene, *Current Microbiol.*, 30, 97-103, 1995.

Milne, P. J., D. D. Riemer, R. G. Zika and L. E. Brand, Measurement of vertical distribution of isoprene in seawater, its chemical fate, and its emission from several phytoplankton monocultures, *Mar. Chem.*, 48, 237-244, 1995.

Moore, R. M., D. E. Oram and S. A. Penkett, Production of isoprene by marine phytoplankton cultures, *Geophys. Res. Lett.*, 21, 2507-2510, 1994.

Swinnerton, J. W. and R. A. Lamontagne, Oceanic distributions of low molecular weight hydrocarbons, *Environ. Sci. Tech.*, 8, 657-663, 1974.

Wanninkhof, R., Relationship between wind speed and gas exchange over the ocean, *J. Geophys. Res.*, 97, 7373-7382, 1992.

Wanninkhof, R. and W. R. McGillis, A cubic relationship between air-sea CO2 exchange and wind speed, Geophys. Res. Lett., 26, 1889-1892, 1999.

Wilson, D. F., J. W. Swinnerton and R. A. Lamontagne, Production of carbon monoxide and gaseous hydrocarbons in seawater: Relation to dissolved organic carbon, *Science*, 168, 1577-1579, 1970.

Dr. David J. Erickson III, Computational Sciences Section, Computer Science and Mathematics Division, Oak Ridge National Laboratory, P. O. Box 2008, Bldg. 6012, Oak Ridge, TN 37831-6367

Jose L. Hernandez, Computational Sciences Section, Computer Science and Mathematics Division, Oak Ridge National Laboratory, P. O. Box 2008, Bldg. 6012, Oak Ridge, TN 37831-6367

Daily Surface Wind Fields Produced by Merged Satellite Data

Abderrahim Bentamy[1], Kristina B. Katsaros[2], William M. Drennan[3], and Evan B. Forde[2]

Scatterometers and radiometers on several polar-orbiting satellites routinely produce oceanic surface wind field data. For this study, we merged the NASA scatterometer (NSCAT) data with scatterometer data from the European Remote Sensing (ERS) satellite 2, and the wind speeds from two of the Special Sensor Microwave/Imagers (SSM/I) and produced daily 1° latitude by 1° longitude gridded wind fields over the global ocean for September 1996 through June 1997. This time period coincides with the lifetime of the NSCAT aboard ADEOS-1. We created these wind fields by utilizing the Kriging technique with its associated variograms, which consider both space and time wind vector structures. The resulting daily wind fields, when compared with moored-buoy wind speed and direction measurements, resulted in a root-mean-square (rms) difference of less than 1.5 m/s. No significant difference was found between statistical parameters estimated over the equatorial zone and middle latitudes. To investigate the global patterns of these new satellite wind fields, comparisons with the National Environmental Prediction Center's (NCEP) re-analysis products have been carried out. The satellite data and the NCEP products have a similar statistical error structure, but the merged wind fields provide complete coverage at much higher spatial resolution. Accurate surface wind speed estimates are an important factor in determining the velocity and magnitude of air-sea gas exchange.

1. INTRODUCTION

This paper studies and analyzes the global gridded wind field calculations derived from satellite wind observations. Three instruments that provide estimates of surface winds over the global oceans at different scales are used: the scatterometer mounted on the European Remote Sensing (ERS) satellite; the NASA scatterometer (NSCAT) on board the Advanced Earth Observing Satellite (ADEOS-1); and the Special Sensor Microwave/Imager (SSM/I) deployed on board the Defense Meteorological Satellite Program (DMSP) satellites F10, F11, F13, and F14. These satellites are in sun-synchronous orbits at about 800-km height with 98° inclination.

Gridded surface wind parameter estimates, including wind vectors, stresses, curl, and divergence, have been made available since August 1991 from ERS-1 and ERS-2 scatterometers, and between September 1996 and June 1997 from NSCAT. These gridded wind fields have been used extensively in global wind studies [*Bentamy et al.*, 1998] and in ocean model forcing [*Grima et al.*, 1999]. Such studies, performed for the tropical Pacific area, indicate that the scatterometer gridded wind fields compare

---

[1]Institut Français de Recherche Pour l'Exploitation de la Mer, Plouzané, France
[2]NOAA/Atlantic Oceanographic and Meteorological Laboratory, Miami, Florida
[3]University of Miami/Rosenstiel School of Marine and Atmospheric Science, Miami, Florida

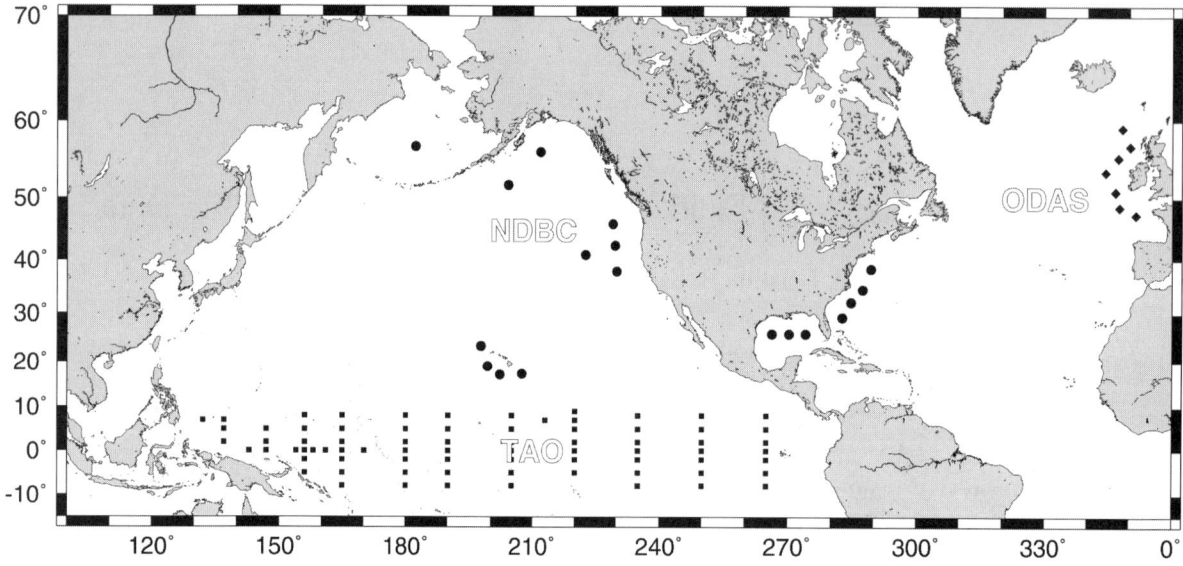

Figure 1. Locations of NDBC (circle), TAO (square), and ODAS (diamond) buoys.

quite well with averaged wind speed and direction estimated from buoy measurements. For instance, the root-mean-square (rms) errors of wind parameters indicate that the statistical differences between weekly buoy and ERS-1 scatterometer wind estimates are about 0.90 m/s for wind speed, 1.50 m/s for the zonal wind component, and 1.60 m/s for the meridional component. Furthermore, the results obtained from experiments forcing an ocean model with gridded scatterometer wind fields show that the correlations estimated from the thermocline ERS-1 simulation and from buoy measurements are statistically significant, and their mean value is greater than 0.70. *Boutin et al.* [1999] verify that satellite derived wind observations can provide quality estimates of the global ocean wind fields, especially as they relate to air-sea gas exchange studies.

This research investigates the improvement of the temporal resolution and the accuracy of gridded wind fields calculated by merging ERS-2, NSCAT, and SSM/I wind observations. Since it is well known that surface wind speed is an important variable for estimating gas transfer velocity [*Wanninkhof*, 1992], this work may also help improve regional and global estimates of air-sea gas fluxes.

This paper is organized into two parts. Section 2 deals with data used in this study and briefly describes the objective method used for global wind analysis. The effects of sampling errors due to the polar satellite-sampling scheme is investigated and results are provided. Some examples of resultant wind fields are shown. Section 3 deals with the estimation of the accuracy of scatterometer wind fields. An objective evaluation of the study's success will be a comparison of our satellite estimates to surface winds derived from moored buoy measurements.

2. DATA AND METHODS

2.1 Satellite Wind Observations

The daily wind fields are calculated from ERS-2 and NSCAT scatterometers and SSM/I wind observations. The calculation is performed during the NSCAT period September 16, 1996 to June 29, 1997. The scatterometers provide near-surface wind vectors (wind speed and direction at 10-m height) over the global oceans with a spatial resolution of 50×50 and 25×25 km^2, respectively, over a swath of 500-km width for ERS-2, and two swaths of 600 km each for NSCAT. SSM/I allows the estimation of surface wind speed with a spatial resolution of 25 km over a swath of 1400-km width. The number of ERS-2, NSCAT, and SSM/I orbits are about 14.3 per day each. They cover the global oceans within two to three days. Furthermore, the ERS scatterometer makes about 79,500 wind observations per day, NSCAT makes 190,000, while the SSM/I makes about 420,000.

The accuracy of ERS and NSCAT retrieval wind speeds and direction and SSM/I wind speeds were determined through a comparison with buoy wind measurements [*Quilfen et al.*, 1994; *Graber et al.*, 1996; *Graber et al.*, 1997; *Bentamy et al.*, 1999]. Three buoy networks were used to estimate the quality of the retrieved scatterometer wind vectors (Figure 1): the National Data Buoy Center (NDBC) buoys off the U.S. Atlantic, Pacific, and Gulf

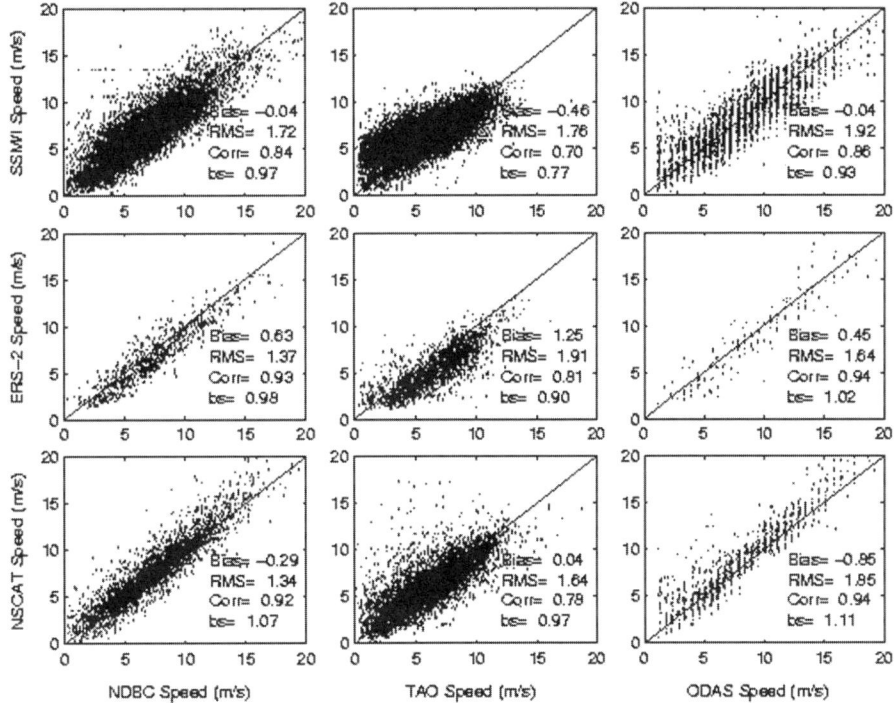

Figure 2. Scatterplots of collocated wind speeds derived from buoy measurements (NDBC, TAO, and ODAS) and from scatterometer observations (SSM/I, ERS-2, and NSCAT).

coasts maintained by the National Oceanic and Atmospheric Administration (NOAA); the Tropical Atmosphere Ocean (TAO) buoys located in the tropical Pacific Ocean and maintained by NOAA's Pacific Marine Environmental Laboratory (PMEL); and the European buoys off European coasts called ODAS and maintained by the United Kingdom's Meteorological Office (UKMet) and Meteo-France. The calculation of buoy wind speed at 10-m height in neutral conditions is performed using the Liu-Katsaros-Businger (LKB) model [*Liu et al.*, 1979]. For the three networks, only hourly buoy wind speed and direction estimates are used in the scatterometer/buoy wind comparisons.

The results obtained by *Graber et al.* [1996] indicated that the ERS-1 scatterometer wind speeds are biased low compared to buoy winds. The bias values derived from ERS-1/NDBC and from ERS-1/TAO comparisons are 0.30 m/s and 1 m/s, respectively. The corresponding rms values are 1.13 m/s and 1.38 m/s. The comparisons between wind direction retrieved from the ERS-1 scatterometer and measured by buoys provided an rms error of 24° for both buoy networks. Using similar collocation procedures, *Graber et al.* [1997] showed that the difference between NDBC and NSCAT wind speeds has mean and rms values of 0.14 m/s and 1.22 m/s, respectively. For the NSCAT wind direction, the rms error is about 24°. The results inferred from NSCAT/TAO comparisons [*Caruso et al.*, 1999] indicated that for wind speed the bias is very low, and the rms difference is about 1.55 m/s, while for wind direction, the rms difference is about 20°. The results obtained from the ERS-2 scatterometer/buoy comparisons are quite similar to those obtained for ERS-1, except that the overall bias of ERS-2 scatterometer wind speed is higher [*Quilfen et al.*, 1999]. The accuracy of SSM/I wind estimates is tested by comparing with NDBC [*Goodberlet et al.*, 1989] and TAO [*Halpern*, 1993; *Bentamy et al.*, 1999] buoy wind measurements. The results indicate that for most wind ranges, the bias between SSM/I and buoy wind speed is low, but it is high for low wind speed (<3 m/s). The standard deviation of the difference is <2 m/s.

Figure 2 shows scatter plots of a comparison of SSM/I, ERS-2, and NSCAT wind speeds with buoy winds at 10 m for NDBC, TAO, and ODAS buoys. Most of the statistical parameters provided within each figure are quite similar to those obtained from previous studies cited above. However, the bias on ERS-2 and SSM/I wind speeds are significant and require correction. To enhance the statistical significance of the retrieved ERS-1/2 scatterometer wind speeds, a collocated data set between ERS-1/2 and NDBC buoy measurements was generated. All valid ERS-1/2 scatterometer measurements obtained within 1 hour and 50 km from buoy measurements during the period March 1992-November 1998 were used to derive a new version of the ERS C-band model [Bentamy et al., 1994], which minimizes the bias seen in Figure 2. The latter is used to retrieve ERS-1/2 scatterometer wind speed observations from measured backscatter coefficients. For SSM/I, the method developed by Bentamy et al. [1999] is used to improve SSM/I wind speed estimates with respect to buoy comparisons. Hence, the ERS-2, NSCAT, and SSM/I merged gridded wind fields are calculated from the ERS-2 and SSM/I corrected wind speeds. SSM/I wind speed (U_{ssmi}) is calculated over each SSM/I cell (25 km) using the following formula:

$$U_{ssmi} = f(T_B) + f(WV) \quad (1)$$

where T_B and WV indicate brightness temperature and water vapor content, respectively. The global analysis of wind direction is performed from the standard ERS-2 and NSCAT wind directions.

2.2 Objective Method

The details of the gridded wind field calculations from remotely-sensed wind observations are provided by Bentamy et al. [1996, 1999]. Briefly, the radar and radiometer wind observations are objectively analyzed using a method based on the Kriging approach [Wackernagel, 1995]. At each grid point of 1° × 1° in longitude and latitude, the estimation of the daily wind vector is obtained consistently by interpolation of wind speeds, zonal components, and meridional components. For each variable, the method determines an estimator at each grid point, X_0, using n observations at point X_i:

$$\hat{U} = \sum_{i=1}^{N} \lambda_i V(X_i), \quad \sum_{i=1}^{N} \lambda_i = 1 \quad (2)$$

where $\hat{U}$ stands for wind speed, zonal component, or the meridional component estimator; $V(X_i)$ are the scatterometer observations at X_i, and X_i represents spatial and temporal coordinates. The weights λ_i are determined as the minimum of the linear system named Kriging:

$$-\sum_{i=1}^{n} \lambda_i \Gamma(i,j) + \Gamma(j,0) - \tau_1 - \tau_2 S(j) + \lambda_j \sigma 2 = 0$$

$$\sum_{i=1}^{n} \lambda_i = 1 \qquad j = 1, n \quad (3)$$

$$\sum_{i=1}^{n} \lambda_i S(i) = S(0)$$

where Γ is the structure function providing the spatial and temporal behavior of the variable to be estimated. The estimation of Γ is obtained from the ERS-1 scatterometer wind speed, zonal component, and meridional component observations. The variable $S(i)$ represents a regionalized variable and is considered as deterministic. Its values are inferred from SSM/I corrected wind speed estimates. Over an ocean basin, scatterometer and SSM/I winds are assumed to be related by: $E(U(i)) = a_0 + b_0 S(i)$. U is the scatterometer wind variable, $E( )$ indicates the first moment, and a_0 and b_0 are constant. Furthermore, the Kriging method provides an expression for variance error, named the Kriging error.

In practice, the calculation of merged satellite gridded wind fields using the Kriging method is based on the following items:

(1) The remotely sensed wind products, including backscatter measurements, brightness temperatures, and retrieved wind vectors, extracted from the Centre ERS d'Archivage et de Traitment (CERSAT)/WiNd Field (WNF) database. Only validated data, according to standard quality controls, are used.

(2) At each ERS-1/2 scatterometer cell (50 km), a new wind speed is estimated from the three backscatter coefficients using the new C-band model function. In this study, the wind direction selected by the operational algorithm (WNF or NSCAT) is chosen. However, for low wind conditions, a comparison between each scatterometer wind direction solution and European Centre for Medium Range Weather Forecasts (ECMWF) wind direction, interpolated in space and time on the scatterometer cell, is performed. The closest scatterometer wind direction from ECMWF is selected. The zonal and meridional wind components are estimated from scatterometer wind speed and direction.

(3) At each SSM/I cell, 10-m surface wind speed is estimated in the absence of rain, using equation (1).

(4) The world oceans are divided into boxes with constant grid spacing in latitude and longitude. In this study, the spatial resolution is 1° × 1° in latitude and longitude. All available scatterometer wind observations at nearly the same time (less than 1 hour) are averaged arithmetically in each box. Similar averaging computation is done for SSM/I winds.

(5) At each grid point, the determination of a neighborhood containing the scatterometer data used to estimate wind vectors (wind speed, zonal, and meridional components) is performed. It is quite a sensitive step. Indeed, due to the highly irregular spatial and temporal arrangement and the density of the scatterometer wind observations, the determination of a local neighborhood is not straightforward. The neighborhood is determined as successive circles centered on the grid point. The radius of these circles correspond to the variogram parameters. The maximum number of observations in a grid point neighborhood

Table 1. Comparison of averaged daily wind speed and direction estimated from NDBC, TAO, and ODAS buoy measurements and from SSM/I, ERS-2, and NSCAT wind observations.

Data Set	Buoy Wind Speed Range (m/s)	Length	Wind Speed					Wind Direction	
			Bias (m/s)	Rms (m/s)	Correlation Coefficient	Slope	Intercept	Bias (deg)	Std (deg)
NDBC/Satellite	0-24	3932	-0.15	1.79	0.90	0.96	0.41	5	25
TAO/Satellite	0-20	13427	0.04	1.61	0.88	0.81	1.40	3	23
ODAS/Satellite	0-24	870	-0.89	2.56	0.82	0.90	1.71	2	30

is 1200. This data set is then sorted by time, and for each hour the closest scatterometer observations from the grid point are used for wind vector estimation.

3. SATELLITE/BUOY AVERAGED WIND COMPARISONS

For each day (September 16, 1996-June 29, 1997), mean values of buoy wind speed and zonal and meridional components are computed arithmetically. Daily means are computed for all NSCAT periods for which at least 12 hour buoy measurements are collected. For each averaging period, the closest scatterometer grid point ($1° \times 1°$) to each buoy location is selected. Therefore, a collocated data set between scatterometer-gridded wind fields (averaging objective method) and buoy-averaged winds are performed for NDBC, TAO, and ODAS buoy networks. Results are then compared using standard statistics.

Table 1 provides the main statistical parameters characterizing wind speed and direction comparisons. The wind speed correlation coefficients range from 0.82 to 0.90 and indicate a good consistency between satellite and buoy-averaged winds. The rms values of the differences of buoy-satellite wind speeds do not exceed 1.80 m/s over NDBC and TAO networks. Results derived from ODAS/satellite comparisons show a higher bias and rms values of 0.89 m/s and 2.56 m/s, respectively. The latter are mainly due to very few comparison data points, and to the high wind variability in the ODAS area (Figure 1). Furthermore, the statistics calculated by several meteorological centers (ECMWF, Commission on Marine Meteorology, UKMet) indicate that ODAS buoy wind speeds tend to be underestimated according to meteorological wind analysis (see ftp://ftp.shom.fr/meteo/qc-stats, site maintained by P. Blouch).

In the NDBC area, buoy and merged satellite wind speeds agree quite closely (slope of 0.96, with intercept 0.41) after the linear regression fit. Comparisons between buoy and satellite winds in the tropical Pacific Ocean give a regression line slope of about 0.80, suggesting an overestimation of low wind speed and underestimation of high wind speed by the merged satellite wind fields compared to TAO winds. The scatterometer wind fields are consistently high compared to the ODAS weekly-averaged wind speeds. Inspection of the calculation of the statistical parameters based on the buoy wind speed ranges (not shown) indicates that their values are increased by some outlier points at low and high wind speeds.

For wind direction, no systematic bias is found, and the overall bias and standard deviation about the mean angular difference are less than 5° and 30°, respectively. These results are consistent with the calibration/validation of the scatterometers against buoy [*Graber et al.*, 1996, 1997; *Caruso et al.*, 1999]. For instance, in the tropical Pacific area where the wind direction is quite steady, the standard deviation calculated for buoy wind speeds higher than 5 m/s does not exceed 17°.

The agreement between scatterometer and buoy-averaged wind fields was investigated through time series comparisons with the NDBC, TAO, and ODAS buoy arrays. Figure 3 shows examples of daily averaged time series of wind speed at five buoy locations, representing various latitudes, in the NDBC array. The comparisons indicate that the matchups are strongly correlated, and their geographical features compare well. For instance, the correlation coefficient is greater than 0.80 at all locations. The lowest correlation values, varying between 0.80 and 0.89, are found in the TAO array. At TAO locations 95°W-0 and 95°W-2°N, the difference between buoy and scatterometer wind speeds is consistent, and the bias reaches 1 m/s. Several error sources can explain the discrepancies between scatterometer and TAO winds. For instance, since the scatterometer measures winds relative to the moving ocean surface, if the current is in the same direction as the wind, the scatterometer winds tend to be lower than buoy estimates of surface wind, relative to the solid earth. In the ODAS array, scatterometer averaged wind speeds are consistently higher than buoy estimates. However, we notice that the bias between buoys and scatterometers tends to be large between October and December 1996. During this period, the correlation coefficient is only about 0.69, which is 22% smaller than for the whole period. The comparisons between NDBC and scatterometer averaged wind speed time series does not exhibit any systematic bias. At some locations, such as 157.8°W-17.2° (Figure 3a) and 177.7°W-57°N) (Figure

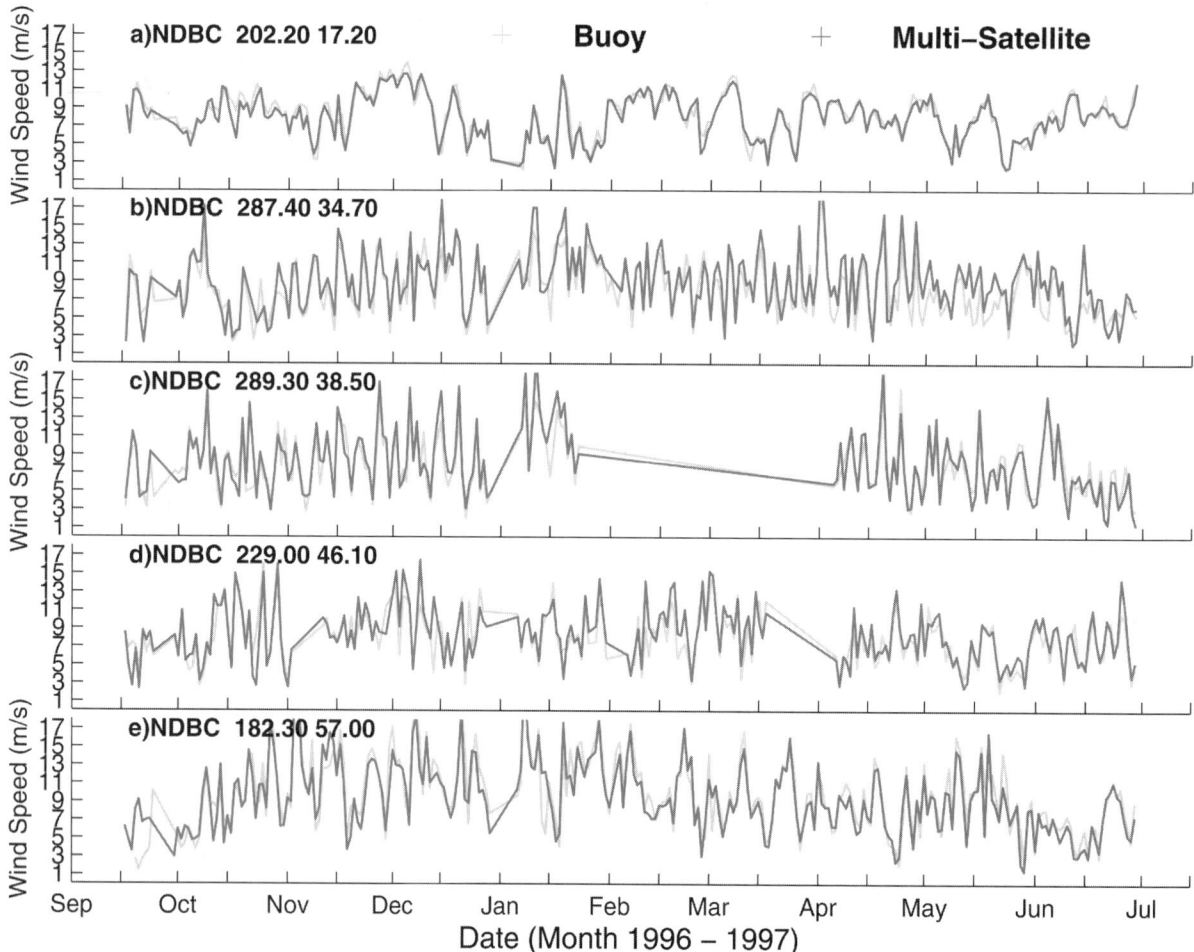

Figure 3. Time series of the daily averaged wind speed derived from merged satellite winds (red line) and from NDBC buoys (green line) at five NDBC locations.

3e), we found a seasonally-related data bias. The bias tends to be slightly positive in winter and negative in summer. This variation may be related to the dependence of the wind speed residuals on buoy wind speed ranges.

A comparison of hourly winds from the merged satellite data with predictions from the NCEP re-analysis indicated good overall agreement. However, significant regional differences are apparent, particularly in the southern oceans (not shown).

CONCLUSIONS

By merging NSCAT data with scatterometer data from the ERS-2 satellite, and the wind speeds from two SSM/Is, we have produced daily 1° latitude by 1° longitude gridded wind fields over the global ocean for September 1996 through June 1997. The overall accuracy of the merged wind fields is excellent. Biases are <0.7 m/s, rms and on the order of 2 m/s. When compared to moored buoy wind speed and direction measurements, our data have a rms difference of <1.5 m/s. No significant difference was found between statistical parameters estimated over the equatorial zone and middle latitudes. Comparisons with NCEP's re-analysis products revealed that our merged satellite data and the NCEP products have similar statistical error structure, but the merged wind fields provide complete coverage at much higher spatial resolution. With the addition of QuikSCAT in 1999 and other satellites scheduled for launch soon, satellite derived global ocean wind fields will continue to be available to produce these types of daily wind field maps in the foreseeable future. Since accurate surface wind speed estimates are an important factor in determining the velocity and magnitude of air-sea gas exchange, our research may also help improve regional and global estimates of air-sea gas fluxes.

Acknowledgments. The ERS-2 data were obtained from CERSAT, SSM/I data from Remote Sensing Systems, Inc., and NSCAT data from the Jet Propulsion Laboratory facility. The work was supported in part by NASA grant UPN261-75 to K.B. Katsaros.

REFERENCES

Bentamy, A., Y. Quilfen, P. Queffeulou, and A. Cavanie, Calibration of the ERS-1 scatterometer C-band model, *IFREMER Tech. Rep.*, DRO/OS-94-01, pp. 72, 1994.

Bentamy, A., Y. Quilfen, F. Gohin, N. Grima, M. Lenaour, and J. Servain, Determination and validation of average wind fields from ERS-1 scatterometer measurements, *Global Atmos. Ocean Syst.*, 4, 1-29, 1996.

Bentamy, A., N. Grima, and Y. Quilfen, Validation of the gridded weekly and monthly wind fields calculated from ERS-1 scatterometer wind observations, *Global Atmos. Ocean Syst.*, 6, 373-396, 1998.

Bentamy, A., P. Queffeulou, Y. Quilfen, and K.B. Katsaros, Ocean surface wind fields estimated from satellite active and passive microwave instruments, *IEEE Trans. Geos. Rem. Sens.*, 37 (5), 2469-2486, 1999.

Boutin, J., J. Etcheto, M. Rafizadeh, and D.C.E. Bakker, Comparison of NSCAT, ERS 2 active microwave instrument, Special Sensor Microwave Imager, and carbon interface ocean atmosphere buoy wind speed: Consequences for the air-sea CO_2 exchange coefficient, *J. Geophys. Res.*, 104, (C5), 11,375-11,392, 1999.

Caruso, M., S. Dickinson, K. Kelly, M. Spillane, L. Mangum, M. McPhaden, and L. Stratton, Evaluation of NSCAT scatterometer winds using equatorial Pacific buoy observations, *Tech. Rep.*, Applied Physics Lab., Univ. of Washington, Seattle, WA, pp. 60, 1999.

Goodberlet, M.A., C.T. Swift, and J.C. Wilkerson, Remote sensing of ocean surface winds with the special sensor microwave/imager, *J. Geophys. Res.*, 94, 14,547-14,555, 1989.

Graber, H.C., N. Ebuchi, and R. Vakkayil, Evaluation of ERS-1 scatterometer winds with wind and wave ocean buoy observations, *Tech. Rep., RSMAS 96-003*, Div. of Applied Marine Physics, Univ. of Miami, FL, pp. 58, 1996.

Graber, H.C., A. Bentamy, and N. Ebuchi, Evaluation of NSCAT scatterometer winds with ocean buoy observations, *Proc., NASA Scatterometer Science Symp.*, November 10-14, 1997, Maui, HI, pp. 106-107, 1997.

Grima, N., A. Bentamy, K. Katsaros, Y. Quilfen, P. Delecluse, and C. Levy, Sensitivity of an oceanic general circulation model forced by satellite wind stress fields, *J. Geophys. Res.*, 104 (C4), 7967-7989, 1999.

Halpern, D., Validation of Special Sensor Microwave Imager monthly mean wind speeds from July 1987 to December 1989, *IEEE Trans. Geos. Rem. Sens.*, 31 (3), 692-699, 1993.

Liu, W.T., K.B. Katsaros, and J.A. Businger, Bulk parameterization of air-sea exchanges of heat and water vapor including the molecular constraints at the interface, *J. Atmos. Sci.*, 36, 1722-1735, 1979.

Quilfen, Y., and A. Bentamy, Calbration/validation of ERS-1 scatterometer precision products, *Proc., Internat. Geosci. and Remote Sens. Symp. (IGARSS'94)*, Pasedana, CA, August 8-12, 1994, pp. 945-947, 1994.

Quilfen, Y., A. Bentamy, K.B. Katsaros, and G. Lorand, Estimation of ocean-atmosphere turbulent fluxes from satellite measurements, *Proc., Internat. Conf. on the Ocean Observing System for Climate (OCEANOBS99)*, St. Raphael, France, October 18-22, 1999, Centre National d'Etudes Spatiales, 2000.

Wackernagel, H., *Multivariate Geostatistics,* Springer-Verlag p. 256, 1995.

Wanninkhof, R., Relationship between gas exchange and wind speed over the ocean, *J. Geophys. Res.*, 97, 7373-7381, 1992.

A. Bentamy, Institut Français de Recherche Pour l'Exploitation de la Mer, B.P. 70, 29280 Plouzané, France.

W.M. Drennan, University of Miami, Rosenstiel School of Marine and Atmospheric Science, 4600 Rickenbacker Causeway, Miami, FL 33149.

E.B. Forde and K.B. Katsaros, NOAA/Atlantic Oceanographic and Meteorological Laboratory, 4301 Rickenbacker Causeway, Miami, FL 33149.

The Effect of Using Time-Averaged Winds on Regional Air-Sea CO_2 Fluxes

Rik Wanninkhof,[1] Scott C. Doney,[2] Taro Takahashi,[3] and Wade R. McGillis[4]

Gas transfer velocities are often related to wind speeds in order to estimate air-sea gas fluxes on regional and global scales. Since climatological averaged winds are frequently used with non-linear gas exchange wind speed relationships, the treatment of wind speed distribution will affect the magnitude of the fluxes if time-averaged winds are used. Commonly, a Rayleigh distribution is assumed for monthly or yearly averaged wind speeds. Although this is a reasonable assumption for global winds, significant regional deviations from this distribution exist. For areas with steady winds such as the trade wind regions and Westerlies in the Southern Ocean, the Rayleigh assumption will overestimate the long-term gas transfer velocities. Using 6-hour National Centers for Environmental Prediction (NCEP) reanalysis winds to determine the actual local distribution of the winds instead of averaged winds with an assumed Rayleigh distribution, the global oceanic CO_2 uptake estimate decreases by 5% if a quadratic dependence with wind speed is assumed and by 26% if a cubic dependence of gas exchange with wind speed is used.

1. INTRODUCTION

In order to estimate oceanic CO_2 uptake on monthly time and basin scales, regional information on air-sea CO_2 fluxes is critical. When such information is incorporated into (inverse) models they, in turn, can assist in determining the cause of interannual variations in the atmospheric CO_2 growth rate. Air-sea CO_2 fluxes, F, have been estimated from monthly mean sea-air pCO_2 differences, ΔpCO_2, and wind speed data according to:

$$\overline{F} = \overline{k_{av} s} \; \overline{\Delta pCO_2} \quad (1)$$

and

$$k_{av} = f(u_{av})(Sc/660)^{-1/2} \quad (2)$$

where k_{av} is the mean gas transfer velocity; "s" is the solubility of CO_2 gas in seawater; ΔpCO_2 is the partial pressure difference of CO_2 between water and air; $f(u_{av})$ is the functional dependence with averaged wind speed; and Sc is the Schmidt number, defined as the kinematic viscosity of the water divided by the molecular diffusivity of the gas in water. The Schmidt number is dependent on temperature and salinity.

Several assumptions are made in this parameterization. First, the cross correlation term between "k" and ΔpCO_2 [$(k s)' \Delta pCO_2'$] is considered negligible on a global scale. Although high winds could increase surface water pCO_2 by stirring sub-mixed layer waters with high pCO_2 into the mixed layer, they also tend to cool the surface water.

---

[1]NOAA/Atlantic Oceanographic and Meteorological Laboratory, Miami, Florida

[2]Climate and Global Dynamics, National Center for Atmospheric Research, Boulder, Colorado

[3]Lamont-Doherty Earth Observatory of Columbia University, Palisades, New York

[4]Department of Applied Ocean Physics and Engineering, Woods Hole Oceanographic Institution, Woods Hole, Massachusetts

Gas Transfer at Water Surfaces
Geophysical Monograph 127
Copyright 2002 by the American Geophysical Union

Hence these effects tend to cancel each other. Accordingly, the cross correlation term is probably small as suggested by an analysis of *Keeling et al.* [1998] who determined that this term contributed about 10% to the total flux for global O_2 fluxes. Because of the buffer effect of CO_2 in seawater, we expect that the cross correlation term would be equal or less for CO_2 than for O_2. Second, the wind speed frequency distribution pattern is implicitly assumed to be the same everywhere over the ocean.

Several different approaches have been taken to develop relationships between gas exchange and wind speed. Controlled studies in wind-wave tanks have yielded tight relationships, but their applicability to the natural environment is unclear. Field studies were, until recently, based on water column mass balance of natural or deliberately injected gases yielding multi-day to multi-year averages. Advances in air-side gradient and co-variance measurements have decreased the time scale to sub-hour. Algorithms relating gas exchange to wind speed are either developed from compilations of field data [*Nightingale et al.*, 2000], from controlled studies at a single field or laboratory site [*Wanninkhof et al.*, 1985; *Watson et al.*, 1991], or a combination thereof [*Liss and Merlivat*, 1986]. Since most relationships between wind speed and gas exchange are non-linear, the relationships will depend on the averaging interval and the distribution of wind speeds. This has been taken into account in some relationships by assuming a Rayleigh distribution of winds and using relationships of the form $k = a\, u^b\, (Sc/660)^{-1/2}$ where the exponent b is 2 or 3 and coefficient "a" is an adjustable constant depending upon if steady winds or long-term ($\geq$ monthly averaged) winds are used [*Wanninkhof*, 1992; *Wanninkhof and McGillis*, 1999].

Field observations and controlled laboratory studies suggest that wind speed alone cannot be used as a definitive predictor for gas transfer since other factors, such as wave age, wave slope, surfactants, and fetch control the transfer in the surface boundary layer (< 1 mm). However, recent compilations of deliberate tracer results show that after applying appropriate adjustments to the different data sets, the gas exchange increases with wind speed with data from different parts of the ocean falling within a broad envelope [*Nightingale et al.*, 2000]. More sophisticated algorithms, including boundary layer stability [*Erickson III*, 1993] and wave slope [*Frew et al.*, 1999; *Jähne et al.*, 1987], have been developed and will be improved in the future.

A significant advance in the quantification of global gas transfer and fluxes is the availability of synoptic wind and wave fields from remote sensing and analysis products. As shown here, these fields will better constrain our flux estimates. The issue of being able to extrapolate non-linear relationships of physical forcing over longer time intervals is germane irrespective of the functional form or environmental forcing parameters used.

Here we investigate the biases in global CO_2 fluxes caused by the assumption that the Rayleigh wind speed distribution is valid for all regions of the ocean when using monthly-averaged winds compared to using real winds. Six-hour wind speeds for 1995 obtained from the NCEP reanalysis [*Kalnay et al.*, 1996] are used to determine the gas transfer velocity and flux for each 4° × 5° grid box. The results are compared to using a Rayleigh wind speed distribution function for the CO_2 flux calculation. The fluxes are determined using Eqn. 1 and the climatological distribution of ΔpCO_2 by *Takahashi et al.* [1999]. The analysis is performed for the quadratic dependence of gas exchange with wind speed as proposed by *Wanninkhof* [1992], W-92, and the cubic dependence proposed by *Wanninkhof and McGillis* [1999], WM-99. The emphasis is on the anomalies for the cubic relation since it yields significantly larger gas exchange rates than the quadratic dependency for the high wind speed regime and, by its strong non-linear nature, it is very sensitive to the assumed wind speed distribution. Since most wind speed/gas exchange relationships are non-linear, the use of time averaged winds will lead to an enhancement.

2. METHODS

To account for the biases caused by non-linearity in the derived relationships of gas transfer with wind speed when using long-term ($\geq$ monthly) averaged winds, a wind speed distribution is assumed to adjust the relationships. W-92 and WM-99 used a slightly modified Rayleigh distribution function that closely represents the global wind speed distribution as shown by *Wentz* [1984].

The modified Rayleigh function differs from the exact Rayleigh distribution by the factor $1/2\pi$ in Eqn. 3 which accounts for the directionality of the wind. The probability distribution function, $P(u)$ can be expressed as:

$$P(u) = u[\exp(-u^2/2\Delta u^2)]/[2\pi\Delta u^2] \qquad (3)$$

in which the standard deviation of the wind, Δu, is related to the average scalar wind:

$$\Delta u = u_{av}(\pi/2)^{-1/2}$$

and where u is the steady (or short-term) scalar wind.

W-92 and WM-99 used long-term constraints on air-sea gas transfer based on natural and bomb ^{14}C inventories in the ocean. These constraints are the basis for relationships of the form of $k_{av} = a\, u_{av}^2$ and $k_{av} = b\, u_{av} + c\, u_{av}^2 + d\, u_{av}^3$, respectively. The long-term relationship of WM-99 can also be expressed as $k_{av} = e\, u_{av}^3$ which does not fit the derived long-term trend from co-variance measurements and a Rayleigh distribution quite as well as a third order polynomial fit. It shows a slight positive bias at intermediate winds (Figure 1) relative to the full cubic polynomial. However, the differences are small and in order to express wind speed distribution anomalies in terms of simple proportionalities, we use this cubic expression in the exercise below. These k_{av} relationships are deconvolved to relationships for short term (steady winds) assuming a quadratic dependency. The coefficient a' can then be determined according to:

$$a' = \left\{ k_{av} \Big/ \sum [P(u) u^b] \right\} \quad (4)$$

The resulting equation is:

$$k = a' u_{10}^b (Sc/660)^{-1/2} \quad \text{(steady/short-term wind)} \quad (5)$$

where the exponent b is 2 in the case of W-92 (quadratic dependence) and 3 for WM-99 (cubic dependence). For a quadratic dependence, the coefficient "a" for steady and long-term winds is 0.31 and 0.39, respectively, a difference of 25%; but for the cubic dependence "a" is 0.0283 for the steady wind case and 0.0615 for long-term wind equation, a factor of 2.17. Thus, the assumptions regarding wind speed distribution are particularly critical when using the cubic dependency to determine gas transfer velocities over longer time intervals.

The global CO_2 fluxes have been determined following Eqn. (1) using the updated monthly distribution of ΔpCO_2 of Takahashi et al. [1999]. This distribution is based on about a half a million observations of atmospheric and seawater pCO_2, and the data are interpolated in time and space using the advective flow field of the Princeton circulation model [Toggweiler et al., 1989]. The monthly mean values are expressed on a 4° × 5° grid. The revised distribution is for a reference year 1995 and shows significantly greater annual uptake compared to the original one referenced to 1990 [Takahashi et al., 1997]. The original air-sea CO_2 uptake estimate for the global oceans was 1.4 Pg C yr^{-1}, while the revised climatology gives 2.2 Pg C yr^{-1} using the long-term W-92 relationship. This difference is due to the increased number of observations made over extensive areas of the global oceans during various seasons, particularly in the southern Indian Ocean which is a significantly greater sink than the original estimate. The increase in the sink is also attributed to changing the reference year from 1990 to 1995, during which the atmospheric pCO_2 increased by about 8 µatm which increased the sink strength at high latitudes.

3. DISCUSSION

3.1 Effects on Global Transfer Velocities and CO_2 Fluxes

Although the Rayleigh distribution function is a good fit for global wind speed distribution, there are significant regional deviations. We express these anomalies in terms of an enhancement factor, R, defined as the ratio of the average of the 6-hour winds squared (cubed) divided by square (cube) of the average winds: $R = (\sum u^b/n)/(u_{av})^b$, where n is the number of observations. For steady winds, $R = 1$ and, as mentioned previously for a Rayleigh distribution, $R = 1.25$ for the quadratic dependence and 2.17 for the cubic dependence. The bias in the average gas transfer velocity introduced by assuming a Rayleigh fit can be estimated from comparing the gas transfer velocities determined by:

$$k_{av}' = R\, a'\, u_{av}^b (Sc/660)^{-1/2} \quad (6)$$

where b is 2 or 3 and a' is 0.31 or 0.0283, compared to our conventional assumption:

$$k_{av} = a\, u_{av}^b (Sc/660)^{-1/2} \quad (7)$$

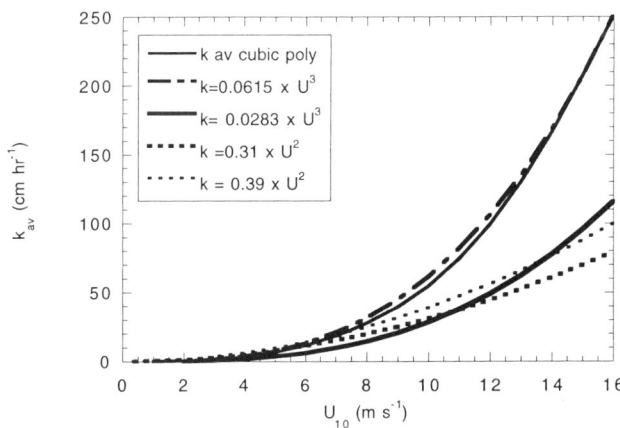

Figure 1. Different parameterizations of gas exchange with wind speed used in this analysis. Dashed lines are the quadratic relationships (lower curve = steady winds, upper curve = long-term averaged winds) while the solid lines are the cubic relationships (lower curve = steady winds, upper curve = long-term averaged winds). The dashed-dotted line is the cubic approximation for the long-term averaged winds.

Table 1. Global sea-air flux computed using a Rayleigh wind speed distribution and 6-hour NCEP wind speeds, and the quadratic and cubic wind speed dependencies of the gas transfer velocity.

	Wind Speeds Rayleigh Assumption[a]	Wind Speeds NCEP 6-hr[b]	Rayleigh/ NCEP 6-hr
$k_{av}(u^2)$[c]	20 cm hr^{-1}	18 cm hr^{-1}	1.11
$k_{av}(u^3)$[d]	24 cm hr^{-1}	17 cm hr^{-1}	1.41
$F(u^2)$[e]	-2.2 Pg yr^{-1}	-2.1 Pg yr^{-1}	1.05
$F(u^3)$[f]	-3.3 Pg yr^{-1}	-2.6 Pg yr^{-1}	1.26

[a] Assuming that the monthly mean wind speeds for each pixel follow a Rayleigh distribution.
[b] Using Eqn. 6.
[c] Average gas transfer velocity using a quadratic dependency (Eqns. 6/7). Note, k includes the Schmidt number dependency which causes the global average to differ from the value of 22 cm hr^{-1} used in W-92.
[d] Average gas transfer velocity using a cubic dependency (Eqns. 6/7). Note, the k includes the Schmidt number dependency which causes the global average to differ from the value of 22 cm hr^{-1} used in W-92.
[e] Average flux using a quadratic dependency.
[f] Average flux using a cubic dependency.

where a is 0.39 or 0.0625 for the quadratic or cubic expression. k_{av}' is the exact formulation and k_{av} is the formulation assuming a Rayleigh distribution.

The bias of using a Rayleigh distribution assumption (Eqn. 7) compared to the exact equation (Eqn. 6) on global CO$_2$ fluxes will depend on the actual distribution of the wind speed and the ΔpCO$_2$ and, to a lesser extent, on the solubility (Eqn. 1).

Non-Rayleigh behavior of wind speed is more pronounced in regions with steady winds such as the Westerlies in the Southern Ocean and the trade wind regions [*Erickson III and Taylor*, 1989]. Globally, the enhancement factor is 1.14 ± 0.07 for the quadratic dependence and 1.45 ± 0.23 (1 σ) for the cubic dependence compared to the Rayleigh factors of 1.25 and 2.17.

The overall effect of the Rayleigh distribution assumption on k_{av} is summarized in Table 1. The Rayleigh approximation causes the k_{av} to be overestimated by 11% compared to the exact solution for the quadratic relationship and by 41% for the cubic relationship. The latter result clearly raises questions about the appropriateness of using the Rayleigh distribution for strong non-linear dependencies. The effect is less pronounced for the global gas fluxes but it is still appreciable for the cubic dependence. The smaller difference of the bias for the fluxes compared to gas transfer velocities is caused by the spatial and temporal distribution of ΔpCO$_2$. For the quadratic gas exchange parameterization, the difference in the exact and Rayleigh assumption is 5 % for the CO$_2$ fluxes but for the cubic dependence the Rayleigh assumption yields a global CO$_2$ uptake that is 26% larger than the exact solution.

3.2 Zonal Anomaly Patterns

The magnitude of the enhancement factor, *R*, and associated difference in flux shows distinct zonal patterns. The zonal wind speed distribution for 1995 is given in Figure 2a. The annual enhancement factor for a quadratic and cubic dependence zonally averaged are shown in Figures 2b and 2c, respectively. The trends are the same but the magnitude of the deviation from a Rayleigh enhancement factor is much larger for a cubic dependence. In this case the Westerlies in the Southern Ocean and trade wind belt show enhancement factors of about 1.4. Mid-latitudes and high northern latitudes show enhancement factors closer to 1.7, but the values seldom reach the value of 2.17 inferred for a Rayleigh distribution.

The annual enhancement factor differences compared to a Rayleigh parameterization do not significantly affect the global fluxes for a quadratic dependency because *R* is close to that of a Rayleigh distribution pattern and because of compensating effects of enhancement factor and source/sink patterns for CO$_2$. Regional differences can be significant such as the decreased outgassing in the equatorial region, which is compensated by less uptake in the Southern Ocean. For the cubic dependence there are large differences in the annual averaged zonal flux using the exact cubic and Rayleigh solution (Fig. 2d). Differences in uptake of up to -1 mol m^2 yr^{-1} are apparent for the entire Southern Ocean and in the high latitude northern oceans with the exact solution showing less uptake. In the equatorial outgassing region, the exact solution shows about 0.3 mol m^2 yr^{-1} less evasion than the Rayleigh representation. As in the case of a quadratic dependence, this will partially compensate the effect on the global fluxes. Nevertheless, the exact solution shows a 26% lower global carbon uptake than the Rayleigh parameterization (Table 1). This decrease brings the global CO$_2$ uptake down to 2.6 Pg C yr^{-1} and within the uncertainty of the general consensus estimate of global uptake of 2 ± 0.8 Pg C yr^{-1} [*Houghton et al.*, 1995].

4. RECOMMENDATION

The many relationships of gas exchange and wind speed and the effect of non-linearity of the relationship in temporal extrapolations results in significant uncertainty in

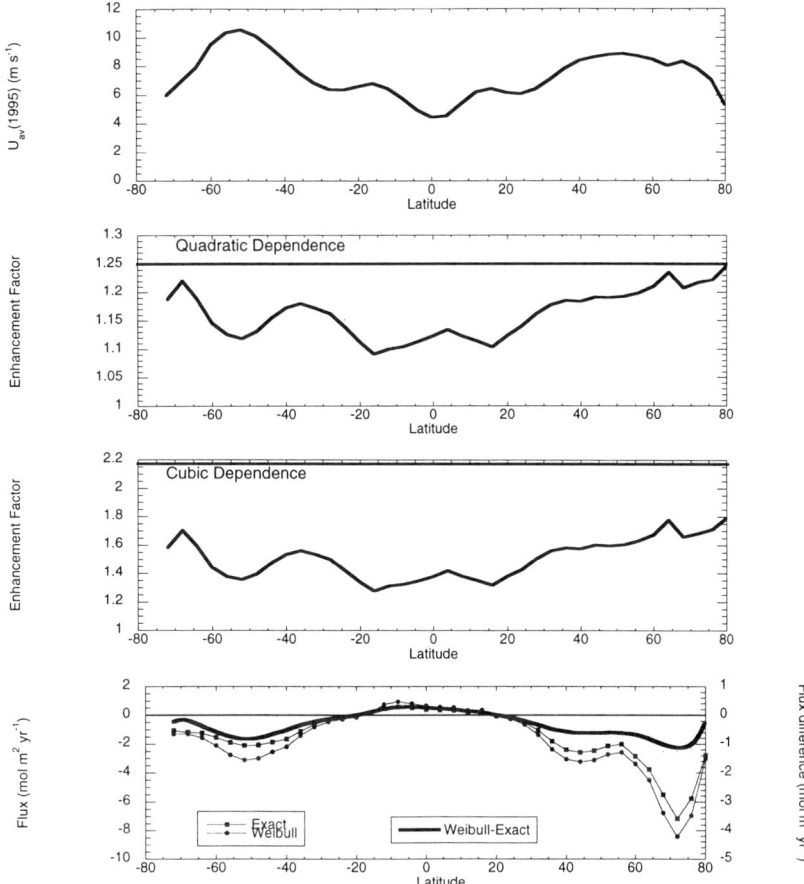

Figure 2 (a) Zonally averaged winds over the ocean for 1995; (b) Zonally averaged enhancement factor for a quadratic dependence. The horizontal solid line is the enhancement factor for a Rayleigh wind speed distribution; (c) Zonally averaged enhancement factor for a cubic dependence with standard deviation. The solid line is the enhancement factor for a Rayleigh wind speed distribution; (d) Zonally averaged fluxes for a cubic dependence. The circles are the fluxes assuming a Rayleigh wind speed distribution. The open squares are the exact solution. The heavy solid line (right axis) depicts the smoothed difference.

global air-sea fluxes of gases. Fluxes cannot be better constrained until we can incorporate a firm mechanistic understanding into parameterizations of gas transfer using proper environmental forcing. The most robust fluxes will be obtained from regional algorithms over short time scales as determined from direct flux measurements such as co-variance measurements combined with near real time synoptic winds from remote sensors, such as QuickSCAT and the ERS scatterometers. The magnitude of the fluxes is dependent on the wind speed product used and the approach by which the products are averaged in time and space. The NCEP winds with 6-hour resolution are believed to miss some of the small scale, high wind events that are captured by satellite observations [*Doney*, 1996].

A research strategy that incorporates field programs to determine the correct parameterization of gas transfer in a variety of regimes is advocated to constrain regional fluxes.

For global models and interpolations, climatological wind fields are used extensively. In these cases, information on wind speed distribution will help improve the estimates, particularly for non-linear dependence of air-sea gas transfer relationships and for skewed wind speed distributions. The validity of a cubic dependence must be explored in more high wind speed environments. Although it has a sound empirical basis related to bubble enhanced gas transfer [*Asher et al.*, 1996; *Monahan and Spillane*, 1984; *Woolf*, 1997] and has better agreement compared to other parameterizations in some mass balances at high latitude [*Signorini et al.*, 2000, *Schneider et al.*, 1999.], evidence of strongly enhanced gas transfer has been lacking in other studies [*Nightingale et al.*, 2000].

The analysis performed here only covers one aspect of the uncertainty in the flux. Sparse coverage of $\Delta p CO_2$, biases in $\Delta p CO_2$ due to near surface gradients (e.g., thermal skin effect) contribute to the uncertainty in regional and global estimates of CO_2 flux.

Our analysis suggests that the long-term relationship, assuming a cubic dependence of gas exchange and a Rayleigh wind speed distribution, significantly overestimates the global air-sea flux, particularly in areas with low enhancement factors such as the regions with Westerlies and trade winds. The bias is less pronounced for a quadratic dependence but will affect regional fluxes. The recommended approach is not to use time averaged winds but rather compute the time-averages of higher moments needed for non-linear relationships from reanalysis or satellite winds.

Acknowledgment. Lisa Dilling, program manager for the Carbon Cycle Program of the Office of Global Programs of NOAA, and the National Science Foundation grant no. OCE-9711218 are gratefully acknowledged for supporting this work. The work benefited from insightful comments of two anonymous reviewers. Formatting and proofreading by Gail Derr of AOML is much appreciated. (Woods Hole Oceanographic Institution contribution number 10335).

REFERENCES

Asher, W.E., P.J. Farley, B.J. Higgins, L.M. Karle, E.C. Monahan, and I.S. Leifer, The influence of bubble plumes on air/seawater gas transfer velocities, *J. Geophys. Res.*, *101*, 12,027-12,041, 1996.

Doney, S.C., A synoptic atmospheric surface forcing data set and physical upper ocean model for the U. S. JGOFS Bermuda Atlantic time series study (BATS) site, *J. Geophys. Res.*, *101*, 25615-26634, 1996.

Erickson, III, D.J., A stability-dependent theory for air-sea gas exchange, *J. Geophys. Res.*, *98*, 8471-8488, 1993.

Erickson, III, D.J., and J.A. Taylor, Non-Rayleigh behaviour observed in a model generated global surface wind field frequency distribution, *J. Geophys. Res.*, *94*, 12,693-12,699, 1989.

Frew, N.M., D.M. Glover, E.J. Bock, C. Goyet, S.J. McCue, and R.J. Healy, Estimation of global air-sea transfer of CO_2 using TOPEX/Poseidon dual-frequency backscatter, *IUGG abstracts, Birmingham U.K.*, 1999.

Houghton, J.T., L.G. Meira Filho, J. Bruce, H. Lee., B.A. Callander, E. Haites, E. Harris, and K. Maskell, *Climate Change 1994: Radiative Forcing of Climate Change and an Evaluation of the IPCC IS92 Emission Scenarios*. Cambridge University Press, Cambridge, 1995.

Jähne, B., K.O. Münnich, R. Bösinger, A. Dutzi, W. Huber, and P. Libner, On parameters influencing air-water gas exchange, *J. Geophys. Res.*, *92*, 1937-1949, 1987.

Kalnay, E., et al., The NCEP/NCAR 40-year reanalysis project, *Bull. Am. Meteorol. Soc.*, *77*, 437-471, 1996.

Keeling, R.F., B.B. Stephens, R.G. Najjar, S.C. Doney, D. Archer, and M. Heimann, Seasonal variation in the atmospheric O_2/N_2 ratio in relation to the kinetics of air-sea exchange, *Global Biogeochem. Cycles*, *12*, 141-164, 1998.

Liss, P.S., and L. Merlivat, Air-sea gas exchange rates: Introduction and synthesis, in *The Role of Air-Sea Exchange in Geochemical Cycling*, edited by P. Buat-Menard, pp. 113-129, Reidel, Boston, 1986.

Monahan, E.C., and M.C. Spillane, The role of oceanic whitecaps in air-sea gas exchange, in *Gas Transfer at Water Surfaces*, edited by W. Brutsaert and G.H. Jirka, pp. 495-503, Reidel, Boston, 1984.

Nightingale, P.D., G. Malin, C.S. Law, A.J. Watson, P.S. Liss, M.I. Liddicoat, J. Boutin, and R.C. Upstill-Goddard, In-situ evaluation of air-sea gas exchange parameterizations using novel conservative and volatile tracers, *Global Biogeochem. Cycles*, *14*, 373-387, 2000.

Schneider, B., K. Nagel, H. Thomas, and A. Rebers, The Baltic Sea CO2 Budget: Do we need a new parameterization of the CO2 transfer velocity?, in *Proceedings of the 2nd International Symposium on CO2 in the Oceans*,, edited by Y. Nojiri, pp. 289-292, Center for Global Environmental Research, NIEST, Tsukuba, JAPAN, January 18-23, 1999.

Signorini, S.R., J.R. Christian, C.R. McClain, C.S. Wong, and P.P. Murphy, Seasonal and interannual variability of phytoplankton, TCO_2, and ΔpCO_2 in the North Pacific (OWS Papa), *EOS Trans. 2000 Ocean Sciences Meeting*, 2000.

Takahashi, T., R.A. Feely, R. Weiss, R. Wanninkhof, D.W. Chipman, S.C. Sutherland, and T.T. Takahashi, Global air-sea flux of CO_2: An estimate based on measurements of sea-air pCO_2 difference, *Proc. Natl. Acad. Sci. USA*, *94*, 8292-8299, 1997.

Takahashi, T., R.H. Wanninkhof, R.A. Feely, R.F. Weiss, D.W. Chipman, N. Bates, J. Olafsson, C. Sabine, and S.C. Sutherland, Net sea-air CO_2 flux over the global oceans: An improved estimate based on the sea-air pCO_2 difference, in *Proc., 2nd Internat. Symp. on CO_2 in the Oceans*,, edited by Y. Nojiri, pp. 9-18, Center for Global Environmental Research, NIEST, Tsukuba, Japan, January 18-23, 1999.

Toggweiler, J.R., K. Dixon, and K. Bryan, Simulation of radiocarbon in a coarse-resolution world ocean model. 1.Steady state prebomb distributions, *J. Geophys. Res.*, *94*, 8217-8242, 1989.

Wanninkhof, R., Relationship between gas exchange and wind speed over the ocean, *J. Geophys. Res.*, *97*, 7373-7381, 1992.

Wanninkhof, R., J.R. Ledwell, and W.S. Broecker, Gas exchange - wind speed relationship measured with sulfur hexafluoride on a lake, *Science*, *227*, 1224-1226, 1985.

Wanninkhof, R., and W.R. McGillis, A cubic relationship between gas transfer and wind speed, *Geophys. Res. Lett.*, *26*, 1889-1893, 1999.

Watson, A.J., R.C. Upstill-Goddard, and P.S. Liss, Air-sea exchange in rough and stormy seas, measured by a dual tracer technique, *Nature*, *349*, 145-147, 1991.

Wentz, F.J., S. Peteherych, and L.A. Thomas, A model function for ocean radar cross sections at 14.6 GHz, *J. Geophys. Res.*, *89*, 3689-3704, 1984.

Woolf, D.K., Bubbles and their role in gas exchange, in *The Sea Surface and Global Change*, edited by P.S. Liss, and R.A. Duce, pp. 173-206, Cambridge University Press, Cambridge, 1997.

A Dynamic Method to Estimate the Reoxygenation Rate at Hydraulic Structures

Badre E. Boumansour, Olivier Dufayt, and Jean L. Vasel

Fondation Universitaire Luxembourgeoise, Arlon, Belgium

Jean M. Hiver

Laboratoire de Recherches Hydrauliques-M.W.E.T, Châtelet, Belgium.

The aim of this study is to present a new dynamic method to estimate the reoxygenation rate "r_{O2}" at hydraulic structures from steady state experimental conditions. The method is based on the creation of an impulse input signal for dissolved oxygen concentration. From this the upstream and downstream signals permit evaluation of the reoxygenation rate. The method has been tested in the laboratory and in the field. Compared to the classical method, this dynamic approach is, in the light of the results obtained in this study, an effective alternative for estimation of "r_{O2}". Moreover it is more rapid and less expensive than the classical static method.

1. INTRODUCTION

Many hydraulic structures used to manage rivers and canals, are characterized by the presence of a waterfall. The sections just downstream from weirs are prime areas for major oxygen transfers that can improve the water quality considerably, especially when the waterfall is located in a stretch with a high oxygen deficit.

Reaeration at hydraulic structures is usually characterized by the "reoxygenation rate," which is defined classically as the ratio between upstream and downstream oxygen deficits [*Avery* and *Novak*, 1978]. This assessment implies constant oxygen concentrations upstream and downstream from the hydraulic structure, hence the need to work at steady state. In practice, this means keeping the upstream oxygen concentration constant during the whole experiment. This procedure is time-consuming, tedious and expensive. For example, *Nakasone* (1987) conducted only one test a day.

In this study we propose what we have called a "dynamic approach" to quantify a waterfall's reoxygenation rate. This approach, which allows one to simulate the downstream signal from the upstream oxygen concentration and other hydrodynamic parameters of the overflow weirs, has been tested on both the laboratory and real scale. Most of the aeration tests were conducted in the laboratory, where the main factors can be varied over a wider range than in the field. The reoxygenation rate values obtained by the dynamic method will also be validated by comparison with the values determined by the conventional method.

2. THEORY

2.1. Conventional Method

The reoxygenation rate at a hydraulic structure is usually expressed as the upstream-to-downstream dissolved oxygen deficit ratio, *i.e.*,

$$r_{O2} = \frac{C_S - C_{up}}{C_S - C_d} \quad (1)$$

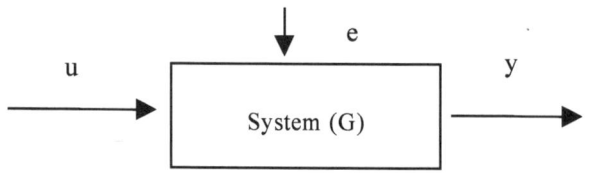

u : input signal
e : disturbance
y : output signal

Figure 1: Principle of the system to study.

where C_{up} and C_d are the dissolved oxygen concentrations upstream and downstream from the hydraulic structure respectively, and C_s is the dissolved oxygen saturation.

2.2. Dynamic Method

In this method we create a non steady-state value and, if possible, try to obtain an impulse input signal of dissolved oxygen, for example by sulfite addition [*Boumansour et al.*, 2000]. The obtained upstream and downstream signals allow one to assess the reoxygenation rate. Indeed, this computation consists of the determination of the transfer function of the waterfall which can be defined as the ratio between Laplace transforms of input and output signals.

The basic theoretical input-output configuration is depicted in figure 1.

Assuming a unit sampling interval, the input signal u(t); output signal are known for y(t) where t= 1, 2, ... , N

Assuming the signals are related by a linear system, the relationship can be written:

$$y(t) = G(q) u(t) + v(t) \quad (2)$$

where q is the shift operator and G(q) u(t) is short for

$$G(q)u(t) = \sum_{k=1}^{\infty} g(k)u(t-k) \quad (3)$$

and

$$G(q) = \sum g(k)q^{-k}...,...q^{-1}u(t) = u(t-1) \quad (4)$$

The numbers {g(k)} are called the impulse response of the system and v(t) the disturbance that can be described as : v(t)= H(q) e(t) where e(t) is white noise.

Rather than specifying the functions G and H in terms of the frequency variable, one can describe them as rational functions of q^{-1} and specify the numerator and denominator coefficients in some way.

A commonly used parametric model is the ARX model which corresponds to

$$G(q) = q^{-nk}\frac{B(q)}{A(q)} \quad \text{and} \quad H(q) = \frac{1}{A(q)} \quad (5)$$

where B and A are polynomials in the delay operator :

$$A(q) = 1 + a_1 q^{-1} + \cdots + a_{na} q^{-na} \quad (6)$$

$$B(q) = b_1 + b_2 q^{-1} + \cdots + b_{nb} q^{-nb+1} \quad (7)$$

Here, the numbers *na* and *nb* are the orders of the respective polynomials. The number *nk* is the number of delays from input to output. The model is usually written:

A(q)y(t) = B(q)u(t − nk) + e(t)   (8)

In our case, the disturbance can be ignored and we get

$$A(q)y(t) = B(q)u(t-nk) \quad (9)$$

In this study we used the Matlab® software to estimate both the parameters a_i and b_i of the ARX model and the transfer function. For this reason we utilized the matlab function arx

$$th = arx(z, [na\ nb\ nk]) \quad (10)$$

where na, nb and nk are the corresponding orders and delays that define the exact model structure. Let's note that this function estimates discrete and continuous models, it means that no transformations between continuous-time and discrete-time model are necessary.

The basic format for representing models in the System Identification Toolbox is called the *theta format*. It stores all relevant information about the model structure used, including the values of the estimated parameters, the estimated covariances of the parameters, and the estimated variance.

More information is given in the Matlab® Systems Identification Toolbox Manual

In our case, the reoxygenation rate of the hydraulic structure, na=1 nb=1 and nk=0, that the model transfer function is in the form (discrete):

$$H(z) = \frac{1}{r_{(t)}} = \frac{a'_1.z + a'_0}{b'_1.z + b'_0} \quad (11)$$

If the input signal is the upstream oxygen deficit and output signal the downstream deficit, from this transfer function we can determine the asymptotic value of H(z), which corresponds to the static gain of the hydraulic structure. Therefore, the inverse of the static gain should be

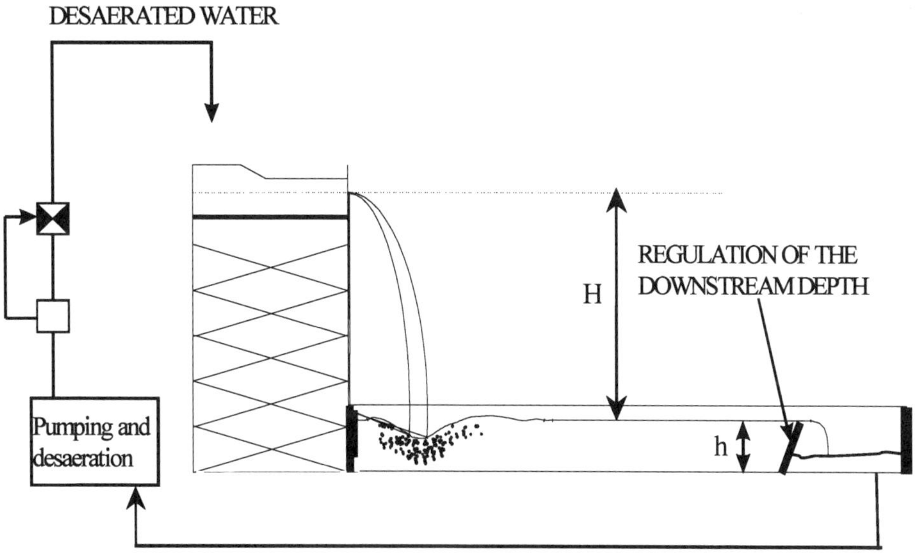

Figure 2: Experimental installation of waterfall.

equivalent to the classical reoxygenation rate determined at steady state (= upstream deficit/downstream deficit).

3. MATERIALS AND METHODS

The experiments were realized on a pilot scale installation built at the Wallon Ministry of Public Works and Transport's (MET's) Hydraulic Research Laboratory at Châtelet (Belgium). It allowed us to vary the main operating factors in a wide operating range (Figure 2).

This installation is composed of an experimental channel 22m long, 1m wide, and 1.20m high. The downstream depth is adjustable from 0 to 1m by a flap gate placed at the channel's outlet. The fall height can be varied between 1m to 4m, the scaffold control bearing the channel where the water flows. Two types of weir crests were tested in this work, that is, "rectangular" and "comb"-shaped crests. The water was pumped from a water tank (100 m³) located at the end of the channel and flowed in a closed circuit. The water flow rate was controlled by an electromagnetic flow meter placed upstream of the waterfall.

Our dynamic method was validated on the test platform under various operational conditions as summarized in Table 1.

We note that deoxygenation was achieved by injecting nitrogen after the water had been pumped up. Previously calibrated dissolved oxygen probes (Orbisphère MOCA, 3600) were placed upstream and downstream from the hydraulic structure to monitor the oxygen concentration continuously. A PC was used for signal acquisition. The data were stored in a file for later determination of the hydraulic structure's reoxygenation rate.

4. EXPERIMENTAL RESULTS

More than 100 aeration tests were carried out in the laboratory scale pilot during this work. Figure 3 gives an example of the instantaneous oxygen deficits monitored upstream and downstream from the waterfall. Using equation (1) the "r_{O2}" value is equal to 5.16.

When we apply the dynamic model explained previously, the transfer function estimation corresponds to the following equation:

$$H(z) = \frac{0.0156 \cdot z}{z - 0.9174} \quad (12)$$

where A = [0.0156   0], B= [1.0000 -0.9174]

Table 1. Operational parameters varied in test platform

Water flowrate (l/s)	Height of fall (m)	Downstream depth (m)	Type of crest
5; 10; 15 and 20	2; 3 and 4	0.4; 0.6; 0.7 and 0.8	Rectangular notch and comb with rectangular notch

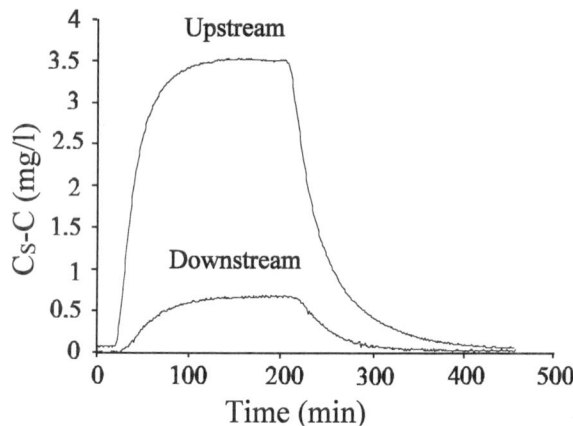

Figure 3: Oxygen deficit evolution upstream and downstream of hydraulic structure. (H=4.107 m, Q= 15 l/s, h= 0.211 m, Cs= 9.34 mg/l)

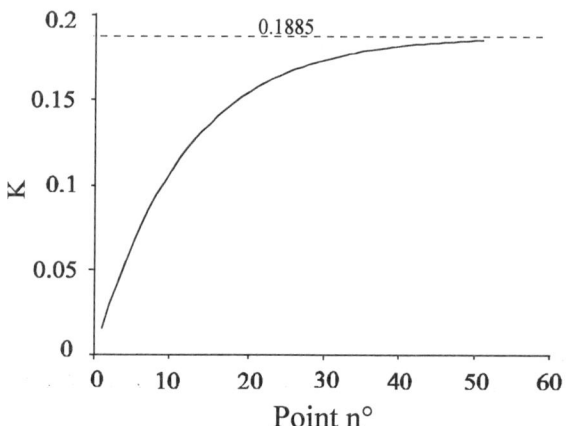

Figure 4: Static gain of the laboratory waterfall.

We note that the sampling period is 0.167 min.

From this model, the static gain "K" of the waterfall has been calculated using the matlab® function [zepo K]= th2zp(th). In this case we obtained K= [1.0000 0.1885 0.0031] where 0.1885 corresponds to the static gain with a standard deviation equal to 0.0031 (figure 4).

From the static gain value, the reoxygenation rate has been calculated as r_{O2}= 1/K, thus r_{O2}= 5.30. Knowing the transfer function of the system, we can apply it to any input signal and simulate the outlet response of the system by deconvolution.

We observe in figure 5 that downstream oxygen deficit curves determined experimentally and by the dynamic model are very close.

In Figure 6 we gathered all the reoxygenation rate values determined by conventional and dynamic methods for both rectangular and comb crests. Moreover, we observe that dynamic and conventional reoxygenation rate methods are well correlated with R^2 in the range of 74% and 79% and slope of linear regression between 0.956 and 0.991.

Other validation tests have been carried out at a real scale facility built on the Vire (a small river in Belgium). Figure 7 shows the oxygen concentrations monitored upstream and downstream of a waterfall. We note that the water desaeration has been conducted by an impulse of sulfite and cobalt solutions in order to work at non steady state.

We note that the error reoxygenation rate determined by the dynamic method is around 4%. This value is lower than those determined for the standard method which is around 17% [*Boumansour et al.*, 2000]. Taking into account those errors as well as error calculus on equation (1), which are greater for the static method, especially for higher r_{O2} values, it appears that the dynamic method can be utilized in a wider r_{O2} range than the static method.

The static gain corresponding to the Vire waterfall is equal to 0.65. The reoxygenation value, which is the inverse of asymptotic value of static gain, is equal to 1.54. We observed that the same value has been obtained when we realized the measurement at steady state [*Boumansour et al.*, 2000].

As the method is validated the next step will be to relate the measured r_{O2} with local gas transfer parameters and dimensionless numbers such as Froude and Reynolds numbers [*Avery and Novak*, 1975]

5. CONCLUSIONS

The dynamic method is a very interesting technique for quantifying hydraulic structures' reoxygenation rates. It

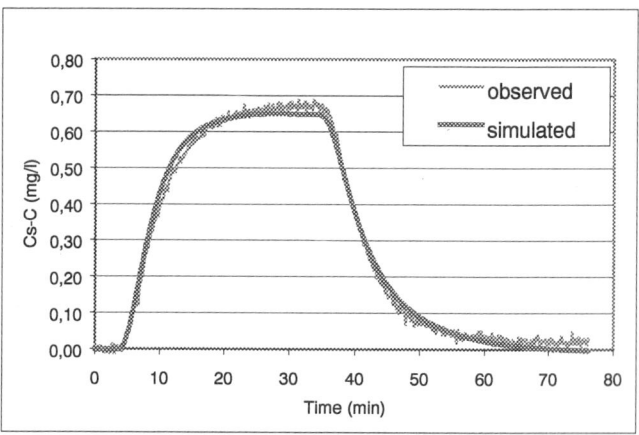

Figure 5 : Downstream oxygen deficit simulated by dynamic method.

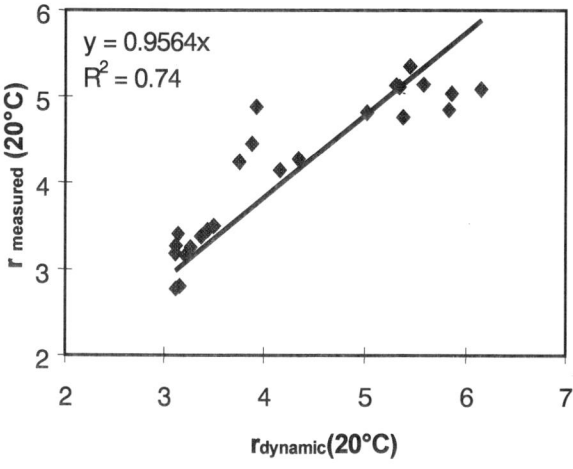

a: rectangular crest.

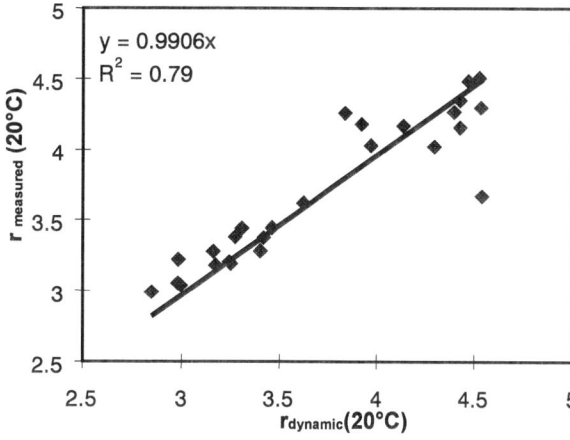

b : Crest: comb with rectangular notch.

Figure 6: Correlation between usual and dynamic oxygenation rate determined in two cases: rectangular and comb crests.

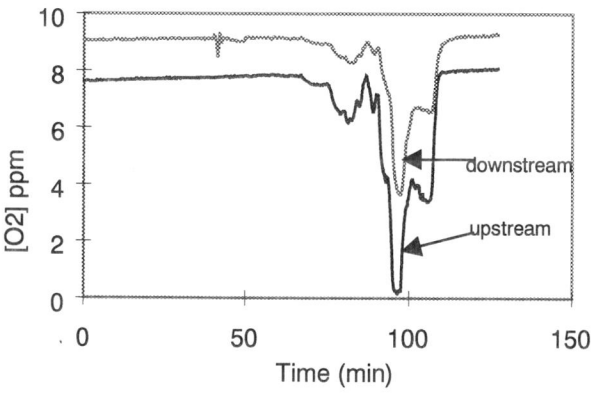

Figure 7 : Dissolved oxygen monitored upstream and downstream of waterfall built on the Vire river (H= 0.8m, h=0.58m, Q= 232 l/s).

has many advantages over the conventional method: it is easier to use non-steady state oxygen concentrations; it is not expensive, and fewer people are needed to carry it out.

Finally, the experiment's shorter run time means that a greater number of more accurate tests can be conducted in a day.

Acknowledgments. The authors gratefully acknowledge the financial support provided for this project by the Wallon Ministry of Public Works and Transport (MET) and the use of facilities at the Hydraulic Research Laboratory of Châtelet (Belgium).

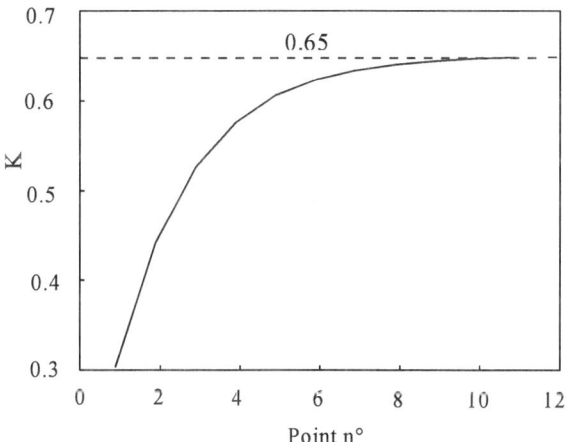

Figure 8 : Static gain of the Vire waterfall (Sampling period 0.5min).

REFERENCES

Avery, S. T.,. Novak, P., Oxygen uptake in hydraulic jumps and over falls, *Proceedings of the XVI Congress*, International Association for hydraulic research, C38, 329-337, 1975.

Boumansour, B. E., Jupsin, H., Praet, E., Vasel, J. L., Comparison of three methods to measure reoxygenation at hydraulic structures, *Fourth International Symposium on Gas Transfer at Water Surfaces*, June 5-8, Miami, 2000.

Matlab® Software Version 5.3, January 21, 1999.

Nakasone, H., Study of aeration at weirs and cascades, *Jour. Envir. Engrg.*, ASCE, *113*, 1, 64-81, 1987.

Mass Transfer in Bubbly Flows: Influence of Physico-Chemical and Hydrodynamic Conditions

Cornelia Lang

Institute for Hydromechanics, University of Karlsruhe, Germany

Results from an experimental study of air bubbles in a finite water volume are presented. The influence of water containing admixtures and of turbulent conditions on bubble gas transfer mechanism has been studied systematically by measurements of oxygen transfer. Bubble characteristics like size distribution, diameter and shaping, quality of air water interface as well as the effect of the energy dissipation in the system are examined and correlated with the measured overall volumetric transfer coefficient $k_L \cdot a$. Both terms, the transfer coefficient k_L and the volumetric interfacial area $a = A/V$ are analyzed and a description of the effects influencing either k_L and/or a is given. The results are related to known theoretical and empirical estimates of bubble size and transfer coefficient.

1. INTRODUCTION

There are many processes in nature or technical systems in which mass transfer occurs between a continuous liquid phase and a dispersed bubble phase. Examples of such processes are aeration of lakes and rivers, water and wastewater treatment, chemical and biochemical processes.

The efficiency of the transfer process is substantially affected by the physico-chemical properties and the hydrodynamic effects of the liquid phase. In order to quantify the influence of these conditions on the key parameters describing the mass flux, we give in this paper a detailed analysis of the bubble transfer mechanisms, illustrated by oxygen transfer in clear water, salt water and water containing surface-active agent.

The knowledge about this influence is important for the prediction of the oxygen transfer rate across the interface, for the controlling of engineering processes, as well as for modeling the transfer process in transport or water quality models [Jirka, 1999].

2. THEORY

2.1 Fundamentals of Gas Transfer at Air-Water Interfaces

It is commonly accepted that the interface transfer of a gas of low solubility (or high Henry's constant, e.g. O_2, CO_2) is controlled by the liquid boundary layer close to the interface itself [Treybal, 1981].

Hence it is the interaction of molecular diffusion and hydrodynamic effect at the liquid side of the interface that determines the kinetics of the gas transfer. The existence of admixtures due to pollution (e.g. surfactants) may affect gas transfer by providing additional resistance to the gas flux [Asher and Pankow, 1986].

The flux $\dot{M}$ of a gas across a clean interfacial area A into the bulk liquid is parameterized by the liquid phase transfer coefficient k_L:

$$\dot{M} = \frac{dM}{dt} = k_L \cdot A \cdot (C_S - C_L). \qquad (1)$$

The quantities C_S and C_L are, respectively, the saturated and bulk concentrations of the dissolved gas.

Measurement of the change of concentration in air-water systems with finite volume V will be analyzed on the base of equation (1) with a kinetic reaction model of the first order:

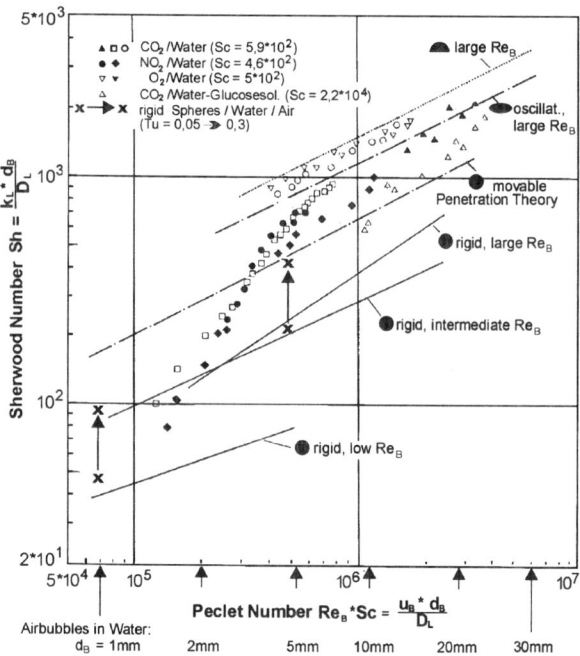

Figure 1. Mass transfer of rigid and fluid particles.

$$\frac{dC_L}{dt} = \frac{dM}{dt} \cdot \frac{1}{V} = k_L \cdot \frac{A}{V} \cdot (C_S - C_L). \quad (2)$$

The overall volumetric transfer coefficient

$$k = k_L \cdot \frac{A}{V} = k_L \cdot a \quad (3)$$

characterizes the attainable gas transfer in this system. The volumetric interfacial area a represents the summation over all interfacial areas A in volume V. For practical transfer processes it is not possible to determine separately both k_L and a. However, to clarify the detailed transfer mechanism and to identify the transfer controlling effects it is necessary to undertake a distinct examination of these parameters.

For the predicative formulation of k_L two forms of conceptual models can be found in the scientific literature:

(a) The classical theories, which are based on the stagnant "film model" [Lewis and Whitman, 1924]: $k_L = D/\delta_L$, where δ_L is the thickness of the diffusive film at the liquid side of the interface, or on the "surface renewal model" [Danckwerts, 1951]: $k_L = D^{1/2} \cdot r^{1/2}$, where r is the mean frequency of random replacement of fluid elements at the interface and D is the molecular diffusion coefficient of the gas.

(b) Hydrodynamic models that consider the dynamic interfacial phenomena as related to the size of the most efficient turbulent eddies. These models may be conventionally separated into the "small-eddy model" [Lamont and Scott, 1970], where the dissipate structures expressed by the turbulent dissipation rate ε is the significant parameter:

$$k_L \sim D^{1/2} \cdot (\varepsilon/\nu_L)^{1/4}, \quad (4)$$

and the "large-eddy model" [Fortescue and Pearson, 1967] using the integral length scale L for time scale

$$k_L \sim D^{1/2} \cdot (u'_L/L)^{1/2}, \quad (5)$$

where u'_L is the velocity scale of turbulence and ν_L is the kinematic viscosity of the liquid phase.

Nakayama [2000] gives a review for open channels on various investigations on hydrodynamic models and modifications of the theories mentioned above.

2.2 Gas Transfer of Bubbles

A dimensionless relationship between transfer coefficient k_L (resp. Sherwood Number Sh) and the bubble diameter d_B (resp. Peclet Number Re_B S_C) is shown in Fig. 1. The presented plot combines data from several experimental and theoretical studies of movable bubbles and rigid bodies [e.g. Calderbank and Lochiel, 1964; Clift, 1978]. In the further discussion, air bubbles in water are considered.

The gas transfer of bubbles depends strongly on the interfacial mobility. Phenomena like internal circulation, deformation of shape, and shape oscillation lead to a significant increase of transfer coefficient k_L for bubbles with diameters d_B = 2 to 8 mm. These irregular ellipsoidal bubbles undergo periodic dilations or random wobbling. Smaller bubbles (d_B < 2 mm) have an approximately fixed spherical shape with a rigid surface that reduces gas transfer. Bubbles with d_B > 8 mm look very similar to spherical or ellipsoidal caps and show an increasing tendency to shape oscillation and deformation resulting in an enhancement of k_L. Experimental results of Galloway and Sage [1968] showing the transfer-enhanced influence of a turbulence level are marked with crosses. The paths of rising bubbles are different of various bubble diameters and shapes: the motion of spherical bubbles is rectilinear, whilst bubbles with a mean ellipsoidal shape rise in a zigzag or spiral path with rocking motion. Large spherical or ellipsoidal caps move more or less rectilinearly.

The efficiency of a single bubble (volume V_B) to gas transfer is also related to the interfacial area a_B (see Eq. (3))

$$a_B = A_B/V_B = 6/d_B \quad (6)$$

where A_B is the bubble surface.

In a finite liquid volume V with a bubble swarm, the volumetric interfacial area a comprises two flow properties:

(a) the mean fractional gas holdup e_G, which is defined as the volume fraction of the gas V_G in the total volume V = $V_G + V_L$ of the two phase mixture:

$$e_G = V_G/(V_G + V_L) \quad (7)$$

(b) the Sauter diameter d_{BS}, which is the equivalent diameter of a sphere with the same volume to surface ratio as the sum over all bubbles

$$d_{BS} = 6 \cdot \Sigma V_B / \Sigma A_B = 6 V_G / A. \qquad (8)$$

Combining Eq. (7) and (8) leads to the definition of A/V

$$A/V = a = 6 \cdot e_G / d_{BS}. \qquad (9)$$

Eq. (9) provides the inverse relationship of bubble size and the significance of all effects increasing the residence time of bubbles, e.g. low rise velocity and long rise path.

With the experimental determination of overall volumetric transfer coefficient k and volumetric interfacial area a all terms of Eq. (3) can be identified separately.

3. EXPERIMENTAL SETUP

The measurements are conducted in a bubble column (closed circuit system). The working cross-section, made of acrylic glass, measured a height H of 2 m, a diameter D of 0.44 m and a finite volume V of 0.3 m³ of water (Fig. 2). A uniform velocity of the liquid phase distribution is obtained by an arrangement of screens in a diffuser. A variation in turbulence conditions can be achieved by additional grids [Baines and Peterson, 1951] at the entrance of the cross-section. The velocity and turbulence measurements are obtained using a LDV system.

Air was injected through 25 nozzles, positioned at each cross of the turbulence grid. Three replaceable nozzle types were constructed to produce bubbles with the following size

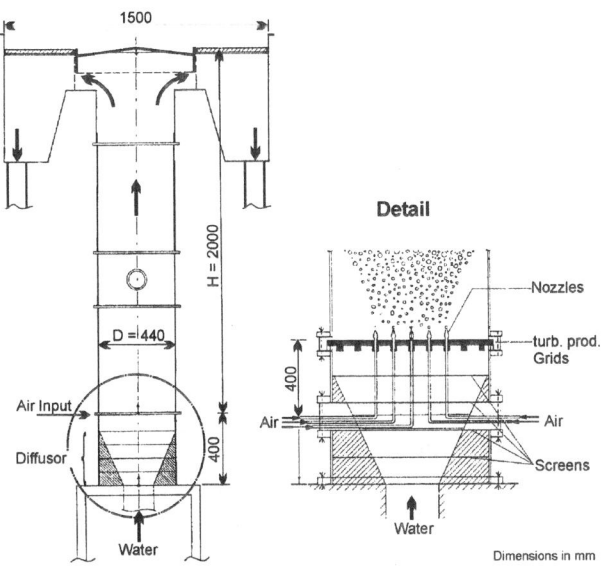

Figure 2. Bubble column with grid-generated turbulence device and air-input system with 25 nozzles.

Organic Admixture: Surfactant (Surface-active agent)[1]			Inorganic Admixture: Salt NaCl	
Surface tension σ [· 10⁻³ N/m]			Salinity [%]	
71[2]	55	40	0.5	1.5
Z_0	Z_1	Z_2	I_1	I_2

[1] Sodium dodecylbenzenesulphonate (anionic detergent)
[2] Clear Water

Table 1. Physico-chemical properties of water containing organic and inorganic admixtures.

ranges: a) d_{BS} = 2-3 mm (small bubbles), (b) d_{BS} = 5-8 mm (medium bubbles), (c) d_{BS} = 8-11 mm (large bubbles).

Bubble sizes and size distributions were measured by an opto-electronical bubbles size test section to estimate the sphere equivalent Sauter diameter d_{BS}.

The operation conditions of the bubble column are: superficial air velocity U_{SG} = 0.07 – 0.66 cm/s, superficial water velocity U_{SL} = 0.07 – 0.46 m/s, void fraction 1 < β = $U_{SL}/(U_{SL} + U_{SG})$ < 10%. Gas holdup e_G was determined by measuring the change in liquid height with and without air.

The properties of the water, artificially contaminated with admixtures, are listed in Table 1.

The overall volumetric transfer coefficient k was determined using an unsteady state chemical method (Na_2SO_3, $CoSO_4$) in accordance with ASCE-Standard (ASCE, 1984).

4. RESULTS AND DISCUSSION

4.1 Physico-Chemical Conditions

Bubble size distribution, rise velocities and the behavior of bubbles in a swarm determine the air volume V_G and therefore gas holdup e_G in the water bulk. Fig. 3a shows the influence of the admixtures on e_G for the three bubbles types. In clear water no significant difference in e_G is observable, though the three pictures in the first line of Fig. 3b indicate differences in bubble size distribution. For all air velocities U_{SG} the Sauter diameter d_{BS} is essentially constant. With the determination of d_{BS} and e_G the volumetric interfacial area a is computable (see Eq. (9)). Due to the decreasing bubble surface with increasing bubble volume the relation of a for small: medium: large bubbles =1: 0.35: 0.25. This proportion is nearly constant for all U_{SG} considered.

The adsorption of surface-active agent (surfactant) or natural impurities at the interfacial area is caused by its asymmetric molecule structure with an hydrophilic and a hydrophobic part. Whereas surfactants reduce the surface tension σ, no significant influence on physical properties of water is ascertainable with natural impurities.

The presence of surfactant affects the gas holdup in two ways: (a) due to the adsorption of the molecules at the interface, the mobility of the bubbles is suppressed. Bubbles of smaller and medium size now are spherical or ellipsoidal

366 INFLUENCE OF PHYSICO-CHEMICAL AND HYDRODYNAMIC CONDITIONS

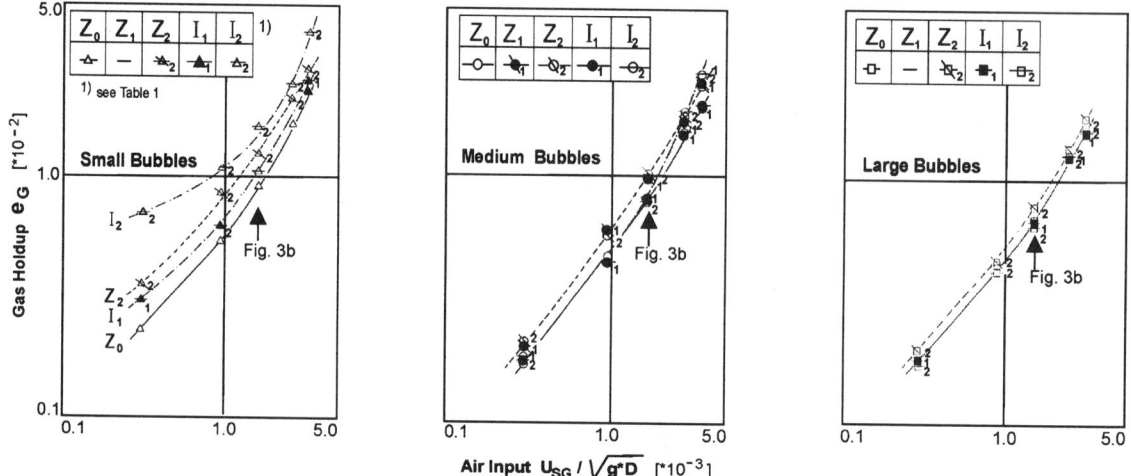

Figure 3a. Gas holdup versus air input: Influence of water containing admixture (surfactant, salt).

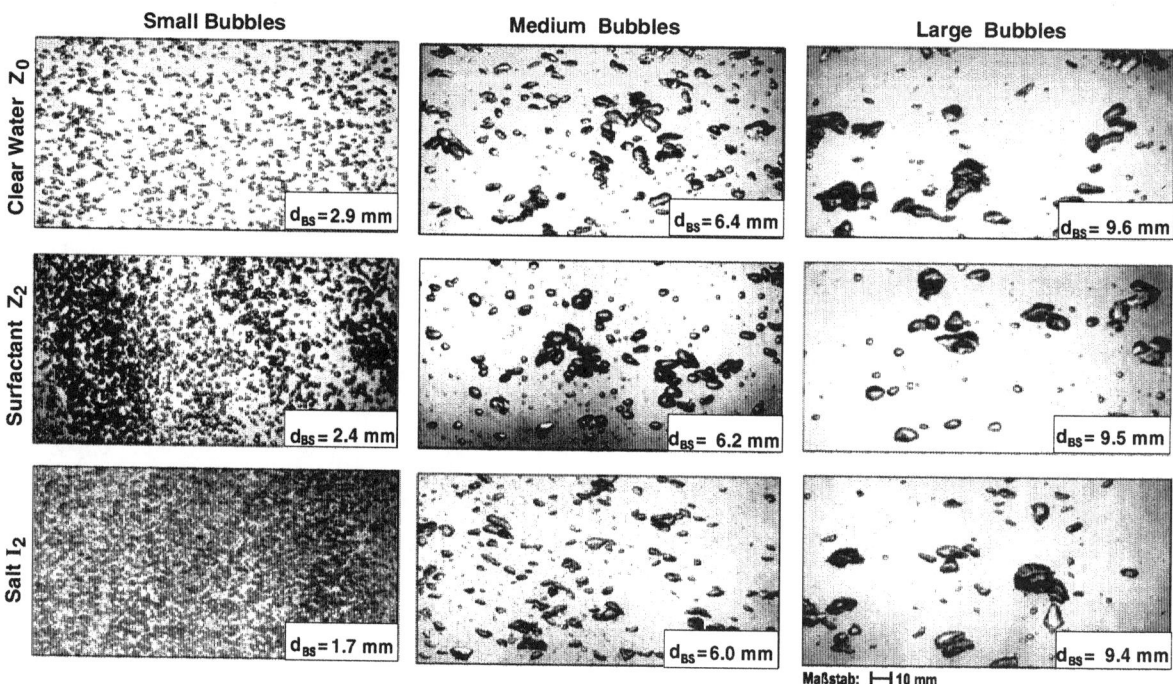

Figure 3b. Photographs of air bubbles in clear water (Z_0) and water containing admixtures (surfactant Z_2, salt I_2).

without any shape oscillation and internal circulation (Fig. 3b, second line). Now the rectilinear rise motion, certainly with a reduced rise velocity, is similar to rigid particles. Large bubbles show a hampered interfacial mobility and a less swaying rise path than in clear water. (b) Gas bubbles in clear liquids tend to coalescence, i.e. if a small bubble approaches closely enough another one, the thin liquid film between them "flows out" and after a period of time the bubbles merge. If the interfacial area is covered with surfactant molecules, this coalescence time increases as a function of concentration [Meusel, 1980] and the coalescence tendency is reduced. The picture for small bubbles in the second line in Fig. 3b (concentration Z_2) shows a slightly decreasing bubble diameter, effecting also the rise velocity.

Both phenomena mentioned above influence the gas holdup e_G. With regard to the interfacial area a for small bubbles in clear water, a is now increasing to the 1.6-fold. The size of large and medium bubbles are nearly unchanged (Fig. 3b, second line) because of the already observed low tendency for coalescence in clear water. Only the reduced

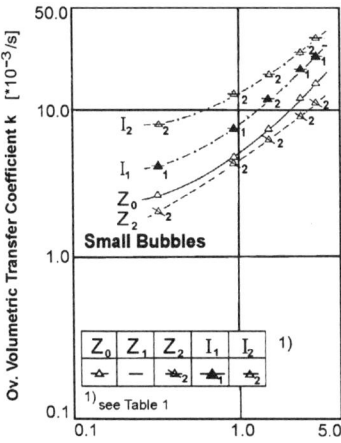 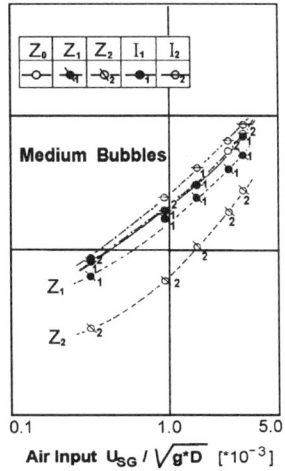 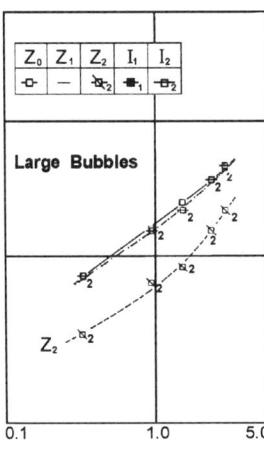

Figure 4. Volumetric transfer coefficient versus air input: Influence of water containing admixtures (surfactant, salt).

bubble mobility leads to a slight increase of e_G and a of approximately 4%.

Water molecules are strong dipoles which form so-called clusters. Interaction between the ions of salt and water leads to a more regular molecular formation. This effect also hampers the coalescence in form of a reduced draining of the fluid film between the bubbles. In recent investigations [Lessard and Zieminski, 1971, Meusel, 1980] a strong increase of coalescence inhibition was detected for salinity I between 0.5% - 1%. An enhancement of I > 1% should have no additional effect.

The pictures in the third line of Fig. 3b show a significant reduction of small bubble diameter d_{BS} with salinity I_2. Already a low concentration of $I_1 = 0.5\%$ shifts the diameter to $d_{BS} = 2.4$ mm. The shape of the bubbles is now spherical and they rise rectilinear. The graph for e_G reflects these effects. Again fixing the value of interfacial area a for small bubbles in clear water (Z_0) at 1, the proportion for Z_0 : I_1 : I_2 = 1 : 1.4 : 3.0. Similar to the results of water containing surfactant, diameter and mobility of bubbles of intermediate and large size are relatively unimpaired.

Fig. 4 shows the results of the oxygen transfer measurements under the same conditions just discussed. It should be mentioned first that the surfactant do not only affect the bubble behavior but also the interfacial transfer mechanism. The molecules adsorbed on the bubble surface first block the diffusion of oxygen through the interface and second reduces the internal circulation of a bubble. Thus the contact time of liquid surface elements with the gas phase may be increased and the transfer will be reduced.

The overall volumetric transfer coefficient k for small bubbles in liquid with concentration Z_2 is similar to clear water. This means that the decreasing of bubble size and the evident increase of the gas holdup e_G compensate the surfactant influence on transfer coefficient k_L. Oxygen transfer is visibly reduced for bubbles of medium and large size, in spite of a slightly increased gas holdup e_G and nearly constant bubbles sizes.

An effect of salinity on the transfer mechanism is not known [Burckhart and Deckwer, 1975, Elster 1979]. Considering that also no influence on bubble mobility was determined, the increasing value of k for small bubbles is mainly caused by extension of the interfacial area. The values for both other bubble types are equal to clear water. With these results the transfer coefficient k_L is computable (see Eq. (3)). The mean values for clear water are $k_L = 4.7 - 5.6 \cdot 10^{-4}$ m/s with the maximum for large bubbles caused by intensive shape oscillations.

The adsorption of surfactant (Z_2) leads to a reduced $k_L = 1.7 - 2.2 \cdot 10^{-4}$ m/s. Tests with low concentration (Z_1) yield a $k_L = 2.8 - 3.2 \cdot 10^{-4}$ m/s. This indicates an interdependence between concentration of surfactant and k_L as mentioned in literature [Meusel, 1980]

Salinity is of no effect on gas transfer, the values of $k_L = 4.7 - 5.4 \cdot 10^{-4}$ m/s are equal to clear water. Only the results for small bubbles (concentration I_2) diminish to $k_L = 3,2 \cdot 10^{-4}$ m/s, the reduced diameter being the reason for them behaving like rigid spheres.

4.2 Hydrodynamic Conditions

The effect of turbulence is significant mainly in the region of intense shear near the grids. This zone is fixed by the meshlength M of the grids [Baines and Peterson, 1951, Lance and Bataille, 1985]. With the power drop Δp across the grids and M the turbulent dissipation rate ε varies in the range of $\varepsilon = 0,2 - 34$ m^2/s^3, correspond with maximum

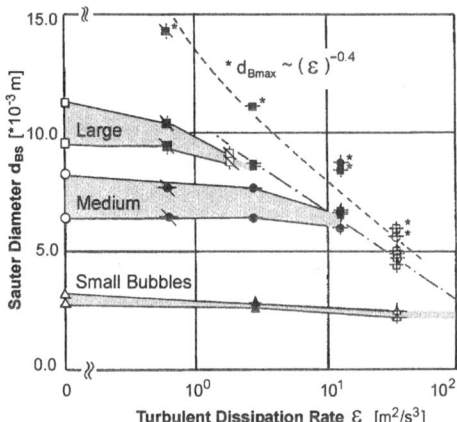

Figure 5. Bubble diameter versus turbulent dissipation rate.

turbulence intensities between 5% to 37%. The injection of air was located in this zone of high energy input. The effects of the grid-induced turbulence on gas transfer process may be the following: (a) additional shape deformation and oscillation lead to an enhancement of surface renewal frequency affecting the transfer coefficient; (b) bubble breakup in the region of intense shear reduces the bubble size; (c) vortexing of bubbles results in an increased residence time in the water bulk respectively an increasing gas holdup e_G. Both latter effects influence the dimension of the interfacial area a.

The bubble diameters d_{BS} for low and relatively high air input are plotted in Fig. 5 as a function of dissipation rate ε. On the left vertical axis of the figure the data for measurement without a grid are listed. In all series of measurement the bubble size distribution of small bubbles is not influenced. The shear forces may deform the bubbles, but do not divide them. A turbulent dissipation rate $\varepsilon > 10$ m^2/s^3 is necessary to break up the bubbles of medium size. Large bubbles will be divided into bubbles of smaller sizes for $\varepsilon > 1$ m^2/s^3.

The results indicate that the turbulent dissipation rate ε determines the final bubble size, independent of the origin bubble diameter. The correlation between ε and the maximum stable bubble diameter d_{Bmax} was found [Grossmann and Mersmann, 1973, Meusel, 1980] to:

$$d_{Bmax} = C \cdot (\sigma/\rho_L)^{0.8} \cdot \varepsilon^{-0.4}. \quad (10)$$

A comparison of the measured data, in Fig. 5 marked with a star, with this equation (dashed line) give an acceptable prediction in the magnitude of maximum bubble diameter d_{Bmax}. As expected the break-up of the bubbles enhance the gas holdup e_G for measurements with bubbles of medium and large sizes, whereas no variation is detected for small bubbles. As discussed in 4.1 the relation of interfacial area a for bubbles of small, medium and large size was denoted to

1: 0.35: 0.25. This proportion shifts to 1: 0.45: 0.45 for measurements with high dissipation rate $\varepsilon = 34$ m^2/s^3.

The combined consideration of d_{BS}, e_G and the measured overall volumetric transfer coefficient k yields the transfer coefficient k_L as a function of turbulent dissipation rate ε, presented in Fig. 6. A linearized regression over all data obtained shows a dependence of $k_L \sim (\varepsilon)^n$ with n = 0.20. Other authors [e.g. Levec and Pavko, 1979, Jun and Jain, 1993] identify in their experiments with turbulent two-phase flows a correlation with n = 0.17 - 0.33. In Fig. 6 the relation for n = 0.20 (dotted line) and n = 0.25 (small-eddy model, see Eq. (4), dashed line) is indicated. Under the given experimental conditions the role of the dominant eddy sizes on gas transfer cannot be clarified exactly. However the strong enhancement of gas transfer with increasing ε is shown.

5. CONCLUSIONS

Systematical measurements of air bubbles in clear water and water containing admixtures (surfactant, salt) and under various turbulent conditions have been carried out to quantify oxygen gas transfer. Based on a comparison of bubbles of different size ranges the results of transfer coefficient k_L are summarized as follows:

k_L is nearly constant for all bubble sizes in clear water and salt water (salinity 0.5%, 1.5%). Gas transfer is hampered in water containing surfactant due to the effect of molecular blocking at the bubble interface and also the suppression of any interfacial mobility. The value of k_L is reduced by as much as 60%.

Turbulent conditions parameterized by the turbulent dissipation rate ε increase the transfer coefficient k_L for all bubble sizes in clear water and k_L can be predicted with approximately $(\varepsilon)^{0.20}$.

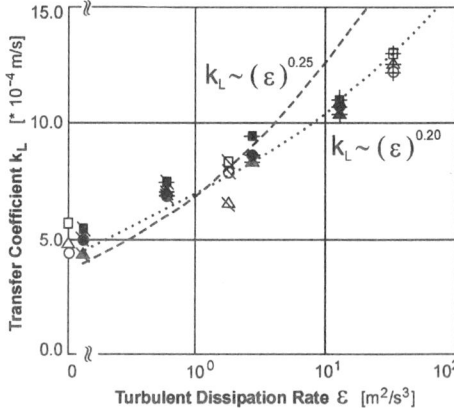

Figure 6. Transfer coefficient versus turbulent dissipation rate.

The dissipation rate ε is found to play an important role affecting the size of medium and large bubbles. The maximum value of bubble diameter can be predicted with $(\varepsilon)^{-0.4}$.

Considering also the dimension of the volumetric interfacial area a it can be shown that small bubbles in clear water are most effective for gas transfer, in spite of its tendency to coalescence. The presence of surfactant or salt increases the coalescence inhibition and therefore the interfacial area. The enhancement of a dependent on the concentration of the admixtures also is discussed. Because of the differences in k_L this leads to an increasing gas transfer for salt water and decreasing values for clear water and water containing surfactant.

Acknowledgments. This work was supported by the Ministry of Education and Research of the Federal Republic of Germany.

REFERENCES

ASCE-Standards, A standard for the measurement of oxygen transfer in clear water, *ASCE*, New York, 1984.

Asher, W.E., Pankow, J.F., The interactions of mechanically generated turbulence and interfacial films with a liquid phase controlled gas/liquid transport process, *Tellus*, 38B, pp. 305-318, 1986.

Baines, W.D., Peterson, E., An Investigation of Flow through Screens, *Trans. ASME*, July 1951.

Buckhart, R., Deckwer, W.D., Bubble size distribution and interfacial areas of electrolyte solutions in bubble columns, *Chem. Eng. Science*, Vol. 30, 1975.

Calderbank, P.H., Lochiel, A.C., Mass transfer coefficients, velocities and shapes of carbon dioxin bubbles in free rise through distilled water, *Chem. Eng. Science*, Vol. 19, 1964.

Clift, R., Grace, J.R., Weber, M.E., Bubble, Drops and Particles, *Academic Press N.Y.*, 1978.

Danckwerts, P.V., Significance of liquid-film coefficients in gas adsorption, *Ind. and Eng. Chem.*, Vol. 43, pp. 1460-1467, 1951.

Elster, F., Oxygen transfer from air to liquid solutions, *Ph.D. Thesis*, University of Dortmund, Germany, 1979.

Galloway, T.R., Sage, B.H., Thermal and Material Transfer from Spheres, *Int. J. of Heat and Mass Transfer*, vol. 11, 1968.

Jirka, G.H., Environmental Fluid Mechanics: Its Role in Solving Problems of Pollution in Lakes, Rivers and Coastal Waters in *Env. Appl. of Mechanics and Computer Science*, edited by G. Bianchi, Ed. Springer, 1999.

Jun, K.S., Jain, S.C., Oxygen transfer in bubbly turbulent shear flow, *J. Hydr. Eng.*, Vol. 119, pp. 21-36, 1993.

Lance, M. Bataille, J., Homogeneous Turbulence in Bubbly Flows, *ASME*, N.Y., 1985.

Lessard, R.R., Zieminski, S.A., Bubble Coalescence and Gas Transfer in Aqueous Electrolytic Solutions., *Ind. Eng. Chem. Fundam.*, 1971.

Levec, J., Pavko, S., Mass transfer in square gas-liquid contractors, *Chem. Eng. Science*, Vol. 34, 1979.

Lewis, W.K., Whitman, W.G., Principles of gas adsorption, *Indust. and Eng. Chem.*, Vol. 16, pp. 1215-1220, 1924.

Meusel, W., Influence of particle coalescence on mass transfer in turbulent gas-liquid-systems, *Ph.D. Thesis*, School of Engineering, Köthen, Germany, 1980.

Nakayama, F., Turbulence and Coherent Structure Across Air-Water Interface and Relationship with Gas Transfer, *Ph.D. Thesis*, University Kyoto, Japan, 2000.

Treyball, T.G., Mass-Transfer Operations, *McGraw Hill*, New York, 1981.

Cornelia Lang, Institute for Hydromechanics, University of Karlsruhe, D-76128 Karlsruhe, Kaiserstrasse 12, Germany.
Email: c.lang@uni-karlsruhe.de

Air-Water Gas Transfer in Uniform Flows With Large Gravel-Bed Roughness

Douglas B. Moog

Dept. of Geological Sciences, Case Western Reserve University, Cleveland, Ohio

Gerhard H. Jirka

Institut für Hydromechanik, Universität Karlsruhe, Karlsruhe, Germany

This study presents laboratory data on oxygen absorption by uniform flow over a gravel bed, and assesses the suitability of an expression derived for uniform flow over a smooth bed. This is a formulation of the small eddy model, a version of surface renewal theory in which the smallest eddies dominate interfacial gas transport. It was earlier found to work for nonuniform, "macrorough" flows, in which resistance is dominated by depth-scale form drag, upon proper scaling of the turbulent energy dissipation rate. While the small eddy model worked well for smooth and small-roughness flows, the same expression was found to underpredict reaeration in uniform flow over gravel with relative roughness over 0.13. More successful was an equation which scaled the turbulent intensity in the dissipation rate expression by the mean velocity rather than friction velocity. This success is believed to reflect the production of turbulence in bed particle wakes, rather than by the mean shear. Studies of turbulence in gravel-bed channels in the laboratory and field support this scaling. The small eddy model has been found to well describe the trend of the gas transfer coefficient with Reynolds number in a wide variety of cases: smooth and small-roughness walls; nonuniform, macrorough flows; and uniform, large-roughness gravel beds. The key is to properly express the near-surface turbulent energy dissipation rate in terms of macro-scale parameters. Results of this study lend confidence to use of the small eddy model in flows with turbulent Reynolds number above 400.

INTRODUCTION

In gravel-bed streams, fish habitat has attracted much recent attention. Dissolved oxygen and its absorption from the atmosphere, known as "reaeration," is an important component. (Bain & Stevenson, 1999). Gravel sizes affect not only spawning and shelter, but reaeration as well, through effects on water turbulence. This study describes laboratory studies on oxygen absorption in a gravel-bed channel and their theoretical implications, for both reaeration and air-water gas transfer theory in general. The results are compared with similar laboratory work [Lau, 1975].

Small Eddy Model

Experiments and analysis [Moog and Jirka, 1999a] found that the small-eddy model, first formulated by Lamont and

Scott [1970], represents an accurate form of renewal theory for uniform flow in channels of small roughness, $k_s^+ \equiv k_s u_*/\nu < 136$, and shear Reynolds numbers $R_* \equiv u_* H/\nu > 400$, where u_* is friction velocity; H is flow depth, ν is kinematic viscosity, and k_s is the equivalent sand-grain diameter. Renewal theory may be expressed as

$$K_L = \sqrt{D_m r} \qquad (1)$$

where K_L is the gas transfer coefficient, D_m is the molecular diffusivity of oxygen in water, and r is the *renewal rate*, the mean frequency at which turbulent motions impinge on and renew the thin region near the free surface which is nearly saturated with oxygen. The small-eddy model approximates r using the Kolmogorov time scale as

$$r \propto \left(\frac{\epsilon}{\nu}\right)^{1/2} \qquad (2)$$

where ϵ is the turbulent energy dissipation rate near the free surface. It is of order

$$\epsilon \sim \frac{(u')^3}{L} \qquad (3)$$

where u' is the mean turbulent intensity and L is the macroscale. For open channels, it is commonly taken as

$$\epsilon \sim \frac{u_*^3}{H} . \qquad (4)$$

Moog and Jirka [1999a] showed this form to work well in the small-eddy model for small-roughness uniform flow. Combining Eqs. 1, 2, and 4 leads to the nondimensional form

$$K_L^+ \propto Sc^{-0.5} R_*^{-0.25} \qquad (5)$$

where $K_L^+ \equiv K_L/u_*$; Sc is the Schmidt number ν/D_m; and the shear Reynolds number R_* is $u_* H/\nu$.

Roughness Regimes

Moog and Jirka [1999b] tested the small-eddy model on *macrorough flow*, nonuniform flow in which irregularities in the bed geometry are large enough to cause significant depth-scale form resistance. They found that the form of the small eddy model used for uniform, small-roughness flow did not work, apparently because Eq. 4 is inaccurate for macrorough flow, for which much more of the energy dissipation should occur near the free surface. Dissipation rates formulated on this basis worked very well.

This report concerns flows which differ from macrorough flows in that they are globally uniform; that is, the depth is reasonably spatially constant. Drag forces may be considered to originate near the bed, with most energy dissipation well below the free surface.

Even though the flows are uniform and not "macrorough," they may be considered to be of "large roughness." They differ from smooth or small-roughness flow, in which turbulence production is dominated by interaction of Reynolds stresses with the mean velocity gradient, which takes a logarithmic form. When the bed particles are large enough, their wakes become important sources of resistance, distorting the logarithmic profile. Ferro and Baiamonte [1994] showed this to occur for a relative roughness $D_{50}^+ \equiv D_{50}/H > 0.13$, where D_{50} is the median grain size. Considering this and other evidence, Moog [1995] adopted this as the range for "large-roughness," uniform flow over gravel beds.

Turbulence Over Gravel Beds

In smooth or small-roughness channels, the mode of turbulent energy production leads u' to scale with u_*. With larger grains, production takes place in particle wakes. Because turbulent intensity in a wake scales with the difference in velocity between the wake and the free stream [Tennekes and Lumley, 1972], one may expect u' to scale instead with the mean stream velocity U.

Turbulence measurements in uniform flow over gravel beds do indicate that u' scales better with U than with u_*. Laboratory measurements over uniform gravel [Wang, 1990; Wang et al., 1993] showed that u'/u_* varied with roughness for $D_{50}^+ < 1$, but u'/U remained constant over the entire range. Over nonuniform gravel in the large-roughness range, Wang and Dong [1996] found that both u'/u_* and U/u_* were independent of roughness above the particle wake layer; hence, u'/U would be independent of roughness in the source region for interfacial transfer. In three New Zealand gravel-bed rivers, Nikora and Smart [1997] found u' profiles to scale better with U than with u_*. (They did not report D_{50}, but judging from D_{90}, it is very likely that the almost all data were taken for $D_{50}^+ > 0.13$.) They reported a similar result from rivers in Georgia. It should be noted that the effective roughness size for a given D_{50} is much greater in a natural stream than in a laboratory flume with a single, immobile gravel layer, owing to particle agglomeration [Whiting & Dietrich, 1990].

Present Study

This paper reports on laboratory reaeration experiments over gravel beds with relative roughnesses varying from small- to large-scale. The objectives are to determine whether the small eddy model is consistent with such flows, and whether it requires different scaling than smooth channels, as did the macrorough case.

EXPERIMENTAL METHODS

Oxygen absorption rates were measured in the Tilting Wind-Water Tunnel (TWWT) at Cornell University, a tilting, recirculating flume of 20 m length and 1 m width. Gravel of median diameter 0.75 cm and 84th percentile of 1.25 cm was glued to canvas sheets in a single layer, and placed on the bed of the flume. The walls remained smooth. Mean flow depths were 2.5, 5, 10, and 20 cm, so that relative roughnesses D_{50}^+ values of 0.0375, 0.075, 0.15, and 0.3 were tested.

Deoxygenated water was pumped into the flume and circulated while it absorbed oxygen from the air. Samples were taken repeatedly at stations separated by 17 m, and analyzed for dissolved oxygen using the Winkler titration method. Absorption rates were determined from the total time derivative of the concentration change between the measurement stations.

Flow rates were obtained from an ultrasonic flowmeter mounted on the return pipe, with flow depth read from rulers on the plexiglass channel walls. Friction velocity was calculated as

$$u_* = \sqrt{gR_hS} \qquad (6)$$

where g is gravitational acceleration, R_h is the hydraulic radius based on a flat bed, and S is the flume slope. Further details of the test conditions and procedures are described by Moog [1995]. The procedures are the same as those reported by Moog and Jirka [1999a] for smooth-bed flows.

EXPERIMENTAL RESULTS AND DISCUSSION

From Eqs. 1 and 2, the small eddy model predicts that K_L varies as $\epsilon^{1/4}$. Given proper estimates for ϵ, a plot of K_L vs. $\epsilon^{1/4}$ should thus form a straight line. Figure 1 depicts such a plot for the gravel-bed tests, compared to the best-fit line for the smooth-channel runs. K_L values were corrected to a standard 20 C water temperature [Moog and Jirka, 1999a]. The dissipation rate is calculated from Eq. 4 and labeled ϵ_s to indicate that it corresponds to the smooth-wall case. Within measurement uncertainty, the small-roughness runs - at 10 cm

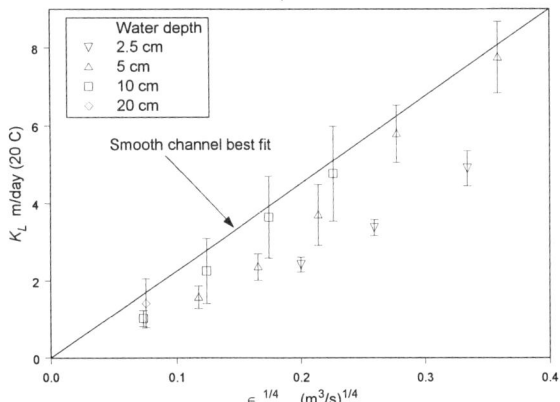

Figure 1. Gravel-bed gas transfer coefficients versus small-eddy model dissipation rate dependence, showing deviation from smooth channel behavior for large relative roughness (low water depth). Error bars indicate 95% confidence intervals.

and 20 cm depths - obey the same small-eddy equation calculated from the smooth-channel runs.

However, the large-roughness flows - at 2.5 cm and 5 cm depth - show K_L values significantly below those predicted by the smooth-channel equation. The same data are shown in non-dimensional form in Figure 2. A line is drawn on this logarithmic plot with slope of -0.25, equal to the small-eddy model's R_* exponent from Eq. 5. Again, the less-rough gravel-bed data, at the highest R_*, agree with the trend of the smooth bed data, but the others fall well below. Shown for comparison are the data of Lau [1975], as reported by Gulliver and Halverson [1989]. Though lower in magnitude (as discussed by Moog and Jirka, 1999a), they show the same invariance of $K_L^+ Sc$ with R_* as do the TWWT gravel-bed data. Lau's small-roughness data (not shown) also resemble those from the TWWT in having a slope of -0.25, in agreement with the small eddy model [Moog and Jirka, 1999a].

Despite the apparent roughness distinction, the gravel-bed results do agree with the small-eddy model upon improved scaling of the dissipation rate ϵ. As discussed in the Introduction, evidence indicates that the turbulent intensity u' in uniform flow over gravel beds is proportional more to U than to u_*. Using the mean velocity to scale ϵ in Eq. 3,

$$\epsilon \propto \frac{U^3}{H} \qquad (7)$$

K_L no longer shows separation by flow depth and relative roughness as in Figure 1. A non-dimensional plot using Eq. 7 and K_L normalized by U (Figure 3) shows slopes close to -0.25 for both the TWWT and Lau [1975] data. It is apparent

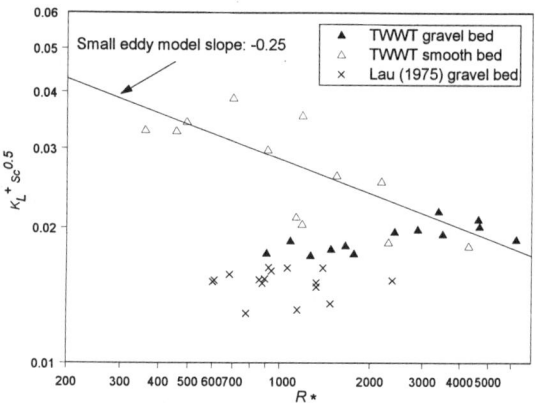

Figure 2. Nondimensionalized gas transfer coefficients showing deviation of gravel-bed data from smooth channel small eddy model when using friction velocity scaling, for large roughness (low R_*).

that the small eddy model, Eq. 5, is consistent with the large-roughness, uniform-flow gravel bed data when the turbulent intensity in Eq. 3 is scaled with U rather than u_*. Together with the success of the macroroughness modification [Moog and Jirka, 1999b], it lends confidence to use of the small eddy model when near-surface dissipation rates are accurately estimated.

The distinction here between small- and large-scale roughness at $D_{50}^+ = 0.13$ is unrealistically abrupt. A better formulation for the dissipation rate would provide a smoother transition, as provided by a linear interpolation between Eqs. 4 and 7:

$$\epsilon_G = (1 - \phi_G)\left(\frac{u_*^3}{H}\right) + \phi_G\left(\frac{(bU)^3}{H}\right) \quad (8)$$

where b is a constant equal to the value of u_*/U for small roughness, about 0.07 for the TWWT data, and ϕ_G is an interpolation factor

$$\phi_G = 1 - e^{-aD_{50}^+} \quad (9)$$

which ranges from 0 under smooth conditions to nearly 1 in large-scale roughness. The exponential form was chosen to provide a sufficiently rapid transition, and a smooth asymptote at large D_{50}^+. Nonlinear regression determined a value of 4.2 for a in the TWWT. As noted earlier, the same D_{50}^+ produces greater roughness in a stream than in the TWWT; Moog [1995] presents an adjustment for this condition which corrects a to 6.8 for natural gravel-bed streams.

Figure 4 shows K_L plotted versus the small-eddy model dependence on the dissipation rate estimate in Eq. 8, similar to Figure 1 with ϵ_G replacing ϵ_S. The smooth-bed and gravel-bed data show identical dependence on ϵ_G. By smoothly switching the scaling of turbulent intensity from friction velocity to mean velocity, ϵ_G generalizes the small eddy model to work for smooth, small-roughness, and large-roughness uniform flow.

CONCLUSIONS

While the small eddy model worked well for smooth and small-roughness flows, the same formulation was found to underpredict reaeration in uniform flow over gravel with relative roughness over 0.13. More successful was an equation which scaled the turbulent intensity in the dissipation rate expression by the mean velocity rather than the friction velocity. This success is believed to result from the production of turbulence in bed particle wakes rather than by the interaction of Reynolds stresses with the mean shear. Previous laboratory and field measurements of turbulence in gravel-bed channels support this scaling.

The small eddy model has been found to well describe the trend of the gas transfer coefficient with Reynolds number in a wide variety of cases: smooth and small-roughness walls; nonuniform, macrorough flows; and uniform, large-roughness gravel beds. The key is to properly express the near-surface turbulent energy dissipation rates in terms of macro-scale parameters. Results of this study lend confidence to use of the small eddy model in flows with turbulent Reynolds number above 400.

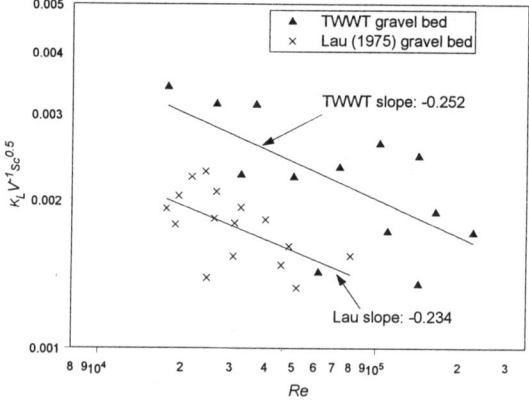

Figure 3. Nondimensionalized gas transfer coefficient with turbulent intensity scaled by mean velocity instead of friction velocity, now showing small eddy model dependence on Reynolds number.

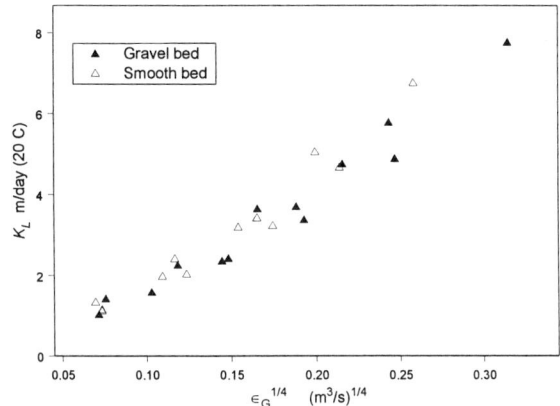

Figure 4. Gas transfer coefficient versus small-eddy model dissipation rate dependence, with scaling of turbulent intensity switching from friction velocity to mean velocity at large relative roughness.

Acknowledgements. The support of the U.S. Geological Survey and the DeFrees Family Foundation is gratefully acknowledged.

REFERENCES

Ferro, V., and G. Baiamonte, Flow velocity profiles in gravel-bed rivers, *J. Hydraul. Eng.*, 120, 60-80, 1994.

Gulliver, J. S., and M. J. Halverson, Air-water gas transfer in open channels, *Water Resour. Res.*, 25, 1783-1793, 1989.

Lamont, J. C., and D. S. Scott, An eddy cell model of mass transfer into the surface of a turbulent liquid, *AIChE J.*, 16, 513-519, 1970.

Lau, Y. L., An experimental investigation of reaeration in open-channel flow, *Prog. in Water Tech.*, 7, 519-530, 1975.

Moog, D. B., *Stream Reaeration and the Effects of Large-scale Roughness and Bedforms*, Ph.D. thesis, Cornell Univ., Ithaca, NY, 1995.

Moog, D. B., and G. H. Jirka, Air-water gas transfer in uniform channel flow, *J Hydraul. Eng.*, 125, 3-10, 1999a.

Moog, D. B., and G. H. Jirka, Stream reaeration in nonuniform flow: macroroughness enhancement, *J Hydraul. Eng.*, 125, 11-16, 1999b.

Nikora, V. I., and G. M. Smart, Turbulence characteristics of New Zealand gravel-bed rivers, *J Hydraul. Eng.*, 123, 764-773, 1997.

Tennekes, H., and J. L. Lumley, *A First Course in Turbulence*, The MIT Press, Cambridge, Mass., 1972.

Wang, J.-J., Distribution of turbulent intensity in a gravel-bed flume, *Expts. In Fluids*, 11, 201-202, 1990.

Wang, J., C. Chen, Z. Dong, and Z. Xia, The effects of bed roughness on the distribution of turbulent intensities in open-channel flow, *J. Hydraul. Res.*, 31, 89-98, 1993.

Wang, J.-J., and Z.-N. Dong, Open-channel turbulent flow over non-uniform gravel beds, *Applied. Sci. Res.*, 56, 243-254, 1996.

Whiting, P. J., and W. E. Dietrich, Boundary shear stress and roughness over mobile alluvial beds, *J. Hydraul. Eng.*, 116, 1495-1511, 1990.

Douglas B. Moog. Dept. of Geological Sci., Case Western Reserve Univ., Cleveland, OH, 44106-7216, USA.

Gerhard H. Jirka, Institut für Hydromechanik, PO Box 6380,Universität Karlsruhe, D-76128 Karlsruhe, Germany.

Large Scale Laboratory Experiments for Water Oxygenation Under Breaking Waves

V.K. Tsoukala, E.I. Daniil and C.I. Moutzouris

Laboratory of Harbor Works, Faculty of Civil Engineering, National Technical University of Athens, Greece

Oxygenation experiments under breaking waves in large-scale experimental facilities are presented. Experiments were performed with a sloping beach at the wind wave flume of Delft Hydraulics and a three-layer rubble mound breakwater at the Schneideberg Wave Flume of Franzius Institut in Universitat Hannover. The scaling law of the phenomenon has not yet been determined and the model for prediction of the transfer coefficient from the wave characteristics needs to be refined by inclusion of a parameter describing the breaking. The apparent transfer coefficients for the large scale experiments were lower than those determined from small scale experiments in the Laboratory of Harbour Works, NTUA. However, the actual oxygen transfer coefficients, as computed using a discretized form of the transport equation and accounting for dispersion, are in the same order of magnitude for small and large scale experiments. Correlations of the oxygen transfer coefficients with dimensionless breaking wave indexing parameters are presented and a modified vorticity based renewal model incorporating the breaking wave Reynolds number is proposed, that describes both small and large scale experimental data well.

1. INTRODUCTION

A research program is conducted by the Laboratory of Harbour Works (LHW), National Technical University of Athens, on the role of breaking waves on seawater oxygenation in coastal zones since 1991. For the transfer of laboratory measurements to field conditions, the scaling law of the phenomenon has to be determined. *Falvey and Ervine* [1988] refer to four different cases of air entrainment and in each case a different dimensionless parameter has to be used for scaling. In the case of oxygenation due to breaking waves no appropriate parameter has been found yet. Scale effects should be further investigated and the model for prediction of the transfer coefficient from the wave characteristics be refined, possibly by the inclusion of a parameter describing the breaking.

Towards this direction, small-scale experimental measurements were initially performed in the wave flume of LHW (27x0.60x1.53m), with waves breaking on a smooth sloping beach with uniformly slope of 1:2.3 and 1:5 and on a rubble mound breakwater with a 1:1.5 sloping front [*Daniil and Moutzouris*, 1994; *Tsoukala*, 2000]. For the investigation of scale effects large-scale experiments were conducted in Delft Hydraulics, The Netherlands, financed by the European Union under the Human Capital Mobility - Large Installations Programme (1994) and at Franzius Institut, Germany, financed by Training and Mobility of Researchers – Access to Large Scale Facilities Programme (1999). Additionally, field data are presently being collected in selected harbour basins [*Daniil et al.*, 2000a].

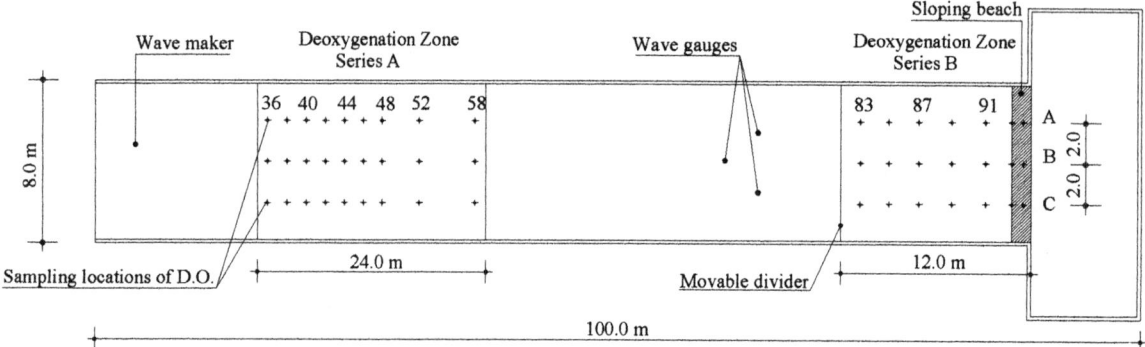

Figure 1. Schematic Plan View and Experimental Setup of Wind Wave Flume in Delft Hydraulics

Herein results from the large-scale experiments are presented.

2. EXPERIMENTAL PROCEDURE

Large-scale oxygenation experiments under breaking waves were performed with a sloping beach at the Wind Wave Flume of Delft Hydraulics (DH) in The Netherlands and with a three layer rubble mound breakwater at the Schneideberg Wave Flume at Franzius Institut (FI) at the University of Hannover in Germany.

The test method is based upon removal of dissolved oxygen (D.O.) from the water volume followed by reoxygenation to near saturation level as described in the *ASCE Standard* [1993] for Measurement of Oxygen Transfer in Clean Water. Water temperature and pressure were recorded for the determination of the saturation concentration. D.O.-time histories were obtained for all experiments and sampling locations. In order to verify the accuracy of readings obtained with the oxygen meters, water samples were also taken and analysed using the azide modification of the iodometric Winkler titration method. The measurements of D.O. concentration commenced immediately upon wave generation and continued until the D.O. value reached ~80% of the estimated saturation level for the specific conditions of temperature and atmospheric pressure.

Wind Wave Flume Delft Hydraulics

The wave flume of Delft Hydraulics is "T" shaped with a total length of 100m and 8m width, while the last 9m are 25m wide. The outline of the flume is shown in Figure 1. The flume is equipped with a hydraulic dual piston wave maker for the generation of mechanical waves. Waves are generated by a computer controlled wave board with adjustable rotation and translation. Waves with heights between 7.5–28.3cm and wave periods between 1.07-1.90sec were produced. The experimental procedure consisted of two distinct series of experiments. In the first series of experiments (AD_4-AD_5, D_{10}-D_{12}) waves were produced without a coastal structure in the flume. In the second series of experiments (D_1-D_9) a concrete structure with a uniform slope of 1:2.3 was placed at the one end of the flume in order to model a sloping beach and initiate wave breaking. The structure was watertight and no water exchange between the two sides of the sloping beach was possible. Three capacitance wave gauges were used for recording the wave characteristics (Figure 1). The water was deoxygenated using sodium sulphite and cobalt chloride for the experiments AD_4-AD_7. About 8kg of Na_2SO_3 and 0.55kg of $CoCl_2 \cdot 6H_2O$ were required for every experiment to deoxygenate an area of 12m length and 8m width. For the experiments D_1-D_{12} the nitrogen stripping method was used. Long elastic tubes with small pores were connected to nitrogen cylinders equipped with regulating pressure valves. The elastic tubes were placed in a 12m long area of the flume, which was isolated during the deoxygenation. The concentration of D.O. in the water body was monitored over time by a portable OXI-196 of WTW Germany oxygen meter, at 21 sampling locations along the flume.

Schneideberg Wave Flume

The Schneideberg Wave Flume of Franzius Institut is ~100m long, 2m wide and 2m deep. For the conducted experiments about half of the flume length was used and it was isolated from the rest of the flume with a vertical watertight wooden wall. A three-layer breakwater model with 1:1.5 slope and 0.50m long crest at 1.20 above bottom elevation was constructed at a distance of 27.75m from the wave generator (Figure 2). The armor layer consisted of stones of mean diameter 3.5cm and mean weight W=100gr. For the under layer and the core, compared to the armor layer, stones of W/15÷20 and W/2000 were used respectively.

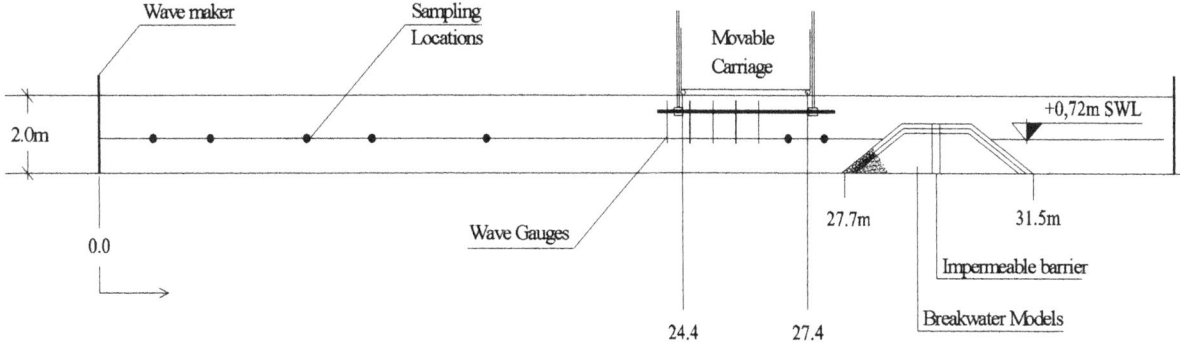

Figure 2. Shematic Cross View and Experimental Setup in Schneideberg Wave Flume at Franzius Institut

Two different series of experiments were conducted. For the first series the breakwater structure was impermeable. A wooden barrier was placed in the middle of breakwater body prohibiting any transport through the structure. In the second series of experiments the barrier was removed but the breakwater remained almost impermeable due to the layering of the structure. Wave height used in the experiments ranged from 9.2 to 16.5cm and wave periods from 1.11 to 1.43sec. Five capacitance wave gauges were used for recording the wave characteristics. The nitrogen stripping method was used for deoxygenation. The water was deoxygenated in front of the breakwater in all experiments and behind the breakwater for the second set of experiments. D.O. concentration at FI was monitored at 15 sampling locations by a YSI Model 95 oxygen meter.

3. DATA ANALYSIS

Determination of the Oxygen Transfer Coefficient K_L

The transfer coefficient is determined indirectly through the mass transport equation, when the rest of the terms are known. The unsteady one dimensional transport equation is usually used for analysis of data from laboratory flumes

$$\frac{\partial C}{\partial t} + U \frac{\partial C}{\partial x} = \frac{\partial}{\partial x}\left(D_x \frac{\partial C}{\partial x}\right) + S \quad (1)$$

where C is the concentration of dissolved oxygen, U is the mean stream velocity in the x-direction (U=0 for the wave flume), D_x is the longitudinal dispersion coefficient in the x-direction, S is source (S>0) or sink (S<0) term per unit volume. The oxygen transfer through the air water interface can be expressed as a source term as:

$$S = K_L(C_S - C)A_S / V \quad (2)$$

where K_L is the oxygen transfer coefficient, t is time, A_S is the average air-water surface area on the horizontal plane and V is the aerated water volume extending from the free surface to the bottom of the channel.

If the only source term is the air-water gas transfer, and horizontal transport and dispersion terms can be neglected, the transport equation is reduced to a first order differential equation. For the initial condition $C=C_0$ at t=0, and constant concentration for all the experiment the solution is:

$$C_i = C_S - (C_S - C_0) e^{-K_{Li}(A_{Si}/V_i)t} \quad (3)$$

Equation (3) can be expressed as a linear function of time by taking algorithms and the transfer coefficient determined using linear regression and the measured DO concentrations. The transfer coefficients computed from Equation (3) for the large-scale experiments reported here were lower than those from the small-scale experiments [*Tsoukala and Moutzouris*, 1997, *Tofa*, 1999].

The above solution is exact for the case of a completely mixed fluid volume. If the source terms are not uniformly distributed, horizontal and diffusion terms are generated, [*Horsch*, 1998 *Daniil et al.*, 2000b] and Equation (3) is not applicable. In this case a more detailed model which includes the effect of all the terms of Equation (1), is required for the calculation of transfer coefficients

For the performed experiments the effect of increased longitudinal dispersion due to wave breaking on the structures, and sometimes by the wave maker, has been considered. Equation (1) is discretized using a control volume approach [*Patankar*, 1980]. Integrating over a control volume of length L_i, volume V_i, cross sectional area equal to W (width of the channel) times d (the water depth), assuming that $\partial C/\partial t$ and S can be represented by their values at point i and a linear profile between grid points for $\partial C/\partial x$, Equation (1) can be written:

$$V_i \frac{\partial C}{\partial t} = \frac{D_{i,i+1}}{L_{i,i+1}} W\, d\, (C_{i+1} - C_i)$$

$$- \frac{D_{i-1,i}}{L_{i-1,i}} W\, d\, (C_i - C_{i-1}) + V_i\, S_i \quad (4)$$

Table 1. Wave Characteristics and Transfer Coefficients.

Exp. Series	Exp. No	T (sec)	H (cm)	Structure	H_b (cm)	L (m)	L_b (m)	Θ (°C)	P (mmHg)	C_S (mg/l)	Sc	apparent $K_{Lap}*10^4$ (m/s)	actual K_L*10^4 (m/s)
DH-A	AD_4	1.90	28.27	-	-	4.37	-	12.7	762.5	10.62	809	0.34	-
DH-A	AD_5	1.55	21.24	-	-	3.30	-	13.2	761.0	10.48	786	0.35	-
DH-A	AD_7	1.55	20.18	-	-	3.30	-	13.9	766.5	10.40	756	0.39	-
DH-A	D_{10}	1.10	9.12	-	-	1.86	-	17.9	759.0	9.46	609	0.57	-
DH-A	D_{11}	1.55	15.27	-	-	3.30	-	17.9	759.0	9.46	609	0.62	-
DH-A	D_{12}	1.30	12.55	-	-	2.50	-	17.9	751.1	9.36	609	0.33	-
DH-B	D_1	1.52	15.31	SB	22.41	3.20	1.15	19.4	757.5	9.15	563	1.28	1.80
DH-B	D_2	1.54	14.41	SB	21.42	3.27	1.09	19.3	760.0	9.20	566	0.83	0.95
DH-B	D_4	1.54	8.18	SB	14.08	3.27	0.72	18.5	764.0	9.41	590	0.20	0.67
DH-B	D_5	1.23	10.07	SB	14.88	2.28	0.76	18.6	757.5	9.31	587	0.68	1.19
DH-B	D_6	1.26	7.06	SB	11.52	2.37	0.59	18.6	754.5	9.27	587	0.34	1.22
DH-B	D_7	1.22	8.06	SB	12.53	2.22	0.64	18.6	751.0	9.23	587	0.38	0.97
DH-B	D_8	1.07	7.02	SB	10.62	1.77	0.54	18.3	752.0	9.30	596	0.25	0.27
DH-B	D_9	1.13	8.04	SB	12.06	1.96	0.62	18.0	759.0	9.44	606	0.34	0.58
FI	F_1	1.11	15.60	IB	21.42	1.91	0.46	13.3	728.0	10.10	782	12.5	*
FI	F_2	1.43	12.68	IB	20.97	2.92	0.45	15.3	711.0	9.35	701	0.67	*
FI	F_3	1.43	15.74	IB	18.46	2.92	0.53	15.5	711.0	9.32	693	0.80	*
FI	F_4	1.11	13.16	IB	18.30	1.89	0.41	15.7	714.0	9.32	686	0.68	*
FI	F_5	1.11	16.45	IB	20.37	1.89	0.47	11.7	724.0	10.33	856	1.04	*
FI	F_6	1.43	13.01	PB	23.96	1.89	0.40	12.9	722.0	10.01	800	0.58	*
FI	F_7	1.43	15.25	PB	23.40	2.92	0.52	13.5	717.5	9.83	774	0.69	*
FI	F_8	1.43	9.22	PB	16.04	2.92	0.35	13.9	715.0	9.70	757	0.76	*
FI	F_9	1.11	13.51	PB	21.37	1.89	0.47	14.5	725.0	9.71	732	0.41	*

SB : Sloping beach with slope 1 : 2.3, IB/ PB : impermeable/ permeable breakwater with a 1 : 1.5 sloping front,
* : Analysis of experiments from FI is presently underway
Water depth d = 0.72m for all experiments, with the exception of F_1 where d= 0.80cm.

where $L_{i,i+1}$, $L_{i-1,i}$ is the distance and $D_{i,i+1}$, $D_{i-1,i}$ the horizontal diffusion coefficient between grid points i, i+1 and i-1, i respectively.

For the experiments in DH in Series A, no longitudinal variation in D.O. concentration was observed. Therefore Equation (3) was used for the determination of the oxygen transfer coefficient. In Series B, where the waves were breaking and longitudinal variation was observed the oxygen transfer coefficients were determined based on the numerical scheme described by Equation (4) [*Tsoukala*, 2000]. Five control volumes were used. Each control volume containing at least a sampling point. The system of equations described by Equation (4) was solved numerically using a T.D.M.A. algorithm for the period of each experiment. The initial concentrations measured at each sampling point were used as inputs. The gas transfer and diffusion coefficients were adjusted till a best fit with observed D.O. values was obtained [*Wanninkhof, R. and L.F. Bliven*, 1994]. The actual oxygen transfer coefficients (Table 1) are in the same order of magnitude for small and large-scale experiments [*Tsoukala*, 2000].

For the FI experiments similar analysis is presently underway and the actual oxygen transfer coefficient are anticipated to be on the order of 25-50% higher than those determined from Equation (3).

Transfer coefficients were translated to 20 °C assuming a square root dependence on the Schmidt number according to the equation given by *Daniil and Gulliver* [1988].

4. RESULTS

Comparison to Vorticity Model

According to the first surface renewal model [*Dankwerts*, 1951] the gas transfer coefficient can be expressed as a function of the rate of surface renewal:

$$K_L \propto \sqrt{D_m r} \text{ or } K_L Sc^{1/2} \propto \sqrt{v r} \qquad (5)$$

where K_L is the gas transfer coefficient, D_m (m²/s) is the molecular diffusivity of the gas in the water, v (m²/s) is the kinematic viscosity of the water, and r (sec⁻¹) is the average surface renewal rate.

As *Moog and Jirka* [1999] noticed, our knowledge of near-surface dynamics is not at present precise enough to warrant the addition of another unknown beyond the surface renewal rate. Thus, from an engineering point of view, the primary research goal should be the prediction of the surface renewal rate based on the wave characteristics. *Daniil and Moutzouris* [1995] presented a vorticity–based renewal model, for gas transfer under breaking waves. The surface renewal rate was expressed as:

$$r = a_r \omega G_r \quad (6)$$

where a_r is a constant of proportionality,

$$\omega = -\frac{4\pi f(\pi H/L)^2}{1-(\pi H/L)^2} \quad (7)$$

is the wave vorticity at the water surface (sec^{-1}), f(sec^{-1}) is the wave frequency, H(m) is the wave height, L(m) is the wave length. The factor Gr was expressed as Gr =(L/d)^2 in order to incorporate the influence of relative depth, where d(m) is the water depth. The following expression was fitted to data from experiments with a sloping beach (slope 1:2.3) and a rubble mound breakwater (1:1.5 sloping front) performed in small-scale facility at LHW [*Daniil et al.*, 1998].Wave heights used in the experiments ranged from 5.6 - 28.0cm and wave periods from 0.75 - 1.75 sec. The water depth ranged from 0.56 - 0.83m.

$$K_L Sc^{1/2} = \alpha (L/d)\sqrt{\nu\omega} - \beta \quad (8)$$

The actual transfer coefficients both for the sloping beach and the breakwater [*Daniil et al.*, 2000] give the same correlation with the wave characteristics (α=3.03 and β=0.00236 m/s).

In Figure 3 the actual transfer coefficients from the experiment in Delft: DH-Series B (sloping beach 1:2.3) and DH-Series A(no structure – non breaking waves), as well as the apparent transfer coefficients from the experiments in Franzius Institut (FI-apparent) are compared with Equation (8). Small-scale laboratory data for sloping beach (slope 1:2.3 and 1:5) are also shown.

It is observed that the actual transfer coefficients for DH-Series B follow Equation (8) quite well, whereas DH-Series A with non-breaking waves is lower, as expected. Apparent transfer coefficients of FI experiments are also lower, but the actual ones are expected to be close to Equation (8) as well.

A Modified Vorticity Based Renewal Model for the Prediction of Transfer Coefficients Under Breaking Waves

The computation of the wave vorticity in Equation (8) is based upon the deep water wave theory (h/L$\geq$0.5) [*Kinsman*, 1984]. It is therefore essential, for the prediction of the gas transfer coefficient under breaking waves, to incorporate in the Gr factor an expression describing the wave breaking. During the present experiments it is observed that the oxygenation is high when the breaking of the waves becomes violent. Therefore a breaker type index that expresses the type of wave breaking should also be incorporated in the models for the prediction of oxygen transfer coefficient.

Dimensionless parameters reported in the literature for indexing of wave breaking [*Irribaren and Nogales* 1949; *Galvin*, 1968; *Zhang and Sunamura*, 1990] were used in the Gr term:

Irribaren $\quad I = \tan^2\alpha / (H_b/L_o) \quad (9)$

Galvin $\quad B_t = H_b / (gT^2 \tan\alpha) \quad (10)$

Reynolds $\quad Re_{wb} = H_b L_b / \nu T \quad (11)$

where H_b is the wave height in the breaking zone T is the wave period, L_o is the wave length in deep water, L_b is the wave length in the wave breaker zone, tanα is the slope of the beach or the coastal structure, ν is the kinematic viscosity of the water and g is gravity acceleration.

For the conducted experiments the wave heights in the breaking zone were calculated from the equation of *Le Mehaute and Koh* [1967]:

$$H_b/H_0 = 0.76(\tan\alpha)^{1/7} \gamma_o^{-1/4} \quad (12)$$

where $\gamma_o = L/d$ is the wave steepness, while the wavelengths, from the relation suggested from the *US Army Corps of Engineers* [1990]:

$$L_b = T(gd_b)^{1/2} \quad (13)$$

where d_b is the water depth in the wave breaking zone. In Figures 4, 5 oxygen transfer coefficients for small and large scale experiments are plotted against the parameters $\sqrt{\nu\omega}$ I and $\sqrt{\nu\omega} B_t$.

The presence of the slope of the structure in the both parameters leads to a pronounced grouping of data based on the slope rather than on the scale.

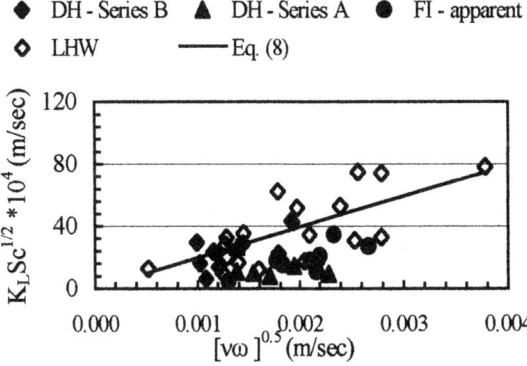

Figure 3. Comparision of vorticity model with large-scale experimental results.

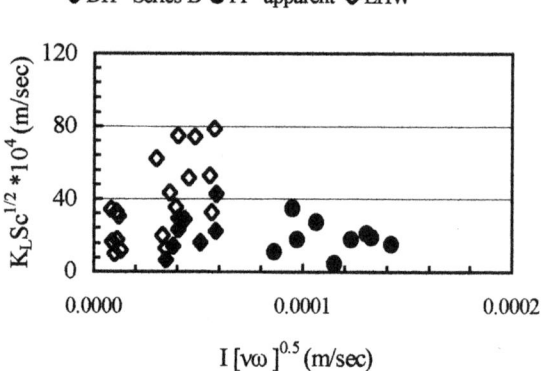

Figure 4. Comparison of transfer coefficients with wave parameter $I[\nu\omega]^{0.5}$

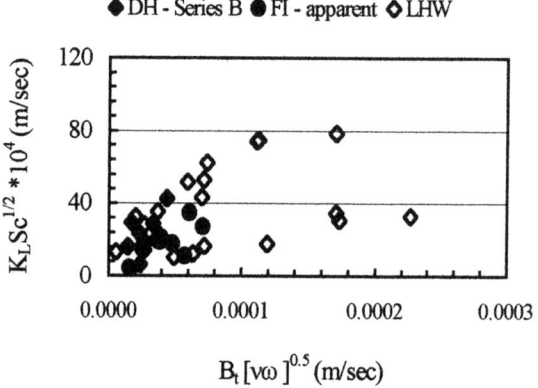

Figure 5. Comparison of transfer coefficients with wave parameter $B_t[\nu\omega]^{0.5}$

In Figure 6 oxygen transfer coefficients are plotted against the parameter $\sqrt{Re_{wb}\,\nu\omega}$. The actual transfer coefficients correlate quite well with this parameter. The vorticity model is thus modified to:

$$K_L Sc^{1/2} = a\sqrt{\nu\omega Re_{ewb}} = a(\omega H_b L_b T)^{0.5} \quad (14)$$

The constant a=0.102 was determined by best fitting the experimental results of LHW and DH-Series B with a very good correlation coefficient (r=0.80) It also describes data from literature fairly well. Data from DH-Series A are also shown in Figure 6 substituting the deepwater wave characteristics H and L in place of the breaking wave characteristics H_b and L_b. They are also plot quite close to Equation (14) [*Tsoukala*, 2000]. Apparent transfer coefficients from FI experiments are below the proposed equation and the corresponding actual transfer coefficients are expected to follow the equation closely as well.

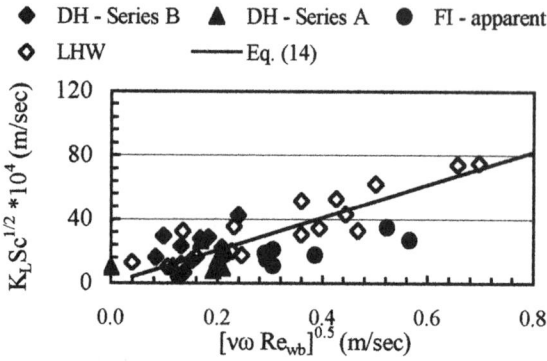

Figure 6. Comparison of transfer coefficients with modified vorticity model.

The proposed equation improves the equation suggested by *Daniil and Moutzouris* [1995] as it eliminates the influence of temperature, which enters in the calculation of transfer coefficient through the kinematic viscosity of the water and can also be used with rather good precision in the case of non-breaking waves, if the breaking wave characteristics H_b and L_b are substituted with the corresponding deepwater characteristics [*Tsoukala*, 2000].

5. CONCLUSIONS

Actual oxygen transfer coefficients, as computed from a discritized form of the transport equation and accounting for longitudinal dispersion, are found to have the same relation with wave characteristics both for small and large scale experimental data. The inclusion of the wave Reynolds number in the expression of the surface renewal rate along with the vorticity parameter was shown to improve the gas transfer coefficient prediction model. The proposed modified vorticity model describes gas transfer also well both for breaking waves when the breaking waves characteristics are used and for non breaking waves when the deep water characteristics are used.

Acknowledgements. This work was partly supported by the European Programme Human Capital Mobility -Access to Large Installation Facilities (1994) and Training and Mobility of Researchers – Access to Large Scale Facilities Programme (1999).

REFERENCES

ASCE Standard, Measurement of oxygen transfer in clean water, ASCE, New York, USA, 1993.

Bennett, J.P., and R.E. Rathbun, Reaeration in open-channel flow, *U.S. Geological Survey*, Professional paper, No. 737, 1972.

Daniil, E.I., and J.S. Gulliver, Temperature dependence of the liquid film coefficient for gas transfer, *J. of Environmental Engineering*, ASCE, Vol. 114, No. 5, pp. 1224-1229, 1988.

Daniil, E.I., and C.I. Moutzouris, A vorticity-based model for gas transfer under breaking waves, *Annales Geophysicae*, EGS, vol. 13, pp.1039-1046, 1995.

Daniil E.I., V.K. Tsoukala and C.I. Moutzouris, Harbour Basin Seawater Oxygenation through Rubble Mound Breakwater Structures, *Journal of Marine Environmental Engineering*, Vol. 4, No 4, pp. 277-300, 1998.

Daniil E.I., V.K. Tsoukala and C.I. Moutzouris, Dissolved Oxygen Measurements for Water Quality in a Harbour Area near Athens, Greece, ASCE, Water Resources Conference, Mineapolis, MN, USA, (*in press*) 2000a.

Daniil E.I., V.K. Tsoukala and C.I. Moutzouris, The Beneficial Role of RubbleMound Coastal Structure on Seawater Oxygenation, *Annales Geophysicae*, EGS, (*in press*) 2000b.

Dankwerts, P.V., Significance of liquid-film coefficients in gas absorption, *Industrial and Engineering Chemistry*, Vol. 43, No. 6, pp. 1460-1467, 1951.

Falvey, H.T., and D.A. Ervine, Aeration in jets and high velocity flows in Model – Prototype Correlation of Hydraulic Structures, *Proc. of the Int. Symposium, ASCE*, pp. 25-55, 1988.

Galvin, C.J., Breaker type classification on three laboratory beaches, *Journal of Geophysical Research*, Vol. 73, No. 12, pp. 3651-3659, 1968.

Horsch, G.M., Steady, diffusive – reactive transport in shallow triangular domain, *Journal of Engineering Mechanics*, ASCE, Vol. 124, No. 10, 1998.

Irribaren, C.R., and C. Nogales, Protection des ports, *XVII, International Navigation Congress*, Section II, Comm. 4, Lisbon, pp. 27-47, 1949.

Kinsman, B., *Wind waves*, Dover Publications, 1984.

LeMehaute, B., and R.C.Y., Koh, On the breaking of waves arriving at an angle to the shore, *Journal of Hydraulic Research*, Vol.5, No.1, 1967.

Moutzouris, C.I., and E.I. Daniil, Water oxygenation in the vicinity of coastal structures due to wave breaking, *Coastal Engineering 1994*, Edge, B.L., (ed.), pp. 3167-3177, 1995.

Patankar, S.V., *Numerical heat transfer and fluid flow*, McGraw Hill, New York, USA, 1980.

Tofa, E., Large-scale experimental measurements at Franzius Instutut at the University of Hanover in Germany, for the oxygenation in the vicinity of breakwater, *Diploma Thesis*, National Technical University of Athens, Athens, 2000.

Tsoukala V.K., E.I. Daniil and C.I. Moutzouris, Experimental and Mathematical Study of harbour area oxygenation, *Proceedings of the 5th and Conference on Science Environmental and Technology*, Mytilini, Greece, Vol. II, pp. 581-588, 1997.

Tsoukala V.K., Wave induced sea water oxygenation at the coastal zone, *Doctoral Dissertation*, National Technical University of Athens, Athens, 2000.

Tsoukala, V.K. and C.I. Moutzouris, Scale effects in oxygenation in the breaker zone of coastal structures, *Coastal Engineering 1996*, ASCE, Vol. 1, pp. 403-414, 1997.

U.S. Army Corps of Engineers, Laboratory study on macro-features of wave breaking over bars and artificial reefs, *Technical Report*, No. CERC-90-12, Coastal Engineering Research Center, Waterways Experiment Station, Vicksburg, Mississippi, USA, 1990.

Wanninkhof, R., and L.F. Bliven, Relathionship between gas exchange, wind speed and radar backscatter in large wind wave tank", *Journal of Geophysical Research*, Vol. 96, No 2, pp2785-2796, 1991.

Zhang, D.P. and T. Sunamura, Conditions for the occurrence of vortices induced by breaking waves, *Coastal Engineering in Japan*, Vol.33, No. 2, pp. 145-155, 1990.

Laboratory of Harbour Works, Faculty of Civil Engineering, National Technical University of Athens, 5 Iroon Polytechniou, 157 73 Zografou, GREECE. e_mail:V.Tsoukala@hydro.ntua.gr, ina@hydro.ntua.gr.